Information Technology, Geospatial Information and Advanced Computing 2018

Transportation Research Record: Journal of the Transportation Research Board Volume 2672, Issue 45

ISBN: 978-1-5108-8251-5

Printed from e-media with permission by:

Curran Associates, Inc.
57 Morehouse Lane
Red Hook, NY 12571

CURRAN ASSOCIATES INC. proceedings.com

Some format issues inherent in the e-media version may also appear in this print version.

Published by Sage Publications USA

ISBN (Print) 978-1-5108-8251-5
ISBN 2018 Printed Set (All Issues) 978-1-5108-7735-1

Printed by Curran Associates, Inc. (2019)

Additional copies of this publication are available from:

Curran Associates, Inc.
57 Morehouse Lane
Red Hook, NY 12571 USA
Phone: 845-758-0400
Fax: 845-758-2633
Email: curran@proceedings.com
Web: www.proceedings.com

Volume 2672 Number 45 December 2018

TRR
JOURNAL OF THE TRANSPORTATION RESEARCH BOARD

Contents

Information Technology, Geospatial Information, and Advanced Computing

Articles

TRB

TRANSPORTATION RESEARCH BOARD

CORE SPONSORS

Transportation Departments of the 50 States and the District of Columbia

Federal Government

U.S. Department of Transportation
- Federal Aviation Administration (FAA)
- Federal Highway Administration
- Federal Motor Carrier Safety Administration (FMCSA)
- Federal Railroad Administration (FRA)
- Federal Transit Administration
- National Highway Traffic Safety Administration (NHTSA)
- Office of the Assistant Secretary of Transportation for Research and Technology (OST-R)

U.S. Air Force Civil Engineer Center (AFCEC)
U.S. Coast Guard (USCG)
U.S. Department of Energy
U.S. Department of the Interior (DOI)
U.S. Environmental Protection Agency (EPA)

Nongovernmental Organizations

American Public Transportation Association (APTA)
Association of American Railroads (AAR)
California Air Resources Board

The National Academies of
SCIENCES · ENGINEERING · MEDICINE

The **National Academy of Sciences** was established in 1863 by an Act of Congress, signed by President Lincoln, as a private, nongovernmental institution to advise the nation on issues related to science and technology. Members are elected by their peers for outstanding contributions to research. Dr. Marcia McNutt is president.

The **National Academy of Engineering** was established in 1964 under the charter of the National Academy of Sciences to bring the practices of engineering to advising the nation. Members are elected by their peers for extraordinary contributions to engineering. Dr. C. D. Mote, Jr., is president.

The **National Academy of Medicine** (formerly the Institute of Medicine) was established in 1970 under the charter of the National Academy of Sciences to advise the nation on medical and health issues. Members are elected by their peers for distinguished contributions to medicine and health. Dr. Victor J. Dzau is president.

The three Academies work together as the National Academies of Sciences, Engineering, and Medicine to provide independent, objective analysis and advice to the nation and conduct other activities to solve complex problems and inform public policy decisions. The Academies also encourage education and research, recognize outstanding contributions to knowledge, and increase public understanding in matters of science, engineering, and medicine.

Learn more about the National Academies of Sciences, Engineering, and Medicine at **www.national-academies.org.**

The **Transportation Research Board** is one of seven major programs of the National Academies of Sciences, Engineering, and Medicine. The mission of the Transportation Research Board is to increase the benefits that transportation contributes to society by providing leadership in transportation innovation and progress through research and information exchange, conducted within a setting that is objective, interdisciplinary, and multimodal. The Board's varied activities annually engage about 7,000 engineers, scientists, and other transportation researchers and practitioners from the public and private sectors and academia, all of whom contribute their expertise in the public interest. The program is supported by state transportation departments, federal agencies including the component administrations of the U.S. Department of Transportation, and other organizations and individuals interested in the development of transportation.

Learn more about the Transportation Research Board at **www.TRB.org**.

TRR
JOURNAL OF THE TRANSPORTATION RESEARCH BOARD

Article

Transportation Research Record
2018, Vol. 2672(45) 1–13

DOI: 10.1177/0361198118790330
journals.sagepub.com/home/trr
SAGE

A Spatial–Temporal Multitask Collaborative Learning Model for Multistep Traffic Flow Prediction

Kun Tang[1], Shuyan Chen[2], and Aemal J. Khattak[3]

Abstract

Traffic flow prediction is a fundamental capability for successful deployment of intelligent transportation systems. Traditionally, multiple related prediction tasks are undertaken individually, without considering the relationships among the tasks. This paper presents a spatial–temporal multitask collaborative learning model for multistep traffic flow prediction. The novel approach learns multiple related prediction tasks collaboratively by extracting and utilizing appropriate shared information across tasks. First, each traffic flow prediction problem is formulated as a supervised machine-learning task. Next, the sparse features shared across multiple tasks are learned by solving a regularized optimization problem. To deal with the non-convex and non-smooth challenges, the optimization problem is then transformed into an equivalent convex problem. Finally, the global optimal solution of the convex problem is found by solving a variation of this problem using an alternating minimization algorithm. The proposed model incorporates both the spatial correlation between different observed stations and the intrinsic relationship between different traffic flow parameters, as well as the coarse-grain temporal correlation between different days in a week and the fine-grain temporal correlation between different prediction steps. Application of the proposed model to a real case study for SR180-E freeway in Fresno, California showed its effectiveness, robustness and advantages for multistep traffic flow prediction.

Growing traffic congestion due to increasing number of automobiles harms transportation system efficiency and the environment (*1*). Traffic flow prediction, which provides future traffic volume, occupancy and speed information, plays a crucial role for many levels of traffic management and transport policy (*2*). Availability of accurate and timely traffic flow information allows individuals to make better travel-related decisions and increase the efficiency of their trips. For traffic management, traffic flow information at the station level can reveal problematic locations where introduction of new or revised traffic control schemes may improve performance. For transport policy, network level traffic flow information provides input for travel demand forecasting and assessments of policy instruments such as congestion pricing.

As short-term traffic flow prediction is an important issue for many traffic management operations, a number of technologies focusing on this topic have been developed in the past few decades (*3–5*). However, most existing approaches solve the traffic flow prediction tasks of different stations and different time intervals individually, without considering correlations amongst these tasks. However, traffic flows at different locations are not independent but constrained by the flows at other locations (*6*). That is, potential mutual interactions couple traffic flows at different locations with each other. Modeling traffic flow prediction from a network perspective and utilizing the interrelationships among different prediction tasks improves prediction performance, especially when the training sample data size is small for each task.

In this paper, we model traffic flow (volume, occupancy and speed) predictions at different observed locations in various time intervals as different prediction tasks from a network perspective and propose a spatial–temporal multitask model to learn these tasks

[1]Jiangsu Key Laboratory of Urban ITS, Jiangsu Province Collaborative Innovation Center of Modern Urban Traffic Technologies, School of Transportation, Southeast University, Nanjing, China
[2]School of Transportation, Southeast University, Nanjing, China
[3]Department of Civil Engineering, University of Nebraska-Lincoln, Lincoln, NE

Corresponding Author:
Address correspondence to Shuyan Chen: chenshuyan@seu.edu.cn

collaboratively by taking advantage of the internal and spatial–temporal correlations. This novel approach first learns a set of data representations (features) that are shared across tasks, and then identifies (selects) the high-quality ones that contribute to improve prediction performance. To the best of our knowledge, this is the first time that both "feature learning" and "feature selection" are incorporated in the multitask collaborative learning for multistep traffic flow prediction. The salient contributions of this research are as follows.

- We model traffic flow from a network perspective, and develop a novel model to learn multiple traffic flow prediction tasks collaboratively by taking advantage of the internal and spatial–temporal correlations.
- We devise a novel algorithm to solve the spatial–temporal multitask collaborative learning model in an alternating optimization manner.
- We present an application of the proposed model to a real case based on freeway SR180-E traffic in Fresno, California. Empirical results from extensive experiments demonstrate the effectiveness, robustness, and advantages of the proposed model.

The paper is organized as follows. The next section presents a review of published literature on short-term traffic flow prediction, followed by a section describing methodology of the proposed model. The ensuing section reports the experimental results of the case study. A section on conclusions and future research conclude this paper.

Literature Review

Multistep short-term traffic flow prediction aims to predict traffic volume, occupancy and speed at observed locations over several time intervals in the future. Time intervals are usually defined as short 5 to 30 min durations. As it is an important component in many intelligent transportation systems (ITS) applications, a large number of models have been developed to improve transportation system efficiency (*7*, *8*). According to Lint and Hinsbergen (*9*), the existing approaches can be classified into the following four categories.

- Naïve models that predict directly based on the data used and exact physical relationships, e.g., historical average. These models are independent of model structures and parameters. While widely used in practice due to easy implementation and low computational effort, the accuracy is generally low. For example, Weijermars and Berkurn investigated the determination of historical traffic patterns by means of Ward's hierarchical clustering procedure (*10*).
- Parametric approaches (*11*), e.g., analytical models and time series models, model traffic data with mathematical or statistical formulations. Such approaches suffer from the underlying model assumptions made to fit the formulations. The performance of these approaches can be unsatisfactory when traffic conditions fluctuate significantly, or road settings are complex. For example, by combining the Hull-White (HW) model with the extended Vasicek (EV) model, Rajabzadeh et al. proposed a two-step approach based on the stochastic differential equations (SDEs) to improve short-term prediction (*12*).
- Non-parametric approaches (*13*) such as data analysis-based methods and neural network techniques, utilize algorithms of machine learning and data mining to provide future traffic predictions. These approaches are data-driven and discover knowledge within the data using the data itself, without explicitly modeling with a dedicated model which heavily relies on the analyst's personal experience and intuition. Non-parametric models are more effective for many transportation applications. For example, based on deep learning, Polson and Sokolov developed an architecture that combined a linear model and a sequence of hyperbolic tangent activation layers to improve the short-term traffic flow prediction (*14*).
- Hybrid approaches (*15*) refer to models that fuse the aforementioned approaches at levels of model or decision, e.g., ensemble learning. Because hybrid approaches take advantage of multiple models, they are usually more effective than the individual approaches. For example, to capture the complex patterns underlying traffic flow, Li et al. proposed a hybrid strategy that is general and can make use of a large number of underlying machine-learning or time-series prediction models (*16*).

In summary, the growing need for real time traffic flow information has resulted in a significant number of traffic flow prediction approaches involving various techniques borrowed from different disciplines. However, most of them treat different traffic flow prediction tasks as individual single task learning (STL) problems, ignoring correlations among these tasks. Recent attention has been shifted to making use of multitask learning (MTL) to overcome this issue (*17*). Since multiple related tasks are learned collaboratively by extracting and utilizing appropriate shared information across tasks, the prediction performance can be effectively improved.

Methodology

In this section, we first formulate the travel time prediction as a supervised machine-learning problem, and then elaborate the proposed spatial–temporal multitask collaborative learning model. Finally, we develop an alternating optimization algorithm to solve the proposed model.

Traffic Flow Prediction

Formulation of the traffic flow prediction problem is as follows. Let $x_{i,t}$ denotes the traffic flow at the observed station i during the time interval t. The prediction task for a special station s at the time interval T is to predict the traffic flow $y = x_{s,T}$, based on the observations

$$\mathbf{x} = [x_{1,T-p}, \cdots, x_{1,T-1}, x_{2,T-p}, \cdots, x_{2,T-1}, \cdots, x_{q,T-p}, \cdots, x_{q,T-1}] \in \mathbb{R}^d, \; d = p \times q \tag{1}$$

where d is the dimensionality of $\mathbf{x}$, p the number of previous time intervals considered, and q the number of stations in the road network.

Given a dataset of m observation samples

$$\{(\mathbf{x}_1, y_1), (\mathbf{x}_2, y_2), \cdots, (\mathbf{x}_m, y_m) | \mathbf{x} \in \mathbb{R}^d, y \in \mathbb{R}\} \tag{2}$$

The objective of traffic flow prediction is formulated as a supervised machine-learning problem as

$$\arg\min_{\mathcal{F}} \left\{ \sum_{i=1}^{m} \|y_i - f(\mathbf{x}_i)\|^2 \right\} \tag{3}$$

where $\|\bullet\|$ is the ℓ_2-norm.

Based on the learned model $f : \mathbb{R}^d \to \mathbb{R}$, the traffic flow prediction for a new input $\mathbf{x}_{new}$ is given by

$$\hat{y} = f(\mathbf{x}_{new}) \tag{4}$$

Spatial–Temporal Multitask Collaborative Learning Model

Given T supervised learning tasks, each task is associated with a specific traffic flows prediction objective. For each task $t = 1, 2, \cdots, T$, the corresponding task is identified by a model $f_t : \mathbb{R}^d \to \mathbb{R}$, and the given dataset of m samples is $\{(\mathbf{x}_{t1}, y_{t1}), (\mathbf{x}_{t2}, y_{t2}), \cdots, (\mathbf{x}_{tm}, y_{tm}) | \mathbf{x} \in \mathbb{R}^d, y \in \mathbb{R}\}$. To model traffic flow from a network perspective, we wish to design an algorithm that uncovers particular relationships across tasks and compute the function f_t, $t = 1, 2, \cdots, T$, based on the data given above. As sparse feature learning and shared representation learning have been extensively studied and proven effective for many applications (*18*, *19*), we study the case that the related tasks share a small set of features. Specifically, we represent the task function as

$$f_t(\mathbf{x}) = \sum_{i=1}^{N} a_{it} h_i(\mathbf{x}), \; t = 1, 2, \cdots, T \tag{5}$$

where $h_i : \mathbb{R}^d \to \mathbb{R}$, $i = 1, 2, \cdots, N$ are the feature learning function, and $a_{it} \in \mathbb{R}$ the regression parameters.

Our goal is to learn the number of features N, the functions h_i and the parameters a_{it}. Inspired by the idea of the unsupervised algorithm, principal component analysis (PCA), we consider the case that the features are linear homogeneous and orthogonal. Formally, the feature learning function is of the form

$$h_i(\mathbf{x}) = \langle \mathbf{u}_i, \mathbf{x} \rangle \tag{6}$$

where $\mathbf{u}_i \in \mathbb{R}^d$ denotes the feature learning vector, and $\langle \bullet, \bullet \rangle$ the inner product of vectors.

Let $\mathbf{U} \in \mathbf{O}^d$ denote the feature learning matrix whose columns are the vectors $\mathbf{u}_i$, where $\mathbf{O}^d$ denotes the $d \times d$ orthogonal matrices. The task function can then be formulated as

$$f_t(\mathbf{x}) = \sum_{n=1}^{d} a_{nt} \langle \mathbf{u}_n, \mathbf{x} \rangle = \langle \mathbf{a}_t, \mathbf{U}^T \mathbf{x} \rangle \tag{7}$$

where $\mathbf{U}^T$ denotes the transpose of $\mathbf{U}$.

As we study the problem of learning a small set of features shared across multiple tasks, the feature selection matrix $\mathbf{A}$, whose columns are the vectors $\mathbf{a}_t$, has many zero rows. We introduce a $\ell_{2,1}$-norm regularization term, which favors a small number of nonzero rows in the matrix $\mathbf{A}$. In addition, the inclusion of a ℓ_2-norm prevents over-fitting. In this fashion, the spatial–temporal multitask collaborative learning model is given by

$$\min \left\{ \mathcal{E}(\mathbf{A}, \mathbf{U}) = \sum_{t=1}^{T} \sum_{i=1}^{m} \mathcal{L}(y_{ti}, \langle \mathbf{a}_t, \mathbf{U}^T \mathbf{x}_{ti} \rangle) + \frac{\lambda}{2} \left(\|\mathbf{A}\|_{2,1}^2 + \|\mathbf{A}\|^2 \right) : \mathbf{A} \in \mathbb{R}^{d \times T}, \mathbf{U} \in \mathbf{O}^d \right\} \tag{8}$$

where $\|\bullet\|_{2,1}$ denotes the $\ell_{2,1}$-norm. The first term is the so called empirical loss term measured by a prescribed loss function $\mathcal{L} : \mathbb{R} \times \mathbb{R} \to \mathbb{R}_+$, where $\mathbb{R}_+$ denotes the set of positive real numbers. For example, the square loss defined as $\mathcal{L}(y, x) = (y - x)^2$ is usually used. The second term is the regularization term, and $\lambda > 0$ is the regularization parameter controlling the number of shared features learned and the magnitude of matrix **A**.

We interpret learning the objective in Equation 8 with respect to matrices **U** and **A** as learning the features shared across tasks and selecting the high-quality ones that contribute to improve prediction performance at most, respectively. The proposed model learns multiple related prediction tasks collaboratively by extracting and utilizing appropriate shared information across tasks. It is an inductive transfer mechanism and the training signals for the extra tasks serve as an inductive bias. As it leverages the domain-specific information contained in the training signals of related tasks, the model generalization performance for traffic flow prediction can be effectively improved.

There are two challenges in solving Equation 8. First, it is a non-convex problem, although matrices **A** and **U** are individually convex. Second, the regularization term is not smooth. Finding the global optimal solution requires transformation of this problem into an equivalent convex problem based on reference (*20*). Specifically, the problem in Equation 8 is equivalent to the convex optimization problem:

$$\min\left\{\begin{array}{l}\mathcal{F}(\mathbf{W},\mathbf{D}) = \sum_{t=1}^{T}\sum_{i=1}^{m}\mathcal{L}(y_{ti},\langle\mathbf{w}_t,\mathbf{x}_{ti}\rangle) + \frac{\lambda}{2}\sum_{t=1}^{T}\langle\mathbf{w}_t,\mathbf{D}^{\dagger}\mathbf{w}_t\rangle + \frac{\lambda}{2}\|\mathbf{W}\|^2 : \\ \mathbf{W}\in\mathbb{R}^{d\times T},\mathbf{D}\in\mathbf{S}_{+}^{d}, trace(\mathbf{D})\le 1, range(\mathbf{W})\subseteq range(\mathbf{D})\end{array}\right\} \quad (9)$$

where $\mathbf{S}_{+}^{d}$ denotes the $d\times d$ positive semidefinite real symmetric matrices, $\mathbf{w}_t$ the columns of $\mathbf{W}$, $\mathbf{D}^{\dagger}$ the pseudoinverse of $\mathbf{D}$, $trace(\mathbf{D})$ the trace (the sum of the elements on the main diagonal) of matrix $\mathbf{D}$, $range(\mathbf{W})$ the set $\{\mathbf{w}\in\mathbb{R}^d : \mathbf{w} = \mathbf{W}\mathbf{z}, \text{ for some } \mathbf{z}\in\mathbb{R}^T\}$, and $\lambda>0$ is the regularization parameter. If $\widehat{\mathbf{W}}$ and $\widehat{\mathbf{D}}$ is an optimal solution of problem in Equation 9, then any $\widehat{\mathbf{U}}$ whose columns form an orthonormal basis of eigenvectors of $\widehat{\mathbf{D}}$ and $\widehat{\mathbf{A}} = \widehat{\mathbf{U}}^T\widehat{\mathbf{W}}$ are an optimal solution of problem in Equation 8.

Alternating Optimization Algorithm

This section presents the algorithm to solve the convex optimization problem in Equation 9. It is a modification of the algorithm developed in Argyriou et al. (*20*), which solves a variation of the problem in Equation 9 in an alternating optimization manner. Specifically, it minimizes a perturbation of the Equation 9 problem with a small parameter ε, as

$$\min\left\{\mathcal{F}_{\varepsilon}(\mathbf{W},\mathbf{D}) = \sum_{t=1}^{T}\sum_{i=1}^{m}\mathcal{L}(y_{ti},\langle\mathbf{w}_t,\mathbf{x}_{ti}\rangle) + \frac{\lambda}{2}trace(\mathbf{D}^{-1}(\mathbf{W}\mathbf{W}^T+\varepsilon\mathbf{I})) + \frac{\lambda}{2}\|\mathbf{W}\|^2 : \mathbf{W}\in\mathbb{R}^{d\times T},\mathbf{D}\in\mathbf{S}_{+}^{d}\right\} \quad (10)$$

where $\mathbf{D}^{-1}$ denotes the inverse of $\mathbf{D}$ and $\mathbf{I}$ represents the identity matrix.

The problem in Equation 10 can be solved using the Algorithm 1 summarized in Figure 1. It comprises of two steps; in the first, the matrix **D** is fixed and the objective is to minimize with respect to **W**, as

$$\min\left\{\sum_{t=1}^{T}\sum_{i=1}^{m}\mathcal{L}(y_{ti},\langle\mathbf{w}_t,\mathbf{x}_{ti}\rangle) + \frac{\lambda}{2}\sum_{t=1}^{T}\langle\mathbf{w}_t,\mathbf{D}^{-1}\mathbf{w}_t\rangle + \frac{\lambda}{2}\|\mathbf{W}\|^2 : \mathbf{W}\in\mathbb{R}^{d\times T}\right\} \quad (11)$$

This minimization can be carried out independently across the tasks since the matrix **W** in the regularization term decouples when **D** is fixed. In the second step, matrix **W** is fixed and the function is minimized with respect to **D**, as

$$\min\left\{\sum_{t=1}^{T}\langle\mathbf{w}_t,\mathbf{D}^{-1}\mathbf{w}_t\rangle + \varepsilon trace(\mathbf{D}^{-1}) : \mathbf{D}\in\mathbf{S}_{++}^{d}, trace(\mathbf{D})\le 1\right\} \quad (12)$$

where $\mathbf{S}_{++}^{d}$ denotes the $d\times d$ positive definite real symmetric matrices. As shown in Argyriou et al. (*20*), the optimal solution of the problem is given by

$$\mathbf{D}_{\varepsilon}(\mathbf{W}) = \frac{(\mathbf{W}\mathbf{W}^T+\varepsilon\mathbf{I})^{1/2}}{trace(\mathbf{W}\mathbf{W}^T+\varepsilon\mathbf{I})^{1/2}} \quad (13)$$

and the optimal value equals $\left(trace(\mathbf{W}\mathbf{W}^T+\varepsilon\mathbf{I})^{1/2}\right)^2$.

Algorithm 1 solves a perturbation of Equation 10. As the number of iterations increases, the algorithm converges to the optimal solution of Equation 10. To obtain the optimal solution of Equation 9, we modify the algorithm by reducing the perturbation parameter ε towards to zero. Specifically, consider a sequence $\{\varepsilon_k>0 : k = 1,2,\cdots\}$ that converges to 0. Let $(\mathbf{W}_k,\mathbf{D}_{\varepsilon_k}(\mathbf{W}_k))$ be the optimal solution of $\mathcal{F}_{\varepsilon_k}$, for every

Algorithm 1: Spatial-temporal Multi-task Collaborative Learning

Input: training sets $\{(\mathbf{x}_{ti}, y_{ti})\}_{i=1}^{m} \in \mathbb{R}^d \times \mathbb{R}$, for every $t = 1, 2, \cdots, T$

Parameters: regularization parameter λ, tolerances ε, tol

Output: $d \times d$ matrix $\mathbf{D}$, and $d \times T$ regression matrix $\mathbf{W} = [\mathbf{w}_1, \mathbf{w}_2 \cdots, \mathbf{w}_T]$

Initialization: initialize $\mathbf{D} = \frac{\mathbf{I}^d}{d}$

While $\|\mathbf{W} - \mathbf{W}_{prev}\| > tol$ do

for $t = 1, 2, \cdots, T$ do

Compute $\mathbf{w}_t = \arg\min_{\mathbf{w}} \left\{ \sum_{i=1}^{m} \mathcal{L}(y_{ti}, \langle \mathbf{w}, \mathbf{x}_{ti} \rangle) + \frac{\lambda}{2} (\langle \mathbf{w}, \mathbf{D}^{-1}\mathbf{w} \rangle + \mathbf{w}) : \mathbf{w} \in \mathbb{R}^d \right\}$

end for

Set $\mathbf{D} = \frac{(\mathbf{W}\mathbf{W}^T + \varepsilon \mathbf{I})^{1/2}}{trace\left((\mathbf{W}\mathbf{W}^T + \varepsilon \mathbf{I})^{1/2}\right)}$

end while

Figure 1. Algorithm for sparse multitask feature learning.

$k = 1, 2, \cdots$. Then any limiting point of the sequence $\{\mathbf{W}_k, \mathbf{D}_{\varepsilon_k}(\mathbf{W}_k)\}$ is an optimal solution to Equation 9.

We explain Algorithm 1 as performing supervised and unsupervised steps alternately. In the supervised step, we learn task-specific functions, i.e., the vectors $\mathbf{w}_t$, based on the data representations that are shared across multiple tasks. In the unsupervised step, we fix the task functions and learn the tasks' shared features in turn.

Case Study

Data

Implementation of the proposed model involved utilization of real traffic data collected from the Caltrans Performance Measurement System (PeMS). This system collects data every 30s from over 43,000 individual detectors deployed on freeway systems across California (*13*). The collected data are aggregated with 5 min interval for each vehicle detector station (VDS). In this paper, the traffic flow data (including volume, occupancy and speed) collected on the freeway SR180-E (as shown in Figure 2*a*) in Fresno, California from June 06–26, 2016 were used for experiments. To demonstrate the effectiveness of the proposed model when the training sample size is small, we train the model using only samples of the first week (one-third) and predict the samples of the remaining two weeks (two-third). As some VDS may fail to work, we removed samples imputed by PeMS itself and predict the real observations in the test stage. Figure 2*b* presents a plot of traffic volume, occupancy and speed over time from a typical VDS (ID: 601204).

Model Specification

The related prediction tasks to be learned are defined first. Since the proposed model accounts for the spatial–temporal correlation and models traffic flow from a perspective of network, the following four categories of related tasks were learnt collaboratively.

1) Different parameters. Traffic flow consists of multiple parameters: volume, occupancy and speed. We treat the predictions of different traffic flow parameters as different tasks. These tasks are related because the traffic flow parameters are intrinsically relevant.
2) Different stations. There are multiple VDS on the freeway SR180-E. We consider all the VDS simultaneously and treat the predictions of different VDS as different tasks. These tasks are related because the VDS on the freeway are spatially correlated.
3) Different steps. We predict the traffic flow of multiple steps ahead, and the time interval for each step is 5 min. The predictions of different steps are treated as different tasks. These tasks are related because the prediction steps are fine-grain temporally correlated.
4) Different days. Traffic patterns vary from day to day during a week, especially between weekdays

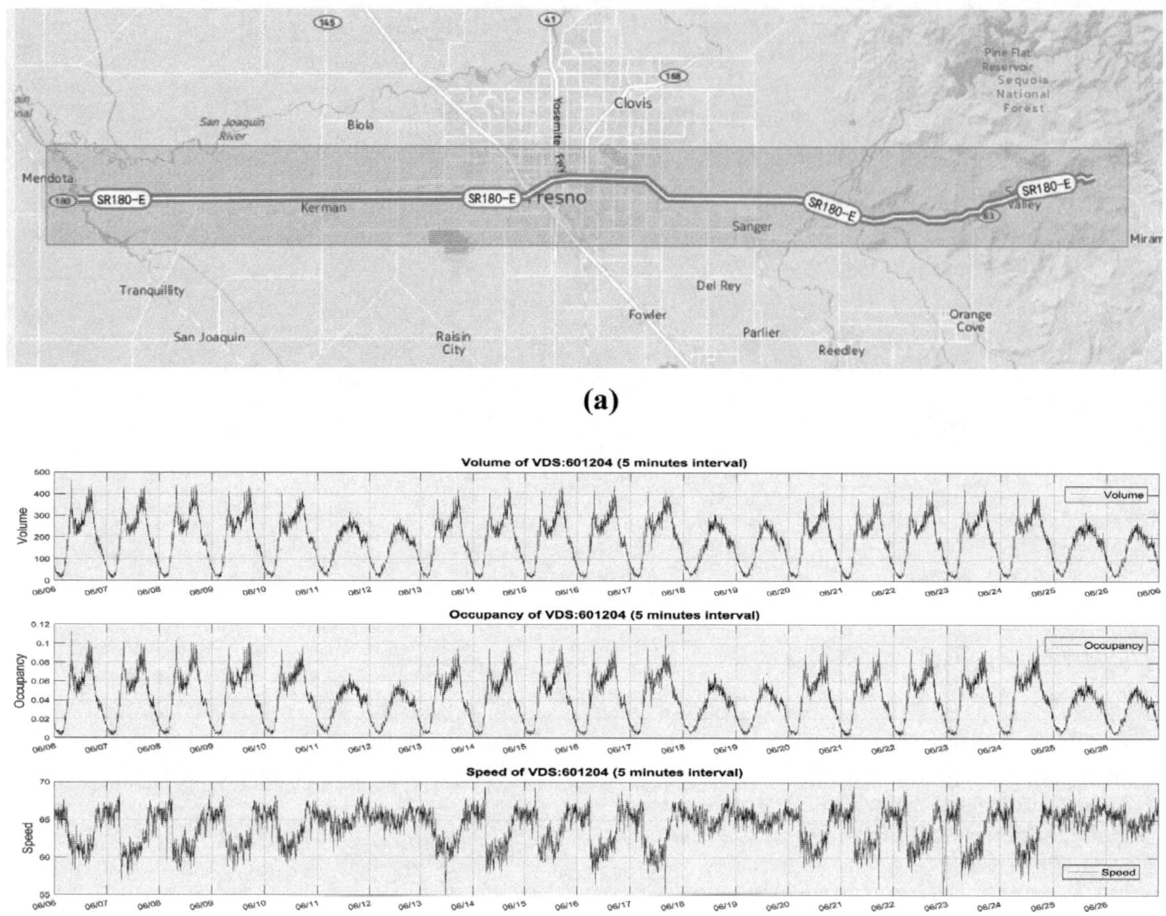

Figure 2. Case Study freeway: (*a*) SR180-E in Fresno, California, and (*b*) the traffic flows of a typical VDS.

and weekends (see Figure 2*b*). We treat the predictions of different days as different tasks. These tasks are related because the days in a week are coarse-grain temporally correlated.

For each task $t = 1, 2, \cdots, T$, we construct the training and test dataset according to Equation 1. Considering the temporal correlation of traffic flow, we predict the traffic flow in a time interval using the traffic flow data at its previous p time intervals. To balance the effectiveness and efficiency, p is set to 4 for experience. Furthermore, to model traffic flow from the network perspective, the data collected from all VDS on the freeway are used and thus the number of considered stations $q = 15$. As a result, the dimensionality of the input for each sample is $d = p \times q = 15 \times 4 = 60$. The proposed model incorporates both spatial and temporal correlations inherently. Since the metrics measuring volume, occupancy and speed are different, we use the command "zscore" in the machine-learning toolbox of Matlab to normalize experimental data. For all experiments, the square loss function is applied to measure the empirical error, the optimal regularization parameter λ is determined from the values $\{10^r : r \in \{-3, -2, \cdots, 3\}\}$ with 5-fold cross-validation. The tolerance parameters were set as $\varepsilon = 10^{-20}$ and $tol = 10^{-6}$, and the matrix $\mathbf{D}$ initialized to $\mathbf{D} = \mathbf{I}^{60}/_{60}$.

Validation Metrics

Four validation metrics are utilized to evaluate the performance of the proposed model. They are mean absolute error (MAE), mean absolute percentage error (MAPE), root mean square error (RMSE), and equal coefficient (EC), defined as

$$MAE = \frac{1}{n}\sum_{i=1}^{n}|x_i - \widehat{x_i}| \quad (14)$$

$$MAPE = \frac{1}{n}\sum_{i=1}^{n}\left|\frac{x_i - \widehat{x_i}}{x_i}\right| \quad (15)$$

$$RMSE = \sqrt{\frac{1}{n}\sum_{i=1}^{n}(x_i - \widehat{x_i})^2} \quad (16)$$

$$EC = 1 - \frac{1}{\sqrt{\sum_{i=1}^{n} x_i^2} + \sqrt{\sum_{i=1}^{n} \widehat{x_i}^2}}\sqrt{\sum_{i=1}^{n}(x_i - \widehat{x_i})^2} \quad (17)$$

where x_i and $\widehat{x_i}$ are the i-th true and predicted travel time respectively, and n is the number of travel times.

Competing Methods

A comparison of the proposed model with the following four methods demonstrates its advantages. Amongst the four methods, the first three are STL methods whereas the fourth is the MTL method.

- Random Walk (RW). RW is a naïve method that predicts traffic flows in the future as equal to the current traffic flows, i.e.

$$x_{s,t+1} = x_{s,t} \quad (18)$$

where $x_{s,t}$ is the traffic flow of the station s in the current time interval t, and $x_{s,t+1}$ is the traffic flow of this station in the next time interval. RW is the simple baseline method amongst these four competing methods.

- Moving Average (MA). MA is a parametric method that predicts traffic data in the future by computing the average of the previous n data, i.e.

$$x_{s,t+1} = \frac{x_{s,t} + x_{s,t-1} + \cdots + x_{s,t-(n-1)}}{n} = \frac{1}{n}\sum_{i=0}^{n-1} x_{s,t-i} \quad (19)$$

where $x_{s,t-(n-1)}$ is the traffic value of the previous $n-1$ time points, and so on.

- Auto-Regressive Moving Average (ARMA). ARMA is a more sophisticated parametric method. It predicts the value in the future in terms of two polynomials, one is the auto-regressive (AR) part that involves regressing the variable on its own lagged values, and the other is the moving average (MA) part that involves modeling the error terms occurring contemporaneously. The usual reference for ARMA is ARMA(p,q) model, where p and q are the order of AR part and MA part, respectively, and formulated as

$$x_{s,t+1} = c + \varepsilon(t+1) + \sum_{i=0}^{p-1}\varphi_i x_{s,t-i} + \sum_{i=0}^{q-1}\theta_i\varepsilon(t-i) \quad (20)$$

where φ_i and θ_i are the AR and MA parameters, respectively. $\varepsilon(t)$ is the white noise in time point t, and c is a constant.

- Multitask Learning Neural Network (MTLNN). MTLNN is a non-parametric method proposed by Jin and Sun (*21*). It is a neural network-based MTL method, which models traffic flows at several contiguous time instants as different nodes in the output layer.

Results

Model Results. Since the proposed model performs well by learning multiple tasks collaboratively, especially when the tasks are highly correlated, computation of correlations between different tasks was undertaken first. The basis for correlation computation was the estimation of a simple linear model for each task independently and then calculation of the correlation coefficients between different models. These coefficients were stored as a correlation matrix, of which each entry denotes the correlation between two tasks. Figure 3 shows a plot of the matrix as an image using the command "imshow" of Matlab. The original tasks correlation graph is plotted in Figure 3*a*. The higher the correlation between the two tasks, the whiter the correlation pixel. The tasks are highly correlated except for a small number of special tasks. The correlation graphs under different thresholds are further provided. The color of the pixel with the correlation larger than the threshold is white, otherwise it is black. Figure 3*b* is completely white, which indicates the correlations between tasks are all larger than 0.6. Only a small number of tasks (black part) are with correlation less than 0.8, as shown in Figure 3*c*. Even when the correlation threshold is 0.9, the majority of tasks show correlation with each other, as shown in Figure 3*d*. Learning these highly correlated tasks collaboratively helps improve prediction performance.

Next was the study of the effectiveness of the proposed model. Table 1*a* summarizes the average accuracy of traffic volume prediction, which is grouped by prediction steps. Results show that 1-step ahead prediction achieves the best accuracies in terms of all validation metrics. The MAE and MAPE are 10.56 and 0.1193, respectively. Accuracies decrease to some extent as the

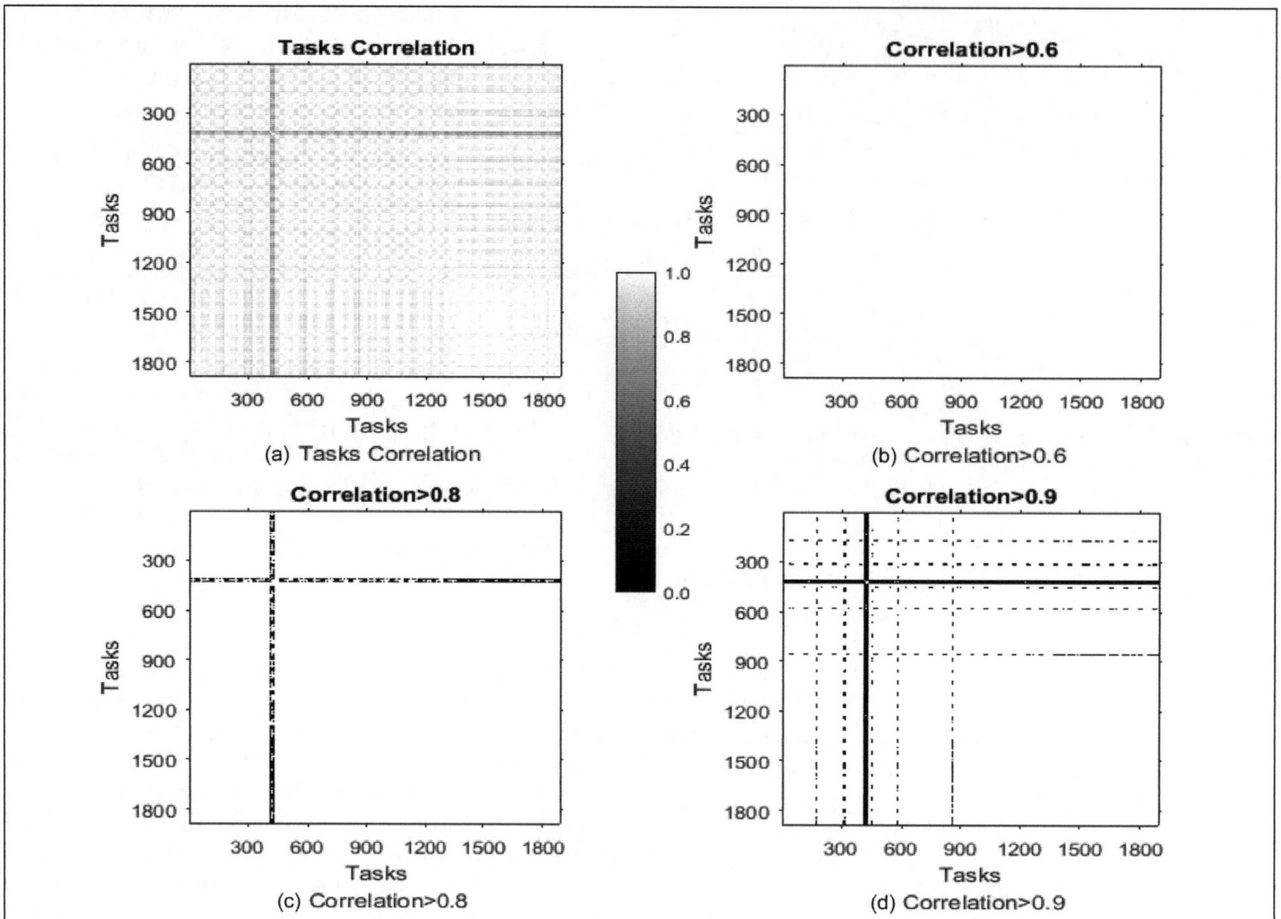

Figure 3. Tasks correlation.

Table 1. Average Accuracy for Traffic Flows Prediction

Metrics	1-step	2-step	3-step	4-step	5-step	6-step
Volume						
MAE	10.56	12.68	14.05	14.79	15.08	15.47
MAPE	0.1193	0.1446	0.1622	0.1762	0.1870	0.2008
RMSE	14.06	17.30	19.32	20.65	21.67	22.10
EC	0.9452	0.9362	0.9280	0.9229	0.9199	0.9170
Occupancy						
MAE	0.0031	0.0036	0.0040	0.0043	0.0045	0.0046
MAPE	0.1359	0.1604	0.1784	0.1944	0.2071	0.2216
RMSE	0.0044	0.0053	0.0058	0.0061	0.0062	0.0064
EC	0.9319	0.9215	0.9137	0.9102	0.9067	0.9041
Speed						
MAE	0.7572	0.8876	0.9572	1.0117	1.0314	1.0413
MAPE	0.0102	0.0118	0.0126	0.0130	0.0132	0.0134
RMSE	0.9895	1.1890	1.2811	1.3215	1.3463	1.3672
EC	0.9925	0.9911	0.9905	0.9900	0.9898	0.9897

number of prediction steps increase. It is consistent with the empirical and theoretical evidence that the knowledge for prediction contained in the training data decreases as the prediction steps increase (*22*). The accuracies of multistep prediction are located in a reasonable range. The MAE and EC of 6-step prediction are 15.47 and 0.9170,

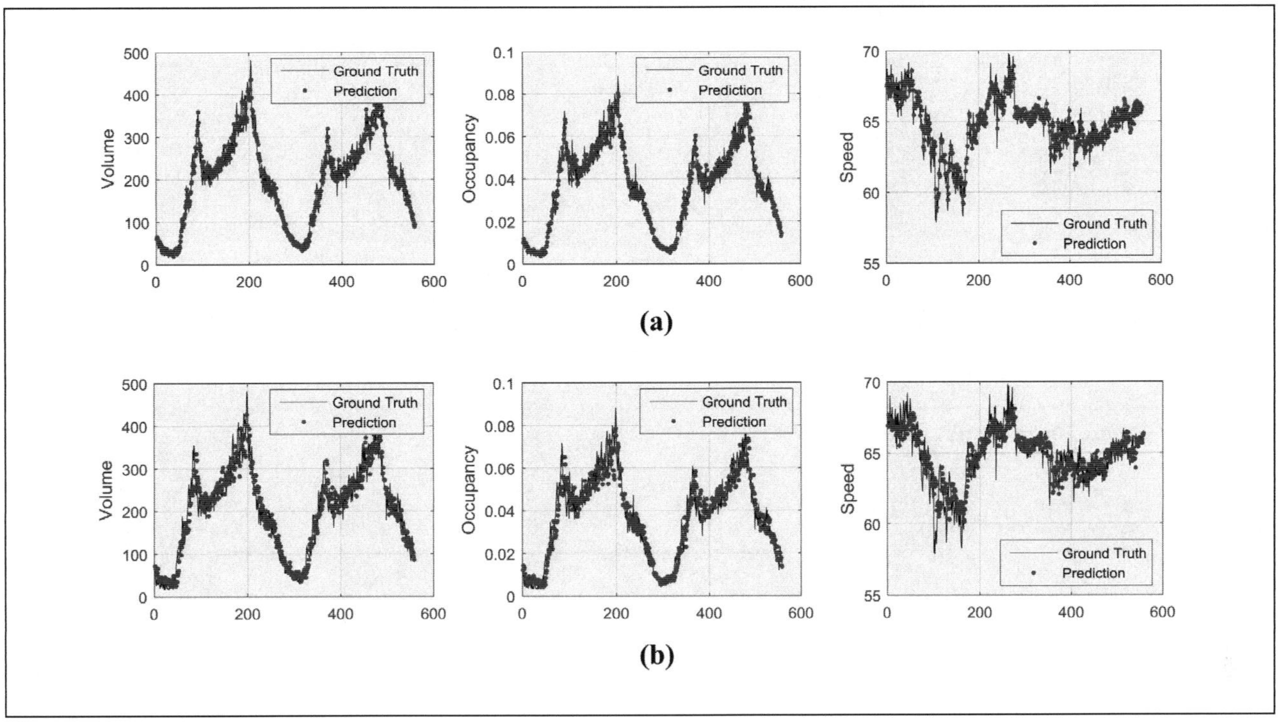

Figure 4. Visual comparison between the ground truth and prediction.
(*a*) Prediction of 1 step ahead.
(*b*) Prediction of 6 steps ahead.

respectively. Table 1*b* and *c* present similar results for traffic occupancy and speed prediction. Speed prediction achieves a relatively higher accuracy. The MAPE of 6-step prediction is 0.0134, which is better than that of volume and occupancy prediction. Note that the model relies only on the traffic flow data and is trained using only the first week data. The promising results demonstrate the effectiveness of the proposed model for multistep traffic flow prediction, even when the training data size is small.

Figure 4 provides the comparison between the true traffic flows and their predictions using the proposed model. For simplicity, the results of 1-step and 6-step ahead prediction for a typical VDS (ID: 601348) on a typical day (Monday) are presented in subgraphs as Figure 4a and b. The three subplots in each subgraph correspond to the result of volume, occupancy and speed, respectively. The ground truths are linked with a black solid line and predictions are denoted with blue points. As shown in Figure 4*a*, the 1-step ahead prediction of volume (left), occupancy (middle) and speed (right) exhibit no apparent bias and have very high correlation with their ground truths. Although some deviations are found as the number of prediction steps increase, the predictions and their ground truths have similar patterns, as shown in Figure 4*b*. This is in agreement with the results in Table 1. The results prove the effectiveness of the proposed model for multistep traffic flow prediction.

To demonstrate the robustness of the proposed model, Figure 5 presents the box plots of MAPE for multistep traffic flows prediction. It is separated by the three traffic flow parameters, and the results with different prediction steps (1~6 on x-axis) are drawn with different colors. Each boxplot is associated with the results of a specific traffic flow parameter with a specific step and summarizes the MAPE scores of all considered VDSs and all days in a week. The median value is denoted as the solid line inside a box, the 25th and 75th percentile values are indicated by the top and bottom edges of the box, and the outliers are shown as red crosses. Among the three traffic flow parameters, speed prediction achieves the best accuracy, while the accuracy of occupancy prediction is relatively poor. With the increase of the number of prediction steps, the prediction accuracies for all traffic flow parameters decrease to some extent. However, they are all located in a reasonable range. Specifically, the medians of MAPE for 6-step traffic flow, occupancy and speed prediction are approximately 0.19, 0.21 and 0.01 respectively. The results prove the robustness of the proposed model for multistep traffic flow prediction.

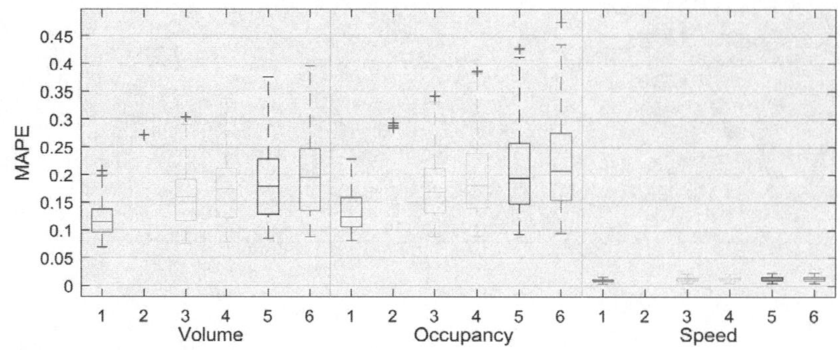

Figure 5. MAPE scores of multistep traffic flows prediction.

Table 2. Average Accuracies of Different Methods

Metrics	Methods	1-step	2-step	3-step	4-step	5-step	6-step
MAE	RW	13.69	15.79	17.62	18.74	19.30	20.17
	MA	13.70	15.22	16.41	17.46	18.56	19.84
	ARMA	13.37	14.91	16.06	16.94	17.71	18.60
	MTLNN	12.74	14.76	15.51	15.95	16.15	16.50
	OA	**10.56**	**12.68**	**14.05**	**14.79**	**15.08**	**15.47**
MAPE (%)	RW	19.19	21.72	24.37	26.26	27.30	28.50
	MA	18.65	21.37	23.05	24.40	25.57	26.83
	ARMA	15.04	16.88	18.71	2022	21.26	22.57
	MTLNN	14.90	16.61	18.05	19.50	20.95	22.23
	OA	**11.93**	**14.46**	**16.22**	**17.62**	**18.70**	**20.08**
RMSE	RW	18.22	21.20	23.73	25.29	26.11	27.31
	MA	18.42	20.58	22.27	23.76	25.34	27.03
	ARMA	17.79	19.90	21.75	23.07	24.16	25.36
	MTLNN	17.02	19.15	20.64	21.16	22.08	23.52
	OA	**14.06**	**17.30**	**19.32**	**20.65**	**21.67**	**22.10**

Results Comparison. The traffic volume prediction accuracy of competing methods is compared in Table 2. Accuracies in terms of MAE, MAPE and RMSE for different prediction steps are given, and OA (our approach) denotes the proposed model. It is clear that the proposed model achieves the best accuracies for all prediction steps and validation metrics. The MAPE of 1-step ahead prediction achieved by the proposed model is 11.93%, which is much better than that achieved by RW (19.19%), MA (18.65%), ARMA (15.04%), and MTLNN (14.90%). For all methods, the accuracy of 1-step ahead prediction is the best. As the number of prediction steps increases, the prediction accuracy of all methods decreases. The RMSE of 1-step and 6-step ahead prediction achieved by the proposed model is 14.06 and 20.81, respectively. The results prove that i) the proposed model outperforms the considered competing methods in multistep traffic flow prediction, and ii) the prediction accuracy of all methods decrease with the increase of the prediction steps.

The traffic occupancy prediction accuracy of competing methods is compared in Figure 6*a*. The box plots of accuracy in terms of EC for different prediction steps are separately given in different subplots. Similar to Figure 5, each boxplot in a subplot summarizes the EC scores of a method for all considered VDSs and all days in a week. Three conclusions can be drawn from this figure. First, the proposed model achieves the best accuracy for multistep occupancy prediction, since the median values (red line in blue box) for different prediction steps are all the highest compared with other methods. Second, the proposed model provides a robust approach for multistep occupancy prediction, since most of the EC scores are located in a relatively small range (blue box). And finally, the proposed model is the most reliable one for multistep occupancy prediction, since the number of prediction outliers (red crosses) is the least compared with other methods.

To further demonstrate the advantages of the proposed model for multistep traffic speed prediction,

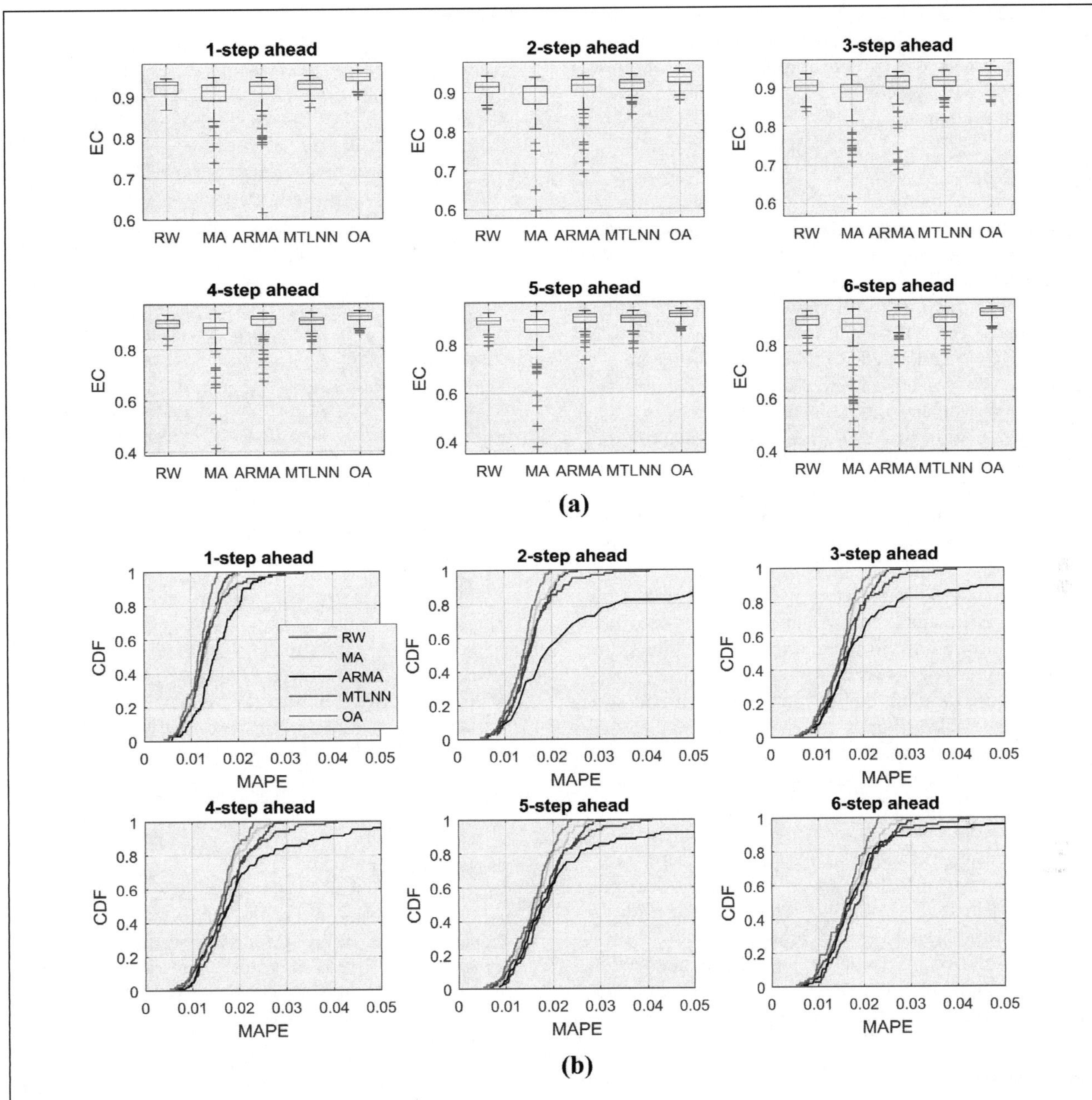

Figure 6. Comparison of different methods for traffic occupancy and speed prediction.
(*a*) EC scores for occupancy prediction with different methods. (*b*) Cumulative distribution functions (CDF) of MAPE for speed prediction with different methods.

Figure 6*b* provides the comparison of prediction accuracy achieved by different methods. The accuracy in terms of MAPE with different prediction steps are separately given in different subplots. Each curve with different color in a subplot is associated with a method and displays the CDF of MAPE scores for all considered VDSs and all days in a week. It is evident that, compared with the methods of RW, MA, ARMA and SVM, the proposed model effectively improves the accuracy of traffic speed prediction. For example, for the 1-step ahead traffic speed prediction, the percentage of MAPE scores of the proposed model less than 0.16 is approximately 100%, which is definitely better than 80% of the methods RW, MA and SVM, and 50% of the method ARMA. It shows the superiority of the proposed model over the competing methods for multistep traffic speed prediction.

Conclusion

This paper presented a spatial–temporal multitask collaborative learning model for multistep short-term traffic flows prediction. The proposed model treats different traffic flow predictions as different supervised machine-learning tasks and learns multiple tasks collaboratively by solving a regularized optimization problem. To address the non-convex and non-smooth challenges, the optimization problem was converted to an equivalent convex optimization problem. Finally, the global optimal solution can be found by solving a variation of the convex optimization problem with an alternating minimization algorithm. The proposed approach models traffic flows from the perspective of network. It incorporates both the spatial correlation between different VDSs and the intrinsic relationships between different traffic flow parameters, as well as the coarse-grain temporal correlation between different days in a week and the fine-grain temporal correlation between different prediction steps.

The proposed model application involved a real case study of the SR180-E freeway in Fresno, California. Empirical results from extensive experiments demonstrate the effectiveness, robustness, and advantages of the proposed model for multistep traffic flow prediction

Recommendations for future research include a study of the correlation between different tasks in depth, further improvement of the prediction accuracy by learning those multiple tasks with stronger relevance, and exploration of other methods to extract and utilize the common information shared across tasks. As well, a need exists for the development and implementation of new algorithms for efficient multi-task learning.

Acknowledgments

This work was sponsored by the National Science Foundation of China (NO. 61374195), the Fundamental Research Funds for the Central Universities and the Jiangsu Province University Graduate Student Research and Innovation Program (NO. KYLX_0180).

Author Contributions

The authors confirm contribution to the paper as follows: study conception and design: Kun Tang, Shuyan Chen, Aemal J. Khattak; data collection: Kun Tang; analysis and interpretation of results: Kun Tang, Shuyan Chen, Aemal J. Khattak; draft manuscript preparation: Kun Tang. All authors reviewed the results and approved the final version of the manuscript.

References

1. Abadi, A., T. Rajabioun, and P. A. Ioannou. Traffic Flow Prediction for Road Transportation Networks With Limited Traffic Data. *IEEE Transactions on Intelligent Transportation Systems*, Vol. 16, No. 2, 2015, pp. 653–662.
2. Kumar, K., M. Parida, and V. K. Katiyar. Short Term Traffic Flow Prediction in Heterogeneous Condition Using Artificial Neural Network. *Transport*, Vol. 30, No. 4, 2015, pp. 397–405.
3. Van Lint, H., W. Schakel, G. Tamminga, P. Knoppers, and A. Verbraeck. Getting the Human Factor into Traffic Flow Models New Open-Source Design to Simulate Next Generation of Traffic Operations. *Transportation Research Record: Journal of the Transportation Research Board*, 2016. 2561: 25–33.
4. Van Wageningen-Kessels, F. Framework to Assess Multiclass Continuum Traffic Flow Models. *Transportation Research Record: Journal of the Transportation Research Board*, 2016. 2553: 150–160.
5. Habtemichael, F. G., and M. Cetin. Short-Term Traffic Flow Rate Forecasting Based on Identifying Similar Traffic Patterns. *Transportation Research Part C: Emerging Technologies*, Vol. 66, 2016, pp. 61–78.
6. Yang, S., S. Shi, X. Hu, and M. Wang. Spatiotemporal Context Awareness for Urban Traffic Modeling and Prediction: Sparse Representation Based Variable Selection. *PLoS One*, Vol. 10, No. 10, 2015. e0141223.
7. Joueiai, M., H. van Lint, and S. Hoogendoorn. Multi-Scale Traffic Flow Modeling In Mixed Networks. *Transportation Research Record: Journal of the Transportation Research Board*, 2015. 2421: 142–150.
8. Tang, J., H. Wang, Y. Wang, X. Liu, and F. Liu. Hybrid Prediction Approach Based on Weekly Similarities of Traffic Flow for Different Temporal Scales. *Transportation Research Record: Journal of the Transportation Research Board*, 2014. 2443: 21–31.
9. Van Lint, J. W., and C. P. Van Hinsbergen. Transportation Research Circular E-C168: Short-Term Traffic and Travel Time Prediction Models. Artificial Intelligence Applications to Critical Transportation Issues. Transportation Research Board of the National Academies, Washington, D.C., 2012. http://www.trb.org/Publications/Blurbs/168134.aspx
10. Weijermars, W., and E. van Berkum. Analyzing Highway Flow Patterns Using Cluster Analysis. *Proc., 2005 IEEE Intelligent Transportation Systems, 2005.*, No. October, 2005, pp. 831–836.
11. Zhao, J., and S. Sun. High-Order Gaussian Process Dynamical Models for Traffic Flow Prediction. *IEEE Transactions on Intelligent Transportation Systems*, Vol. 17, No. 7, 2016, pp. 2014–2019.
12. Rajabzadeh, Y., A. H. Rezaie, and H. Amindavar. Short-Term Traffic Flow Prediction Using Time-Varying Vasicek Model. *Transportation Research Part C: Emerging Technologies*, Vol. 74, 2017, pp. 168–181.
13. Lv, Y., Y. Duan, W. Kang, Z. Li, and F. Y. Wang. Traffic Flow Prediction with Big Data: A Deep Learning Approach. *IEEE Transactions on Intelligent Transportation Systems*, Vol. 16, No. 2, 2015, pp. 865–873.

14. Polson, N. G., and V. O. Sokolov. Deep Learning for Short-Term Traffic Flow Prediction. *Transportation Research Part C: Emerging Technologies*, Vol. 79, 2017, pp. 1–17.
15. Zhang, Y., Y. Zhang, and A. Haghani. A Hybrid Short-Term Traffic Flow Forecasting Method Based on Spectral Analysis and Statistical Volatility Model. *Transportation Research Part C: Emerging Technologies*, Vol. 43, 2014, pp. 65–78.
16. Li, L., S. He, J. Zhang, and B. Ran. Short-Term Highway Traffic Flow Prediction Based on a Hybrid Strategy Considering Temporal–Spatial Information. *Journal of Advanced Transportation*, Vol. 50, No. 8, 2016, pp. 2029–2040.
17. Huang, W., G. Song, H. Hong, and K. Xie. Deep Architecture for Traffic Flow Prediction: Deep Belief Networks with Multitask Learning. *IEEE Transactions on Intelligent Transportation Systems*, Vol. 15, No. 5, 2014, pp. 2191–2201.
18. Yang, S. On Feature Selection for Traffic Congestion Prediction. *Transportation Research Part C: Emerging Technologies*, Vol. 26, 2013, pp. 160–169.
19. Zhang, X., E. Onieva, A. Perallos, E. Osaba, and V. C. S. Lee. Hierarchical Fuzzy Rule-Based System Optimized with Genetic Algorithms for Short Term Traffic Congestion Prediction. *Transportation Research Part C: Emerging Technologies*, Vol. 43, 2014, pp. 127–142.
20. Argyriou, A., T. Evgeniou, and M. Pontil. Convex Multi-Task Feature Learning. *Machine Learning*, Vol. 73, No. 3, 2008, pp. 243–272.
21. Jin, F., and S. Sun. Neural Network Multitask Learning for Traffic Flow Forecasting. *2008 IEEE International Joint Conference on Neural Networks*, 2008, pp. 1897–1901.
22. Cai, P., Y. Wang, G. Lu, P. Chen, C. Ding, and J. Sun. A Spatiotemporal Correlative K-Nearest Neighbor Model for Short-Term Traffic Multistep Forecasting. *Transportation Research Part C: Emerging Technologies*, Vol. 62, 2016, pp. 21–34.

The Standing Committee on Artificial Intelligence and Advanced Computing Applications (ABJ70) peer-reviewed this paper (18-02554).

Article

TRR
JOURNAL OF THE TRANSPORTATION RESEARCH BOARD

Transportation Research Record
2018, Vol. 2672(45) 14–22

DOI: 10.1177/0361198118775839
journals.sagepub.com/home/trr
SAGE

3-D Data Processing to Extract Vehicle Trajectories from Roadside LiDAR Data

Yuan Sun[1], Hao Xu[1], Jianqing Wu[1], Jianying Zheng[2], and Kurt M. Dietrich[3]

Abstract

High-resolution vehicle data including location, speed, and direction is significant for new transportation systems, such as connected-vehicle applications, micro-level traffic performance evaluation, and adaptive traffic control. This research developed a data processing procedure for detection and tracking of multi-lane multi-vehicle trajectories with a roadside light detection and ranging (LiDAR) sensor. Different from existing methods for vehicle onboard sensing systems, this procedure was developed specifically to extract high-resolution vehicle trajectories from roadside LiDAR sensors. This procedure includes preprocessing of the raw data, statistical outlier removal, a Least Median of Squares based ground estimation method to accurately remove the ground points, vehicle data clouds clustering, a principle component-based oriented bounding box method to estimate the location of the vehicle, and a geometrically-based tracking algorithm. The developed procedure has been applied to a two-way-stop-sign intersection and an arterial road in Reno, Nevada. The data extraction procedure has been validated by comparing tracking results and speeds logged from a testing vehicle through the on-board diagnostics interface. This data processing procedure could be applied to extract high-resolution trajectories of connected and unconnected vehicles for connected-vehicle applications, and the data will be valuable to practices in traffic safety, traffic mobility, and fuel efficiency estimation.

High-resolution vehicle data including continuous location, speed, and direction are important to connected-vehicle applications, micro-level traffic performance evaluation, and adaptive traffic control. Connected-vehicle technologies, applications and potential benefits have been studied since 2003 when the U.S. Department of Transportation (U.S. DOT) initiated the Vehicle Infrastructure Integration (VII) program (*1*). The connected-vehicle system provides longer detection distance for drivers (or autonomous vehicles) and pedestrians to "see" around corners or "through" other vehicles so that threats can be perceived earlier. A long list of connected-vehicle applications (*2–4*) was intended to be developed to test technical functionality of the connected-vehicle system, but the testing has not well demonstrated the effectiveness or end-user benefits of the identified applications in the real world. This problem was caused by the scope of proof-of-concept, and the current connected-vehicle system is highly reliant on information broadcast by each vehicle. The maximum safety benefits of current connected-vehicle deployment would need all vehicles to have connected-vehicle devices and broadcast their information in real time. However, mixed traffic with connected-vehicles and unconnected-vehicles will exist in the next decades or even longer. Therefore, supplemental data to help the connected-vehicle deployment needs to be considered.

High-resolution micro traffic data can be collected by conventional probe vehicles with the GPS logging function. However, probe vehicles provide only sample data of the traffic fleet on roads, while the connected-vehicle system needs the data of all road users. The traditional traffic sensors such as loop detectors, video detectors, Bluetooth sensors and radar sensors mainly provide macro traffic data such as traffic flow rates, average speeds and occupancy (*5*) or spot vehicle speed, so the existing sensors cannot provide the continuous micro traffic data needed by connected vehicles. Even the popular crowd-sourced data, such as real time travel time data from Wave, is still macro-level traffic information. A new method to collect high-resolution micro traffic data for the connected-vehicle system is needed to help the current connected-vehicle deployment and future connected-

[1]Department of Civil and Environmental Engineering, University of Nevada, Reno, NV
[2]Soochow University, Suzhou, China
[3]Traffic Engineering & Operations, Reno, NV

Corresponding Author:
Address correspondence to Hao Xu: haox@unr.edu

vehicle applications. The high-resolution micro traffic data will also change existing traffic safety engineering and traffic operation. For example, the micro-level trajectories of vehicles and pedestrians at intersections can be used to analyze intersection traffic safety and signal performance with much more detail than the traditional safety and performance analysis.

A light detection and ranging (LiDAR) sensor transmits and receives laser signals at a high frequency to obtain the point clouds of surrounding objects (*6*). The LiDAR sensor is better than video image processing for its insensitiveness for light variation. The LiDAR sensor usually has a finer resolution compared with Radar. Moreover, the existing Radar sensors deployed in the U.S. only report the spot speed and spot traffic volumes, which cannot support applications of connected vehicles. LiDAR can scan a 360º horizontal range and is able to detect the direction of arrival as well as the number of vehicles accurately. However, it has a relatively high cost compared with other sensors, so its usage was limited. The popularity of autonomous vehicles brought opportunities to the LiDAR sensor. The LiDAR sensor is widely adopted for object perception and recognition in the autonomous vehicle field. As an indispensable component for autonomous vehicles, the price of LiDAR sensors has decreased significantly in recent years. For example, the 2017 LeddarTech Vu-8, which has 215 m of detection range, 100º of horizontal sensing range, and weighs 107 g is now priced at $750. The LiDAR sensor used in this research paper was Velodyne's VLP-16; it was $7,999, with 100 m of detection range, 360º of horizontal sensing range, and weighed 830 g. This trend enables traffic engineers to utilize cheaper LiDARs to obtain high-resolution datasets.

Extensive research has been conducted to apply LiDAR and optical sensors for detecting and tracking vehicles in autonomous sensing systems. The most widely adopted approach to pedestrian and vehicle detection is to combine LiDAR features and vision-based features (*7*). The advantage of using a LiDAR associated with a vision system is that the laser sensor is not very sensitive under extreme weather changes, and the distance or depth measurement is accurate, while at the same time the vision system includes rich information. In Cristiano's research (*8*) for LiDAR data processing, a 15-dimensional LiDAR-based feature vector, Histogram of Oriented Gradients and Covariance image descriptors are extracted. Himmelsbach (*9*) proposed a fast response 3D object perception procedure which could be applied to autonomous vehicles. The uniqueness of his method is that the segmentation was done in a 2.5 D occupancy grid, and classification was done in 3D point clouds. These literature reviews suggest that the LiDAR data processing procedures for autonomous vehicles provide good references for processing the roadside LiDAR sensor data. However, the LiDAR application for roadside traffic surveillance is different from LiDAR application on autonomous vehicles. First, the LiDAR sensors are installed at a fixed place instead of a moving vehicle. Secondly, the system should be able to constantly detect and track vehicles even without a good shape of a vehicle (shaped by 3D LiDAR points). This requires the data processing algorithm to be scalable and robust to track the vehicle with fewer LiDAR points. Thirdly, the LiDAR sensor should be able to track the speed without help of other types of sensors. More sensors, especially different types of sensors, mean more maintenance labor and higher budgets, which may cause unacceptable cost for cities or states. Therefore, a data processing procedure was developed specifically for roadside LiDAR sensor data, which could track vehicles with low LiDAR point density and without the support of other optical sensors. These trajectories could be applied to traffic performance evaluation, safety countermeasures, and traffic operation. The developed procedure is especially important for filling the data gap of existing connected-vehicle deployment.

This paper is organized as follows. The major data processing procedure is presented first. Next, is demonstrated the testing results with LiDAR data collected at different intersections and roads. Finally, all the findings are summarized, and future research is introduced. The data processing procedure proposed in this paper could automatically extract high-resolution vehicle trajectories from LiDAR data.

Procedure of Extracting High-Resolution Trajectories

The LiDAR sensor used in this research has the following specifications: 16 channels, 300,000 points per second, 360º horizontal field of view (FOV) as well as a 15º vertical FOV. The data processing procedure proposed in this paper includes preprocessing of the raw data, statistical outlier removal, a Least Median of Squares (LMedS) based ground estimation method to remove the ground points accurately, vehicle data clouds clustering, a principle component-based oriented bounding box method to estimate the location of vehicle, and a geometrically-based tracking algorithm.

Raw Data Processing

The sensor manual suggested not using the points within a 2-m radius. Hence the points in the 2-m-radius range to the sensor were excluded first. The region of interest (ROI) was then determined based on the road locations and geometry. Four reference points have been selected.

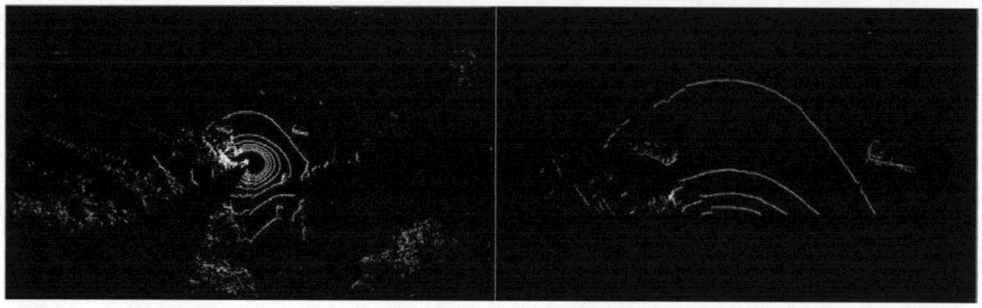

Figure 1. A raw data frame (left) and the same data frame after preprocessing (right).

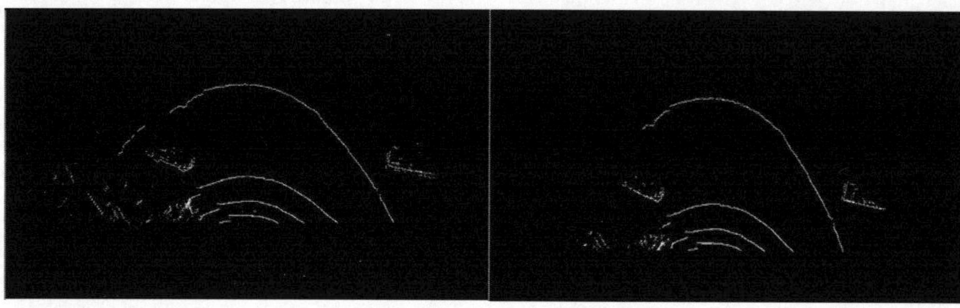

Figure 2. Data frame before outlier remover (left) and after outlier remover (right).

Linking these four points would yield a cuboid area. The space outside of this cuboid area was considered a region of no interest. By trimming out the region of no interest, the buildings, trees, traffic signs and other irrelevant objects would be filtered out from the dataset. Filtering of irrelevant objects decreases the data size for processing, so it increases the efficiency of the follow-up steps. This step also reduces the influence of uninteresting objects. Figure 1 compares a raw data frame and a preprocessed data frame.

Statistical Outlier Removal

This step removes the noise data points near the objects. These noises in each data frame complicate the vehicle identification and tracking and introduce errors of vehicle locations and speeds. A statistical outlier remover algorithm (*10*) was applied at this step. This sparse outlier removal algorithm is based on the computation of the distribution of point to its K neighbor's distances in the input data frame. For each point, the algorithm will compute the mean distance from it to all its neighbors. By assuming that the resulted distribution is Gaussian, all points whose mean distances are outside an interval defined by the global distances mean and standard deviation can be considered as outliers and trimmed from the data frame. Figure 2 presents the comparison of before and after the outlier removal step.

Ground Plane Identification and Segmentation

Correctly and accurately identifying and segmenting the ground plane is an essential step for the proposed data processing procedure. The ground plane here is defined as a group of concentric circle points when LiDAR scanning the ground. Therefore, these sparse noises are subject to being identified and removed. The ground plane is recognized as the noise of the data frame, thus random sample consensus (RANSAC) (*11*), an iterative method to estimate the parameters of mathematical models, was adopted. The input of RANSAC method contains three parts: the data to be processed, a candidate mathematical model fit into the data frame outliers, and parameters related to the estimation of the model. Here a mathematical model was considered as a surface function which could represent the ground plane. There are two key parameters. The first parameter is the number of iterations. The iteration limit was set to 200 by experiments regarding efficiency and successful rate (*12*). The second allowed error parameter is the allowed deviate angle. This deviate angle was set to 15º, which means the direction of the estimated surface could not deviate 15º away

Table 1. Ranking Different Sample Consensus Methods by Average Segmentation Time and Accurate Frame Numbers Against 300 Test Frames

Name	RANSAC	MSAC	MLESAC	rRANSAC	rMSAC	PROSAC	LMedS
Time	1 μs	1 μs	1 μs	2 μs	2 μs	2 μs	1 μs
Accurate frame numbers	231	242	25	187	158	236	291

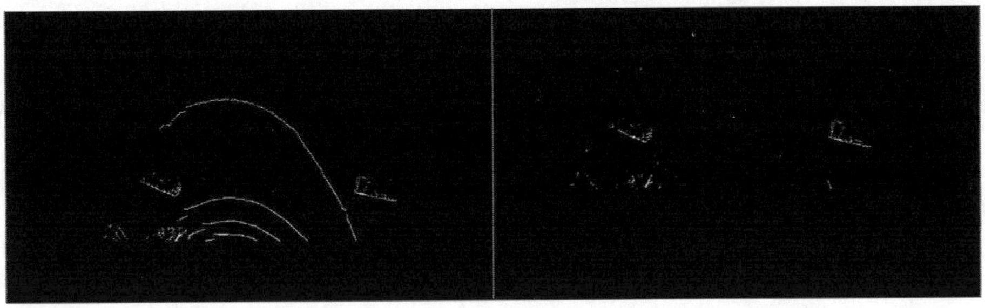

Figure 3. Data frame before ground plane remover (left) and after ground plane remover (right).

from the vertical vector (0, 0, 1). If the estimated surface has a degree larger than this, it will automatically trigger failure segmentation.

The RANSAC iteration method contains four steps. A random subset of the original data was selected first. This subset of data was called hypothetical inliers. Then, the predefined model was fitted to the set of hypothetical inliers. Next, all other data are tested against the fitted model. Those points that fitted this model well would be incorporated into this model. This larger estimation model is considered as a good model if there are enough points in the model. Finally, if the model is not good, another subset of data frame was randomly selected and estimated. RANSAC can estimate the surface function parameters even with significant noises presented in the data frame, but it has no guarantee of delivering optimal solutions within limited iterations. This suggests that by computing a greater number of iterations the optimized solution was approximated. Other variants of RANSAC (*13*) were also tested. RANSAC solves the selection problem as an optimization problem with a bounded loss function. The loss function is defined as a summation of geometric distance for all the points to the estimated plane. When a point is within allowed distance, this point is considered to belong to the estimated plane hence the loss value for this point is set to zero.

LMedS is a modified RANSAC method. LMedS method does not need any tuning variable because it tries to minimize median squared error (*14*). Hence it is considered as a robust estimation. Robust here means fast and insensitive to noises of a data frame. Different modified RANSAC methods were tested against the same data frames; the LMedS model was found to achieve a high accuracy while only consuming a limited computation time, as presented in Table 1. Therefore, it was adopted for ground plane segmentation method in this research. The accurate frame in the table was defined as a frame whose ground plane was perfectly removed and there is no scanning line on the ground left for this frame after ground plane segmentation.

It should be noted that none of the mentioned methods could guarantee a perfect segmentation for all tested data frames. In this research, a follow-up cleaning algorithm was designed. This algorithm removes a 0.2 m thick cuboid by utilizing the plane model parameters, instead of just removing points in a plane. This follow-up algorithm deletes residual ground points around the plane while keeping the vehicle clouds points integrated and completed. Figure 3 indicates the effect of the ground removal step.

Vehicle Clustering

A clustering method is needed to cluster points of different vehicles. The algorithm applied in this research was developed by Rusu (*15*). This clustering algorithm was named Basic Clustering Techniques. The key part is letting the system understand what an object point cluster is, and what makes it different from another. A threshold is required as the input to this algorithm, which is the minimal distance between two cloud sets. Rusu proposed to approximate nearest neighbors via the kd-tree structure (*16*). First, an empty list of clusters was set up, and an empty queue was set up. Then for a random point P

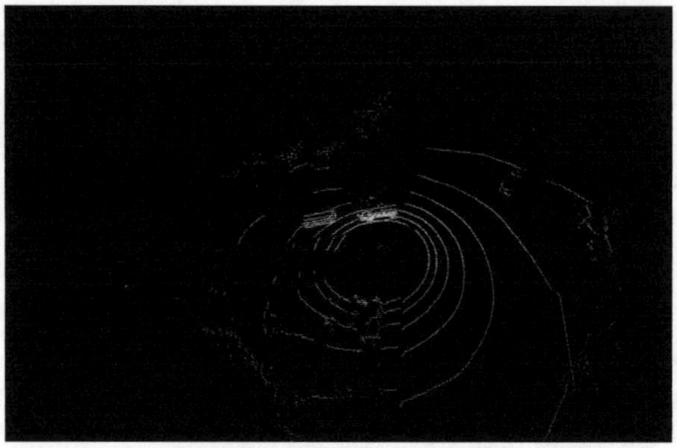

Figure 4. Results of LiDAR points clustering: red points indicate raw data; blue points indicate data after preprocessing; green points indicate clustered vehicles.

in a LiDAR data frame, P is added to the current queue. The algorithm then searches points within a user-defined radius from point P. These neighbor points are marked as processed points and will not be added to another queue. After the search for point P and adding all its neighbor points into the current queue, the corresponding queue will be saved, then a new one is created. The next candidate point is randomly selected, and its neighbor points are searched in the same method. The algorithm terminates when all points have been stored in queues. After clustering, the points are grouped into individual objects and subject to extraction of the useful information. Figure 4 demonstrates clustered result.

Vehicle Identification

Because the LiDAR sensor would penetrate the window of vehicles, or simply the scanning could not cover every corner of the vehicle, the vehicle identification step is to identify the location and shape of vehicles accurately. The space center of a cluster is first identified to represent the location of a vehicle. However, the center of a vehicle shifts frame by frame due to the incomplete structure of objects. The LiDAR sensor can only detect a part of a vehicle body in a frame, so the detected part changes when the vehicle moves. Thus, a reconstruction method is used to restore the shape of vehicles before identification of the actual location of vehicles. It is to find the minimum-volume oriented bounding box (OBB) of a given finite set of N points. There are several methods to identify the OBB for a given finite set of points: O'Rourke's Algorithm, brute-force methods, principal components analysis (PCA) based methods and some other method like $(1 + \varepsilon)$-Approximation (*17*). In this research, a min-max PCA-based OBB method was used to identify vehicle boundaries. This method first calculates the principles (eigenvectors), and then a reference coordinate system is constructed based on these eigenvectors. The original vehicle cluster is translated into this new reference coordinate system, and a bounding box is constructed for the vehicle. This bounding box is then translated back to the original coordinate system and used as the optimum bounding box for this vehicle. The front corner point and back corner point, which are closest to the sensor, are identified as a key data pair to represent the location of this vehicle. Figure 5 indicates the identified vehicle boundary and the points used to track a vehicle.

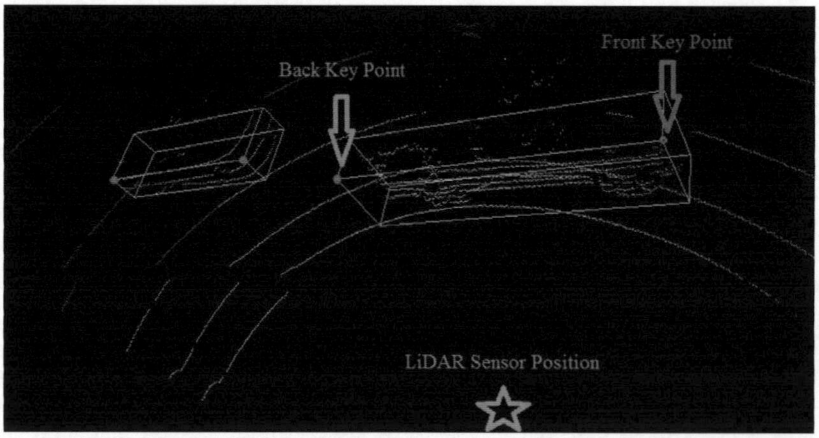

Figure 5. Identified vehicle boundary and key data pair.

Vehicle Tracking

Finally, an object tracking algorithm was applied to track vehicle trajectories. This tracking algorithm utilizes the geometric location information of vehicle key data pair to identify key points in different frames belonging to same vehicles. The algorithm tracks the front key point when the vehicle is approaching the sensor and tracks the back corner key point when the vehicle is leaving the sensor.

Figure 6 describes the whole data processing procedure.

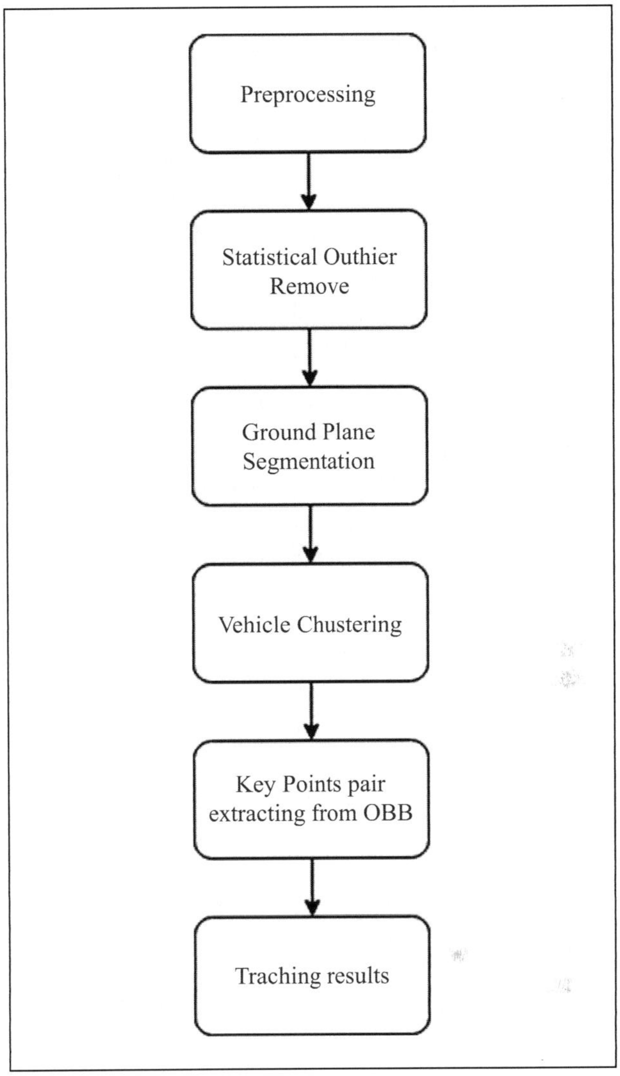

Figure 6. Flow chart of vehicle speed tracking with roadside LiDAR.

Testing Data Sets and Results

The data processing procedure has been tested with data collected at three sites for different traffic scenarios. To compare the tracking results and actual vehicle speeds, the authors installed a logging system at a testing vehicle. The logging system read and stored vehicle speeds from the on-board diagnostics (OBD) interface of the vehicle. This dataset was collected at the University of Nevada, Reno's northern parking lot. Figure 7 presents the comparison of the extracted speeds from LiDAR data and the logged speeds of the testing vehicle. The tracking results of both directions match to the OBD speed very well. This indicates that this data processing procedure is precise and able to track vehicles in both directions.

Another dataset was collected at the intersection of Evans Street and Enterprise Road in Reno, Nevada, which is a two-way-stop-sign intersection with four legs. Trajectories of two vehicles are shown in Figure 8. The leading vehicle stopped at the eastbound Enterprise Road, waited for a gap, and then crossed the intersection. The follow-up vehicle decreased speed and waited for the leading vehicle to pass the street, as the frame index 4000–4500 indicates. The follow-up vehicle then accelerated and approached to stop bar. When there was a gap, it accelerated sharply and crossed the intersection. This experiment demonstrates the effectiveness of data processing framework in a multiple-vehicles passing environment.

The data processing procedure was then applied to the data collected on Kietzke Lane. Kietzke Lane is an arterial road with a speed limit of 40 mph in Reno, Nevada. Figure 9 shows the tracking results. It should be noted that some vehicles slowed down when approaching the sensor. It is possibly the data collection that attracted drivers' attention and caused the speed reduction. The flat speed profile suggests higher speed tracking is also effective and stable.

To test whether this data processing procedure obtains accurate traffic volumes, the research group tested this procedure against another additional dataset collected at the intersection of 15th Street and Virginia Street, Reno Nevada. Vehicle information from 1000 frames of data was extracted by hand and was considered as valid references for comparison. Out of 36 passing vehicles, 35 vehicles were accurately tracked. One vehicle was missing because a near-sensor vehicle proceeded at the same speed with the missing vehicle, and this near-sensor vehicle blocks the missing vehicle's whole body all the way.

Another finding from this research is that the tracking speed of vehicles passing the sensor using this data processing procedure might suddenly drop or increase. The scanning shape of a vehicle may not be complete when this vehicle is close to the sensor, so when tracking the vehicles, the front key point would be cut short. This would cause the drop of the tracking speed. A sudden increase of speed is also possible because the sensor saves the next frame's vehicle data into the current frame, which would cause a sudden extension of the front

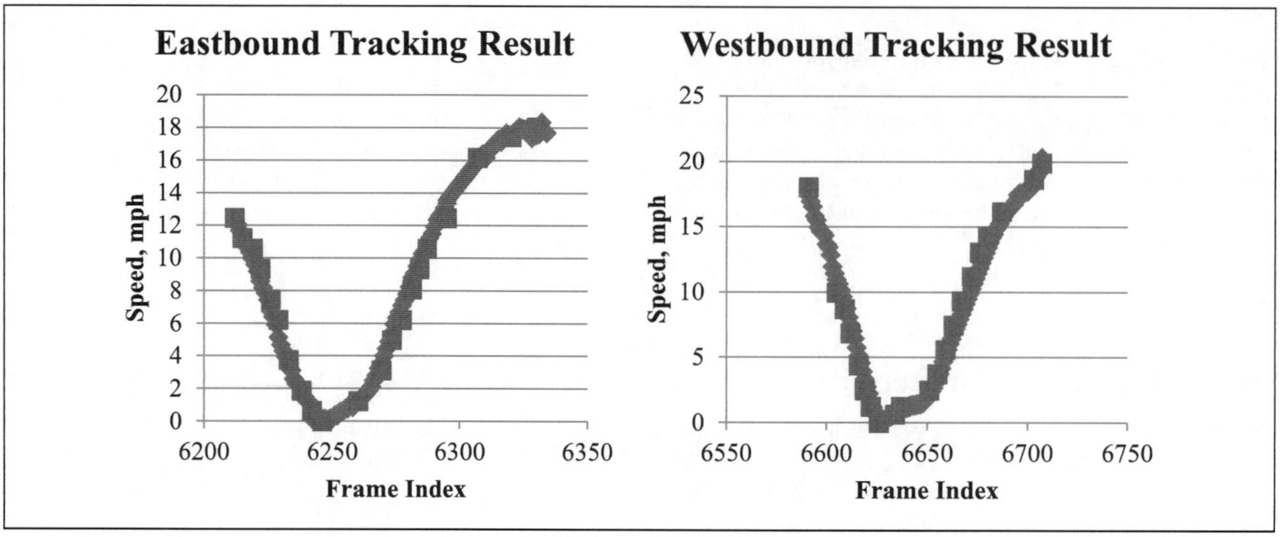

Figure 7. Parking lot tracking results: blue dots indicate tracking result; red dots indicate OBD speed.

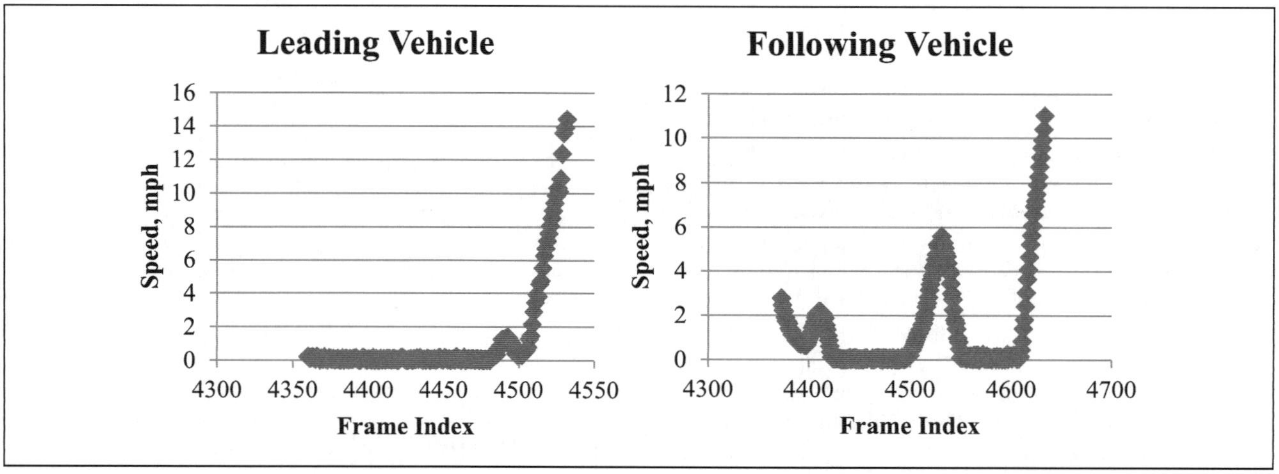

Figure 8. Two vehicle tracking results at the intersection of Evans street and Enterprise road (two-way-stop-sign scenario).

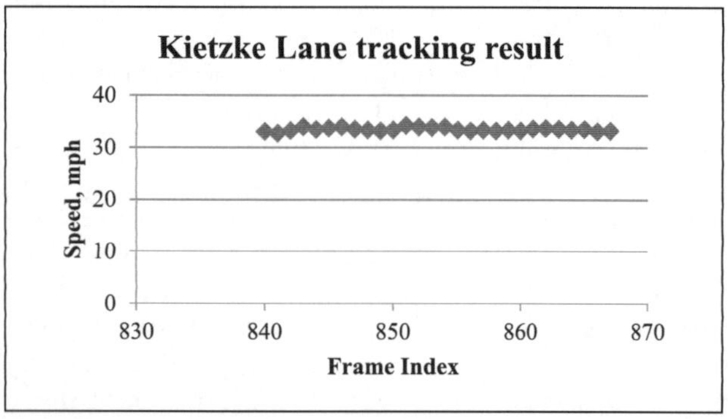

Figure 9. One vehicle tracking result on Kietzke lane (free-flow scenario).

tracking point. The first situation is not likely to happen when there is a distance, preferably 3 m, between the sensor and vehicles. These 3-m distances create space for LiDAR to obtain the complete shape of the vehicles. The second situation is caused by VLP-16 LiDAR itself therefore could not be avoided. Although these scenarios are not common, they can be identified, smoothed, and reported before generating the actual tracking speed profiles in the developed method.

Conclusions and Future Research

This paper presented a data processing procedure to extract vehicle trajectories from roadside LiDAR data.

The procedure includes preprocessing of the raw data, statistical outlier removal, an LMedS based ground estimation method to accurately remove the ground, vehicle data clouds clustering, a PCA-based OBB method to estimate the location of a vehicle, and a geometrically-based tracking algorithm. The documented data processing procedure has been applied to a two-way-stop-sign intersection and free-flow traffic in the Reno area. The results have been compared and verified with speed data from a logger-equipped testing vehicle. A total of 18,000 frames data were collected and tested. The result suggests that the proposed data processing procedure could be used to track vehicle speed at stop-sign intersections and arterial roads, as well as parking lots. This research could be applied to other roads and intersections, and be extended to tracking vehicles in other types of traffic segments, such as four-way-stop-sign intersections or freeways.

This data processing procedure can be applied to extract high-resolution trajectories for various traffic applications. First, by utilizing the roadside LiDAR data processing procedure, connected vehicles could perceive unconnected vehicles. Secondly, traffic safety analysis could be conducted because high-resolution LiDAR data would reveal more details on the roads. Thirdly, traffic counts could be directly extracted from the data processing procedure regardless of daytime or nighttime. This would save time and effort for traffic engineers. Fourthly, this research would benefit operation performance evaluation. Signal evaluation, delay time evaluation or stop time evaluation all will benefit from this research. Finally, the self-detected device could be designed. The ordinary Rectangular Rapid Flash Beacon (RRFB) could be upgraded to detect and predict pedestrian behavior automatically. In summary, the proposed data processing procedure would benefit connected-vehicle applications, traffic safety analysis, traffic mobility analysis, operation performance evaluation and self-detected traffic device design.

Future research will investigate how to integrate multiple LiDAR sensors to extend the detection range. The extension of detection range requires additional algorithms to fuse multiple datasets, while real-time tracking demands the procedure be fast enough to generate and display result within 0.1 seconds. In particular, this paper has not researched bad weather datasets, such as heavy rain and snow, which will introduce noises to the dataset. It is even possible for them to blind the sensor. How to filter snow or rain noise, and still achieve a satisfactory detection range, speed, and accuracy remains to be researched.

Acknowledgments

This research was funded by the SOLARIS Institute, a Tier 1 University Transportation Center (UTC) under Grant No. DTRT13-G-UTC55 and matching funds by the Nevada Department of Transportation (NDOT) under Grant No.P224-14-803/TO #13. The authors gratefully acknowledge this financial support. This research was also supported by engineers with the Nevada Department of Transportation, the Regional Transportation Commission of Washoe County, Nevada, and the City of Reno.

Author Contributions

The authors confirm contribution to the paper as follows: study conception and design: Yuan Sun, Hao Xu; data collection: Hao Xu, Jianqing Wu, Jianying Zheng, Kurt M. Dietrich; analysis and interpretation of results: Yuan Sun, Hao Xu; draft manuscript preparation: Yuan Sun, Hao Xu. All authors reviewed the results and approved the final version of the manuscript.

References

1. Farradyne, P. *Vehicle Infrastructure Integration (VII) Architecture and Functional Requirements, Version 1.1.* FHWA US Department of Transportation, 2005.
2. Lee, J., and B. Park. Development and evaluation of a cooperative vehicle intersection control algorithm under the connected vehicles environment. *IEEE Transactions on Intelligent Transportation Systems*, Vol. 13, No. 1, 2012, pp. 81–90.
3. Goodall, N., B. Smith, and B. Park. Traffic signal control with connected vehicles. *Transportation Research Record: Journal of the Transportation Research Board*, 2013. 2381: 65–72.
4. Rios-Torres, J., A. Malikopoulos, and P. Pisu. Online Optimal Control of Connected Vehicles for Efficient Traffic Flow at Merging Roads. *Proc., 2015 IEEE 18th International Conference on Intelligent Transportation Systems (ITSC)*, Canary Islands, Spain, IEEE, 2015, pp. 2432–2437.
5. Mimbela, L. E. Y., and L. A. Klein. *Summary of Vehicle Detection and Surveillance Technologies Used in Intelligent Transportation Systems.* FHWA Office of Highway Policy Information, 2007.
6. Sun, Z., G. Bebis, and R. Miller. On-road vehicle detection: A review. *IEEE Transactions on Pattern Analysis and Machine Intelligence*, Vol. 28, No. 5, 2006, pp. 694–711.
7. Premebida, C., G. Monteiro, U. Nunes, and P. Peixoto. A LiDAR and Vision-based Approach for Pedestrian and Vehicle Detection and Tracking. *Proc., 2007 IEEE Intelligent Transportation Systems Conference*, Seattle, WA, IEEE, 2007, pp. 1044–1049.
8. Premebida, C., O. Ludwig, and U. Nunes. LIDAR and Vision-based Pedestrian Detection System. *Journal of Field Robotics*, Vol. 26, No. 9, 2009, pp. 696–711.
9. Himmelsbach, M., A. Müller, T. Luettel, and H.-J. Wünsche. LIDAR-based 3D object perception. *Proc., 1st International Workshop on Cognition for Technical Systems*, Munich, Vol. 1, 2008, pp. 1–7.
10. Rusu, R. B., Z. C. Marton, N. Blodow, M. Dolha, and M. Beetz. Towards 3D point cloud based object maps for

household environments. *Robotics and Autonomous Systems*, Vol. 56, No. 11, 2008, pp. 927–941.
11. Fischler, M. A., and R. C. Bolles. Random Sample Consensus: A Paradigm for Model Fitting with Applications to Image Analysis and Automated Cartography. *Communications of the ACM*, Vol. 24, No. 6, 1981, pp. 381–395.
12. Zuliani, M. *RANSAC for Dummies with Examples Using the RANSAC Toolbox for Matlab and More*, University of California at Santa Barbara, November 2008.
13. Choi, S., T. Kim, and W. Yu. Performance evaluation of RANSAC family. *Journal of Computer Vision*, Vol. 24, No. 3, 1997, pp. 271–300.
14. Simpson, D. Introduction to Rousseeuw (1984) Least Median of Squares Regression. In *Breakthroughs in Statistics*, Springer, New York, NY, 1997, pp. 433–461.
15. Rusu, R. B. Semantic 3d Object Maps for Everyday Manipulation in Human Living Environments. *KI-Künstliche Intelligenz*, Vol. 24, No. 4, 2010, pp. 345–348.
16. Bentley, J. L. Multidimensional Binary Search Trees Used for Associative Searching. *Communications of the ACM*, Vol. 18, No. 9, 1975, pp. 509–517.
17. Chang, C.-T., B. Gorissen, and S. Melchior. Fast Oriented Bounding Box Optimization on the Rotation Group SO (3, ℝ). *ACM Transactions on Graphics (TOG)*, Vol. 30, No. 5, 2011, p. 122.

The Standing Committee on Information Systems and Technology (ABJ50) peer-reviewed this paper (18-00110).

TRR
JOURNAL OF THE TRANSPORTATION RESEARCH BOARD

Article

Transportation Research Record
2018, Vol. 2672(45) 23–35

DOI: 10.1177/0361198118786645
journals.sagepub.com/home/trr
SAGE

Integrating Remote Sensing Data in Decision Support Systems for Transportation Asset Management

Andrea Vaccari[1], Tamal Batabyal[1], Nazia Tabassum[1], Edward J. Hoppe[2], Brian S. Bruckno[3], and Scott T. Acton[1]

Abstract

The objective of the research summarized in this manuscript is two-fold: to develop and make available a set of novel analysis tools, fully integrated in ArcGIS, that leverages the rich information provided by satellite-based remote sensing data to detect and characterize geohazards of interest to the transportation community; and to provide a modern web-based decision support system (DSS) in which these novel analysis products can be seamlessly integrated with existing datasets. Specifically, interferometric synthetic aperture radar (InSAR) and its derivatives are employed in this work. Each tool is designed to target a specific asset of interest to the transportation community: sinkhole detection, bridge differential settlement detection, slope stability assessment, and pavement distress evaluation. Furthermore, to allow interested parties to test the introduction of remote sensing data into their decision-making process without requiring the full integration with existing workflows, a modern web-based DSS is introduced. Thanks to the large coverage and the high update frequencies offered by satellite-based remote sensing data, datasets augmented with these new products can offer a near real-time overview of the network status and lead to better support in decisions about resources allocation, existing infrastructure monitoring, new infrastructure planning, and post-construction assessment.

Satellite-based interferometric synthetic aperture radar (InSAR) techniques have proven to be invaluable tools to accurately monitor minute displacements of the Earth's surface over large areas (*1*). Detecting developing geohazards (*2*), monitoring current (*3*) and historical (*4*) displacements of the transportation infrastructure, and optimizing the design and placement of critical infrastructures (*5*) are just a few examples of the potential applications.

There are two challenges associated with these novel datasets. The first is the development of analysis tools that can synthesize them into products meaningful to the transportation community; the second is the seamless integration of both data and products with existing transportation datasets and within transportation agencies' workflows to offer a clear overview of the current status and support in the management of transportation assets.

This study addresses both challenges by introducing a new set of tools that leverages these novel datasets to generate products providing detection and characterization of pavement motion, bridge settlement, and subsidence development. Furthermore, it introduces an open source web-based decision support system (DSS) in which these products are merged with existing Virginia DOT (VDOT) pavement condition datasets to provide an enhanced network level overview of the transportation infrastructure status over two regions in Virginia.

Background

Remote Sensing Technology

Synthetic aperture radar (SAR) is a remote sensing technique that allows operators to achieve high ground resolution by combining the echoes of rapid radar pulses emitted by the satellite (*6*). Thanks to the coherent nature of the radar emissions, each SAR image contains both amplitude and phase information about each ground resolution unit within the imaged area, with the

[1]Charles L. Brown Department of Electrical and Computer Engineering, University of Virginia, Charlottesville, VA
[2]Virginia Transportation Research Council, Charlottesville, VA
[3]Virginia Department of Transportation, Staunton, VA

Corresponding Author:
Address correspondence to Andrea Vaccari: vaccari@virginia.edu

phase representing the time delay of the echo received from that location. The difference between the phases of two SAR images, taken at different times and/or vantage points, generates an "interferogram" (InSAR) (*7*). Variations in intensity in the interferogram represent a combination of both changes in the topographical profile and potential crustal deformation within the imaged region. A technique known as differential InSAR (DInSAR) takes advantage of a digital elevation model (DEM) of the imaged area to remove the topographic effects and quantify the deformation along the line-of-sight (LOS) of the satellite. Although this technique has been successfully used to measure large volcanic deformations (*8*), the achievable resolution is ultimately limited by the phase errors introduced by the atmospheric effects. To resolve this problem, an advanced DInSAR technique called permanent/persistent scatterer InSAR (PSInSAR) was introduced (*9*). Leveraging the difference in nature between the atmospheric phase interference (random) and the phase changes introduced by ground deformation (e.g. linear, quadratic, etc.), PSInSAR can remove the former and preserve the latter. For those locations (permanent/persistent scatterer – PS) that have a large signal-to-noise ratio across several SAR images, PSInSAR can deliver deformation measurements with an accuracy of less than 1 cm (0.4 in.). For each PS, the deformation rate and acceleration, elevation, as well as the time series of the deformation are provided. PS are typically found in correspondence with *locations* that provide excellent reflection of radar signals (e.g. rocky outcrops, lamp post, crash barriers, transmission towers, fences, buildings, etc.). In rural areas, the density of PS can be extremely low. This issue was greatly reduced by the introduction of an improved version of PSInSAR in which an additional layer of statistical analysis is used to also identify *regions* providing a consistent and traceable response to the radar pulses over time. This new algorithm, SqueeSAR (*10*), was used for the analysis of the SAR data in this study.

Areas of Interest (AOI)

This study focused on two regions (Figure 1*a*): (AOI1), centered over the town of Middlebrook in central Virginia, and (AOI2), including Northern Virginia and a section of Washington, D.C.

AOI1 was chosen because of the diversity of the geological conditions (*2*). Over 100 lane miles of interstate and over 200 lane miles of primary and high-volume secondary routes as well as approximately 600 bridges, culverts, and major transportation structures coupled with the strong karst component of this AOI and its historic recurrence of subsidence events, provide the ideal test bed for the analysis tools developed in this study.

By comparison AOI2, including most of the Washington, D.C. metropolitan area, with its high density of urban infrastructures and many transportation features including numerous bridges and areas with slope stability issues, as well as some of the highest volume traffic routes in Virginia, is an ideal location to evaluate roads' response to high traffic loading.

Together these two areas provide an invaluable coverage of transportation infrastructures and potential geohazards that was key in evaluating the efficacy of the proposed tools and the value of the integrated and synthesized analysis offered by the DSS.

Datasets

SqueeSAR Dataset. The AOI1 dataset was generated by applying the SqueeSAR algorithm to 62 SAR images acquired by the Cosmo-SkyMed (CSK) satellite constellation (*11*) from 29 August 2011 to 24 November 2014 approximately every 16 days (with a 232-day interval between 25 October 2012 and 30 June 2013). The images, each covering a 40 km × 40 km (24.85 mi × 24.85 mi) with a ground resolution of 3 m × 3 m (9.84 ft × 9.84 ft), were acquired along an ascending orbit with a LOS vector forming a 24.23° angle with the geographic north and a 12.52° angle with the vertical. The resulting dataset contained about 800 k scatterers with an average density of 497/km^2 (1,287/mi^2) and an average displacement rate accuracy of ± 0.45 mm/year (± 0.02 in./year).

Similarly, the AOI2 dataset was generated from 28 SAR images acquired by CSK from 28 March 2014 to 19 June 2015. The images, with the same ground resolution as those over AOI1, were acquired along a descending orbit with a LOS vector forming a 12.87° angle with the geographic north and a 24.03° angle with the vertical. The resulting dataset contained about 3.6 M scatterers with an average density of 2,278/km^2 (5,899/mi^2) and an average displacement rate accuracy of ± 2.3 mm/year (± 0.09 in./year).

The SqueeSAR datasets for the two AOIs are illustrated in Figure 1, *b* and *c*.

Pavement Quality Dataset. The pavement quality datasets, collected by the VDOT's contractor, Fugro-Roadware Inc., are acquired each year according to different schedules depending on the type of road: entire interstate, primary highway systems, and approximately 20% of secondary systems of highway network (requiring 5 years to provide full coverage) (*12*). These datasets contain multiple indices characterizing the properties of each 0.1 mi (0.16 km) of the pavement. Among the indices, all fully accessible through the DSS, this study focused on the critical condition index (CCI) and the international roughness index (IRI) (*12*). The CCI covers the 0–100

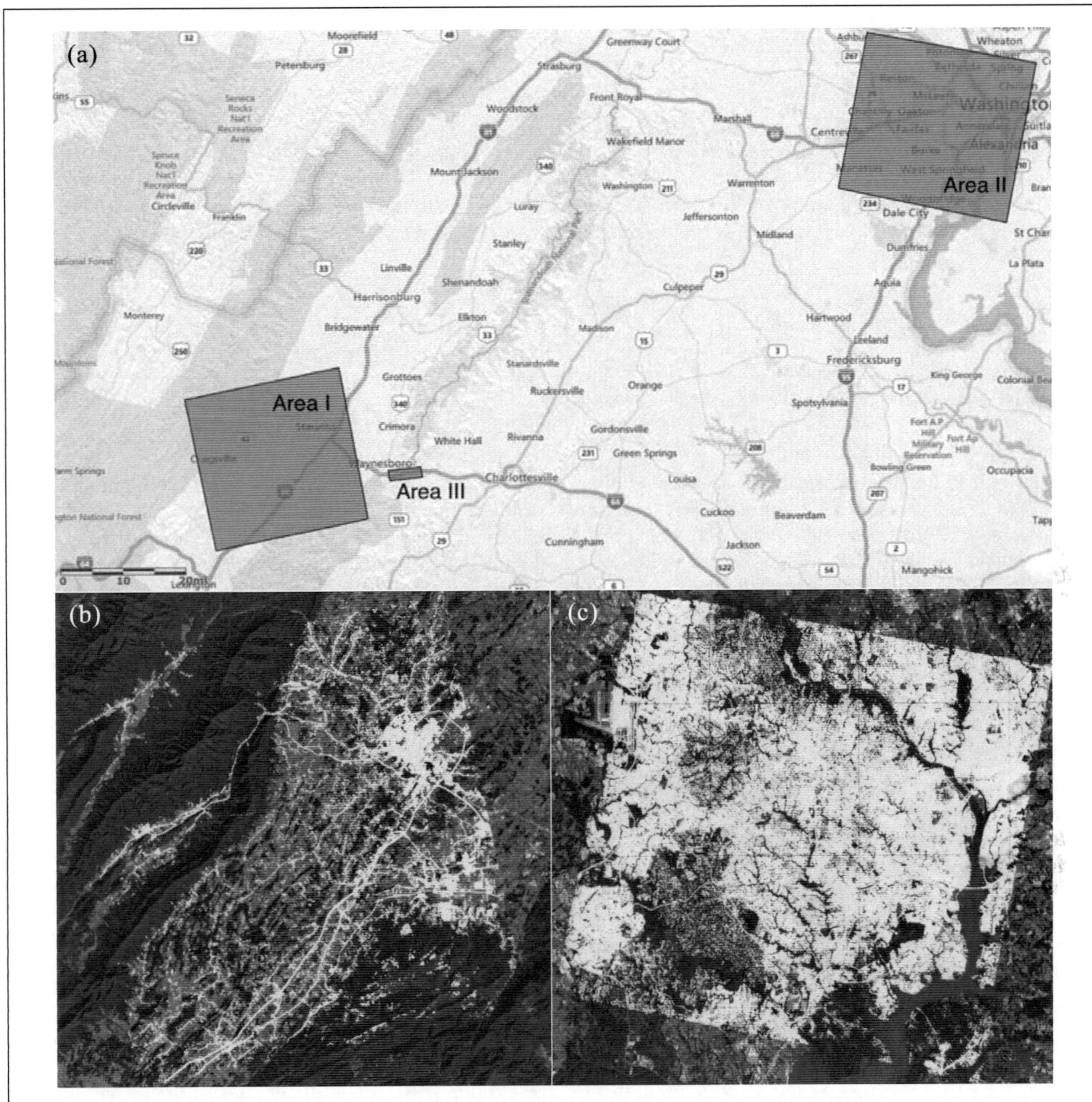

Figure 1. (*a*) Areas of interest. AOI1 and AOI2. SqueeSAR dataset: (*b*) AOI1, (*c*) AOI2. Each point represents a scatterer and is color-coded based on the average displacement rate (red to blue with red representing negative values).

range, it is calculated based on several other measurements, and it is the main index used by VDOT to guide the decision on maintenance and rehabilitation actions. There are five classes: 0–49 (very poor), 50–59 (poor), 60–69 (fair), 70–89 (good), and 90–100 (excellent). Stretches of road with CCI values less than 60, are considered for rehabilitation.

The IRI, on the contrary, provides a direct measurement of the pavement roughness by integrating the pavement surface irregularities (displacements) over distance and reporting the values in in./mi or cm/km. There is not a direct correlation between IRI and CCI but tables linking the two for different types of road have been created (*13*). In general, for interstates and primary roads, IRI

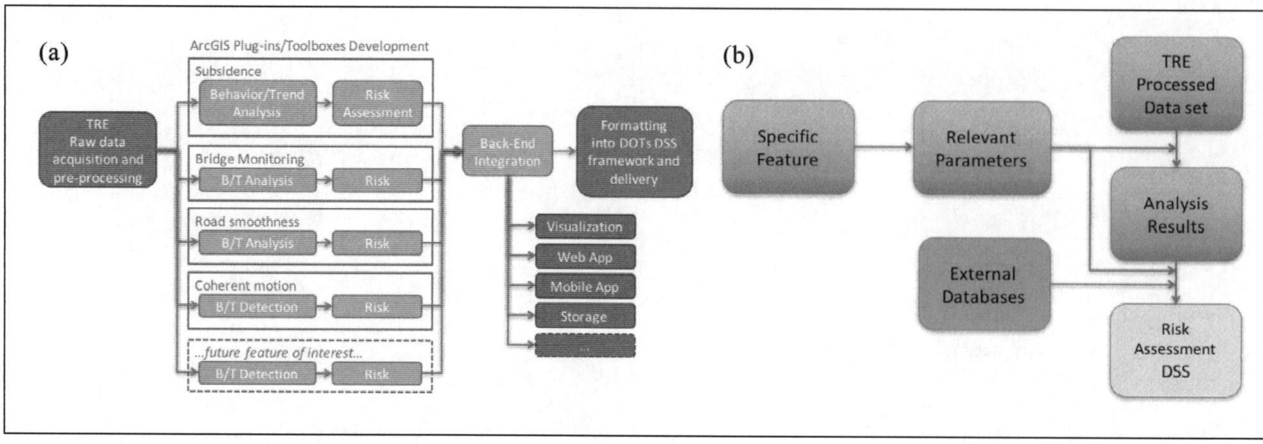

Figure 2. (*a*) Processing pipeline and (*b*) general plug-in/toolbox flow.

values above 140 are considered to be of poor quality while this value is increased to 220 for secondary roads.

The datasets covering the two AOIs were converted into georeferenced shapefiles and merged with the SqueeSAR data creating a unique dataset providing a direct comparison during the period of the SAR acquisitions (2011–2015).

Analysis Tools

Processing Pipeline

One of the goals of this study was to develop new tools that leverage remote sensing datasets and easily integrate with agencies' existing workflow (Figure 2*a*). Each tool requires pre-processed data. For this study, this step was performed by TRE Altamira using their proprietary algorithm SqueeSAR but any permanent scatterer interferometry algorithm that provides the times-series of the displacements could be utilized. Based on these datasets, several decision support tools were developed, with each tailored to the detection and analysis of a specific transportation asset:

- Subsidence analysis for sinkholes.
- Bridge motion analysis for bridge differential settlement.
- Coherent motion analysis for slope stability.
- Road smoothness analysis for pavement distress evaluation.

For each feature of interest listed above, key parameters were identified. For example, for subsidence: location, size, and growth rate. Based on these parameters, a model of the feature was created and used to analyze the pre-processed dataset to identify regions where the behavior of the data matched the model or triggered either warning or critical thresholds defined by the model. The result of the analysis is a "model-matching" or "threshold-triggering" map that can be merged with existing datasets to further refine the results and provide the final assessment (Figure 2*b*).

Two of the main advantages offered by the remote sensing datasets are coverage and frequency. New CSK SAR images are acquired every 16 days because of the orbital design, at which point the entire analysis pipeline can be repeated resulting in updated products. Comparing this to a typical pavement or bridge monitoring schedule, it becomes apparent that being able to access this data in parallel with the existing datasets would provide a valuable infrastructure asset management tool.

All of the tools, except the subsidence detection, were implemented as user-configurable ArcGIS python toolboxes allowing for easy integration in agencies' existing workflows. The subsidence analysis was implemented as a standalone application (because of computational requirements) generating geo-referenced TIFF images (geoTIFF) easily importable into ArcGIS. Although ArcGIS was chosen as the primary production and integration backend, every tool generates outputs that can be easily ported to any other modern GIS environment. To this point, a proof of concept web-based DSS application (described later) that integrates the tools' output with the VDOT pavement quality databases and Google Street View was developed.

Subsidence Detection

Subsidence is a geohazard that is relevant not only for transportation, as it can bring the closure of major highways, but also for the population at large because of the often-devastating consequences brought on by the development of sinkholes.

Leveraging the point cloud nature of the SqueeSAR datasets, a spatiotemporal approach merging the key concepts of a matched filter (*14*) and Hough transform (*15*) was embedded in a tool able to detect and track subsiding regions. The method can be summarized as:

1. Define a spatiotemporal model describing the expected changes in crustal elevation because of a developing subsidence.
2. Identify the key parameters that regulates the model behavior: location, size, and growth rate.
3. Quantize and define the limits of the parameter space.
4. For each set of parameters in the parameter space, generate a template from the model.
5. Search the dataset for regions (group of scatterers) whose spatiotemporal behavior matches the template.
6. Quantify the matching level (0 – no match, 1 – perfect match).

For this study, the developed model was based on measurements performed on existing sinkholes and defined as a Gaussian of fixed width linearly growing in time:

$$g(x,t) = \alpha t \exp\left[-(x - x_0)^2/2\sigma^2\right]$$

where

x = location currently under analysis (e.g. m [ft]);

x_0 = location of the center of the developing subsidence (e.g. m [ft]);

α = growth speed of the subsidence (e.g. mm/month [in./month]);

t = time since the first SAR image (e.g. month);

σ = width of the detected subsidence (e.g. m [ft]).

Different examples of the tool use are illustrated in Figure 3. The top 32 detections in AOI1 (Figure 3*d*) were selected for ground validation. Based on the observed infrastructure deformation and geomorphology at the inspected locations, VDOT geologists determined that 78% of these detections presented strong evidence of subsidence (*2*).

Originally designed in MATLAB and C++ , this application, currently available for both Windows- and OS X-based (Macintosh) machines, was designed to allow fine control of the analysis parameter and to produce results in a geoTIFF format.

Bridge Motion Analysis

Monitoring bridges is a critical task for state DOTs. Currently at VDOT this is performed on a fixed two-year schedule as costly physical inspection by personnel. Nevertheless, slow deterioration of the internal structure can sometimes go undetected increasing the risk of weakening to the point of collapse when under stress. The bridge motion analysis tool was designed to leverage the time series of the remote sensing data and provide frequent and accurate routine monitoring of all the bridges within an imaged area to quickly identify dangerous changes and trends.

The toolbox takes two shapefiles as input: the pre-processed remote sensing data and a shapefile identifying the bridges' location and extents. For each bridge, the relevant scatterers are isolated and undergo a first screening where the median height is extracted and only points within a user-specified distance from this value are preserved. This insures that only scatterers measuring the bridge are used by the algorithm. Users can select a warning and a critical threshold. After consulting with a VDOT bridge maintenance program manager, the default thresholds were set to 0.5 in./month (12.7 mm/month) (critical) and 1.0 in./year (25.4 mm/year) (warning). The displacement time series of each scatterer is interpolated to evaluate the average displacement rate of the corresponding location for each month. If any of the calculated rates exceeds the critical threshold, a flag is raised and the bridge is marked accordingly. Monthly rates are aggregated by evaluating their median value which is then compared against the warning threshold.

For each bridge, the analysis result is stored within the original shapefile as additional attributes seamlessly integrating the analysis results with the existing data. Figure 4,*a* and *b* illustrates an example of this analysis for two bridges on I81. When integrated in ArcGIS, the height and average displacement rate of each scatterer associated with the bridge are displayed on the map, and the bridge shape is colored according to the analysis results (Figure 4*b*).

Coherent Motion Analysis

The coherent motion analysis toolbox is based on the Theil–Sen trend estimator and it was developed to detect regions, within the user-selected analysis area, showing indications of strong linear motion: grooves/ruts on a road/pavement, collapsing culverts, creeping slopes and, in general, any event expected to generate displacements with strong linear behavior in time (*16*).

The Theil–Sen estimator is applied to the time series of the displacements to robustly evaluate the median of the displacement rates effectively reducing the contribution of potential contribution from the SAR additive and multiplicative noise. These rates are then checked against the user-selected warning and critical thresholds (given in mm/day) and each scatterer is colored accordingly. For

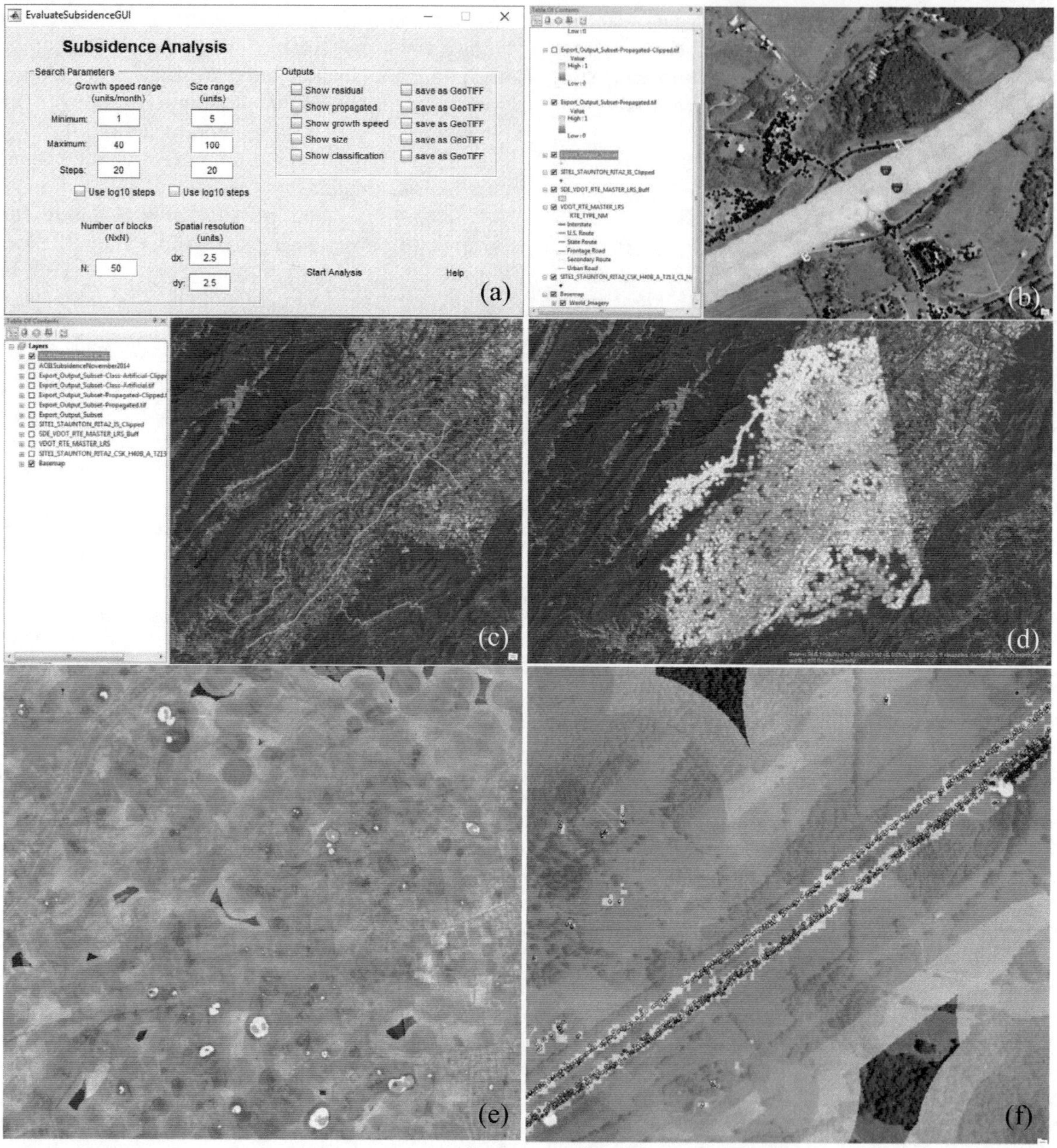

Figure 3. Subsidence analysis. Warmer colors indicate potential subsidence development. (*a*) User interface. (*b*) Analysis of a section of I81N where subsidence activity was identified. (c) Network level analysis (interstates and primaries) in AOI1. (*d*) Region analysis (entire AOI1). (e) Detail of post processing within the ArcGIS environment. The gray scale image represents the results of the subsidence analysis where the darker color indicates a higher degree of match with the subsidence model. The blue lines identify primary roads in the region while the yellow, orange, and red areas indicate the top matches of the subsidence algorithm. (*f*) Example of multi-layer integration providing a wide set of focused information about a specific location. Notice the yellow and orange subsidence detections in the bottom left and top right corner. Also notice the agreement between the displacement rates (red identifies negative rates) of the scatterers (dots), and the subsidence detections. By combining several layers, it was possible to clearly determine the presence of active subsidence on the shoulder of the northbound lane of I81.

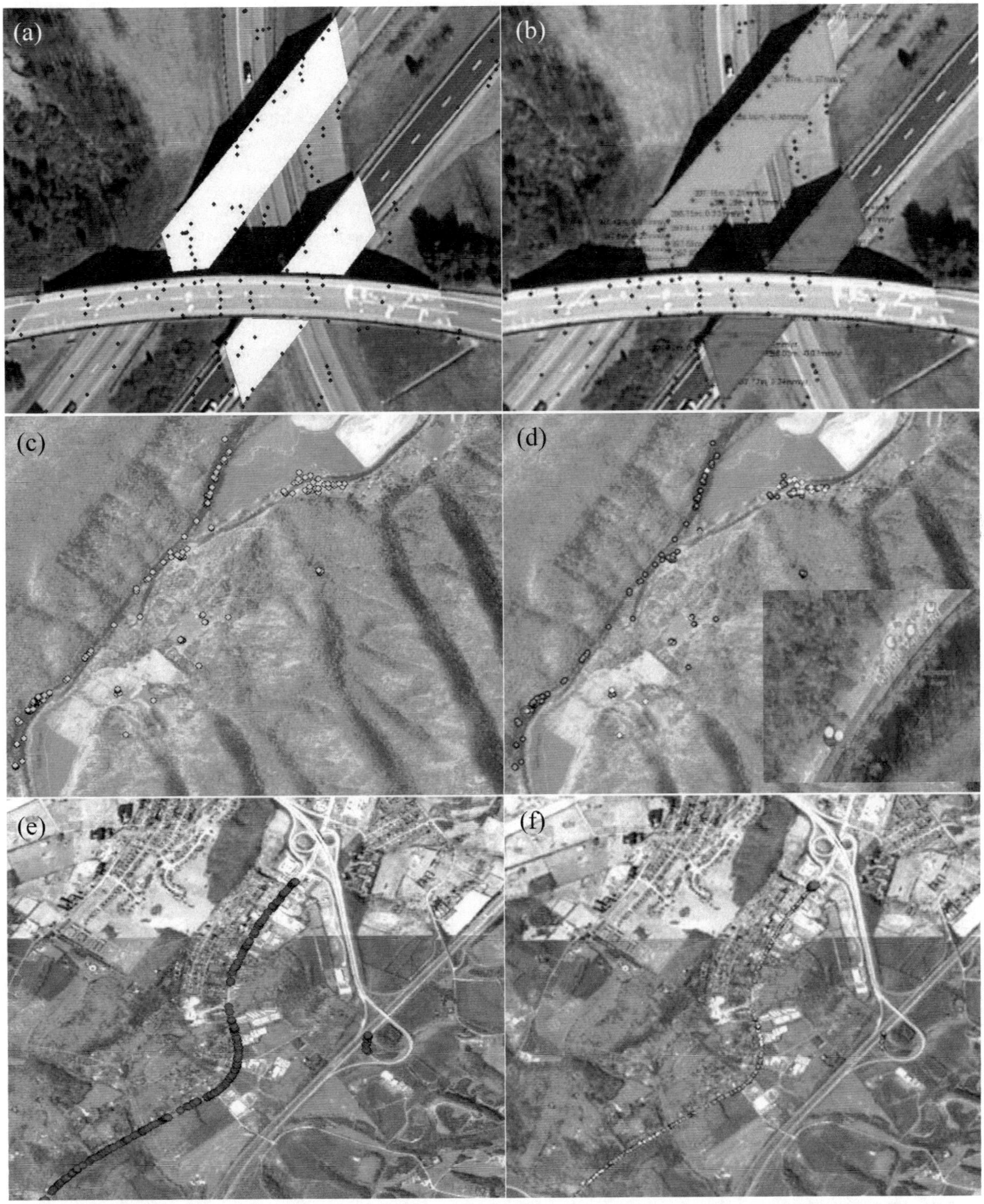

Figure 4. (*a, b*) Bridge motion analysis. (*a*) The two inputs shapefiles: bridges (light green) and SqueeSAR (purple dots). (*b*) The output of the toolbox in which the bridges were marked according to the analysis results: ok (green), warning (yellow), critical (red). (*c, d*) Coherent motion analysis applied to an active slope on Route 600. (*c*) Scatterers along the road (blue dots). (*d*) Scatterers colored according to the results of the analysis. Several active regions were detected with critical linear trend identified at the active slope (bottom left). Detail of slope (inset). (*e, f*) Road smoothness analysis on a road section. (e) Original scatterers (purple dots). (*f*) The scatterers have been colored (green, yellow, and red) corresponding to the results.

this study, the thresholds were defined after discussions with VDOT pavement engineers.

The results are once again stored as additional attributes within the original SqueeSAR shapefile. Figure 4, *c* and *d* illustrates an example of the analysis applied to a region including an active slope along Route 600 in western Virginia.

Road Smoothness Analysis

Smoothness is a critical road parameter as it directly affects the riding perception. Furthermore, roughness almost always inversely correlates with pavement quality. On scales, larger than typical IRI measurements and determined by the density of scatterers along the road, it is possible to define the normalized smoothness index (NSI) computed at each scatterer location and for each satellite acquisition time. This index is calculated using a graph theoretical approach and measures the smoothness of a signal defined over a graph (*17, 18*). In this case, for each (i-th) scatterer, the graph vertices are defined by the locations of all the (N_s) scatterers within a certain user-defined distance or scale (s). The graph edges are obtained by connecting each vertex to all the other vertices and their weights are defined as a function of the physical distance between the vertices they connect. The graph signal (f_t) is defined by the displacements measured at the vertices for a given acquisition time (t). Every graph is then represented by its graph Laplacian (L_i). When L_i is applied to a signal it behaves similarly to the differential operator and can be used to provide an aggregated measure of the discrepancy between the displacements of neighboring connected scatterers.

The hypothesis behind this approach is that regions of the road surface displaying large differential displacements between neighboring scatterers (large NSI) are prone to pavement deterioration. The difference in scale between the NSI and IRI measurements means that direct correlation has only been observed for rare stretches of road with very high PS densities. Nevertheless, NSI can be helpful in identifying locations with potential large-scale stresses. The NSI in the toolbox is implemented as:

$$NSI(s, i, t) = \frac{1}{N_s}\left(\frac{f_t^T L_i f_t}{f_t^T f_t}\right)$$

where

N_s = number of scatterers within a selected distance, or *scale* (s), from the (i-th) scatterer;

L_i = *Laplacian* of the graph with vertices defined by the N_s scatterers;

f_t = displacements measured at the vertices of the graph at a given acquisition time (t).

Figure 4,*e* and *f* illustrates the results of this tool applied to a small section of road. As with the other tools, the results are stored as an additional attribute within the SqueeSAR shapefile. Furthermore, if clusters of scatterers are identified as being in the warning or critical range, an extra shapefile is generated by finding the smallest region that encloses all the identified scatterers.

Web-Based Network Level Decision Support System

To showcase the portability of the analysis results and provide interested transportation agencies with a framework to test the introduction of the remote sensing data and products described above into their decision-making process without requiring the full integration with existing workflows, a DSS web application, based on open source libraries as well as widely used GIS data standards was developed. The application was written in JavaScript (JS) and leverages the following open source libraries: *leaflet* (*http://leafletjs.com*) the leading JS library for mobile-friendly interactive maps, *jQuery* (*https://jquery.-com*) a fast, small, and feature-rich JS library providing optimized access to HTML document manipulation, event handling, animation, and asynchronous Java access (Ajax) to large datasets, and *Chart.js* (*http://www.chartj-s.org*) a simple and flexible JS charting library.

The current version of this portable and modular framework runs in all modern web-browsers (a live demo is available at *http://viva-lab.ece.virginia.edu/viz/viz.html*) and delivers a clear summary of the overall network status (Figure 5*b*) by allowing the users to integrate the following datasets:

- Pavement quality data
- Pavement motion analysis (coherent motion analysis tool)
- Bridge motion analysis (bridge motion analysis tool)
- Subsidence motion analysis (subsidence detection tool)

A menu, typically hidden until selected, provides simple access to various visualization options for each one of the datasets that are independently handled by standalone classes within the code allowing for simultaneous visualization (Figure 5*a*).

The base map is displayed using the standard tiling approach used by most mapping applications, therefore it is easily extended to include any base map the user might need. The data format used by the application is *GeoJson* (*http://geojson.org*) a geospatial data interexchange standard based on JS Object Notation (RFC7946). As both GeoJson and shapefile are

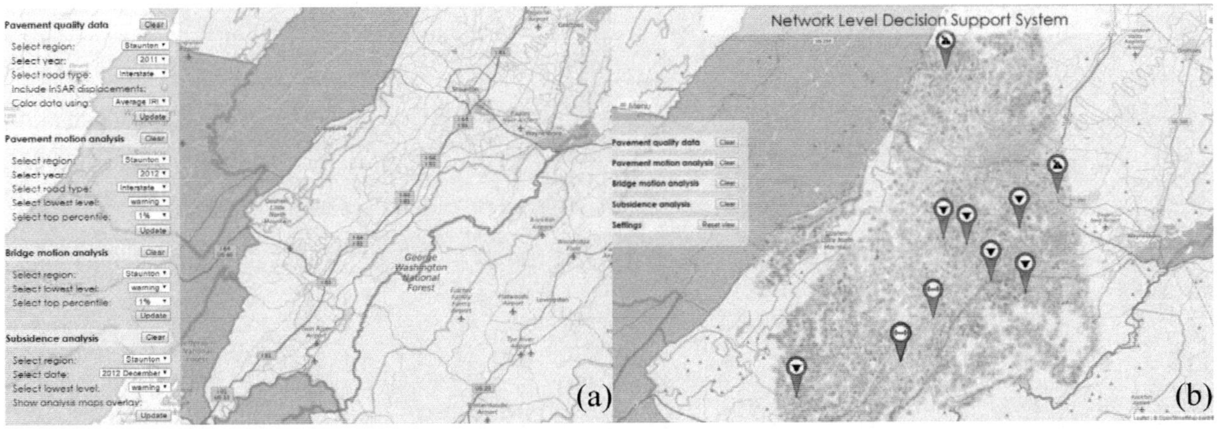

Figure 5. (*a*) Detail of fully expanded DSS user interface. (*b*) Example of top 10% critical detection of each toolbox visualized simultaneously with the pavement condition and the subsidence analysis overlay (gray).

standards in the geospatial community, several tools exist to convert the default shapefile output to the GeoJson format. For example, the basic license for ArcGIS for Desktop includes the *FeaturesToJSON_conversion*() function within the *conversion toolbox* that would allow a potential user to convert their existing datasets for easy inclusion within the DSS.

Pavement Quality Data

This section offers a modern interface to the visualization of the VDOT pavement quality data. The user can apply several filters to select between the regions covered by the satellite (AOI1 – Staunton and AOI2 – NOVA), the years in which satellite data was available (2011–2015), the type of road (interstate, primary, secondary, all) and two options for data coloring (IRI and CCI). All these parameters can be easily extended to include more data and more options depending on the user's needs. The ability to visualize several years of pavement quality data allows users to quickly identify regions where pavement quality indices are changing. In Figure 6, *a* and *d* it is possible to clearly notice that most of the warm colors are slowly turn to green between 2011 and 2014 indicating a successful paving campaign. The same approach could be used to quickly identify roads where the pavement is deteriorating.

All of the information stored by VDOT is still available and can be visualized as a popup window by simply clicking on one of the 0.1 mi sections of road. Furthermore, the user can visualize the Google Street View for the selected location easily (Figure 6*e*).

To allow for a direct comparison with the remote sensing data, the user can visualize the merged pavement and SqueeSAR dataset. When using this merged dataset, the DSS will visualize the scatterers identified along the roads and, for each of them, both VDOT pavement data and full SqueeSAR information are available. Furthermore, the user has the option to visualize a chart showing the time series of the measured displacement as an overlay chart (Figure 6*f*). This feature could be used, for example, to check if the settlement of a newly constructed road is within specifications and track its evolution over time.

Pavement Motion Analysis

For this and the following analysis datasets, the DSS allows the user to simply visualize those locations that triggered either the warning or the critical threshold using a semaphore-like color coding (yellow for warning and red for critical detections) and a different icon for each of the datasets, thus providing a visual summary of the current network status, (Figure 5*b*). Depending on the available data, different visualization options are going to be available to the user.

The pavement motion analysis dataset was obtained using the coherence motion analysis toolbox. For this dataset, the user can select the region of interest, the year for which the analysis was performed, the type of road on which the analysis was performed, the triggered threshold (warning or danger), and the top percentile of detections to visualize (1%, 5%, 10%, 25%, or 100%), allowing for a fine tuning of the desired visualization. Figure 7*a* illustrates an example where all the warning and above detections over AOI1 primary roads for the year 2014 were selected. For each shown detection, it is possible to open a Google Street View visualization and,

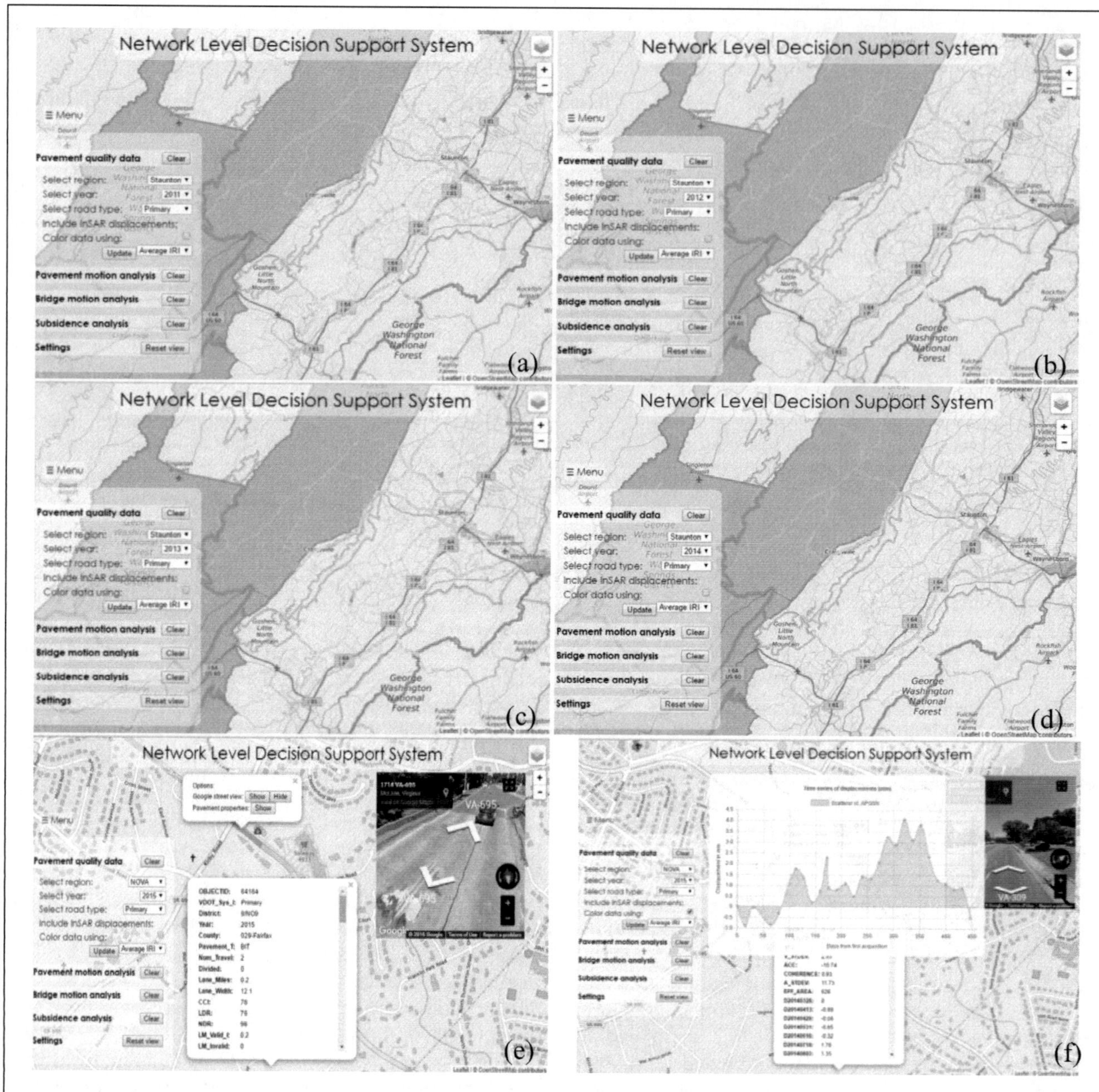

Figure 6. (*a–d*) Example of evolution of primaries pavement IRI in AOI1 (red corresponds to higher values): (*a*) 2011, (*b*) 2012, (*c*) 2013, and (*d*) 2014. (e) Example of detailed pavement information and Google Street View for a section of poor quality pavement in AOI2. (*f*) Example displacement chart and detailed information when using the joint pavement and SqueeSAR datasets.

when used in conjunction with the pavement quality and SqueeSAR dataset, all the additional information described in the previous section are available as well. This provides a very powerful tool that allows DOT engineers to study potential correlations between the pavement quality indices, the SqueeSAR data, and the detected motion while simultaneously providing a quick visual identification of the location thanks to the navigable Google Street View inset.

Bridge Motion Analysis

When visualizing the results of the bridge motion analysis toolbox, the same set of parameters, semaphore-style

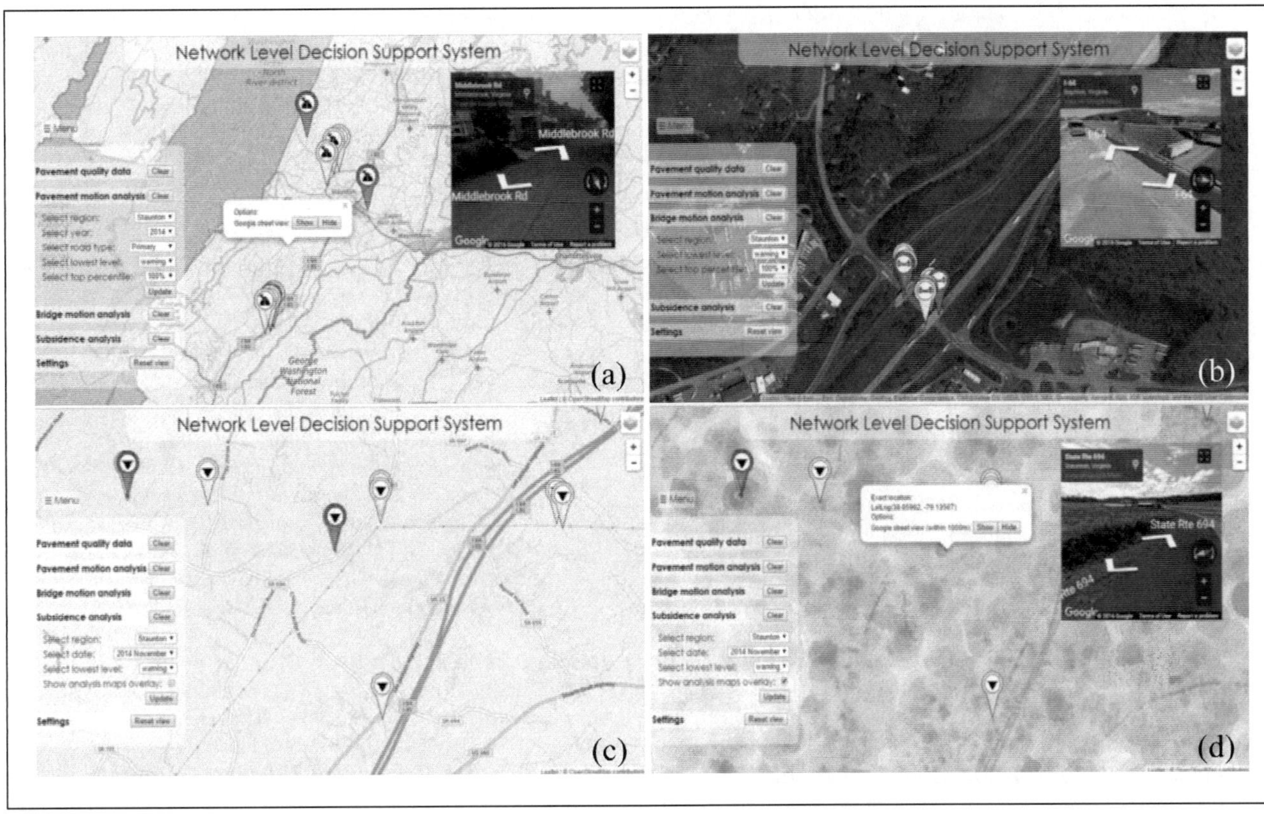

Figure 7. (*a*) Visualization of the pavement motion analysis results for 2014 data over primary roads in AOI1. (*b*) Example of bridge motion analysis results displayed over a satellite view base map. (*c*) Semaphore visualization of locations and contours of the detected subsiding regions within a subset of AOI1. (*d*) Subsidence analysis visualization including full analysis overlay, popup coordinates information for one of the detections as well as the closest Street View available within 1km from the detection location.

pins, and navigable Google Street View inset are available (Figure 7*b*).

Subsidence Analysis

The subsidence analysis module allows the visualization of the potential geohazards detected by the subsidence detection standalone tool previously described.

A script was created that compares the values in the geoTIFF output, produced by the subsidence analysis tool, with thresholds that were defined based on values that have been observed to correctly identify developing sinkholes (*2*). This comparison is used to generate two shapefiles: one that provides the location and severity of the detections (again semaphore-style), and the other that identifies contours of the detected region. An example of the visualization of these two files for a small subset of AOI1 is shown in Figure 7*c*.

A second output of the script is a tiled version of the original geoTIFF output showing the results of the subsidence analysis. The user can choose to visualize this as a grayscale map overlay in the DSS. Currently, the darker colors indicate regions where the detected motion more closely matches the subsidence model previously described, whereas the brighter regions indicate areas where no subsidence matching the model was detected.

When one of the detection pins is selected, a popup window will display its exact coordinates (in latitude and longitude) as well as the option to toggle the closest Google Street View within 1km (Figure 7*d*).

The ability to quickly generate an overview of potential active subsiding regions in relation to the location of a transportation asset provides an incredibly powerful tool in the planning of a new asset location as illustrated in Bruckno et al. (*5*).

Conclusions

- Satellite-based InSAR technology provides continuous large area coverage with a high update frequency.
- When combining several SAR images, InSAR-derived products can measure the Earth's surface

with an average accuracy of the order of ± 0.45 mm/year (± 0.02 in./year) allowing for mm-range displacement monitoring over time.
- A set of novel GIS tools that leverages these high-resolution time series of displacements to provide an accurate analysis of assets of interest to the transportation community (sinkhole detection, bridge differential settlement detection, slope stability assessment, and pavement distress evaluation) was developed.
- Integrating these remote sensing products, specifically designed to analyze the transportation infrastructure asset, with existing datasets provides an augmented overview of the network status and leads to an improved decision-making workflow.
- A modern web-based DSS that can summarize and integrate several remote sensing products with existing transportation datasets was developed and introduced to provide an up-to-date and complete overview of the transportation infrastructure condition.

Acknowledgments

The authors would like to acknowledge the Office of the Assistant Secretary for Research and Technology of the U.S. Department of Transportation for supporting the two projects (RITARS-11-H-UVA and RITARS-14-H-UVA) through which the tools described in this study were developed and the data acquired and analyzed. Rob Minford of VDOT and Dan Widner of the Virginia Information Technology Agency (VITA) for their help in converting the existing pavement quality datasets into shapefiles. Tod Springer, bridge maintenance program manager at VDOT for his direction in selecting the warning and critical thresholds for the bridge movement analysis. Adrian Bohane, Dr. Giacomo Falorni, and Vicky Hsiao of TRE Altamira, who provided data analysis and technical assistance on the remote sensing part of the study.

Author Contributions

The authors confirm contribution to the paper as follows: study conception and design: Andrea Vaccari, Edward J. Hoppe, Brian S. Bruckno, and Scott T. Acton; data collection: Edward J. Hoppe, Brian S. Bruckno (and TRE Altamira); analysis and interpretation of results: all authors; draft manuscript preparation: Andrea Vaccari. All authors reviewed the results and approved the final version of the manuscript.

References

1. Ferretti, A. *Satellite InSAR Data: Reservoir Monitoring from Space*, 1st ed. EAGE, The Netherlands, 2014.
2. Vaccari, A., M. Stuecheli, B. Bruckno, E. Hoppe, and S. T. Acton. Detection of Geophysical Features in InSAR Point Cloud Data Sets Using Spatiotemporal Models. *International Journal of Remote Sensing* Vol. 34, No. 22, 2013, pp. 8215–8234.
3. Hoppe, E., B. Bruckno, E. Campbell, S. Acton, A. Vaccari, M. Stuecheli, A. Bohane, G. Falorni, and J. Morgan. Transportation Infrastructure Monitoring Using Satellite Remote Sensing. In *Materials and Infrastructures 1* (J.-M. Torrenti, and F. La Torre, eds.), Wiley-ISTE, 2016, pp. 185–198.
4. Hoppe, E. J., Y.-J. Kweon, B. S. Bruckno, S. T. Acton, L. Bolton, A. Becker, and A. Vaccari. Historical Analysis of Tunnel Approach Displacements with Satellite Remote Sensing. *Transportation Research Record: Journal of the Transportation Research Board*, 2015. 2510: 15–23.
5. Bruckno, B., A. Vaccari, E. Hoppe, S. Acton, and E. Campbell. Integration and Delivery of Interferometric Synthetic Aperture Radar [Insar] Data into Storm Water Planning Within Karst Terranes. *Proc., 14th Multidisciplinary Conference on Sinkholes and the Engineering and Environmental Impacts of Karst*, Rochester, MN, 2015.
6. Sherwin, C. W., J. P. Ruina, and R. D. Rawcliffe. Some Early Developments in Synthetic Aperture Radar Systems. *IRE Transactions on Military Electronics*, Vol. 1051, No. 2, 1962, pp. 111–115.
7. Rosen, P. A., S. Hensley, I. R. Joughin, F. K. Li, S. N. Madsen, E. Rodriguez, and R. M. Goldstein. Synthetic Aperture Radar Interferometry. *Proceedings of the IEEE*, Vol. 88, No. 3, 2000, pp. 333–382.
8. Lu, Z. InSAR Imaging of Volcanic Deformation Over Cloud-prone Areas–Aleutian Islands. *Photogrammetric Engineering & Remote Sensing*, Vol. 73, No. 3, 2007, pp. 245–257.
9. Ferretti, A., P. Claudio, and R. Fabio. Permanent Scatterers in SAR Interferometry. *IEEE Transactions on Geoscience and Remote Sensing*, Vol. 39, No. 1, 2001, pp. 8–20.
10. Ferretti, A., A. Fumagalli, F. Novali, C. Prati, F. Rocca, and A. Rucci. A New Algorithm for Processing Interferometric Data-stacks: SqueeSAR. *IEEE Transactions on Geoscience and Remote Sensing*, Vol. 49, No. 9, 2011, pp. 3460–3470.
11. Agenzia Spaziale Italiana. *Cosmo-skymed System Description & User Guide*. Publication ASI-CSM-ENG-RS-093-A, ASI, 2007.
12. Papagiannakis, A., N. Gharaibeh, J. Weissmann, and A. Wimsatt. *Pavement Scores Synthesis*. Publication FHWA/TX-09/0-6386-1. Federal Highway Administration, U.S Department of Transportation, Washington, D.C., 2009.
13. Virginia Department of Transportation, Maintenance Division. *State of the Pavement 2016*. www.virginiadot.org/info/state_of_the_pavement.asp. Accessed 21 July 2017.
14. Turin, G. An Introduction to Matched Filters. *IRE Transactions on Information theory*, Vol. 6, No. 3, 1960, pp. 311–329.
15. Hough, P. V. C. Machine Analysis of Bubble Chamber Pictures. *Proc., 2nd International Conference on High-Energy Accelerators*. Geneva, Switzerland, 1959.
16. Sen, P. K. Estimates of the Regression Coefficient Based on Kendall's Tau. *Journal of the American Statistical Association*, Vol. 63, No. 324, 1968, pp. 1379–1389.

17. Shuman, D. I., et al. The Emerging Field of Signal Processing on Graphs: Extending High-Dimensional Data Analysis to Networks and Other Irregular Domains. *IEEE Signal Processing Magazine*, Vol. 30, No. 3, 2013, pp. 83–98.
18. Batabyal, T., A. Vaccari, and S. T. Acton. LaWeCo: Active Region Detection in Non-uniformly Sampled Data Using Laplacian-weighted Covariance. *2016 IEEE Southwest Symposium on Image Analysis and Interpretation*, Santa Fe, NM, 2016, pp. 129–132.

The Standing Committee on Information Systems and Technology (ABJ50) peer-reviewed this paper (18-00489).

The views, opinions, findings and conclusions reflected in this presentation are the responsibility of the authors only and do not represent the official policy or position of the US Department of Transportation/Office of the Assistant Secretary for Research and Technology, or any state or other entity.

Article

TRR
JOURNAL OF THE TRANSPORTATION RESEARCH BOARD

Transportation Research Record
2018, Vol. 2672(45) 36–44
© National Academy of Sciences:
Transportation Research Board 2018
Article reuse guidelines:
sagepub.com/journals-permissions
DOI: 10.1177/0361198118775876
journals.sagepub.com/home/trr
SAGE

Performance of Hyperspectral Imaging with Drone Swarms

Raj Bridgelall[1], James B. Rafert[2], and Denver D. Tolliver[3]

Abstract

The ongoing proliferation and diversification of remote sensing platforms offer greater flexibility to select from a range of hyperspectral imagers as payloads. The emergence of low-cost unmanned aircraft systems (drones) and their launch flexibility present an opportunity to maximize spectral resolution while scaling both daily spatial coverage and spatial resolution simultaneously by operating synchronized swarms. This article presents a model to compare the performance of hyperspectral-imaging platforms in their spatial coverage and spatial resolution envelope. The authors develop a data acquisition framework and use the model to compare the achievable performance among existing airborne and spaceborne hyperspectral imaging vehicles and drone swarms. The results show that, subject to cost and operational limitations, a platform implemented with drone swarms has the potential to provide greater spatial resolution for the same daily ground coverage compared with existing airborne platforms.

Hyperspectral imaging has the potential to identify land cover materials, obstacles, and dynamic targets that would otherwise be difficult or impossible to discern with conventional image sensors (*1*). However, the trade-offs between spatial resolution, spectral resolution, and daily coverage for existing hyperspectral imaging platforms pose practical limitations. Applications such as multimodal transportation infrastructure monitoring place high demands on performance in terms of spectral resolution, spatial resolution, and ground coverage rate during daylight hours (coverage per day) (*1*). Important emerging applications in transportation include condition monitoring of railroads, pipelines, and roadways (*2*). Complementary applications such as traffic-flow modeling, post-disaster evaluation, and environmental health monitoring place similarly high demands on performance (*3*). Insufficient spatial or spectral resolution could result in high false positives that lead to unnecessary expense from conducting verification missions or field inspections in hostile terrain. Practitioners, therefore, will benefit from a comparative framework for hyperspectral image acquisition that can provide practical guidance for the budgeting, acquisition, and configuration of appropriate equipment and sensor packages.

Each so-called hyper-pixel of a hyperspectral image sensor captures the spectrum from radiated or reflected electromagnetic (EM) energy, which includes invisible and visible light (*4*). The device optically separates the EM spectrum from a scene so that dozens of physical pixels from an image sensor can sample the dispersed energy into closely spaced wavelength bins (*5*). The two main methods of capturing a hyperspectral scene during the same sensor integration time are the sweeping optical slit (so-called pushbrooming) and snapshot hyperspectral imaging. The pushbrooming method sweeps a linear array of imaging pixels across the scene such that each row of a two-dimensional (2D) planar image sensor captures the same line of the scene in different wavelength bands (*6*). In contrast, full-frame hyperspectral image sensors use optical filters to subdivide the available 2D planar pixel array into groups of pixels (hyper-pixels), such that each pixel within a group simultaneously integrates energy from a different wavelength band (*7*). Both techniques require image stitching to assemble complete hyperspectral scenes after a mission. Various methods of geometric and orthographic corrections for image stitching are available (*8*, *9*). Subsequently, given the same 2D sensor and assuming that the scene changes relatively slowly (*10*), the pushbrooming approach can achieve both higher spatial and spectral resolutions. Therefore,

[1]SMARTSe[SM], Upper Great Plains Transportation Institute, North Dakota State University, Plano, TX
[2]Department of Physics, North Dakota State University, Golden, CO
[3]Upper Great Plains Transportation Institute, North Dakota State University, Fargo, ND

Corresponding Author:
Address correspondence to Raj Bridgelall: raj.bridgelall@ndsu.edu

this article focuses on characterizing the performance of sweeping optical slit methods across different platforms.

Given a desired spectral resolution, constraints in the design of optical systems for hyperspectral image acquisition force an inherent trade-off in their daily ground coverage (square-kilometers per day) and the spatial resolution (square-meters per pixel). Spaceborne platforms typically move faster than airborne platforms and cover a broader swath, but the spatial resolution can be several-fold larger. On the other hand, airborne platforms can schedule flight paths and missions with the flexibility needed to achieve higher spatial resolution, but the daily ground coverage per aircraft is substantially less. Handheld spectrometers are available to conduct field inspections by foot to achieve even higher spatial resolution, but the remoteness and ruggedness of some terrain limit their accessibility. Practitioners are, therefore, currently limited to selecting an appropriate combination of different platforms to achieve the desired trade-off (Figure 1).

Satellites that carry hyperspectral imager payloads are rare and relatively inaccessible for continuous scanning missions. As of 2017, there were only three such satellites, and researchers must complete an application process several months in advance and allow several weeks for data delivery after its collection (*11*). The growing availability of cube satellites (CubeSat) promises to reduce the cost of spaceborne hyperspectral image acquisition. CubeSats are miniaturized satellites that various manufacturers design to a set of standard specifications so that a single vehicle can launch several of them at once to reduce deployment costs (*12*). However, their size and power constraints limit the availability and performance of hyperspectral imagers. Manned aircraft equipped with hyperspectral imagers require nearby airports, and scheduling them can result in data collection lags of days or even weeks (*13*).

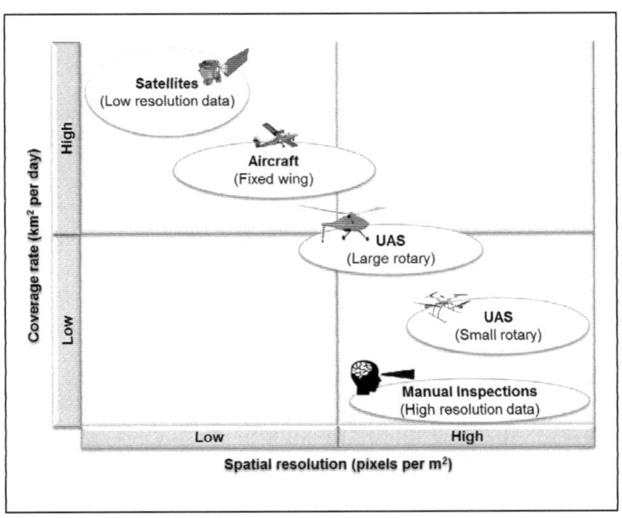

Figure 1. Performance trade-off for hyperspectral imaging platforms.

The emergence of small UAS (sUAS) or small drones and lightweight, low-power hyperspectral imagers for commercial applications promises greater flexibility and a wider range of mobility and spatial resolution options (*14*). The steady pace of cost reduction and the relaxation of regulations are creating many new application opportunities. In particular, the authors of this article have been experimenting with concepts in swarming unmanned aircraft system (UAS) configurations to reconstruct hyperspectral scenes in three-dimensions (*15*). One aspect of that development incorporates rapid hyperspectral image classification to adjust the flight altitude adaptively (*16*). That is, sensor parameters and flight paths change in response to target tracking so that spatial resolution and signal quality improve only for selected scenes, thereby reducing the data load and processing requirements.

Swarm robotics incorporates sensing, artificial intelligence, and local communications to achieve collective behavior directed toward a common goal (*17*). As swarming operations evolved from the field of artificial swarm intelligence, their implementation involves many individual rules, cost minimization functions, and many constraints that require significant computing capabilities to mimic social insect behaviors in the natural world. Communications among swarm members involve a system of constant feedback. To focus this article, the authors do not cover the enabling technologies or the related issues of practical swarm operations such as flight-path management, airspace-access planning, or aircraft refueling (including recharging) strategies. Rather the authors point to a diverse set of important research by others who are advancing those solutions for more effective drone operations (*18*, *19*).

The main contribution of this article is a model to evaluate the performance of sweeping aperture hyperspectral imaging platforms in their daily spatial coverage and spatial resolution envelope. The authors develop a data acquisition framework and use the model to compare the achievable performance among existing airborne and spaceborne hyperspectral imaging vehicles and drone swarms.

The organization of this paper is as follows: the next section presents fundamentals of the optical geometry and the image acquisition methods that lead to the model for evaluating the performance of sweeping aperture hyperspectral-imaging platforms. Section 3 develops a framework for gathering and translating the available data for each platform into the model parameters. Section 4 presents results from applying the model to

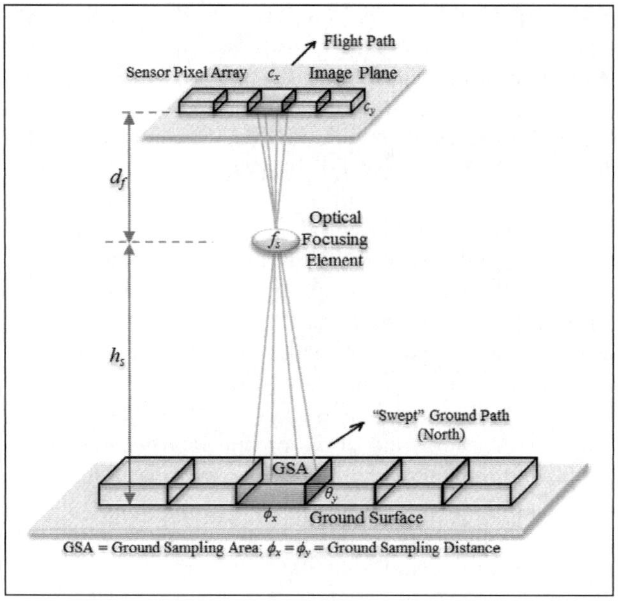

Figure 2. Pushbroom method of image acquisition.

compare the achievable performance among platforms. The final section offers conclusions that could benefit others who utilize the model.

Methods

This section develops the model by applying fundamentals of the optical geometry and the methods of image acquisition and scene assembly that require a sweeping aperture (Figure 2). To follow the steps in the model development, it is convenient to summarize all of the parameters and their units of measure (Table 1).

A hyperspectral scene consists of points $S_{\theta\phi h}$ where the triplet $\{\theta, \phi, h\}$ are the latitude, longitude, and elevation of the geospatial position on the surface S. Each point on the surface radiates or reflects electromagnetic (EM) energy with intensity $R_\lambda(S_{\theta\phi h}, t)$ where t is the integration sample interval and λ is wavelength span of the EM energy spectrum where

$$\lambda = \sum_i \lambda_i \tag{1}$$

and λ_i is a discrete wavelength band i within the spectral range. Hyperspectral image sensors map the aggregate EM reflectance from a ground sample area (GSA) to each pixel $\boldsymbol{p}_{x,y,i}$ of the image sensor at time window $t[n]$ such that:

$$p_{x,y,i}[n] = R_{\lambda_i}\left(G_{x,y}, t[n]\right) \tag{2}$$

$G_{x,y}$ is the center of the GSA with dimensions ϕ_x and θ_y. Therefore, the spatial resolution of the GSA is $\phi_x\theta_y$ meters.

Optical Geometry

The sensor captures image frame n at discrete times $t[n]$ with an integration interval of Δ_t. Each planar photosite $p_{x,y}$ of the image sensor is associated with a radiometric vector

$$\vec{R}_\lambda = R_{\lambda_1}, R_{\lambda_2}, R_{\lambda_2}, \ldots \tag{3}$$

with each element $R_{\lambda i}$ representing the charge accumulation associated with the EM reflectance integrated across the GSA, within the individual spectral bands λ_i.

The focusing element of the generalized optical system has a focal length of d_f and is positioned at a distance of d_s above the image plane. The subject in focus will be a distance of h_s away where, for a relatively thin lens, the approximation is

$$h_s \cong \frac{1}{d_f^{-1} - d_s^{-1}}. \tag{4}$$

For pixels of dimension $c_x c_y$ where $c_x = c_y$, the GSA G_A will be

$$\begin{aligned} G_A = \phi_x^2 &= \left[\frac{c_x}{d_s} h_s\right]^2 = \left[\frac{c_x}{d_s}\frac{1}{d_f^{-1} - d_s^{-1}}\right]^2 \\ &= \left[\frac{c_x}{d_s/d_f - 1}\right]^2 = p_a \frac{1}{(d_s/d_f - 1)^2} \end{aligned} \tag{5}$$

Therefore, the GSA is directly proportional to the pixel area p_a and inversely proportional to the ratio of the image-to-focal plane distance. Hence, the optical area magnification factor M_η is

$$M_\eta = (d_s/d_f - 1)^{-2} \tag{6}$$

This magnification depends on the separation distance δ between the image plane and the focal plane where

$$\delta = d_s - d_f \tag{7}$$

Therefore, the magnification factor is

$$M_\eta = (d_f/\delta)^2 = \left(\frac{d_f}{d_s - d_f}\right)^2 \tag{8}$$

The image plane distance constrains at least one dimension in the overall size of the sensor system. The size also depends on the volume required to accommodate optical path folding elements along the focal length. Given an optically symmetric system and square pixels the area magnification factor must be

$$M_\eta = \left(\frac{\phi_x}{c_x}\right)^2 \tag{9}$$

Table 1. Parameters of the Performance Evaluation Model

Parameter	Units	Description
θ	degrees	Latitude
ϕ	degrees	Longitude
h	meters	Elevation
$S_{\theta\phi h}$	unit-less	Point in 3D space
t	seconds	Energy integration sample interval
λ	meters	Wavelength span
R_λ	Joules	Electromagnetic energy
$\mathbf{p}_{x,y,i}$	unit-less	Hyper-pixel identification number
$G_{x,y}$	unit-less	Center of ground sample area corresponding to hyper-pixel
ϕ_x	meters	Lateral dimension of the ground sample area
θ_y	meters	Vertical dimension of the ground sample area
Δ_t	seconds	Energy integration interval
$R_{\lambda i}$	Coulombs	Spectral charge integration vector of each hyper-pixel
d_f	meters	Focal length of the optical system
d_s	meters	Distance of focusing element above the image plane
δ	meters	Separation between the image plane and the focal plane
h_s	meters	Distance of focused subject in the ground sample area
c_x	meters	Lateral dimension of the image sensor pixel
c_y	meters	Vertical dimension of the image sensor pixel
p_a	sq-meters	Pixel area
G_A	sq-meters	Ground sensing area (GSA)
M_η	unit-less	Optical magnification factor
S_x	meters	Image sensor length
N_x	unit-less	Number of pixels per row of the 2D planar sensor
γ	unit-less	Image overlap fraction
L_y	meters	Sample interval along the ground track
L_G	meters	Length of ground (nadir) path
V_G	m s^{-1}	Ground speed
T_f	seconds	Flying time
N_F	unit-less	Number of frames needed to cover the flight path
F_r	frames s^{-1}	Frame rate
F_{rm}	frames s^{-1}	Minimum required frame rate
G_y	meters	Swath height
A_n	unit-less	Number of aircrafts in the swarm
I_V	degrees	Instantaneous field-of-view
W_a	sq-meters	Ground area coverage

Therefore, given a swath width of G_x, the sensor length S_x must be

$$S_x = G_x M_\eta^{-1/2} \tag{10}$$

The number of pixels per row of the sensor N_x is

$$N_x = \frac{S_x}{c_x} \tag{11}$$

Equating Equations 6 and 9 and solving for the image plane distance gives

$$d_s = d_f\left(\frac{c_x}{\phi_x} + 1\right) \tag{12}$$

Image Acquisition

For a desired image overlap of γ the sample interval L_y along the ground track must be

$$L_y = \theta_y(1 - \gamma) \tag{13}$$

The length of the ground (nadir) path L_G is a function of the aircraft ground speed V_G and flying time of T_f seconds such that

$$L_G = V_G T_f \tag{14}$$

in units of m s^{-1}. Therefore, the number of frames N_F needed along the flight path is

$$N_F = \frac{L_G}{L_y} + 1 \tag{15}$$

Note that there will be two frames if the flight path length is identical to the ground sample interval such that $L_G = L_y$. Given the maximum frame rate F_r and counting the initial frame at the starting position, this is equivalent to

$$N_F = F_r T_f + 1 \quad (16)$$

Combining Equations 13, 15, and 16 and solving for the minimum required frame rate F_{rm} yields

$$F_{rm} = \frac{V_G}{c_y \sqrt{M_\eta}(1 - \gamma)} \quad (17)$$

The aperture of a pushbroom type of imaging system typically limits the field-of-view (FOV) to one row of the image sensor or to a linear image sensor array such that the swath height G_y is

$$G_y = \theta_y = c_y M_\eta^{1/2}. \quad (18)$$

Therefore, a single sensor payload would capture images that cover a ground area W_a (square-meters) of

$$W_a = N_x G_x (L_G + G_y). \quad (19)$$

This model is extensible to full-frame image sensor types that utilize a larger sweeping aperture for scene assembly.

Reflecting the ground sampling distances to the image sensor pixel sizes yields

$$W_a = N_x c_x M_\eta^{1/2} \left(L_G + c_y M_\eta^{1/2}\right). \quad (20)$$

Given the ground speed and flying time of a single platform, and substituting Equation 14 into Equation 20, the expression becomes

$$W_a = N_x c_x^2 M_\eta \left[1 + \frac{V_G T_f}{c_x M_\eta^{1/2}}\right]. \quad (21)$$

If the minimum frame rate is achievable, then this expression provides the ground coverage area as a function of the sensor, optical, and flight parameters.

These expressions describe the fundamental trade-offs in sensor pixel area, pixel density, optical system package size, the image capture rate and overlap percentage, the flight time, and the equivalent ground speed of the platform. For the same swath width, the flight altitude of a single platform is directly proportional to the ratio of the image plane distance from the focusing element and the image sensor size such that

$$h_s = G_x \frac{d_s}{S_x} \quad (22)$$

Therefore, by decreasing the size of the optical system, the image plane distance will decrease. The focal length must also decrease to maintain the same magnification factor. Consequently, the aircraft must fly at a lower altitude to maintain the swath width and GSD. Given the instantaneous field of view (IFOV) denoted I_V, the GSD is

$$\phi_x = 2h_s \tan\left(\frac{I_V}{2}\right) \quad (23)$$

where h_s is the flight altitude. For small angles, $\tan(\theta) \approx \theta$, therefore $\phi_x \approx h_s \times I_V$. The frame rate F_r of the optical system is in frames per second and the maximum speed of the platform is in m s^{-1}. Therefore, the daily ground coverage W_s for swarms containing A_n aircrafts with γ_s percentage lateral overlap is

$$W_s = W_a A_n (1 - \gamma_s). \quad (24)$$

This model guides the design of a lateral swarming configuration with overlapping fields-of-view, using identical aircraft and cameras. Given the scarcity of publicly available data, the simplicity and linear scalability of this model can guide design and budgeting decisions before the acquisition of multiple expensive aircraft and sensor packages to collect hyperspectral data, and to validate performance expectations for specific application objectives.

Data

The data available for each type of hyperspectral imaging platform that currently exists is scattered throughout the literature. A review process compiled, calculated, and validated the values of the relevant parameters (Table 2). The next section describes the dataset and explains the derivations of the model parameters needed.

Table 2. Parameters Values of the Dataset

	h_s	N_x	c_x	ϕ_x	I_V	F_r	γ	V_G	η_l	W_a
Platform	(km)		(μm)	(m)	(mr)	(s^{-1})	(%)	(km h^{-1})	(%)	(km^2)
Hyperion	705	256	60	30	0.043	220	0.6	23,760	2.7	14,600
AVIRIS-ER2	17	677	200	17	1.0	12	1.5	730	16.9	9292
AVIRIS-TO	4	677	200	4	1.0	12	25	130	16.9	387
UAS	4	1600	6.5	1	0.3	250	30	630	58.7	2,417
sUAS	0.12	1600	6.5	0.03	0.3	250	30	19	94.8	3.5

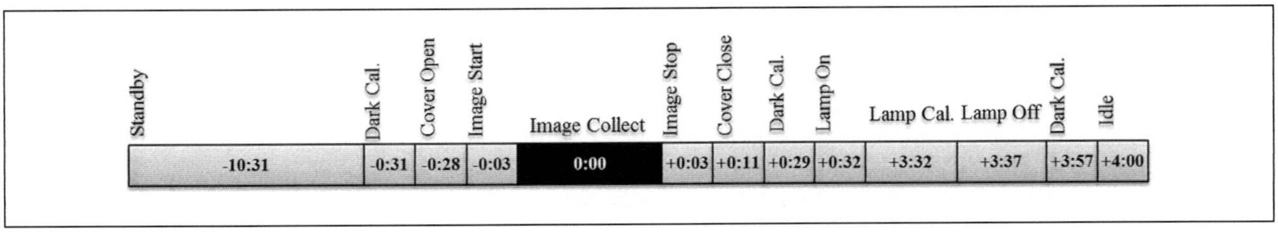

Figure 3. Timing of a Hyperion data collection event.

EO-1 Satellite

The Hyperion imager aboard the Earth Observer-1 (EO-1) satellite provides 220 spectral channels that range from 400 to 2500 nanometers (*20*). The typical revisit for an area is 16 days. The optical system splits the incoming electromagnetic energy into two beams, each focusing on 256 spatial pixels with 60-micrometer sides. The GSD is nominally 30 meters; therefore, the IFOV is 42.55 microradians (Equation 23). The system records images at the rate of 220 frames per second. At an orbit of 705 km, the effective ground path velocity is 23,760 km h^{-1}. At the maximum capture rate of the imaging system of 223.4 Hz, the image overlap will be 1.5% (*21*).

Each data collection event (DCE) produces five files that include calibration and image data. The calibration files are pre-image dark collect, post-image dark collect, internal calibration lamp collect, and post-lamp dark collect (*22*). The timing of a Hyperion DCE before and after the image capture event is summarized in minutes and seconds (mm:ss) format (Figure 3). The total overhead time is 14 minutes and 31 seconds. The maximum image capture time is 24 seconds. Therefore, the daily coverage efficiency, which is the ratio of the image capture time to the total camera operation time, is only 2.6%.

Approximately 10 minutes of standby time is required to temperature stabilize the analogue signal processors for signal-to-noise ratio calibration. Spacecraft maneuvering is necessary to orient the optical system for the solar calibration. The maneuvering, calibration, standby, and temperature stabilization constraint of every DCE limit the time efficiency of the image acquisition to 2.7%. The EO-1 can schedule a maximum of 11 DCE per day (*22*). Therefore, the system has an ability to image 14,600 square-kilometers per day. However, the system has the potential to provide up to 96 DCE per 24-hour period with a sun synchronous orbit. At the maximum acquisition rate, the platform could capture up to 114,000 square kilometers of imagery per 24-hour period.

Manned Aircraft Systems

The Airborne Visible/Infrared Imaging Spectrometer (AVIRIS) sensor is one of the most popular. It provides 224 spectral channels that range from 360 to 2,500 nanometers (*23*). The flight altitude and the optical geometry of the sensor dictate whether the frame rate or the aircraft speed becomes the bottleneck in daily ground coverage. Operators determine the flight altitude based on barometric conditions, aircraft capabilities, cloud cover, sun angle, and other factors such as local regulations and government policies.

The AVIRIS optical system splits the incoming light into four paths that focus the light onto four individual linear image sensor arrays. The nominal dimensions of the charge-coupled detector (CCD) elements are 200 micron × 200 micron (*24*). Three 64-element and one 32-element detector provide the 224 spectral channels. The device utilizes rotating elements in the optical path to scan a field of view across the ground trajectory while focusing the image on one spatial pixel at a time (*25*). The scanning operation provides an equivalent of 677 pixels of lateral spatial resolution. The IFOV of the optical system is 1 milliradian. Therefore, when flown at a nominal altitude of h_s = 17 kilometers on the NASA ER-2 aircraft, the GSD is 17 meters (Equation 23). The ER-2 flies at a nominal ground speed of 730 km/h to accommodate a line scan rate of 12 Hz. From Equation 17, the maximum aircraft speed provides less than 1% of image overlap.

The on-board tape memory capacity of the AVIRIS sensor limits the amount of image capture and recording time to approximately 1.1 hours. Given the maximum flight duration of 6.5 hours, the calculated AVIRIS image capture efficiency is approximately 17%. Therefore, the AVIRIS system is capable of producing a maximum of 9,292 km^2 of hyperspectral scene per day. When flown on a Twin-Otter (TO) aircraft at an altitude of 4 km, the AVIRIS will provide a much higher spatial resolution of 4 meters but the daily ground coverage will be reduced to 387 km^2, given the same set of optical system constraints.

Small Unmanned Aircraft Systems

From Equation 17, the optical system configuration, maximum frame rate, and maximum flight altitude will constrain the maximum ground speed to

$$V_G = F_{rm} c_y \sqrt{M_\eta}(1 - \gamma). \quad (25)$$

The maximum frame rate of a compact hyperspectral imager that is suitable for deployment on sUAS is 250 frames per second (*26*). Given the aperture (*f*/2.5) and slit length (10.5 mm), the effective focal length is 26 millimeters. The image sensor pixel sizes of 6.5 micrometers and the focal length defines the IFOV as 0.3 milliradian (mr). At the maximum flight altitude of 121.2 meters, the GSD is 3 centimeters (Equation 23). Therefore, the ground swath is 48 meters. With a desired frame-overlap of 30% to enable frame stitching, this optical geometry and the maximum frame rate of the imager limits the ground speed to 5.3 m s^{-1} (19 km h^{-1}).

The daily coverage efficiency of a UAS that operators can launch from a mobile platform at the target area will be higher than that for one that relies on the availability of an airport or an airstrip nearby. The additional time overhead includes lifting the payload, reaching a stable velocity at the start of image acquisition, time to descend, and the duration lost to battery replacement. A small rotary aircraft has the greatest agility and could ascend and descend without traveling significant distances horizontally to reach the beginning of their image acquisition coordinates. A suitable rotary aircraft carrying a payload of 1 kg (2.2 lbs.) and using a 22.2-volt battery with 16,600 milliamp-hour (mA-h) charge capacity will provide a maximum flight time of approximately 35 minutes (*27*). The selected reference provides a conservative estimate because battery life will likely improve over time. With an average lift speed of 5 m s^{-1}, the rotary craft will reach the maximum allowable flight altitude in 24 seconds. Experienced operators can change batteries in less than one minute. If a launch vehicle can follow the ground path to accommodate rapid descent and battery changes, then the daily coverage efficiency for the rotary craft of this capability will be 94.8%. Even further improvements are achievable if the launch vehicle carries pre-charged vehicles to alternate missions.

The frame rate of the image sensor and the optical geometry limits the maximum ground speed for this platform to 19 km h^{-1}. Therefore, the sUAS will image approximately 0.5 km^2 per mission. The sun will be at a suitable range of angles to provide proper irradiance for approximately four hours (*4*). Therefore, operators can fly approximately seven data collection missions per aircraft. Consequently, with the sensor, flight altitude, and battery life constraints, the sUAS platform can provide 3.5 km^2 of hyperspectral images per aircraft per day. If placed on a larger UAS that can fly at the same altitude as the Twin-Otter, the image sensing system will provide a spatial resolution of one meter and a daily ground coverage of 2,417 km^2.

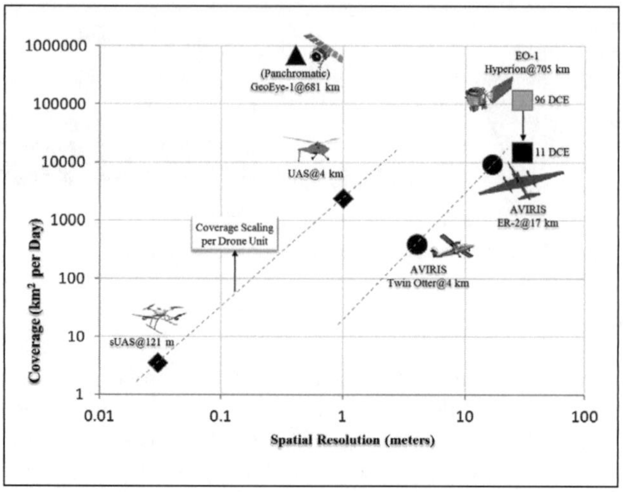

Figure 4. Bounds in the coverage-resolution trade-off for existing platforms.

Results and Discussion

Applying the derived parameters from the dataset to the model for each platform produces their daily coverage shown in the last column of Table 1. The altitude (h_s) is in kilometers (km), the pixel dimension in micrometers (μm), the GSD in meters, the IFOV in milliradians (mr), the frame rate is in frames per second, the ground velocity in km h^{-1}, the image capture efficiency η_I, and the daily ground coverage W_a is in square kilometers. The daily ground coverage and the spatial resolution achievable by each platform is summarized graphically (Figure 4). The figure shows that the smaller IFOV typical of lightweight imaging systems has the potential for providing greater spatial resolution for the same daily ground coverage of existing airborne platforms. The arrow shown indicates the scaling direction of the coverage-resolution bound per Equation 24, for each additional sUAS added to the swarm. The figure also includes the daily ground coverage and spatial resolution of a panchromatic imaging platform aboard the GeoEye-1 satellite for comparison. Hyperspectral imagers force a lower coverage-resolution bound because, given an image sensor, they trade off spatial resolution for spectral resolution. The time overhead needed to begin the actual image acquisition during available natural lighting conditions tends to limit substantially the daily ground coverage of existing aerospace platforms. Therefore, system designers must carefully consider their operational details and logistics to minimize the time overhead between data collection events.

Drone swarms can have a distinct advantage if they incorporate mobile launch and refueling/recharging platforms at the deployment site to minimize the time overhead. Positioning the platform at the deployment site before the time of optimal sun angle will minimize the overhead to begin a data capture mission. Furthermore, the launch vehicle could potentially carry two drone fleets to accommodate launches during the time that the previous fleet is landing. This approach will minimize the time between data capture events and increase the daily coverage efficiency per launch system. These considerations offer important insights into the design optimization of swarming aircraft systems for high-efficiency operations. Given their advantages, practitioners must also consider the precautions required when using drones. Considerations include compliance with federal and state regulations to ensure airspace and public safety. According to the Federal Aviation Administration (FAA) website at the time of this publication, operators must register drones that weigh between 0.55 and 55 pounds, and are not flying under the special rule for model aircraft. The authors recommend that potential practitioners of the method do regularly visit the FAA website because the situation and rules evolve steadily.

Conclusions

Many applications are emerging that demand simultaneously vast ground coverage, high spatial resolution, high spectral resolution, and high data-collection efficiencies during available lighting conditions. The continuous cost and size reduction of high-performance electronic and optical systems will lead to a proliferation and diversification of aerospace platforms that afford greater flexibility in trading off performance bounds. Nevertheless, a fundamental trade-off between spatial resolution and daily coverage for hyperspectral imaging acquisition sets a practical limitation that constrains their use. Therefore, using the framework for data acquisition and the performance evaluation model developed in this article can help other researchers and practitioners in their work.

The finding that drone swarms have the potential to provide greater spatial resolution for the same daily ground coverage of existing airborne hyperspectral imaging platforms points to their potential for use in new applications that demand high-performance imaging. Given an image sensor of some spectral resolution, the flight height, fight speed, and image capture efficiencies fundamentally constrain spatial coverage and spatial resolution. Therefore, an ability to scale economically the swarm size will push the daily spatial coverage dimension without diminishing the spatial or spectral resolutions. Furthermore, the potential for flight-path adaptation based on target identification can further improve performance in reducing false positives. However, this potential comes with the caveat that practical and high-performance methods must be in place to stitch hyperspectral image frames into hyperspectral scenes, and to enable the refueling and alternating of swarm fleets, without the need for manual interventions. These are all important and ongoing areas of research that this article does not address.

Future research will use the models to design a variety of swarming configurations to enable the three-dimensional reconstruction of hyperspectral scenes for various applications in transportation.

Acknowledgments

A grant from the United States Department of Transportation supported this research under Grant DTRT13-G-UTC38.

Author Contributions

The authors confirm contribution to the paper as follows: study conception and design: Raj Bridgelall; data collection: Raj Bridgelall; analysis and interpretation of results: Raj Bridgelall, James Rafert, Denver Tolliver; draft manuscript preparation: Raj Bridgelall. All authors reviewed the results and approved the final version of the manuscript.

References

1. Bridgelall, R., J. B. Rafert, and D. D. Tolliver. *Remote Sensing of Multimodal Transportation Systems*. Mountain-Plains Consortium, North Dakota State University - Upper Great Plains Transportation Institute, Fargo, ND, 2016.
2. Bridgelall, R., B. J. Rafert, and D. D. Tolliver. Hyperspectral Applications in the Global Transportation Infrastructure. *Proc., European Association for Signal Processing (EUSIPCO)*, Nice, France, 2015, pp. 739–743.
3. Bridgelall, R., J. B. Rafert, D. D. Tolliver, and E. Lee. Resolution agile remote sensing for detection of hazardous material spills. *Transportation Research Record: Journal of the Transportation Research Board*, 2016. 2547: 49–56.
4. Goetz, F. Three Decades of Hyperspectral Remote Sensing of the Earth: A Personal View. *Remote Sensing of Environment*, Vol. 113, 2009, pp. S5–S16.
5. Hagen, N., and M. W. Kudenov. Review of Snapshot Spectral Imaging Technologies. *Optical Engineering*, Vol. 52, No. 9, 2013, pp. 090901–090901.
6. Barbieux, K., D. Constantin, and B. Merminod. Correction of Airborne Pushbroom Images Orientation Using Bundle Adjustment of Frame Images. *International Archives of the Photogrammetry, Remote Sensing & Spatial Information Sciences*, Vol. 41, No. B3, 2016, pp. 813–818.
7. Bareth, G., H. Aasen, J. Bendig, M. L. Gnyp, A. Bolten, A. Jung, R. Michels, and J. Soukkamäki. Spectral Comparison of Low-weight and UAV-based Hyperspectral Frame Cameras with Portable Spectroradiometer Measurements. *Proc., Workshop on UAV-based Remote Sensing Methods for Monitoring Vegetation*, Cologne, Germany, 2015, pp. 1–6.

8. Jinmeng, H., and S. Zhong. Research on Image Stitching Algorithm for UAV Ground Station Terminal. *Proc., Asian Simulation Conference*, Springer, Singapore, 2016, pp. 207–215.
9. Stark, F., M. Rosenberger, P.-G. Dittrich, R. Celestre, M. Hänsel, and G. Notni. Geometrical and Spectral Data Stitching for Combining Hyperspectral Imaging Systems. *at-Automatisierungstechnik*, Vol. 65, No. 6, 2017, pp. 381–395.
10. Lippitt, D., and D. A. Stow. Remote Sensing Theory and Time-Sensitive Information. In *Time-Sensitive Remote Sensing* (C. D. Lippitt, D. A. Stow, and L. Coulter, eds.), Springer, New York, 2015, pp. 1–10.
11. Union of Concerned Scientists. *Satellite Database*. Union of Concerned Scientists, Cambridge, MA, 2017.
12. California Polytechnic State University. *CubeSat Design Specification*, Revision 13, California Polytechnic State University, San Luis Obispo, CA, 2015.
13. Kramer, H. J. *Observation of the Earth and Its Environment: Survey of Missions and Sensors*, Springer, New York, 2002.
14. Colomina, I., and P. Molina. Unmanned Aerial Systems for Photogrammetry and Remote Sensing: A Review. *ISPRS Journal of Photogrammetry and Remote Sensing*, Vol. 92, 2014, pp. 79–97.
15. Bridgelall, R., J. B. Rafert, and D. D. Tolliver. Hyperspectral Range Imaging for Transportation Systems Evaluation. *Proc., SPIE Smart Structures and Materials: Nondestructive Evaluation and Health Monitoring*, Las Vegas, NV, Vol. 9803, 2016, p. 98032Y.
16. Bridgelall, R., J. B. Rafert, D. D. Tolliver, and E. Lee. Rapid Hyperspectral Image Classification to Enable Autonomous Search Systems. *Journal of Spectral Imaging*, Vol. 5, No.1, 2016, pp. 1–8.
17. Hassanalian, M., and A. Abdelkefi. Classifications, Applications, and Design Challenges of Drones: A Review. *Progress in Aerospace Sciences*, Vol. 91, 2017, pp. 99–131.
18. Valavanis, K. P., and G. J. Vachtsevanos. *Handbook of Unmanned Aerial Vehicles*. Springer Publishing Company, Inc., New York, 2014.
19. Zhu, X., Z. Liu, and J. Yang. Model of Collaborative UAV Swarm Toward Coordination and Control Mechanisms Study. *Procedia Computer Science*, Vol 51, 2015, pp. 493–502.
20. Lee, P., S. Carman, C. K. Chan, M. Flannery, and M. Folkman. *Hyperion: A 0.4 Micrometer-2.5 Micrometer Hyperspectral Imager for the NASA Earth Observing-1 Mission*. TRW Space and Electronics Group, Redondo Beach, CA, 2001.
21. Folkman, M. A., J. Pearlman, L. Liao, and P. J. Jarecke. EO-1/Hyperion Hyperspectral Imager Design, Development, Characterization, and Calibration. *Proc., SPIE*, Bellingham, Washington, Vol. 4151, 2001, pp. 40–52.
22. Barry, P. *EO-1/Hyperion Science Data User's Guide: Level 1B*. TRW Space, Defense & Information Systems, Redondo Beach, CA, 2001.
23. Coulter, D., P. L. Hauff, and W. L. Kerby. Airborne Hyperspectral Remote Sensing. *Proc., 5th Decennial International Conference on Mineral Exploration*, Toronto, Canada, 2007, pp. 375–378.
24. Green, R. O., M. L. Eastwood, C. M. Sarture, T. G. Chrien, M. Aronsson, B. J. Chippendale, and J. A. Faust. Imaging Spectroscopy and the Airborne Visible/Infrared Imaging Spectrometer (AVIRIS). *Remote Sensing of Environment*, Vol. 65, No. 3, 1998, pp. 227–248.
25. Kruse, F. A. Comparison of AVIRIS and Hyperion for hyperspectral mineral mapping. *Proc., 11th JPL Airborne Geoscience Workshop*, Pasadena, California, Vol. 4, No. 8, 2002, pp. 1–11.
26. Hill, S. L., and P. Clemens. Miniaturization of high spectral spatial resolution hyperspectral imagers on unmanned aerial systems. *Proc. SPIE 9482, Next-Generation Spectroscopic Technologies VIII*, Baltimore, Maryland, Vol. 9482, 2015, p. 94821E.
27. Aerialtronics International. *Altura Zenith ATX8 System Overview*. Aerialtronics International, Katwijk, The Netherlands, 2017.

The Standing Committee on Information Systems and Technology (ABJ50) peer-reviewed this paper (18-00750).

TRR
JOURNAL OF THE TRANSPORTATION RESEARCH BOARD

Article

Transportation Research Record
2018, Vol. 2672(45) 45–54

DOI: 10.1177/0361198118776153
journals.sagepub.com/home/trr
SAGE

WiFi Scanner Technologies for Obtaining Travel Data about Circulator Bus Passengers: Case Study in Obuse, Nagano Prefecture, Japan

Arief Hidayat[1], Shintaro Terabe[1], and Hideki Yaginuma[1]

Abstract
Recently, a variety of technologies, such as the development of Bluetooth, WiFi, and smart cards, have been used to investigate the movements of passengers or vehicles. This study describes the use of a WiFi scanner that was installed on a bus and circulated around the bus's route 14 times. The method presented here involves the use of WiFi and GPS to analyze passengers' movements while the bus is running and while stopped at a bus stop. Nine steps were used to derive travel data from the raw data to estimate the number of movements of the passengers. The results of this study describe the travel data collected between bus stops no. 1 and no. 9 and compare the observer data with the WiFi data.

Data about movements in transit systems in a city or region are becoming increasingly important. For example, it is necessary to know the number of passengers boarding or alighting at a bus stop to be able to implement effective strategies for alleviating congestion and improving the quality of the transportation service. In this case, the collection of transportation data is very useful when investigating the current supply and demand for transit to be able to formulate better transit strategies in the future. The data of interest for this application are commonly called origin–destination (OD) data, which reflect a person's movements from one place to another. However, traditional methods to record travel data rely on manual counts by on-board observers on each trip. The process of manual counting is very time-intensive, high cost, and has the potential for significant errors.

Recently, there has been an increasing interest in advancing collection techniques for travel data by incorporating novel technologies to reduce the survey burden and accurately collect passenger information. However, it is relatively expensive to implement the necessary equipment required for these emerging techniques, such as sensors, smart cards, and smartphones. Many researchers have developed methods for collecting travel data using WiFi, Bluetooth, loop detectors, image processing, and various other technologies. However, several of these methods are high cost and are difficult to apply to further research. WiFi is one of the most widely used communication networks for smartphones, laptops, tablets, and other devices that are currently in high demand around the world and, therefore, WiFi can be used as a tool to observe movement. Consequently, WiFi is one of the most useful options for obtaining travel data. Therefore, the objectives of this study are to gather travel data based on WiFi use with a relatively inexpensive tool and to interpret the transit behavior using these data.

The challenge in this research is how to collect travel data or passenger data from a small bus and a small operator with no smart card and a hop on–hop off system. Developing new methods to collect travel behavior via WiFi remains a challenge for better decision-making based on bus-travel data. This study therefore seeks to achieve this goal by investigating the utility of WiFi-sensing technologies for affordable collection of travel data. This paper's value is to provide some procedures for distinguishing travel data from raw WiFi data (media access control [MAC] address).

Related Work

Research into techniques for estimating OD data is gaining momentum. Previous research investigations have used the method of monitoring the GSM (*1*). Another

[1]Department of Civil Engineering, Tokyo University of Science, Urban and Transportation Planning Laboratory, Noda, Japan

Corresponding Author:
Address correspondence to Arief Hidayat: ariefhidayat06@hotmail.com

approach based on a real-time transit-tracking technique that utilizes mobile phone networks has been proposed, which can count bicycles, buses, and other drivers (*2*). Bluetooth and wireless detectors have evolved enough to be of use in the transportation sector as a method for collecting travel data. Use of a Bluetooth detector and WiFi on a bus is more predictable because the bus users and the passenger-carrying capacity varies so the process of analyzing OD data for each bus stop is very diverse (*3–5*). Some approaches combine wireless mesh networks with Bluetooth capabilities by detecting Bluetooth devices attached to cars equipped with wireless networking at specific locations (*3*, *6*, *7*) or by the installation of detection equipment at certain locations such as public facilities (*8*) or terminals and bus stops. This enables the measurement of travel data and vehicle images (*9*) during movement in either direction at a travel node (*10*). Other research has been done to combine the use of a loop detector with Bluetooth that produces multi-mode data in large areas (*11*). However, the tools used in previous research are numerous and diverse and are therefore impractical when it comes to funding, operation, and maintenance. For example, a loop detector is placed on the road and a Bluetooth sensor is a large, wireless link sensor that requires a large amount of solar power. In addition, on smartphones and laptops, Bluetooth is more commonly turned off than WiFi, and Bluetooth capabilities are quite limited by its short range. Data collection via wireless or WiFi is relatively effective because of the current network ubiquity, especially with mobile phones equipped with WiFi being more and more widely used by people of all ages. WiFi signals travel over a relatively large range according to previous research and are low cost. Some studies have investigated the use of WiFi on a bus by placing a WiFi-detector tool on board (*12*, *13*). Bluetooth and wireless tools have been used not only to determine the number of vehicles or gather traffic data from the road but also to calculate pedestrian data around pedestrian paths that are equipped with Bluetooth sensors (*14*). However, there is still a bias due to the small number of Bluetooth users among pedestrians in travel nodes, such as bus terminals and bus stops, which hinders the collection of data about the behavior of pedestrians in public facilities (*15*, *16*).

Method and Field Experiment

The location of this study is the town of Obuse in the Kamatakai District in the Nagano Prefecture of Japan. Obuse is one of the top tourist destinations in Japan. It attracts 764,000 visitors every year. Obuse is unique for its chestnut processing industry and is a historic city with a variety of tourist attractions (*17*). In October 2016, the town had an estimated population of 10,698 and a population density of 560 people per km^2. Its total area is 19.12 km^2 (*18*).

Test Field

Obuse has a circulating shuttle bus called the "Romango." The Romango Bus is a hop on–hop off bus, which means the bus stops at each location for only a few minutes and allows passengers to travel around the city with a single tour ticket for an entire day. The fare for a day-trip ticket is 300 Japanese yen, which is equivalent to approximately US$3. Obuse operates two circular buses every Saturday and Sunday, which connect nine bus stops. Visitors can reach several tourist attractions by using these buses.

Route. The Romango Bus traverses seven segments in its circulation from bus stop no. 1 (BS1) to bus stop no. 9 (BS9) and passes nine bus stops: BS1 (Obuse Highway Oasis Park), BS2 (Obuse Station), BS3 (Hokusai Museum), BS4 (Obuse Museum), BS5 (Matsumura Town Parking), BS6 (Obuse Hot Springs), BS7 (Floral Garden), BS8 (Jyokoji Temple), and BS9 (Ganshoin Temple). The circular route is approximately 15 km and the overall route length is 8.8 km. The longest segment is 2.7 km from BS1 to BS2 and the shortest is 0.3 km from BS4 to BS5.

Vehicles (Bus Fleets). There are two Romango buses, no. 1 and no. 2, seen in Figure 1. Each bus accommodates about 30 passengers.

Schedule. Each bus starts every 30 minutes from 9:50 a.m. to 5:50 p.m. It takes 50 minutes for a round trip. Bus no. 1 operates from 9:50 a.m. to 5:10 p.m. and bus no. 2 operates from 10:20 a.m to 5:50 p.m.

Equipment

WiFi Technology. WiFi technology periodically transmits a signal, called WiFi, to all information devices around it. When a device receives a WiFi signal, it sends a query, called a probe request, and the access point returns a reply, called a probe response, which includes the SSID. Although the probe request takes between 15 seconds and up to several minutes (and takes around one minute on average), these data also include the MAC address that identifies the transmitting device.

WiFi Scanner. Here, a WiFi scanner (shown in Figure 2) was placed on the bus and used to acquire data. This scanner contains a miniature computer, Raspberry Pi 2 B V1.1 (*19*), to control a WiFi antenna and a GPS antenna.

Figure 1. "Romango" circulating buses no. 1 and no. 2.

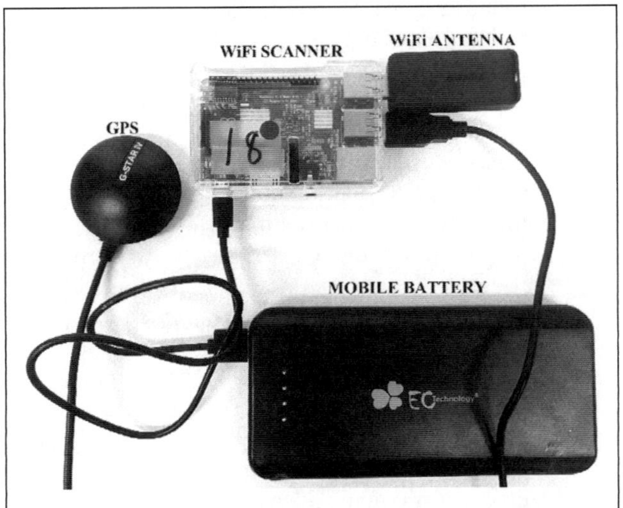

Figure 2. WiFi scanner with GPS antenna and mobile battery.

It records log files including the record date and time, the longitude and latitude from the GPS antenna, and the MAC addresses of the mobile devices of bus passengers and people near the bus from the WiFi antenna. It has a micro SD port for storing data and a micro USB power source (*20*). By collecting and analyzing data, it is possible to determine the spatial flow and distribution of the users of information devices.

Experiment

A WiFi scanner was placed on each of the buses (Figure 3) and a researcher was also stationed on each bus to manually count the number of boarding or alighting passengers at each bus stop. The number of passengers on the bus was estimated for each segment between bus stops. The WiFi scanner was mounted on bus no. 1 near the left window and on bus no. 2 above the driver during the period of 09:50–17:50 on Sunday, October 30, 2016. The installation of the tool is simple and it is placed so that it does not interfere with the bus driver and passengers. The initial plan was to install both scanners above the driver's seat. However, no appropriate place could be found in bus no. 1, which was a different model from bus no. 2. Due to the differences in the inner layouts at the front of the bus, it was not possible to place the scanner in the same position. Therefore, the WiFi scanner was installed near the front-left window on bus no. 1 and above the driver's seat on bus no. 2. However, these different positions had no influence on the detection results as the differences in WiFi-scanner coverage were negligible.

The measured number of boarding or alighting passengers was collected from the WiFi scanner and compared with the manually counted results. The WiFi scanner, which was developed by the author as a prototype, scans for mobile devices that are emitting management packets every five seconds. It can detect people who have devices, such as smartphones and tablets, within an approximately 200 m radius. This scanner records the unique identification codes of the detected mobile devices. The log files were then analyzed to obtain the duration for which each identification code was detected and the beginning and end of each of these periods were interpreted as the boarding or alighting times of the passenger, respectively.

The WiFi scanner also detected identification codes of non-passengers, such as pedestrians and drivers outside of the bus. Several rules of discrimination were applied to filter the log data to include only passengers based on the nature of the user's behavior: (1) the passenger boards the bus at a bus stop, (2) the passenger alights at another bus

Figure 3. Installed WiFi scanners on buses no. 1 and no. 2.

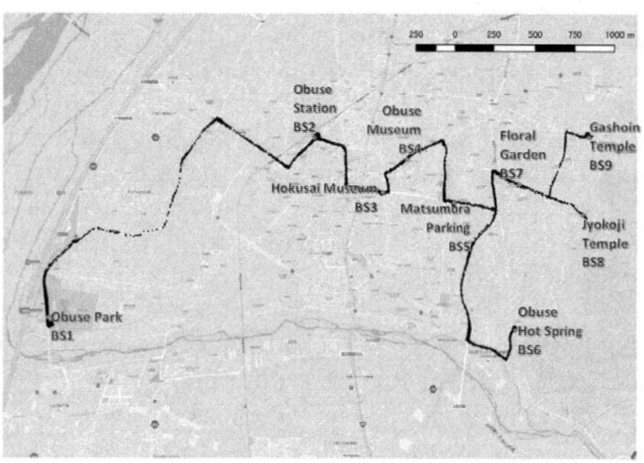

Figure 4. MAC ID point data for two bus circulations.

stop, and (3) the passenger can be detected between these two stops. The time stamps and longitude and latitude of the devices are also recorded by the GPS antenna and the logger integrated into the WiFi scanner. A bus stop visited by the bus was identified by matching the collected longitude and latitude with the map of the bus route.

Data Processing

The resulting raw data include the GPS log and the WiFi log (Figure 4). The GPS log contains time, latitude and longitude information while the WiFi log contains time and MAC ID data. The MAC is a unique identification or address assigned to each network device and can therefore be used to identify each device. The raw data needed to be re-filtered and analyzed before being interpreted. Nine steps were used to filter the raw data into travel data, as follows:

- The first step is to combine the raw data from the WiFi and GPS logs to obtain WiFi data correlated with latitude and longitude data by using the time data. Latitude and longitude location data were correlated with the MAC data from the WiFi data such that the WiFi log became the primary data set.
- The second step is to unify the same data (based on the MAC ID, latitude, longitude, and time) to get a unique data set. The resulting data set is assigned a new column containing "1" for each row to indicate that one row is one datum. After creating the attribute, the data were analyzed by looping (pivoting) to find out how much of the data was repeated.
- The third step is to convert the latitude and longitude data to the UTM (universe transverse mercator) WGS 84 Zone 54N in accordance with the real field conditions based on a WGS 84 decimal degree datum map. Therefore, the output data contained X-, Y-data in the UTM form.
- The fourth step is to define new attributes or labels to each unique identifier to indicate the speed. The data that already contain X-, Y-coordinates in the UTM form are re-analyzed with a new attribute, speed, which is calculated by the formula,

$$V_i = \frac{\sqrt{(X_{i+1} - X_i)^2 + (Y_{i+1} - Y_i)^2}}{T_{i+1} - T_i}. \tag{1}$$

i = Time frame of data
V_i = Speed at the ith time frame
X_i = Longitude for the ith MAC ID in the rows of the tabular data
Y_i = Latitude for ith MAC ID in the rows of the tabular data
T_i = Time for ith MAC ID in the rows of the tabular data

- The fifth step is to justify each identifier as a moving or stopped based on its speed attribute. Setting a threshold for speed as 5 m/s, it is assumed that a device is moving if its speed is greater than the threshold and it is assumed to be stopped if it is less than the threshold. Each identifier is distinguished as moving or stopped according to its speed. The threshold 5 m/s was used to distinguish a moving or stationary MAC ID. The value of 5 m/s was based on the authors' assumption that 5 m/s was reasonable as it was based on the bus's minimum travel speed. Lacking prior knowledge to guide judgment as to the appropriate threshold values, a trial-and-error procedure was used to choose reasonable threshold values that were moderately conservative but not so restrictive as to eliminate potentially viable observations.
- The sixth step is to determine the bus stop locations based on the latitude and longitude positions. Each bus stop location is assumed to be a buffer zone, a region up to 10 meters around the bus stop. Therefore, the data attribute will be assigned new labels: "bus stop," "stop," and "moving." "Stop" was introduced to distinguish being stopped at a bus stop from being stopped at a traffic signal.
- The seventh step is separating the categorized data from step six according to the circulation number out of seven round trips made by the Romango Bus. To do this, an original timetable with data about the arrival and departure times at BS1 was used. Many identified stops located between the bus stops can be assumed to be either a red light or a stop due to a slowdown. In these cases, the "stop" attribute is reassigned to "moving."
- The eighth step is to create an OD matrix for each bus circulation with the MAC IDs, which are observed more than once between two bus stops. It is assumed that any MAC ID included more than once in the data is a device that moves. For each circulation, labels are added to each row of data, such as from M1-2 or M2-3 meaning "moving" (estimated number of passengers) the segment from BS1 to BS2 or the segment from BS2 to BS3, respectively.
- The ninth step is to combine the data from both buses (no. 1 and no. 2). For each bus, there are seven circulations, so after this step, there are 14 circulations in the data set. The data were coded with circulation numbers (CNs) 1–14. The data were then divided again by time into CN1–6 as the a.m. (before noon) data and CN7–14 as the p.m. (after noon) data to facilitate a time-based analysis.

All of the data processing was carried out using an Anaconda 1.5 Jupiter Notebook 5.0 with Python 2.7, Microsoft Excel, and QGIS open source software. Total lines of raw data were 71,630 and the raw data were processed into 200–300 lines of data for each circulation.

Results and Discussion

The results obtained after data processing generally show significant differences between the trends in the number of passengers in trip segments BS1-BS2-BS3, BS3-BS9-BS3, and BS3-BS2-BS1. The number of passengers between BS1-BS2- BS3 and BS3-BS2-BS1 is high as these routes connect to the entrance of Obuse and provide mobility toward Obuse Station (BS2). This data may be overestimated because of the relatively high incidence of non-passenger data being detected from pedestrians or vehicles around the bus. Congestion and high traffic volume on the road caused the WiFi scanners to detect or catch. On the other hand, the BS3-BS9-BS3 segment data were found to be relatively stable in every circulation. According to the data, the change in travel between circulations was low; this is because the high travel volume in the morning is directed toward Obuse Station. In the second through the fifth circulations, there was a higher level of passenger travel because many people travel in the time period of 10:00–15:00 and the number of passengers began to fall in the afternoon.

CN1–6 (Before Noon)

The estimated OD data shown in Figure 5 are based on CN1 to CN6 data from the morning until noon. The pattern of tourist travel is dominated by museums (BS3, BS4), hot spring (BS6), garden (BS7), and historical places (BS8, BS9). Based on the data, the variation was high on segments from M1-2 to M2-3 and from M3-2 to M2-1 and there was high passenger activity on BS2 and BS1. A medium variation occurred in segments from M3-4 until M6-9, from M9-5 until M4-2, indicating stable movement. There was low variation in M6-9, indicating that the tourist attraction destinations, such as temples, are long distances away from the station.

CN7–14 (After Noon)

The estimated OD data in Figure 6 for CN7-14 in the afternoon show a pattern that is almost the same as that in the morning. The relative movement in segments from M3-4 until M4-2 are consistent with the morning data, showing that bus users start to spread evenly throughout the bus stops to reach the tourist destinations. However, compared with the morning travel data, the movement in segments from M1-2 to M3-4, and from M4-2 to

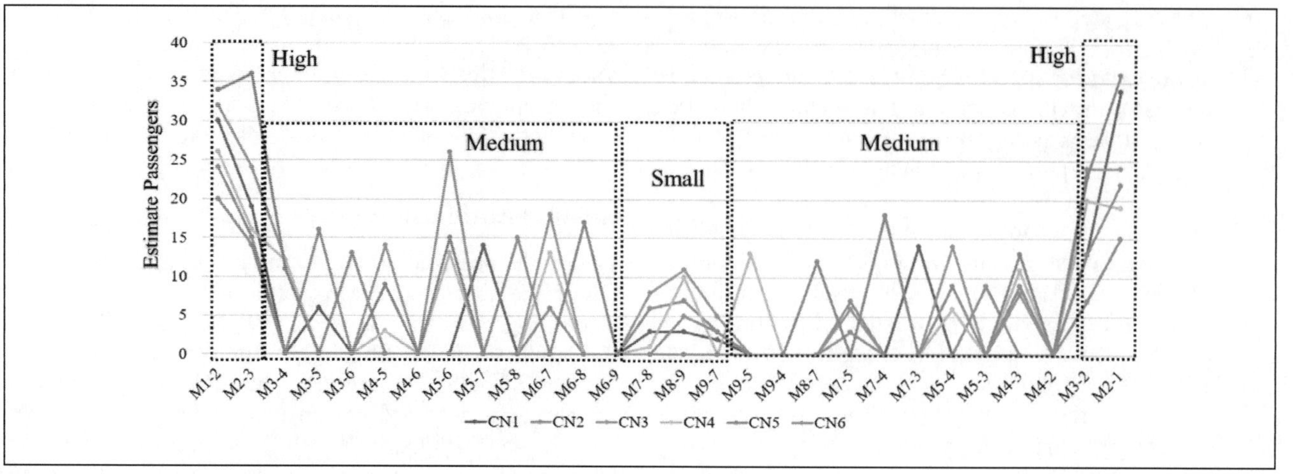

Figure 5. Estimated number of passengers for travel data from CN1–6 (AM).

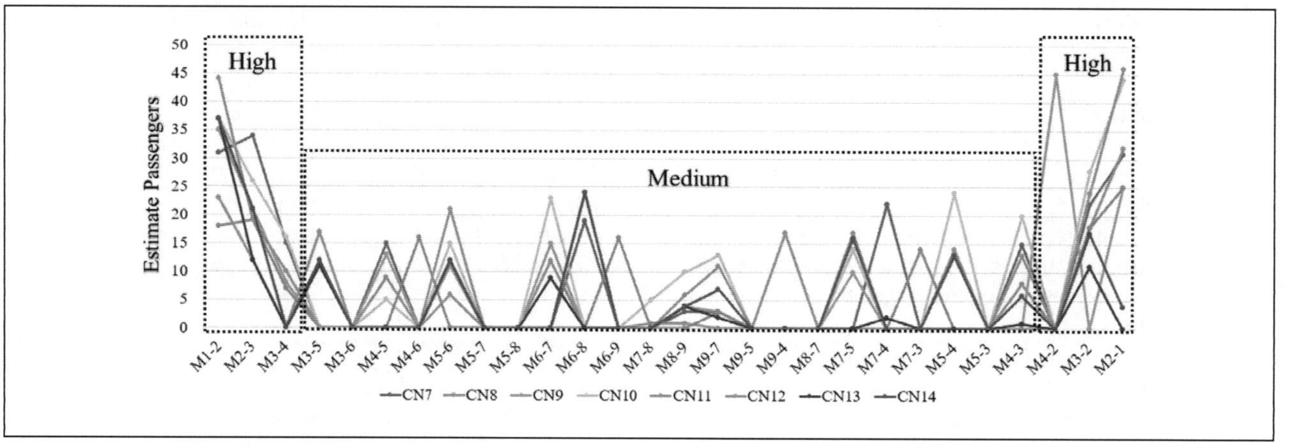

Figure 6. Estimated number of passengers for travel data CN7–14 (PM).

M2-1, were higher in the afternoon due to the end of visiting periods.

Comparison between Manual Counting and Wifi Approaches

The observer count data were then compared with the results of the proposed WiFi-based method collected simultaneously. The observer data are the number of passengers boarding and alighting at each bus stop. The manual count data were only available for CN3, CN5, CN8, CN10, and CN12.

Figure 7 shows the difference between the manually counted (observer) data and WiFi data in five circulations: CN3, CN5, CN8, CN10, and CN12. In general, the data show reliable WiFi data compared with the observer data for CN3 in segments M2-3, M5-6 until M7-5, and M4-3; for CN5 in segments from M3-4 until M4-3; for CN8 in segments from M5-6 until M8-9; for CN10 in segments M4-5 and M7-8; and for CN12 in segments M3-4, M4-5, and M8-9. A striking difference between manually counted and WiFi data, indicating unreliable data, was found for the other segments listed above.

The data were analyzed based on the correlation analysis to determine which WiFi data were related and which were unrelated in each circulation by comparison with the observer data. The correlation analysis for WiFi data compared with the observer data for CN3 reported an r-value of 0.390, indicating a low (positive) relationship. The CN5 data had an r-value of 0.671, which is quite a high (positive) relationship. The CN8 data had an r-value of 0.639, which is quite a high (positive) relationship. The CN10 data had an r-value of 0.633, which is quite a high (positive) relationship. The CN12 data had an r-value of 0.530, which is quite a high (positive)

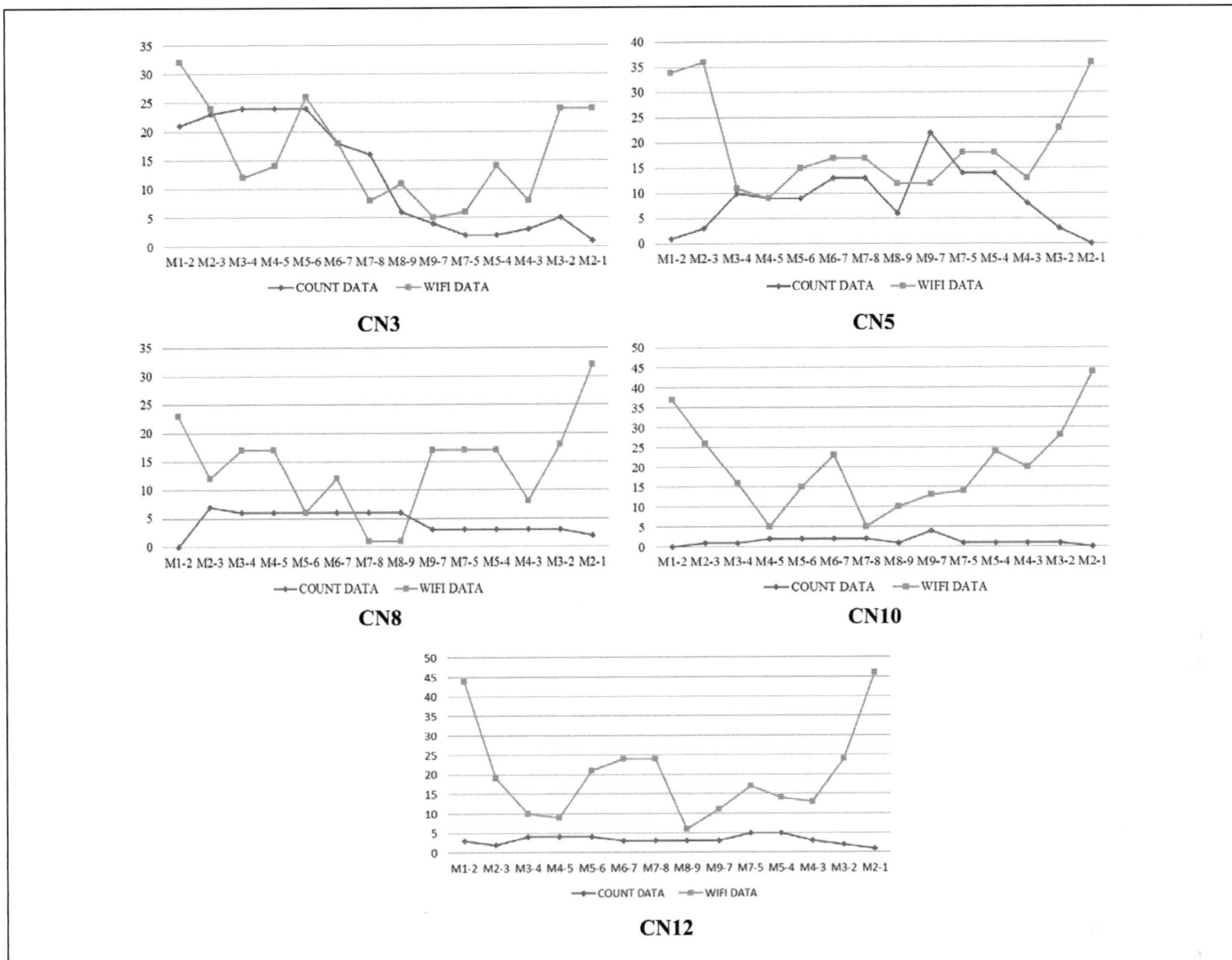

Figure 7. Comparison between the WiFi data and observer data.

relationship. Based on correlation analysis, the observer data and WiFi data generally have a correlation line indicating relation because there was such a small difference in the two data sets.

However, if the data are split based on the movement between each bus stop, there is a small difference between the observer and WiFi data that can be considered reliable. All of the circulation data for BS1 and BS2 were associated with a high traffic volume because it is the main route in Obuse, because the main activity center is Obuse Station (BS2), and because the tourist entrance is at Obuse Park (BS1). CN3, CN5, CN8, CN10, CN12 at BS4, BS5, BS6, and BS7 is a daytime route, which most pedestrians tend to walk so they can see some of the interesting places alongside the route. To assess the details of each circulation and each bus stop, the observer data and WiFi data were analyzed by the ratio of WiFi over estimated (RWOE) to the observer data. The values of RWOE greater than 8 indicate the unreliable route segments. For values >8 (RWOE) was used based on the authors' assumption that if the data difference was between 1 and 7, it was acceptable. RWOE analysis was employed to illustrate the differences between the WiFi data (MAC address) and the traffic count. According to Table 1, CN3, CN5, and CN8 had more reliable data than C10 and C12. For more details, please refer to Table 1.

Conclusion

The analysis and interpretation of the WiFi data yielded travel data about passengers' travel from one place to another. This paper presented the application of a new method of collecting data about boarding or alighting passengers at each bus stop. The results of this paper are as follows:

Table 1. Comparison of Observer and WiFi Data for Each Trip Segment in Five Circulations

M	CN3			CN5			CN8		
	MCD	WD	RWOE	MCD	WD	RWOE	MCD	WD	RWOE
M1-2	21	32	11.0	1	34	33.0	0	23	23.0
M2-3	23	24	1.0	3	36	33.0	7	12	5.0
M3-4	24	12	-12.0	10	11	1.0	6	17	11.0
M4-5	24	14	-10.0	9	9	0.0	6	17	11.0
M5-6	24	26	2.0	9	15	6.0	6	6	0.0
M6-7	18	18	0.0	13	17	4.0	6	12	6.0
M7-8	16	8	-8.0	13	17	4.0	6	1	-5.0
M8-9	6	11	5.0	6	12	6.0	6	1	-5.0
M9-7	4	5	1.0	22	12	-10.0	3	17	14.0
M7-5	2	6	4.0	14	18	4.0	3	17	14.0
M5-4	2	14	12.0	14	18	4.0	3	17	14.0
M4-3	3	8	5.0	8	13	5.0	3	8	5.0
M3-2	5	24	19.0	3	23	20.0	3	18	15.0
M2-1	1	24	23.0	0	36	36.0	2	32	30.0

M	CN10			CN12		
	MCD	WD	RWOE	MCD	WD	RWOE
M1-2	0	37	37.0	3	44	41.0
M2-3	1	26	25.0	2	19	17.0
M3-4	1	16	15.0	4	10	6.0
M4-5	2	5	3.0	4	9	5.0
M5-6	2	15	13.0	4	21	17.0
M6-7	2	23	21.0	3	24	21.0
M7-8	2	5	3.0	3	24	21.0
M8-9	1	10	9.0	3	6	3.0
M9-7	4	13	9.0	3	11	8.0
M7-5	1	14	13.0	5	17	12.0
M5-4	1	24	23.0	5	14	9.0
M4-3	1	20	19.0	3	13	10.0
M3-2	1	28	27.0	2	24	22.0
M2-1	0	44	44.0	1	46	45.0

Note: M = moving; CN = circulation number; MCD = manual count data; WD = WiFi data; RWOE = ratio WiFi over estimated.

1. This paper developed a data processing procedure to combine WiFi raw data and GPS log data into travel data, which show the number of passengers boarding or alighting at a bus stop and OD information.
2. The travel data were analyzed to be divided into two data: before noon and after noon. These two results have characteristics of different travel behavior.
3. The results of comparison between WiFi data and observer data show similar travel patterns among many bus stops; however, because of the non-passenger data, a direct comparison cannot be made between two bus stops.

The results of the conducted experiments indicate not only the benefits of the developed equipment but also the challenges to be addressed in future work. The WiFi scanner could be used for several other purposes:

1. They could assist small bus operators who conduct their own surveys.

2. The proposed method is appropriate for the long-term data collection of daily variations.
3. There is no need to communicate with passengers when collecting data.
4. It is possible to know movement demand (origin and destination) between bus stops, which would allow bus operators to estimate hourly and daily demand and determine the capacity between bus stops. For example, bus demand sometimes increases in certain periods such as the high autumn season and festivals. The WiFi-scanner data can be used to estimate capacity at certain seasons, allowing bus operators to add or reduce the number of operating buses to accommodate the changing demand. They can also make timetable changes to adjust to high demand at certain times.
5. The WiFi-scanner data can also provide information about busy areas such as bus stops, shops, parking lots, and bus terminals. If the WiFi scanners detect an area that has many visitors, these data could inform operators to increase capacity. For local governments, WiFi-scanner data could be used to improve the capacity of facilities and provide input to urban planning related to transportation facilities and public transportation and could also be used to improve the marketing of tourism support facilities.
6. Outside the bus, the WiFi-scanner technology can be used to detect and distinguish vehicles, pedestrians, and buildings (non-passenger data).

The results of the conducted experiments indicated not only the benefits of the developed equipment but also the challenges to be addressed in future work. However, at present, there are several limitations. The WiFi scanner can produce inaccurate results when there is less use of WiFi-enabled mobile devices: there are false WiFi readings because the device is out of range; the MAC address changes (upgrading system devices); and because of slight GPS inaccuracies. As there have been few case studies, the technology needs to be tested in several different places with larger survey areas and longer survey times to improve the analysis. Future work will focus on passenger travel behavior modeling, passenger travel times (how many minutes from origin to destination based on WiFi and GPS data), and non-passenger data analysis such as pedestrians, vehicles, and buildings.

Acknowledgments

The authors would like to thank all those who have contributed to this paper. Thanks also go to members of the Transportation Planning Laboratory, Tokyo University of Science who have provided survey data and much assistance in supporting this paper. Thanks also go to the corresponding author's workplace Universitas Teknologi Sulawesi, Ministry of Research, Technology and Higher Education and Indonesia Endowment Fund for Education (LPDP), Ministry of Finance, Indonesia.

Author Contributions

The authors confirm contributing to the paper as follows: study conception and design: Hidayat, Terabe, Yaginuma; data collection: Terabe, Yaginuma; analysis and interpretation of results: Hidayat, Terabe, Yaginuma; draft manuscript preparation: Hidayat. All authors reviewed the results and approved the final version of the manuscript.

References

1. Caceres, N., J. P. Wideberg, and F. G. Benitez. Deriving Origin Destination Data from a Mobile Phone Network. *IET Intelligent Transport Systems*, Vol. 1, No. 1, 2007, pp. 15–26.
2. Thiagarajan, A., J. Biagioni, T. Gerlich, and J. Eriksson. *Cooperative Transit Tracking Using Smart-Phones. Proc., 8th ACM Conference on Embedded Networked Sensor Systems, SenSys 2010*, Zürich, Switzerland, ACM, New York, 2010, pp. 85–98.
3. Haseman, R. J., J. S. Wasson, and D. M. Bullock. Real-Time Measurement of Travel Time Delay in Work Zones and Evaluation Metrics Using Bluetooth Probe Tracking. *Transportation Research Record: Journal of the Transportation Research Board*, 2010. 2169: 40–53.
4. Kostakos, V., T. Camacho, and C. Mantero. Towards Proximity-Based Passenger Sensing on Public Transport Buses. *Personal and Ubiquitous Computing*, Vol. 17, 2013, pp. 1807–1816. doi:10.1007/s00779-013-0652-4.
5. Kostakos, V. *Towards Sustainable Transport: Wireless Detection of Passenger Trips On Public Transport Buses.* International Energy Outlook, 2007. United States Department of Energy, Washington, D.C.
6. Ahmed, H., M. El-Darieby, Y. Morgan, and B. Abdulhai. A Wireless Mesh Network-Based Platform for ITS. *Proc., IEEE Vehicular Technology Conference, VTC Spring 2008*, Singapore, IEEE, New York, United States Department of Energy, Washington, D.C., 2008, pp. 3047–3051.
7. Bachmann, C., M. J. Roorda, B. Abdulhai, and B. Moshiri. Fusing A Bluetooth Traffic Monitoring System with Loop Detector Data for Improved Freeway Traffic Speed Estimation. *Journal of Intelligent Transportation Systems*, Vol. 17, 2013, pp.152–164. doi:10.1080/15472450.2012.696449
8. Jackson, S., A. Lesani, and L. F. M. Moreno. Towards A WiFi System for Traffic Monitoring In Different Transportation Facilities. Presented at 93rd Annual Meeting of the Transportation Research Board, Washington, D.C., 2014.
9. Lozano, J. J. F., M. M. Guzmán, J. M. Ávila, and A. G. Cerezo. A Wireless Sensor Network for Urban Traffic Characterization and Trend Monitoring. *Sensors Journal*, Vol. 15, 2015, pp. 26143–26169.
10. Oransirikul, T., R. Nishide, I. Piumarta, and H. Takada. Measuring Bus Passenger Load by Monitoring Wi-Fi Transmissions from Mobile Devices. *Procedia Technology*, Vol. 18, 2014, pp. 120–125. http://dx.doi.org/10.1016/j.protcy.2014.11.023.

11. Perttunen, M., V. Kostakos, J. Riekki, and T. Ojala. *Urban Traffic Analysis through Multi-Modal Sensing*. *Personal and Ubiquitous Computing*, Vol. 19, 2015, pp. 709–721.
12. Dunlap, M., Z. Li, K. Henrickson, and Y. Wang. Estimation of Origin and Destination Information from Bluetooth and Wi-Fi Sensing for Transit. *Transportation Research Record: Journal of the Transportation Research Board*, 2016. 2595: 11–17.
13. Jiang, M., X. Fan, and F. Zhang. Characterizing On-Bus WiFi Passenger Behaviors by Approximate Search and Cluster Analysis. *Proc., 7th International Conference on Cloud Computing and Big Data (CCBD2016)*, Macau, China, IEEE, New York, 2016. https://www.researchgate.net/publication/308449230.
14. Malinovskiy, Y., N. Saunier, and Y. Wang. Analysis of Pedestrian Travel with Static Bluetooth Sensors. *Transportation Research Record: Journal of the Transportation Research Board*, 2012. 2299: 137–149.
15. Poucin., G., B. Farooq, and Z. Patterson. Pedestrian Activity Pattern Mining in WiFi-Network Connection Data. Presented at 95th Annual Meeting of the Transportation Research Board, Washington, D.C., 2016.
16. Shlayan, N., A. Kurkcu, and K. Ozbay. Exploring Pedestrian Bluetooth and WiFi Detection at Public Transportation Terminals. *Proc., IEEE 19th International Conference on Intelligent Transportation Systems (ITSC)*, Rio de Janeiro, Brazil, IEEE, New York, 2016, pp. 229–234.
17. Obuse Culture and Tourism Association. In Obuse Tour Guide. http://www.obusekanko.jp/download/pdf/english_s.pd. Retrieved June 1, 2017.
18. Obuse, Nagano. In Wikipedia: The Free Encyclopedia. https://en.wikipedia.org/wiki/Obuse,_Nagano. Retrieved June 1, 2017.
19. Raspberry Pi 2 Model B. In Raspberry Pi Foundation. https://www.raspberrypi.org/products/raspberry-pi-2-model-b/. Retrieved June 5, 2017.
20. Raspberry Pi 2 Model B. In Raspberry Pi 2 Model B 1GB Technical specifications. https://www.element14.com/community/docs/DOC-73827/l/raspberry-pi-2-model-b-1gb-technical-specifications#techspecs. Retrieved June 5, 2017.

The Standing Committee on Information Systems and Technology (ABJ50) peer-reviewed this paper (18-01015).

Article

TRR
JOURNAL OF THE TRANSPORTATION RESEARCH BOARD

Transportation Research Record
2018, Vol. 2672(45) 55–63

DOI: 10.1177/0361198118796068
journals.sagepub.com/home/trr
SAGE

Performance of Wireless Magnetometers for Stop Bar Detection at Signalized Intersections

Jidong J. Yang[1] and Bashan Zuo[2]

Abstract

Wireless magnetometers have been considered as a practical alternative to inductive loops and suitable for large intersections, where span wire is generally used for traffic signal support. In this paper, wireless magnetometers are evaluated for stop bar vehicle detection at signalized intersections. High-resolution detector data were collected in the field subjected to various weather and environmental conditions. Conditional inference trees were used to correlate detection errors with weather and environmental factors that potentially affect the performance of wireless magnetometers. The study results indicated that the wireless magnetometers are fairly robust to various environmental conditions, such as wind, lighting, and visibility. Frequent passing of heavy vehicles, common at large intersections, can cause communication interruption between in-pavement sensors and the access point. This likely increases false and stuck-on call errors, which could be aggravated by adverse weather (e.g., rain, fog, or snow). This communication interruption issue can be mitigated by proper installation of additional repeaters. Provided the interruption issues are site-specific, professional judgment and field test are required for proper system setup, which is critical to delivering accurate and reliable detection for the wireless magnetometer system.

Vehicle detection technologies have been rapidly evolving over the past decade because of the advancement of sensors and wireless communication technologies and increasing deployment of traffic-responsive and adaptive traffic control systems, which heavily rely on robust vehicle detection. The traditional inductive-loop detector was introduced in the early 1960s and since then has become the dominating vehicle sensor in modern traffic signal control systems. With a long history of deployment, inductive-loop detectors have exposed many problems. They are susceptible to pavement cracking, labor intensive for maintenance over time, and cause traffic disruption for repair. Additionally, resurfacing of roadways or utility repairs may require reinstallation of these types of sensors. Because of those practical concerns, less intrusive or nonintrusive vehicle detection technologies have emerged. However, at large intersections, which typically require span wire for signal support, vehicle detection options are rather limited besides the inductive loop. Among those, wireless magnetometers have become a popular alternative to inductive loops, and have been extensively used in the U.S. With no need for cabling from the cabinet to the in-pavement sensors, they are much easier to install and maintain, and cause less disruption to traffic for installation and repair as compared with inductive loops. Given the small size of in-pavement sensors, they are less likely to malfunction because of pavement cracking. Unlike inductive loops, wireless magnetometers can be reused when the pavement is resurfaced. Furthermore, the directional sensors can reduce the false calls from vehicles traveling in different directions. However, because they use the unlicensed frequency band (2400–2483.5 MHz) for communication between the in-pavement sensors and access point or repeaters, communication disruption might occur for various reasons. Additionally, latency might be an issue depending on the field conditions and setup. Previous studies of this technology have been focused on its application for vehicle counting (*1–4*). More recently, Day et al. conducted a study focusing on the latency and recommended optimal spacing of magnetometers to minimize missed calls (*5*). Medina et al. evaluated the performance of wireless magnetometers under adverse

[1]Department of Civil and Construction Engineering, Kennesaw State University, Marietta, GA
[2]Emory University, Atlanta, GA

Corresponding Author:
Address correspondence to Jidong J. Yang: jyang12@kennesaw.edu

weather conditions by simple comparison (*6*). In this paper, wireless magnetometers were evaluated on the basis of high-resolution (100 ms) detector data collected at a test intersection: South Marietta Parkway and Technology Parkway in the City of Marietta, Georgia, shown in Figure 1.

Test Site Setup

At the time of this study, the City of Marietta was in the process of replacing inductive loops with wireless magnetometers at the study intersection. This detection technology transition from inductive loops to wireless magnetometers allowed the research team to use the existing inductive loops as a benchmark to evaluate wireless magnetometers. Note that regardless of practical drawbacks previously described, inductive loops have been deemed as the most accurate and reliable vehicle detectors if they are properly functioning.

Specifically, three new wireless magnetometers were installed in the northbound left turn lane within the confine of the existing loop as illustrated in Figure 1. Both inductive loops and wireless magnetometers were checked after installation and were monitored over the study period to ensure they were functioning. The northbound approach was purposely selected for this study because (1) it is farther away from the existing antenna of the access point as shown in Figure 1 and (2) frequent passing of heavy vehicles on South Marietta Parkway can potentially interrupt communication between in-pavement magnetometers and the access point, as shown in Figures 2 and 3.

The field observation confirms that frequent heavy trucks often move slowly and are sometimes stalled in the middle of the intersection because of extended queues at the downstream intersection of South Marietta Parkway and Cobb Parkway during peak periods of traffic. Those large trucks do interrupt the communication between the in-pavement magnetometers and the antenna mounted on the strain pole at the northeast corner of the intersection. Some large stuck-on calls were recorded during the initial test period. Later, a repeater was added to the strain pole at the southwest corner to improve communication. The effect of the repeater was also evaluated as part of this study. The field setup for the wireless magnetometers is illustrated in Figure 3.

Data Description

Detection errors have generally been defined as discrepancies from ground truth conditions. Four types of detection errors have been commonly referenced: missed

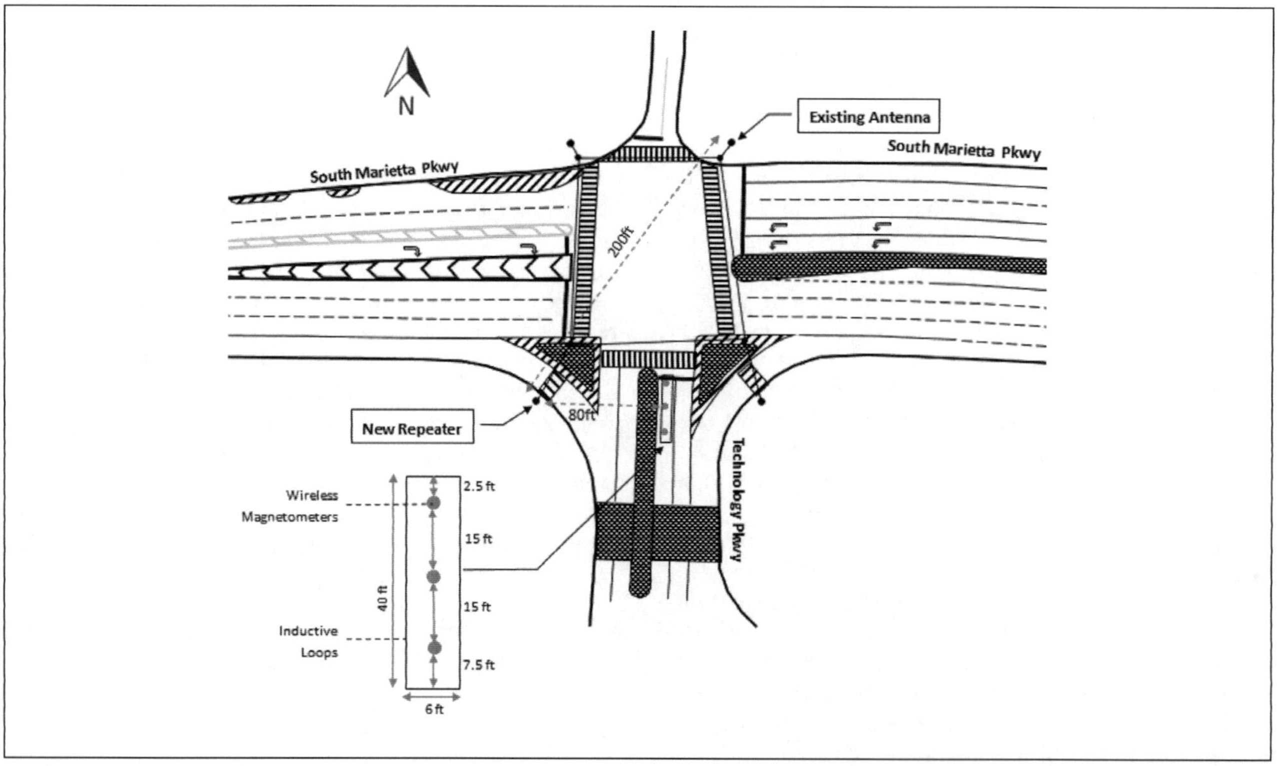

Figure 1. Illustration of the test site.

Figure 2. Northbound approach—potential signal blocking because of frequent passing of heavy trucks on South Marietta Parkway.

Figure 3. Field setup of wireless magnetometers.

call, false call, stuck-on call, and dropped calls. A similar definition by Medina et al. (*6*) was adopted in this study by referencing inductive loops as ground truth. As such, the detection error is defined as a discrepancy between the test device (wireless magnetometers) and the inductive loop conditional on the immediate prior state of detection, as shown in Table 1.

As can be seen, a missed call is defined as the loop detector registering a call (status = 1) when the test device did not (status = 0), implying a vehicle was present at the stop bar but was not detected by the test device. A dropped call is defined as the test device erroneously releasing a previously registered call (status change from 1 to 0) while the call was still held by the loop detector (status = 1), implying a vehicle was still present at the stop bar but was erroneously dropped by the test device. A false call is defined as the test device registering a call but the loop detector not doing so, implying a wrong call was placed by the test device. A stuck-on call is defined as the loop detector having released a previously registered call while the call was still being held by the test device, meaning the vehicle had departed from the stop bar but the call was still erroneously held by the test device. Given the cyclic error patterns, the average error per cycle for each type of erroneous call was computed and used as a performance indicator.

Table 1. Definition of Erroneous Calls (Detection Errors)

	Type of erroneous calls							
	Missed call		False call		Stuck-on call		Dropped call	
Status change	Loop	Test device	Loop	Test device	Loop	Test device	Loop	Test device
Before (*t*–)	0	0	0	0	1	1	1	1
After (*t* +)	1	0	0	1	0	1	1	0

Note: 1 = detector status if on; 0 = detector status if off.

Table 2. Field Test Periods and Weather Events

	Wireless magnetometer	
Test device	Without repeater	With repeater
Period of field test		
From	2/13/2015 10:10:28 a.m.	3/11/2015 1:19:43 p.m.
To	2/20/2015 1:31:08 p.m.	3/20/2015 1:15:05 p.m.
Weather event		
Clear	Y	Y
Partly cloudy	Y	Y
Scattered clouds	Y	Y
Mostly cloudy	Y	Y
Overcast	Y	Y
Light rain or drizzle	Y	Y
Mist	na	Y
Fog	na	Y
Haze	na	Y
Light Snow	Y	na

Note: Y indicates presence of the weather event; na = not applicable.

Table 3. Variables and Data Coding

Factor	Variable	Description	Unit/Value
Wind speed	Wspeed	Actual wind speed	mph
Weather event	Cond	Clear	0
		Cloudy or overcast	1
		Light rain or drizzle	2
		Mist, fog, or haze	3
		Light snow	4
Lighting	Night	Day or night	0 = day; 1 = night
Visibility level	Visibility	Distance	mi
Repeater	RP	A repeater was added	0 = without repeater 1 = with repeater

Concurrent weather information for the test periods was extracted from the closest weather station through the weather underground website (wunderground.com), including wind speed, visibility, and various weather events. Weather events experienced at the test site during the test periods are presented in Table 2.

For analysis purposes, similar weather events were grouped together and assigned an ordinal number by referencing clear weather as baseline. A higher number indicates a more severe weather event. The coding for various weather events and other variables is presented in Table 3.

Table 4. Summary of Descriptive Statistics

Variable	Unit/value	Freq.	Mean	SD	Median	Min.	Max.
Wspeed	mph	5,229	9.3	4.8	8.1	0.0	23.0
Visibility	mile	5,229	8.6	2.8	10.0	0.5	10.0
Weather group	0	637	na	na	na	na	na
	1	3,927	na	na	na	na	na
	2	615	na	na	na	na	na
	3	15	na	na	na	na	na
	4	35	na	na	na	na	na
Night	N	3,493	na	na	na	na	na
	Y	1,736	na	na	na	na	na
RP	N	2,182	na	na	na	na	na
	Y	3,047	na	na	na	na	na

Note: Freq. = frequency; SD = standard deviation; Min. = minimum; Max. = maximum; na = not applicable.

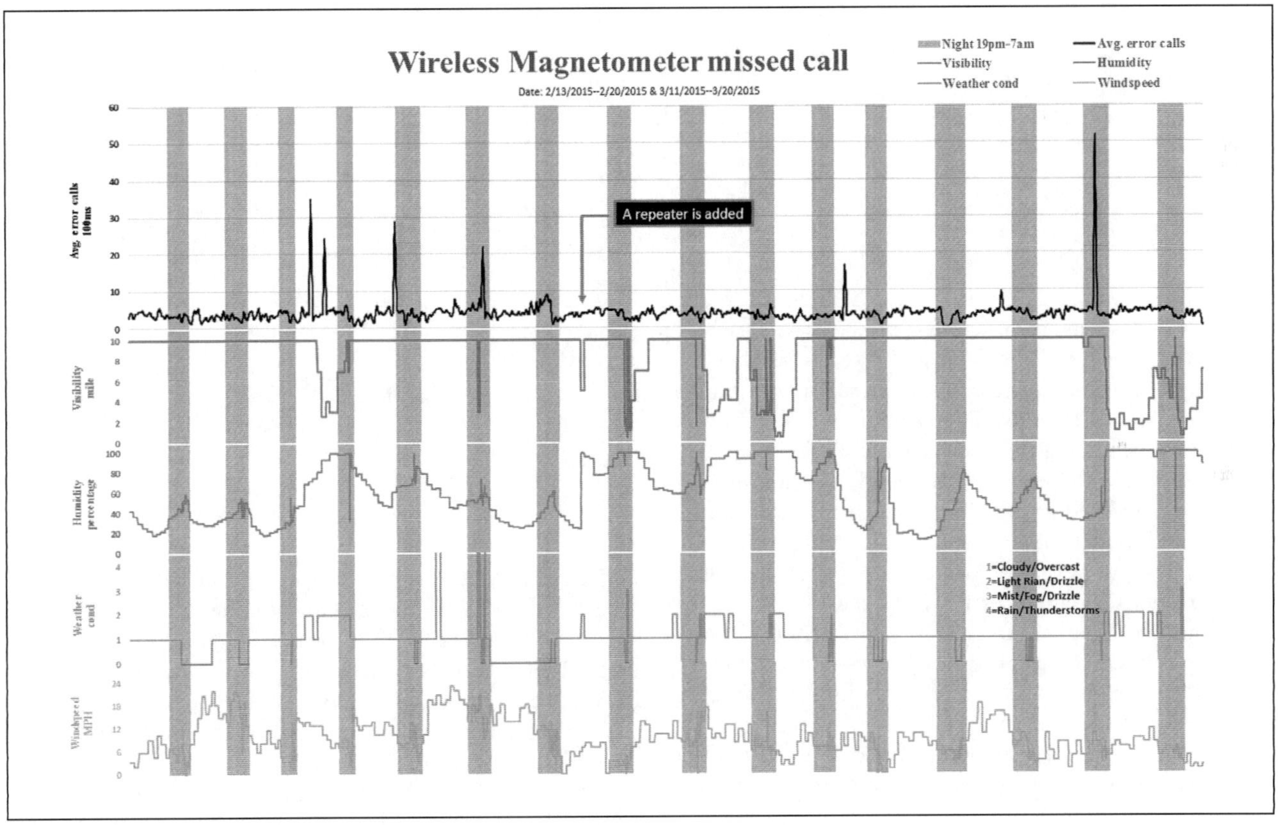

Figure 4. Temporal plots of missed call errors versus weather and environmental conditions.

Data Analysis

Descriptive Statistics and Temporal Plots

The descriptive statistics of the variables are summarized in Table 4. The temporal variation of detection errors was graphed against concurrent weather and environmental conditions. As shown in Figures 4–7, no systematic patterns can be visually discerned except that a sudden reduction of false and stuck-on calls becomes evident after the new repeater had been installed.

Conditional Inference Trees

Background. Recursive partitioning is a fundamental tool in data mining. It helps explore the structure of data, while developing easy-to-visualize decision rules for

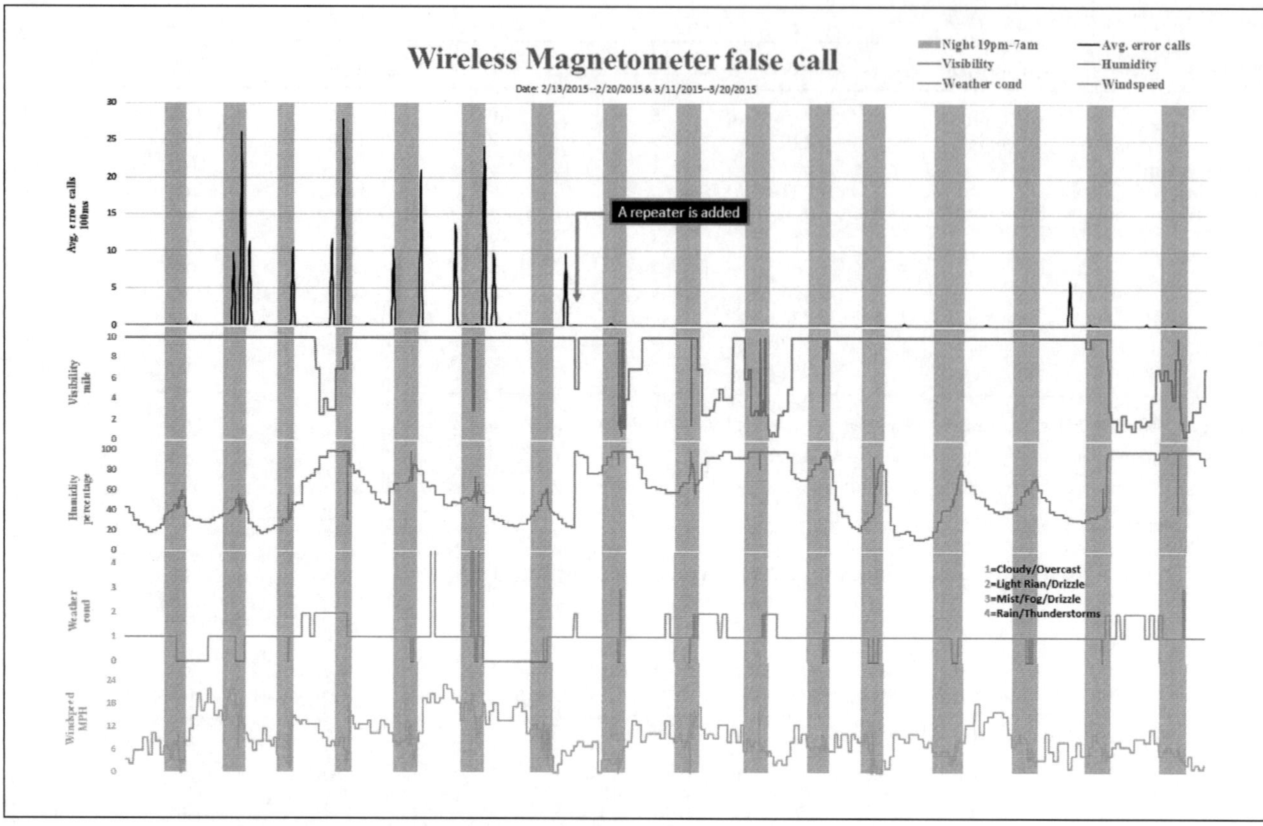

Figure 5. Temporal plots of false call errors versus weather and environmental conditions.

predicting a categorical (classification tree) or continuous (regression tree) outcome. The conditional distribution of statistics measuring the association between responses and covariates is the basis for unbiased selection among covariates measured at different scales (*7*). In this study, conditional inference trees (ctree) in the R party package (*8*) were used to reveal associations between detection errors (responses) and potential influential factors or conditions (covariates). Details on the methodology of conditional inference trees were given by Hothorn et al. (*7*). A brief introduction to conditional inference trees is provided below for reference.

By assuming that the conditional distribution $D(Y|X)$ of the response variable Y given the covariates X depends on a function of X, $f(X)$,

$$D(Y|X) = D(Y|f(X))$$

The covariate space is partitioned into finite disjoint cells using the regression relationship to be fitted based on a learning sample. To determine whether there is any information about the response variable covered by any of the m covariates, a global hypothesis test is used. Given m covariates, the global hypothesis of independence is formulated in terms of the m partial hypotheses as

$$H_0 = \bigcap_{j=1}^{m} H_0^j$$

where

$$H_0^j : D(Y|X_j) = D(Y), j = 1, \ldots, m$$

When H_0 cannot be rejected at a prespecified level, the recursion stops. If the global hypothesis is rejected, the association between Y and each of the covariates X_j is measured by test statistics or p-values indicating the deviation from the partial hypotheses H_0^j.

In application, a generic algorithm that recursively partitions a sample is formulated using non-negative integer valued case weights. Each node of a tree is represented by a vector of case weights, which have nonzero elements when the corresponding observations are elements of the node and are zero otherwise. The algorithm involves two steps: 1) variable selection, and 2) splitting. In Step 1, the covariate of strongest association with the response (i.e., with the minimum p-value) is selected for splitting. In Step 2, a permutation test framework (*9*) is used to find the optimal binary split for the selected covariate in Step 1. The goodness of a split is evaluated by a two-sample linear statistic that measures the discrepancy between the samples. The two steps are repeated

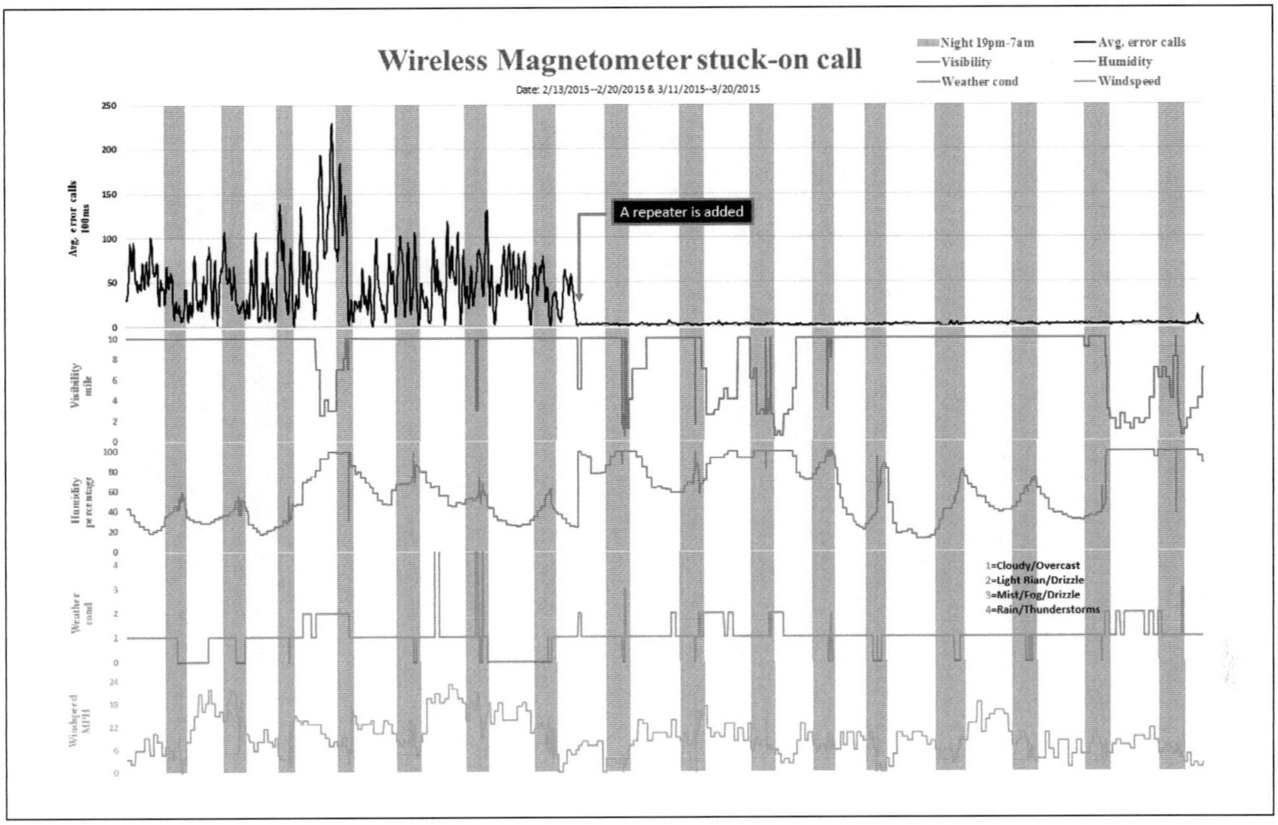

Figure 6. Temporal plots of stuck-on call errors versus weather and environmental conditions.

recursively until the global null hypothesis of independence between the response and any of the covariates cannot be rejected at a prespecified level, say 0.05. Bonferroni-adjusted p-value was used in this case.

Estimation Results. To identify factors contributing to the variation of detection errors for wireless magnetometers, conditional inference trees were estimated. The conditional inference trees for the false call error and stuck-on call error are shown in Figures 8 and 9. For the missed call error and dropped call error, no trees were estimated, implying that adverse weather and environmental conditions appears to not affect the missed call or dropped call errors for wireless magnetometers. The addition of the repeater did not significantly reduce missed call or dropped call errors. It should be noted that false calls and stuck-on calls are similar in that they both indicate that a vehicle is not detected by inductive loops (ground truth) but detected by the wireless magnetometers. On the other hand, missed calls and dropped calls are similar because they reveal that a vehicle is detected by inductive loops (ground truth) but not detected by the wireless magnetometers.

As shown in Figure 8, "y" indicates the mean detection error in milliseconds and "n" indicates the number of observations for that leaf. As a result, the only significant factor to explain the variance of false call errors is whether the repeater is present. As seen, the addition of the repeater significantly reduces the false call error from an average of 0.883 ms to 0.028 ms.

Similarly, Figure 9 shows that the addition of the repeater results in a drastic reduction in the stuck-on call error. Larger stuck-on call errors are correlated with adverse weather conditions (Cond $>$ 1, including light rain, drizzle, mist, fog, haze, or light snow) as compared with fair weather conditions (Cond$<=$1, including clear, cloudy, or overcast). The largest stuck-on call error (103.749 ms) occurred under more adverse weather conditions (Cond $>$ 1) when there was no repeater. After installing the repeater, the stuck-on call error was reduced to 2.486 ms. The smallest stuck-on call error (1.997 ms) occurred under more favorable weather conditions (Cond $\leq$ 1) with the repeater being installed. The repeater appears to significantly mitigate the effect of adverse weather conditions as seen by the much smaller difference in mean error (2.486 − 1.997 = 0.489 ms) as

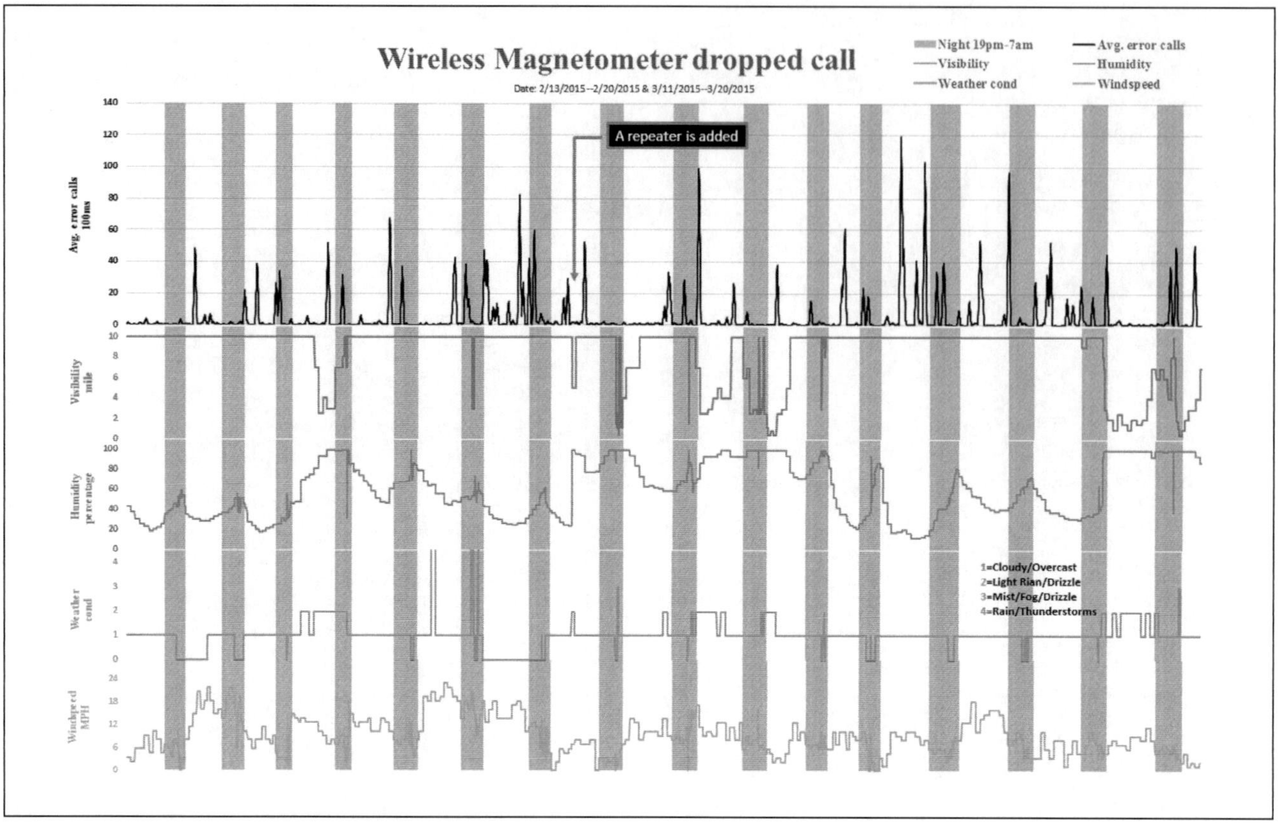

Figure 7. Temporal plots of dropped call errors versus weather and environmental conditions.

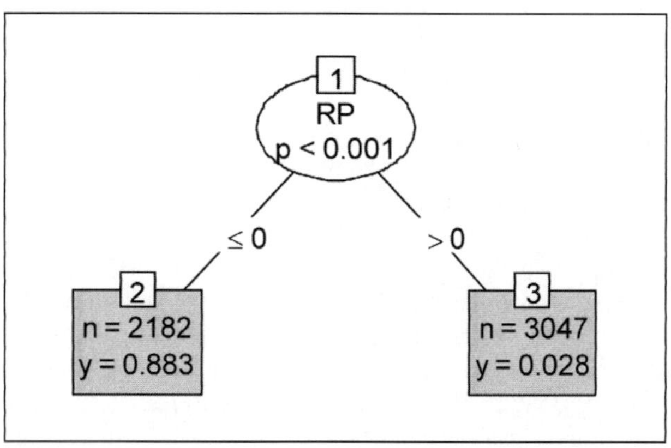

Figure 8. Conditional inference tree for false call errors.

compared with (103.749 – 45.003 = 58.746 ms) in the absence of the repeater.

Note that the estimated conditional inference trees did not include environmental factors, such as wind speed, visibility, lighting conditions (day or night), indicating that wireless magnetometers are fairly robust to those environmental conditions.

Conclusion

Technologies for replacing inductive loops at large intersections with span wire are rather limited. Wireless magnetometers are considered as a practical alternative to inductive loops that have been widely deployed in the U.S. Based on the results of this study, the wireless magnetometers appear to be fairly robust to various environmental conditions, such as wind, lighting, and visibility conditions. Adverse weather (e.g., rain, fog, mist, haze, or snow) likely increases stuck-on call errors. Frequent passing of heavy vehicles, which is common at large intersections, could interrupt wireless communication between the access point and in-pavement magnetometers as demonstrated in this study. This communication interruption can increase false or stuck-on call errors, resulting in inefficient operation of traffic signals. To resolve this interruption issue, additional repeaters could be installed. As demonstrated in this study, the installation of the additional repeater significantly reduces stuck-on and false call errors at the study site. Provided the interruption issues are site-specific, professional judgment and field test are required for proper

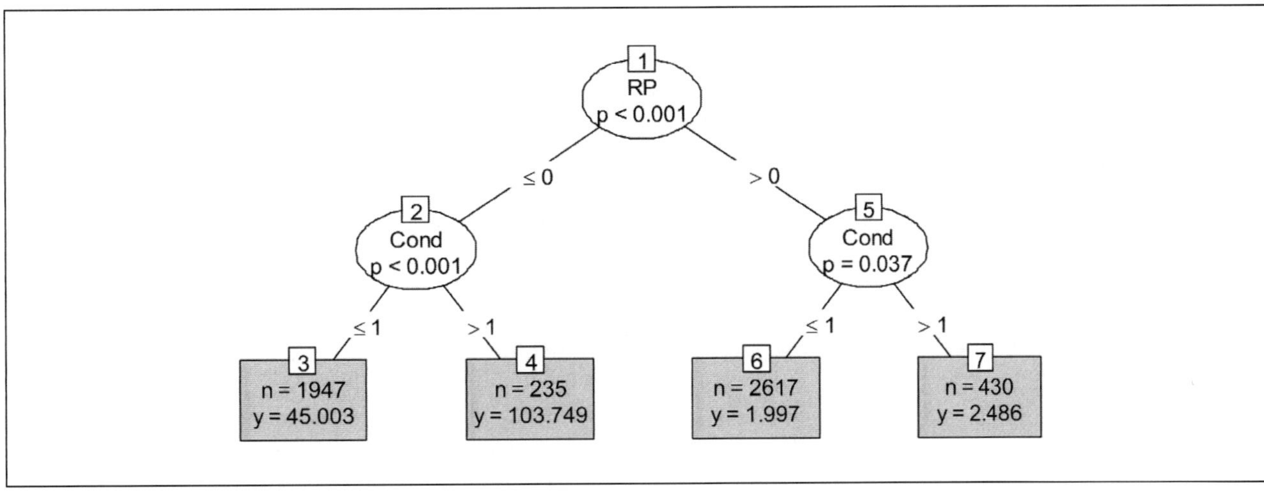

Figure 9. Conditional inference tree for stuck-on call errors.

system setup, which is critical to delivering accurate and reliable detection for the wireless magnetometer system.

Acknowledgments

This work is part of a research project sponsored by the Georgia Department of Transportation. The study was conducted with the Sensys wireless magnetometer detection unit.

Author Contributions

The authors confirm contribution to the paper as follows: study conception and design: Jidong Yang; data collection and processing: Bashan Zuo; analysis and interpretation of results: Jidong Yang; draft manuscript preparation: Jidong Yang, Bashan Zuo. All authors reviewed the results and approved the final version of the manuscript.

References

1. Cheung, S. Y., S. Coleri, B. Dundar, S. Ganesh, C.-W. Tan, and P. Varaiya. Traffic Measurement and Vehicle Classification with Single Magnetic Sensor. *Transportation Research Record: Journal of the Transportation Research Board*, 2005. 1917: 173–181.
2. Haoui, A., R. Kavaler, and P. Varaiya. Wireless Magnetic Sensors for Traffic Surveillance. *Transportation Research Part C: Emerging Technologies*, Vol. 16, No. 3, 2008, pp. 294–306.
3. Margulici, J. D., S. Yang, and C.-W. Tan. *Evaluation of Wireless Traffic Sensors by Sensys Networks, Inc.* California Department of Transportation, Sacramento, 2006.
4. Cheung, S.-Y., and P. Varaiya. *Traffic Surveillance by Wireless Sensor Networks: Final Report.* UCB-ITS-PRR-2007-4. University of California, Berkeley, 2007.
5. Day, C., H. Premachandra, T. M. Brennan, Jr., J. R. Sturdevant, and D. M. Bullock. Operational Evaluation of Wireless Magnetometer Vehicle Detectors at Signalized Intersection. *Transportation Research Record: Journal of the Transportation Research Board*, 2010. 2192: 11–23.
6. Medina, J. C., R. F. Benekohal, and A. Hajbabaie. *Evaluation of Sensys Wireless Vehicle Detection System: Results from Adverse Weather Conditions.* Research Report ICT-11-81. Illinois Center for Transportation, 2011.
7. Hothorn, T., K. Hornik, and A. Zeileis. Unbiased Recursive Partitioning: A Conditional Inference Framework. *Journal of Computational and Graphical Statistics*, Vol. 15, No. 3, 2006, pp. 651–674.
8. R Core Team. *R: A Language and Environment for Statistical Computing.* R Foundation for Statistical Computing, Vienna, Austria, 2014.
9. Strasser, H., and C. Weber. On the Asymptotic Theory of Permutation Statistics. *Mathematical Methods of Statistics*, Vol. 8, 1999, pp. 220–250.

The Standing Committee on Information Systems and Technology (ABJ50) peer-reviewed this paper (18-01756).

Any opinions, findings, and conclusions or recommendations expressed in this paper are those of the authors and do not necessarily reflect the views of the sponsoring agency or other individuals.

Article

TRR
JOURNAL OF THE TRANSPORTATION RESEARCH BOARD

Transportation Research Record
2018, Vol. 2672(45) 64–75

DOI: 10.1177/0361198118796711
journals.sagepub.com/home/trr
SAGE

Transportation Cloud Service Composition Based on Fuzzy Programming and Genetic Algorithm

Weibin Zhang[1], Haifeng Guo[2], Ziqiang Zeng[3], Yong Qi[4], and Yinhai Wang[3]

Abstract

The push toward smarter transportation management and decision-making has increased significantly in recent years. Toward this end, it is constructive to establish a traffic intelligence platform leveraging web services, big data analytics, and cloud computing. The developers of any such platform always face the challenge of selecting or creating composition plans from among numerous possible plans, often with unclear requirements, that satisfy their quality-of-service (QoS) requirements. Typical QoS properties associated with a web service are execution cost and time, availability, successful evaluation, usage frequency, and accuracy. Most of these factors are vague and/or difficult to quantity. To address the need to handle vague inputs, a constraint satisfaction-based web service composition algorithm is proposed, which is based on fuzzy multi-objective linear programming. A genetic algorithm with Pareto optimization evaluation with weighted standardized Euclidean distance is used that demonstrates good performance as the quantity of service candidate increases. The algorithm is capable of finding a satisfactory solution under input of vague QoS requirements, which shows better capability than multiple-objective linear programming. The applicability and performance of the proposed algorithm are validated in an online transportation analytics platform environment.

Big data is currently a hot topic in a broad range of science and engineering fields both in research and, increasingly, in practice (*1*, *2*). In particular, applications for big data concepts and methodologies have become a major focus in transportation research in recent years. Vast quantities of data are available through data acquisition and communications technologies including sensors, cameras, mobile location services, and others, which when combined with emerging computing and analysis methodologies can lead to more informed and proactive management and decision making (*3*).

In the near future, the information required for making optimal travel and operations decisions will be nearly universally accessible, and available in whatever form is most convenient for any given application. For example, when traffic congestion can be monitored and accurately predicted in real time, driving routes will automatically be optimized, updated, and customized for individual users. Such applications are foundational to the concept of a smart city, and as time goes on more and more people will benefit from intelligent infrastructure, mobile computing, and the integration of intelligent applications in the virtual world into the physical infrastructure. The closely related concepts of intelligent transportation and smart cities are defined by the effective integration of intelligence embedded into cloud service platforms, access to real-time data and information streams, and the physical world and the people therein.

To achieve this wealth of opportunity, a standard framework for the effective utilization of data and computing resources, such that these data serve the needs of users, has become a pressing need in the field of transportation (*3*). The challenges associated with this type of

[1]School of Electronic and Optical Engineering, Nanjing University of Science and Technology, Nanjing, China
[2]College of Information Engineering, Zhejiang University of Technology, Hangzhou, China
[3]Smart Transportation Application and Research Laboratory, Department of Civil and Environmental Engineering, University of Washington, Seattle, WA
[4]School of Computer Science and Engineering, Nanjing University of Science and Technology, Nanjing, China

Corresponding Author:
Address correspondence to Yong Qi: 790815561@qq.com

framework primarily stem from the need for machines to collaborate in an efficient way while extracting information from heterogeneous data.

Some previous studies have developed web service-based platforms which organize data to provide domain-specific analytics services (*4*, *5*). However, there are a number of unique challenges associated with constructing such a platform to address the needs of the transportation field, and research in this area is lacking. In this paper, several key questions associated with building a transportation intelligence platform to support the analytical methods, large and diverse datasets, and quality-of-service (QoS) requirements in the transportation field are studied.

In determining how to serve user traffic demand with big data, there are three key issues which need to be addressed:

1. User requirement identification from vague/ambiguous concepts and input.
2. Web service mapping, evaluation, and composition under blurring conditions in an efficient way.
3. Transportation data structure modeling and cloud service provisioning.

Here the term "user" indicates the persons who use the traffic platform service that is discussed in this study.

This paper focuses primarily on how to compose services and match solutions to vague user requirement in an efficient way. The term "user requirement" means traffic service related requirement from users, such as route planning.

The proposed traffic intelligence system is based on service-oriented architecture, providing a loosely coupled computing framework under which traffic function and data-access web services are composed. It is clear that, due to the growth of cloud computing and software as a service (SaaS) (*6*), the number and variety of services provided on the internet will only increase in the future.

In certain scenarios, a simple web service may not meet a desired functional requirement, then a composition of a set of proper component web services may be a possible alternative to fulfill the requirement (*6*, *7*). Web service composition integrates individual, outsourced web services to a new service (*8*, *9*), thereby providing value-added composite services beyond the capabilities of individual atomic web services. This design paradigm requires a method to organize component services to fulfill user requirements, something similar to a workflow model. Consider the following example, which describes an individual making travel plans using various web services. The following steps are involved: First, the individual would contact transportation agency and weather web services to acquire current road limitations and weather forecasts. Other services may then be requested, including car rental, insurance, and hotel related to the trip. After the relevant services have been identified and selected, they are coordinated and executed, and the results approved by the user. This example illustrates a common scenario in which, by combining resources offered from multiple providers, a composite web service can provide greater value and ease of use compared with single web service.

There are some challenges in selecting the optimal individual services. One challenge is how to identify specific user requirements from vague input (*10*). Another, possibly greater challenge is service composition from different service providers with different QoS attributes that is thought of as an NP-hard optimization problem (*4*). Previous studies (*11*, *12*) have spent a great deal of effort on service composition and algorithm performance improvement in NP-hard problems, but the question of how to handle vague requirement is essentially ignored. In this study, a fuzzy linear programming algorithm based on the Pareto-optimal genetic algorithm is proposed to address QoS constraint mapping under vague user input. The Pareto-optimal genetic algorithm is used for solution search, which is validated as an effective and efficient method.

Traffic Intelligence as Cloud Service

Problem Description

Service composition is a useful paradigm that has been proposed by researchers and applied by cloud platform providers. This paper deals mainly with web service composition from the perspective of QoS and mapping between QoS and user requirements. Selecting appropriate services from a service pool, addressing service composition restrictions, determining the important QoS parameters, and understanding the fuzzy characteristics of the problem are some important issues that must be addressed to ensure the service users' satisfaction. QoS is becoming a very important measure to evaluate the services provided (*12*). Typical QoS parameters associated with a web service are execution cost and time, availability, successful execution rate, reputation, and usage frequency (*13*). In addition, recognizing user requirements and mapping them to QoS metrics is another significant challenge.

Because multiple QoS dimensions are involved in service-level agreement (SLA) contracting, and because conflicts exist between dimensions when performing optimization, it is a typical multi-objective optimization problem. Furthermore, considering the often unclear of boundaries of client requirements, a fuzzy preference model is introduced to help service composers elicit and specify preference information upon different QoS

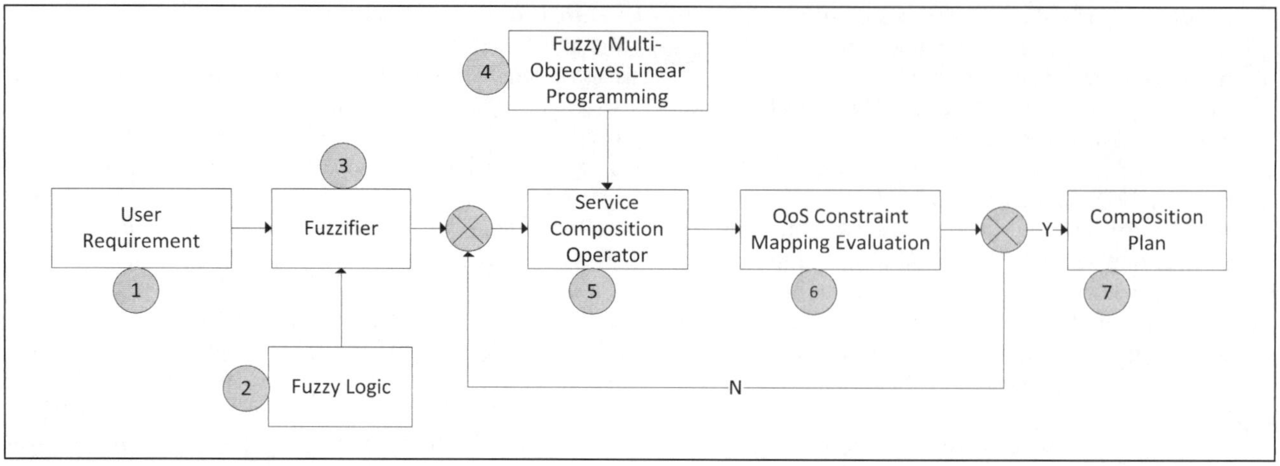

Figure 1. Framework for service composition.

dimensions. Fuzzy logic was selected to represent vague requirements provided by the users of the system in the QoS evaluation.

As shown in Figure 1, user requirements are collected (step 1) and converted to fuzzy membership in the fuzzifier step (step 2 and 3). Service composition (step 5) is based on the fuzzy multi-objective linear programming algorithm (step 4). The composition result is evaluated (step 6). If the result does not meet the QoS criteria, another round of service composition search is conducted. Otherwise, the composition plan is produced (step 7).

Proposed Framework of Intelligent Traffic in Cloud Service

The term "cloud service" refers to a service which is implemented and hosted on a cloud computing platform. The term "traffic intelligence" refers to a set of methods, processes, architectures, and technologies that can be organized automatically to process and transform a collection of datasets into meaningful and useful information for traffic decision support. Thus, traffic intelligence as a cloud service means the implementation of traffic intelligence services on a cloud computing platform.

A web service is generally defined as a collaboration of three components: service requester, service provider, and service broker (*14*). The definitions and links between components follow standard protocols, such as simple object access protocol (SOAP), web services description language (WSDL), or universal description discovery and integration (UDDI). The service broker plays the role of bridge between a service requester and a service provider by providing functions including service registration, search, and others. In the framework proposed in this paper, the fundamental triangle structure is maintained, but the service broker component is improved through intelligent service mapping (see Figure 2). In addition to service registration and search functions, the intelligent service mapping component provides automatic service mapping under a set of cost and QoS constraints.

Intelligent Service Composition

In this approach, web service developers choose outsourced services automatically according to the intent or application in the target composition rather than manual composition methods (*15*). Artificial intelligence planning and software synthesis techniques are an effective way to plan web service composition without human intervention (*4*). Some different automation techniques providing basic flow control processes such as sequence, split/join, iterate, and choice operations to organize atomic process into more complex services have been proposed. Examples of this approach include high-level process description models such as OWL-S (*16*), and state transition machine (*17*).

An important issue related to web service composition is the need to reflect client QoS requirements in synthesizing outsourced web services (*12*). The client's QoS requirements are usually represented as a permissible or constraint value for each QoS property. Given a workflow of atomic processes, the problem to be solved is finding the best combination of outsourced web services that does not violate the QoS constraints set by the client. QoS evaluation of a composition plan is needed to automatically give the client a desired composition plan in terms of QoS, as shown in Figure 3. This problem is an optimization problem based on vague QoS requirements.

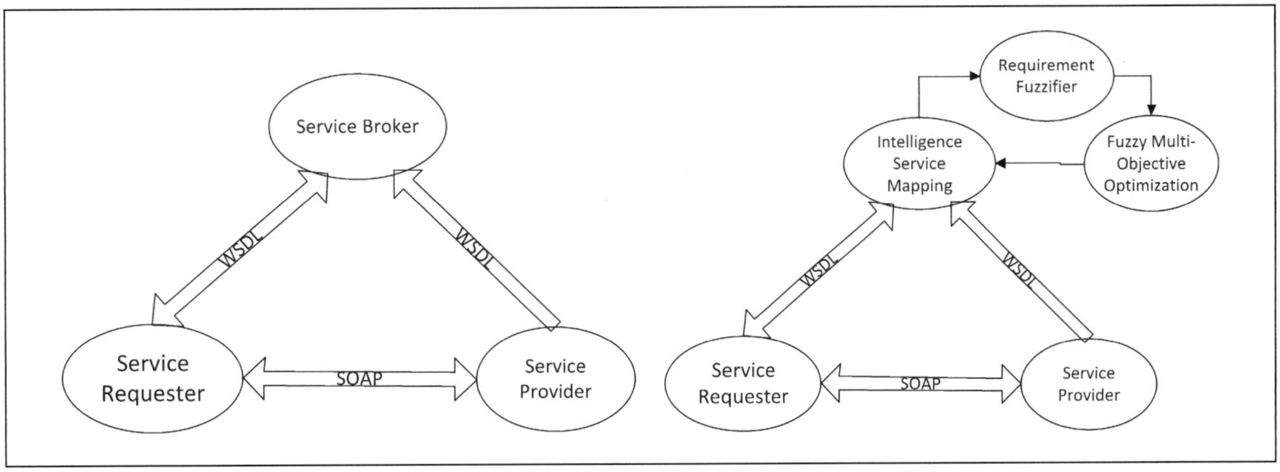

Figure 2. Web service composition concepts.

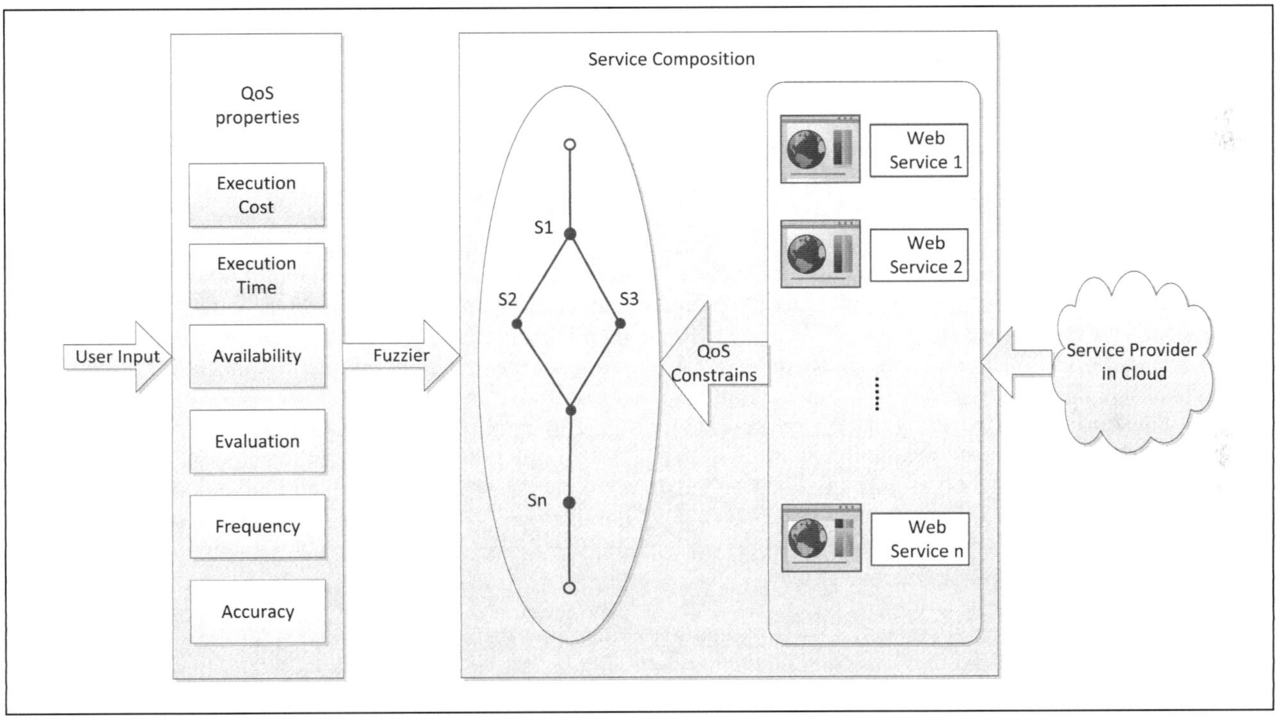

Figure 3. QoS constraint service composition model.

QoS-Oriented Web Service Composition Based on Fuzzy Multi-Objective Linear Programming

QoS Properties and Fuzzifier

To implement service composition operator, the capability of mapping vague input and service provision is necessary. Fuzzy multi-objective linear programming (FMOLP) is able to deal with matching capacity to demand in service composition (*18*, *19*). FMOLP aims to i) set overall service levels for each QoS property to meet fluctuating or uncertain demand, and ii) set decisions and policies concerning QoS properties, like execution time, execution cost, availability, successful execution rate, evaluation, frequency, and accuracy etc. Fuzzy multi-objective programing is one of the effective functions to process uncertainty.

In this paper, we propose a new QoS-oriented web service composition algorithm based on fuzzy multi-objective programming (*20*, *21*). For this purpose, we first normalize QoS properties with fuzzy logic because they are measured in different scales and often inexact.

The composition algorithm generates a composition plan by organizing outsourced web services into the composition schema in an optimal way. Moreover, the composition problem is treated as one of constraint satisfaction, a topic that has been studied in both optimization and artificial intelligence fields (*22*). In this framework, each constraint is associated with a QoS property. The proposed algorithm reflects the degree of QoS constraint violation in the objective function and continues to search the plan space in order to locate an optimal composition plan that minimizes the objective function.

In this study, six QoS properties are selected for solving the service composition problem. It should be noted that most decisions encountered in practical problems must consider a range of additional factors including production costs, skills of workers, product life cycle, etc. However, this discussion is beyond the scope of this study.

The six web-service QoS properties used in this study are defined below, some of which have been emphasized in previous research work (*13*, *23*):

- Execution cost: The cost to execute a web service.
- Execution time: The average time expected for executing a web service.
- Availability: The fraction of time that a web service is available for use. It is measured as uptime/(uptime + downtime) and updated by the service provider.
- Evaluation: The average evaluation score of web service evaluated by the clients. The individual evaluation scores are likely to be subjective, but the average score becomes more trustworthy as the total number of usages increases (*24*).
- Frequency: The number of times clients use web service. High frequency as well as high evaluation indicates that the service is trusted by users.
- Accuracy: An inquiry is often answered by analyzing large volumes of collected data. The capability to extract useful information from the large volumes of data in a limited time interval is important for an online analysis platform. Accuracy is rated by the platform operation administration according to service quality metrics, which indicate the quality or correctness of the results obtained.

Mathematical Formulation

FMOLP. Conventional multiple objective decision-making is concerned with the optimization of multiple, conflicting objective functions to satisfy constraints representing the availability of limited resources and user requirements. The multi-objective linear programming (MOLP) problem is specified by linear functions that are to be maximized subject to a set of linear constraints. The standard form of an MOLP can be written as:

$$\text{Min}\, f(x) = Cx \;\; s.t. x \in X_x, X_d = \{x | Ax \leq b, x \geq 0 \qquad (1)$$

where C is an $r \times n$ objective function matrix, and A is an $m \times n$ constraint matrix. The parameter b is an m-vector, and x is an n-vector of decision variables.

Due to the uncertainty and "fuzziness" in real-world decision-making, the function parameters and constraints may be more appropriately represented as fuzzy numbers. Moreover, the constraints may be expressed by a fuzzy inequality relation. In this case, the problem (1) becomes an FMOLP model as:

$$\text{Min}\, \tilde{f}(x) = \tilde{C}x \quad s.t. x \in X_x, X_d = \{x | \tilde{A}x \tilde{\leq} \widetilde{b \leq} 0, x \geq 0 \qquad (2)$$

where C, A, b are the matrices of the fuzzy numbers. r1 $\tilde{\leq}$ r2 means r1 is approximately smaller than or equal to r2.

Approach for FMOLP. The original MOLP model can be converted to an FMOLP model using the piecewise linear membership function (*25*) to represent the fuzzy objectives of the decision-making, together with the fuzzy decision-making (*26–28*).

With multiple objective functions, it is often the case that no globally optimal solution exists in which all objective functions reach their maximum. Therefore, a compromise solution is required such that each of the objective functions is as large as possible within the solution constraints. Fuzzy mathematical programming can be carried out for each of the objective functions, which transforms the multi-objective problem into a single-objective problem, the solution to which constitutes a fuzzy optimal solution to the multi-objective problem.

More formally, the MOLP problem (2) can be described as:

$$\max Z = Cx \qquad (3)$$

with constraint:

$$\begin{cases} Ax \leq b \\ x \geq 0 \end{cases} \qquad (4)$$

Next (3) and (4) can be transformed into multiple single-objective linear programming problems:

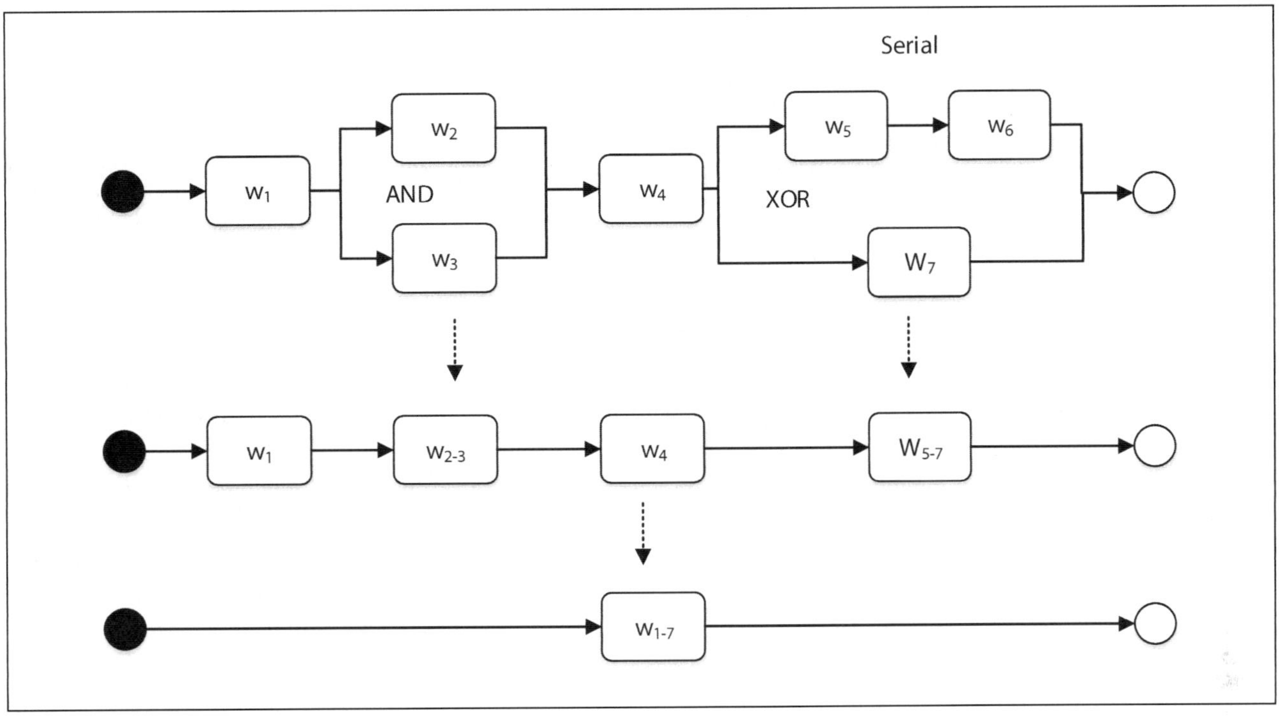

Figure 4. Aggregation process example of QoS properties.

$$\begin{cases} \max Z = \lambda \\ \sum_{j=1}^{n} c_{ij}x_j - d_i\lambda \geq Z_i^* - d_i, i = 1, 2, \cdots r \\ \sum_{j=1}^{n} a_{kj}x_j \leq b_k, k = 1, 2, \cdots, m \\ \lambda \geq 0, x_1, x_2, \cdots, x_n \geq 0 \end{cases} \tag{5}$$

where Z_i^* is the maximum and Z_i^- is the minimum of single objects $Z_i, i = 1, 2, \cdots r. d_i = Z_i^* - Z_i^-, i = 1, 2, \cdots r$ is the scope factor.

Aggregation of QoS Property Values. Generally, composition plans can be composed of serial, AND-parallel, and XOR-parallel execution dependency patterns. The QoS property values of component services connected by a serial pattern are aggregated to create a virtual service with the aggregated QoS property values. For all serial, AND-parallel, and XOR-parallel patterns, the aggregation process are performed, and virtual services are created appropriately. The aggregation steps are iterated until only one virtual service is obtained. Figure 4 demonstrates a typical aggregation procedure for service composition from multiple patterns. As shown in Figure 4, the composition plan consists of seven atomic processes, and their execution dependency is a combination of AND-parallel, XOR-parallel, and serial patterns from web service w1−w7. As the result, the aggregated QoS property values of service s8 correspond to those of the composition plan.

Weighted Standardized Euclidean Distance. For optimization, Euclidean distance is often used to represent the distance in the objective function. Euclidean distance is most easily understood from the Euclidean distance formula between two points in space. In practice, however, Euclidean distance often has the following two disadvantages: i) the dimension of each component (scale), often with differing unit, is treated in the same way; ii) the distribution of the individual components (expectation, variance, etc.) is not considered, and may be different. For example, cost and time are measured on differing scales and with differing distributional characteristics and are not directly comparable. Because of this, in this paper the weighted standardization Euclidean distance is used in the optimization process.

Standardized Euclidean distance can overcome the two previously described drawbacks associated with simple Euclidean distance. Firstly, a sample set can be standardized by the mean m and standard deviation s of the sample, as:

$$X^* = \frac{X - m}{s} \tag{6}$$

This means the standardized value (X^*) equals the raw observation (X) minus the mean of the components (m) divided by the standard deviation of the component (s). From this, the standardized Euclidean distance d_{st} between two n-dimensional vectors, $a(x_{11}, x_{12}, \ldots, x_{1n})$ and $b(x_{21}, x_{22}, \ldots, x_{2n})$ can be obtained from:

$$d_{st} = \|a - b\|_{st} = \sqrt{\sum_{k=1}^{n} \left(\frac{x_{1k} - x_{2k}}{s_k}\right)^2} \tag{7}$$

For a multi-objective optimization problem in the transportation domain, users often show a preference toward one or more optimization objectives. For example, cost and time are often considered the highest priority among optimization goals. Therefore, weight w_i is introduced to express different priorities of optimization objectives. The weights can be obtained by a simple user survey before optimization search, by assigning target priority according to empirical preference. Historical search statistics also can be used to infer weights if no weight input is available.

Objective Function. It is assumed that the binary decision variable x_{ij} is set to 1 if the jth candidate service s_{ij} is chosen for the P_i i_{th} atomic process, otherwise, it is set to 0. Since only one web service should be chosen to satisfy P_i, the following constraint should hold:

$$\sum_{j=1}^{m_i} x_{ij} = 1 \quad \forall i \tag{8}$$

Each QoS property should associate with at least one constraint. If we denote the permissible value of the k_{th} QoS property as a fuzzy crisp number Q_k (k = 1,2,. . ., 6), and we denote the aggregated value A_k (k = 1,2,. . ., 6), then each constraint can be represented as QoS constraints:

$$\begin{cases} A_k \leq Q_k, & \text{if } k = 1, 2 \\ A_k \geq Q_k, & \text{if } k = 3, 4, 5, 6 \end{cases} \tag{9}$$

Equation 9 indicates that execution cost and execution time should be less than or equal to their maximum values, and that availability, evaluation, frequency, and accuracy should be larger than or equal to their minimum values.

For the six QoS constraints in Equation 9, the constraint can be appraised according to the aggregated QoS property values A_k (k = 1,2,. . .,6). The following objective functions are proposed to find an optimal composition plan that minimizes the constraint violation using the FMOLP approach:

$$\begin{aligned} Z = & \left(\sum_{i=1}^{2} w_i \cdot \max\|A_i - Q_i\|_{st}\right) \\ & + \left(\sum_{i=3}^{6} w_i \cdot \max\|Q_i - A_i\|_{st}\right) \end{aligned} \tag{10}$$

$$S_i = \min(Z_i - Z_i^*) \tag{11}$$

where Z_i^* is maximum of single objects, and S_i is the optimal solution. w_i is the weight of the ith QoS constraint, which is a user-defined parameter.

Genetic Algorithm and Pareto Optimality. The genetic algorithm is a search or problem-solving approach commonly used in artificial intelligence applications which is based on the evolutionary process of natural selection. This heuristic is often used to generate solutions to optimization problems, particularly in cases with a large or complex search space (*27*).

An important consideration for multi-objective optimization by genetic algorithms is how to determine the fitness of individuals based on multiple objectives. An approach based on Pareto optimization is one method that has proven effective in past research. Within the scope of multi-criteria decision-making theory, Pareto-optimality is a necessary condition in order to guarantee the rationality of a decision (*19*). Therefore a "reasonable" solution to an MOLP problem should be Pareto-optimal and, as FMOLP is an approach used to solve MOLP problems, any optimal solution of an FMOLP problem should also be Pareto-optimal.

Definition. A decision plan $xo \in X$ is said to be a Pareto-optimal solution to the FMOLP problem (1) if and only if there does not exist another $y \in X$ such that $zi(y) \leq zi(xo)$ for all i and $zj(y) < zj(xo)$ for at least one j.

The above ideas can be integrated into a genetic algorithm to obtain a Pareto-optimal solution to the FMOLP problem. The steps of the proposed algorithm can be described as follows:

Step 1: Formulate FMOLP model from MOLP model:

(1) Assign the degree of membership for several values of each objective function Z_i (i = 1,2,…,k).
(2) Determine the piecewise linear membership function.
(3) Formulate the linear equations for each of the piecewise linear membership functions $f_i(z_i)$ (i = 1,2,…,k).
(4) Introduce the auxiliary variable L; the problem can be transformed into the equivalent conventional LP problem. The variable L can

be interpreted as representing an overall degree of satisfaction with the multiple fuzzy goals of decision making.

Step 2: Solve the FMOLP model with Pareto-optimal genetic algorithm:

(1) In the case that the optimal solution is unique:
   (a) If all the satisfaction degrees are strictly less than 1, the solution is Pareto-optimal and it can be chosen. In this case, go to step 5.
   (b) If some satisfaction degrees are equal to 1, i.e., if one or more targets are fully achieved, the solution may not be Pareto-optimal. In this case, go to step 4.

(2) In the case that there exist multiple optimal solutions, go to step 3.

Step 3: Solve the second phase model, i.e., maximize the sum of satisfaction degrees without making the achievement degrees obtained in the previous step worse:

(1) If all the satisfaction degrees are strictly less than 1, the solution is Pareto-optimal and it can be chosen. In this case, go to step 5.

(2) If some satisfaction degrees are equal to 1, i.e., if one or more targets are fully achieved, the solution may not be Pareto-optimal. In this case, go to step 4.

Step 4: Maximize the sum of negative deviations for targets that are fully achieved, without making the values obtained in the previous step worse. The solution is Pareto-optimal, go to step 5.

Step 5: Evaluate the decision making. If it is not satisfied with the initial solution, return to step 1 and iterate until a satisfactory solution is found.

Following the steps above, the Pareto-optimal genetic algorithm is proposed to search for a solution in an FMOLP problem. Clearly, this is a time-exhausted task. The algorithm takes a composition schema, six QoS permissible values, and a vector of QoS weights as inputs, uses weighted standardized Euclidean distance as distance metric, and evaluates the fitness according to the Pareto-optimality rule for genetic calculation. As the result, a composition plan is obtained that is able to satisfy the client's QoS requirement.

The flow process of the algorithm is demonstrated in Figure 5.

Validation and Discussions

In this section, experiments to verify the correctness and efficiency of the proposed algorithm are conducted. Firstly, a composition schema is created to meet a typical travel process. The experiment involves the physical intelligent transportation platform DRIVE Net (Digital Roadway Interactive Visualization and Evaluation Network), and web service providers from internet. Comparison between the proposed FMOLP algorithm and MOLP is carried out to validate the applicability of FMOLP. Secondly, a Monte Carlo simulation was performed to evaluate performance of the proposed FMOLP algorithm. The simulation environment is developed in Matlab 2015 on a PC server with one Intel Core i7-4790 @3.6 GHz CPU, 32 GB RAM, and Windows 7 operating system.

Example of QoS Constraint Composition Plan

DRIVE Net is developed and maintained by STAR Lab (Smart Transportation Applications and Research Laboratory) at the University of Washington. This is an on-line transportation platform aimed at data sharing, integration, visualization, and analysis (*28*). The system provides users with the capability to store, access, and manipulate data from anywhere, and requires only an active internet connection and web browser on the client side. DRIVE Net automates data fusion and data management details, and provides a functioning archive service capable of collecting detector data from all Washington State Department of Transportation (WSDOT) regions and incorporating third party data from WITS, INRIX and weather databases. A number of useful web services are integrated into DRIVE Net, a few of which are listed below:

- Transportation Information Service: providing real-time information describing commercial vehicle restrictions, border crossings, mountain pass conditions, etc.
- Ferry Terminal Traffic Service: providing ferry schedule, ferry position, and traffic prediction integrated with weather forecasting.
- Freeway Travel Time Service: providing travel time reliability analysis based on both historical traffic data and real-time traffic volume.
- Weather Information Service: providing severe weather notification and forecasting information.

Vacation travel is a common activity, but many details need to be tracked before or during the travel for optimal decision making. For example, consider a travel plan from San Francisco to Olympic National Park in Washington State. A great deal of preparation and information gathering might be completed before such a trip, including flight booking, hotel booking, car rental, route selection, weather forecast retrieval for safe driving, road information retrieval to avoid congestion, food service, etc. Clearly, it is quite complex and labor intensive to identify all sources of information, retrieve said information, and make travel plans accordingly. Many travelers

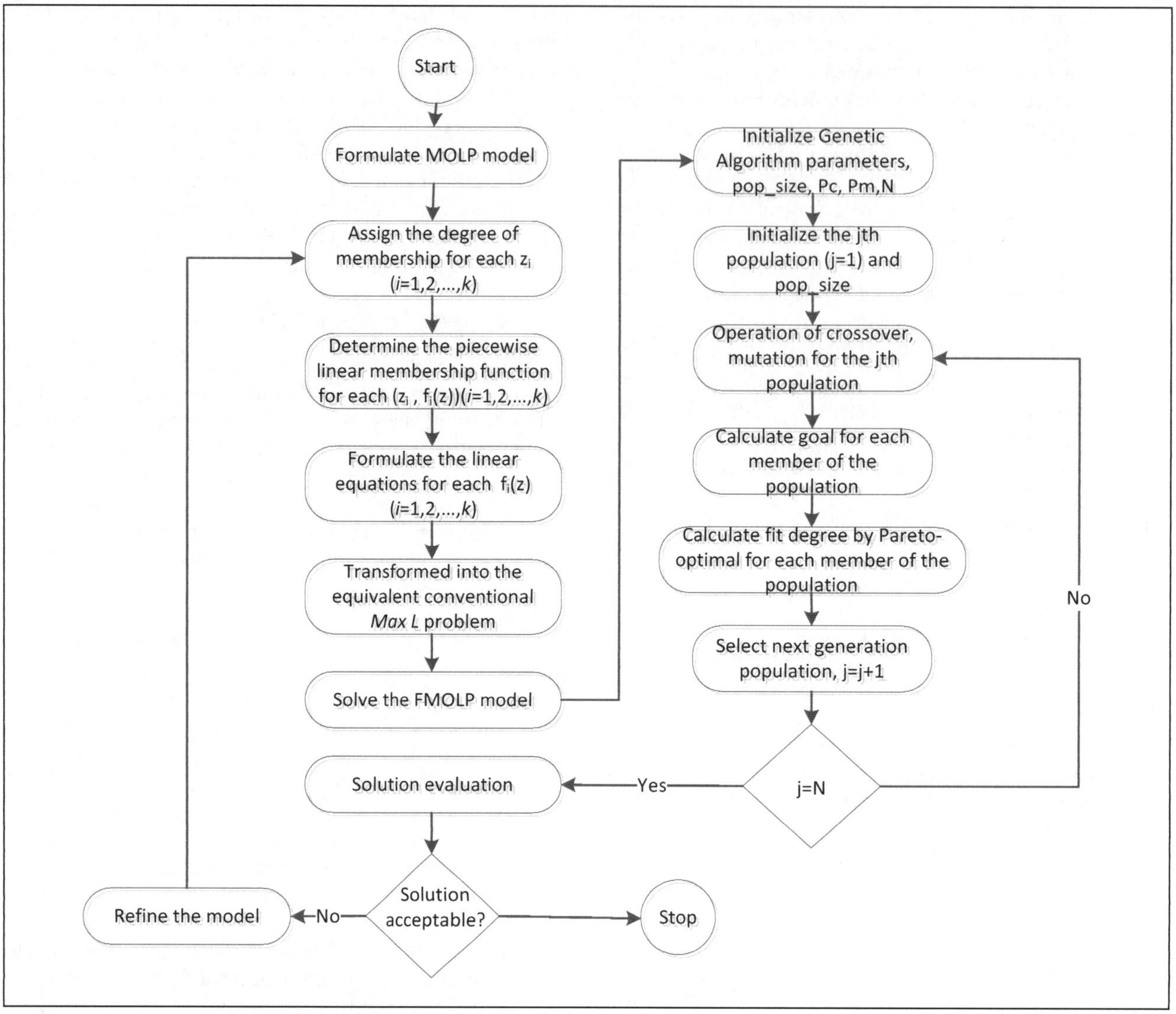

Figure 5. FMOLP search solution using genetic algorithm.

will not know how to access much of the required information, and their degree of familiarity really depends on personal experience.

In the experiment, a composition schema that consists of atomic processes is assumed to accomplish a common travel procedure. The procedure aggregates seven atomic processes including flight booking, hotel booking, car rental, travel time estimation, ferry traffic prediction, food services, and weather forecasts, etc. In each atomic process, there are two to five candidate web services available. The web service for transportation information comes from DRIVE Net, WSDOT, or google maps. Others, such as flight, hotel, and car rental, come from some common web services elsewhere on the internet. The QoS properties are rated by user survey. The QoS requirements are typically vague, because individual travelers will nearly always want a cheaper, safer, and faster travel plan, but may not express their preferences in terms of multiple conflicting objectives. The QoS requirements for cheaper, faster, better service quality, higher reputation, and better accuracy degree corresponding to the QoS properties of execution cost, execution speed, availability, frequency, evaluation, and accuracy are fuzzified by the fuzzy logic method.

A comparison is conducted between FMOLP and MOLP in the experiment to compare the ability of each algorithm to find an acceptable solution. Table 1 shows the experiment results. The match degree in Table 1 is calculated by the combination of AND-parallel, XOR-parallel, and serial patterns from aggregated membership

Table 1. Comparison of FMOLP and MOLP

	FMOLP			MOLP		
Experiment	Successful	Time cost (sec)	Match degree (%)	Successful	Time cost (sec)	Match degree (%)
1	Yes	21.15	73.23	Yes	11.65	0
2	Yes	21.17	96.32	Yes	11.68	100
3	Yes	21.15	68.56	No	11.65	0
4	Yes	21.11	77.82	No	11.62	0
5	Yes	21.13	98.56	Yes	11.64	100
6	Yes	21.12	53.26	No	11.65	0
7	Yes	21.16	65.58	No	11.67	0
8	Yes	21.12	99.37	Yes	11.65	100
9	Yes	21.17	58.73	No	11.63	0
10	Yes	21.12	97.62	Yes	11.65	100

Table 2. TOC Results by Different Quantity of Atom Process and Service Candidate without Genetic Algorithm

		Quantity of web service candidate in each atom process			
		2	4	6	8
Quantity of atom process in composition schema	2	0.02	0.06	0.18	0.29
	4	0.06	0.29	2.43	13.52
	6	0.25	1.43	26.51	121.04
	8	0.93	8.67	85.83	389.42

functions of all atom processes, indicating the extent to which the results met the QoS requirements.

From Table 1, we can see that:

1. MOLP has better performance in time cost compared to FMOLP when the QoS constraints can be met. The average time cost of MOLP is 11.65 sec, while the average time cost of FMOLP is 21.12 sec.
2. In some QoS constraints, the MOLP solution search failed. MOLP searches achieved a success rate of only 40% in the experiments, while FMOLP is always able to find a solution regardless of the QoS constraints. Further, FMOLP produces results comparable to MOLP in cases where MOLP was able to find a solution.

Generally, the MOLP algorithm does not guarantee that all of the conditions can be met when finding an optimal solution. In some QoS constraints, the absolute optimal solution does not exist. This means it is impossible to meet all constraint conditions simultaneously. On the other hand, FMOLP is capable of finding a satisfactory solution even if the match degree is not high.

Performance Experiment

As indicated, service composition is thought of as an NP-hard optimization problem because it exposes a very large number of similar single services from different service providers in the cloud. An experiment to evaluate the performance of the proposed algorithm is conducted in this section. For this experiment, a virtual composition schema is assumed which is composed of two to eight atomic processes. Each process has two to eight atomic service candidates. The QoS properties of these atomic service are simulated using the Monte Carlo method.

In Table 2, the TOC (time of calculation) is listed for each composition plan with different quantities of atomic processes and web service candidates. Figure 6 illustrates the result graphically.

From the results shown in Table 2 and Figure 6, we can see that the genetic algorithm provides efficient solutions to the programming problem and demonstrates good performance as the quantity of service candidate increases. The maximum calculation time is 389.42 sec when both the quantity of service candidates and atomic processes reach eight, which constitutes a very large solution search space.

Discussion

From the comparison experiment above, it is clear that the FMOLP algorithm is an effective method capable of handling vague user requirements. FMOLP consistently outperforms the MOLP algorithm in the experiment cases and produces acceptable solutions even when it is impossible to meet all constraint conditions simultaneously. Therefore, these experiments suggest that the

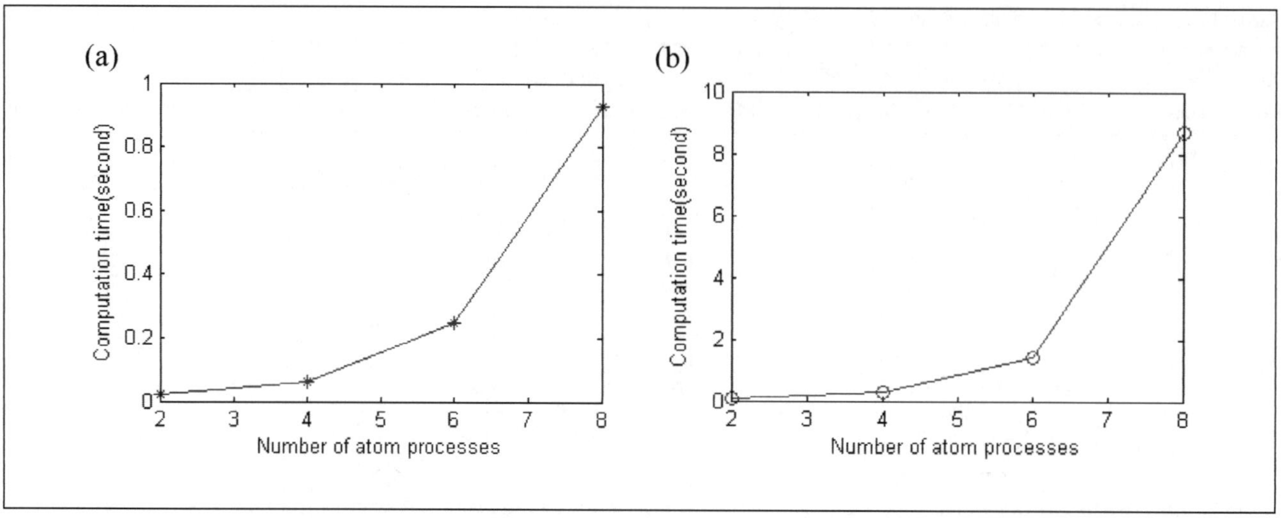

Figure 6. TOC results for service quantity comparison showing quantity of service candidates is (*a*) two and (*b*) four.

FMOLP algorithm embedded in an intelligent traffic framework will be capable of dealing with unclear requirements in real-world applications.

The service composition is an NP-hard problem that is time-consuming. This point is effectively illustrated in the experiment, and the Pareto-optimal genetic algorithm provides an effective resolution to the programming problem. Although the performance of the proposed algorithm has been verified under a certain number of component services, for situations of rapid increment of the service quantity further research related to genetic algorithms still needs to be conducted to improve the efficiency.

Another issue that bears further improvement is how to resolve vague QoS requirements more effectively. User requirements are always unclear, resulting in a great deal of ambiguity for solution finding in real-world applications. Though the proposed algorithm shows ability to handle the fuzzy input, the input in this study actually includes only six properties, including execution cost, execution time, availability, evaluation, frequency, and accuracy. As the limitation, these inputs have to be transformed into a fuzzy number with fuzzy operations. In fact, the model's ability to deal with uncertainty still needs to be enhanced to capture various possible inputs.

Conclusions

This study addresses a key consideration in the design of a traffic intelligence framework integrating web service resources into a single platform for real-world decision support. In principle, such a platform combines web services offered by a variety of entities, real-time data access and query, and value-added intelligence while consistently meeting user QoS requirements. As the mechanism coordinating services to meet QoS requirements in a cost-effective way, a web service composition algorithm is one of the core components of this type of system. However, in the real world, user requirements are often vague, and handling vague or ambiguous requests is a key challenge addressed by this research.

To address the need for a transportation-oriented service composition approach, as well as the challenge of vague user input and service quality priorities, a service composition algorithm based on FMOLP is proposed. The algorithm is able to handle vague QoS requirements and obtains an optimal solution by solving the optimization problem with fuzzy multiple objective programming. A genetic algorithm with Pareto optimization is adopted to solve the programming problem. In a validation procedure conducted using the DRIVE Net platform, the algorithm shows superior capability in handling uncertainty compared with conventional MOLP.

Finally, a traffic intelligence framework used in the real world must be capable of extracting actionable instructions from vague input. Therefore, one obvious way to improve on the results presented here is to enhance the capability of the system in terms of handling greater ambiguity in user input. Additionally, the optimal solution search is an NP-hard problem, and algorithm efficiency could be further improved. These questions will be studied in the further research.

Acknowledgments

This research is sponsored by National key research and development program: Key projects of international scientific and technological innovation cooperation between governments (No. 2016YFE0108000), and PacTrans at the University of Washington.

Author Contributions

The authors confirm contribution to the paper as follows: study conception and design: WZ, YQ, YW; data collection: ZZ, YW; analysis and interpretation of results: WZ, HG, YQ; draft manuscript preparation: WZ, HG. All authors reviewed the results and approved the final version of the manuscript.

References

1. Ibrahim, A. T. H., I. Yaqoob, N. B. Anuar, S. Mokhtar, A. Gani, and S. U. Khan. The Rise of "Big Data" on Cloud Computing: Review and Open Research Issues. *Information Systems*, Vol. 47, 2015, pp. 98–115.
2. Christian, E., M. Ficco, F. Palmieri, and A. Castiglione. A Knowledge-Based Platform for Big Data Analytics Based on Publish/Subscribe Services and Stream Processing. *Knowledge-Based Systems*, Vol. 79, 2015, pp. 3–17.
3. Dobre, C., and F. Xhafa. Intelligent Services for Big Data Science. *Future Generation Computer Systems*, Vol. 37, 2014, pp. 267–281.
4. Jong, M. K., C. Ouk Kim, and I.-H. Kwon. Quality-of-Service Oriented Web Service Composition Algorithm and Planning Architecture. *The Journal of Systems and Software*, Vol. 81, 2008, pp. 2079–2090.
5. Hussain, A.-A., L. Liu, R. Hill, and N. Antonopoulos. Cloud BI: Future of Business Intelligence in the Cloud. *Journal of Computer and System Sciences*, Vol. 81, 2015, pp. 85–96.
6. Hassina, N., N. Djebari, H. Slimani, and D. Aissani. A Distributed Authentication Model for Composite Web Services. *Computers & Security*, Vol. 70, 2017, pp. 144–178.
7. Zhi-Zhong, L., D.-H. Chu, Z.-P. Jia, J.-Q. Shen, and L. Wang. Two-Stage Approach for Reliable Dynamic Web Service Composition. *Knowledge-Based Systems*, Vol. 97, 2016, pp. 123–143.
8. Amin, J., E. Sundararajan, and Z. Othman. Cloud Computing Service Composition: A Systematic Literature Review. *Expert Systems with Applications*, Vol. 41, 2014, pp. 3809–3824.
9. Chia-Feng, L., R.-K. Sheu, Y.-S. Chang, and S.-M. Yuan. A Relaxable Service Selection Algorithm for QoS-based Web Service Composition. *Information and Software Technology*, Vol. 53, 2011, pp. 1370–1381.
10. Bo, Z., Q. Zhang, Q. Shi, Q. Yang, P. Yang, and Y. Yu. 2017. Measuring Web Service Security in the Era of Internet of Things. *Computers and Electrical Engineering*, Vol. 66, 2018, pp. 305–315.
11. Victor, C. The Business Intelligence as a Service in the Cloud. *Future Generation Computer Systems*. Vol. 37, 2014, pp. 512–534.
12. Xin, Z., L. Shen, X. Peng, and W. Zhao. Toward SLA-constrained Service Composition: An Approach Based on a Fuzzy Linguistic Preference Model and an Evolutionary Algorithm. *Information Sciences*, Vol. 316, 2015, pp. 370–396.
13. O'Sullivan, J., D. Edmond, and A. T. Hofstede. What's in a Service? *Distributed and Parallel Databases*, Vol. 12, No. 2–3, 2002, pp. 117–133.
14. Rajesh, K., F. Khendek, and R. H. Glitho. A Novel Architecture for Web Service Composition. *Journal of Network and Computer Applications*, Vol. 35, 2012, pp. 787–802.
15. Juric, M. B., B. Mathew, and P. Sarang. *Business Process Execution Language for Web Services*. Packt Publishing, Birmingham, UK, 2004.
16. Sirin, E., B. Parsia, D. Wu, J. Hendler, and D. Nau. HTN Planning for Web Service Composition Using SHOP2. *Journal of Web Semantics*, Vol. 1, No. 4, 2004, pp. 377–396.
17. Berardi, D., D. Calvanese, G. De Giacomo, M. Lenzerini, and M. Mecella. Automatic Services Composition Based on Behavioral Descriptions. *International Journal of Cooperative Information Systems*, Vol. 14, No. 4, 2005, pp. 333–376.
18. Heinrich, R. Fuzzy Linear Programming and Applications. *European Journal of Operational Research*, Vol. 92, 1996, pp. 512–527.
19. Mariano, J., and A. Bilbao. Pareto-optimal Solutions in Fuzzy Multi-objective Linear Programming. *Fuzzy Sets and Systems*, Vol. 160, 2009, pp. 2714–2721.
20. Edward, L. H. Linear Programming with Multiple Fuzzy Goals. *Fuzzy Sets and Systems*, Vol. 6, 1981, pp. 235–248.
21. Young-Jou, L., and C.-L. Hwang. A New Approach to Some Possibilistic Linear Programming Problems. *Fuzzy Sets and Systems*, Vol. 49, 1992, pp. 121–133.
22. Hentenryck, V. P., and L. Michel. *Constraint-Based Local Search*. MIT Press, Cambridge, MA, 2005.
23. Mani, A., and A. Nagarajan. *Understanding Quality of Service for Web Services*, White Paper, IBM, 2002. http://www.ibm.com/developerworks/webservices/library/ws-quality.html.
24. Jøsang, A., R. Ismail, and C. Boyd. A Survey of Trust and Reputation Systems for Online Service Provision. *Decision Support Systems*, Vol. 43, No. 2, 2007, pp. 618–644.
25. Gulcin, D. Y., and N. Erginel. Fuzzy Multi-objective Programming Algorithm for Vehicle Routing Problems with Backhauls. *Expert Systems with Applications*, Vol. 42, 2015, pp. 5632–5644.
26. Ozan, K., and R. Keskin. A Novel Portfolio Selection Model Based on Fuzzy Goal Programming with Different Importance and Priorities. *Expert Systems with Applications*, Vol. 42, 2015, pp. 6898–6912.
27. Mitchell, M. *An Introduction to Genetic Algorithms*. MIT Press, Cambridge, MA, 1996.
28. Ma, X. L., Y. R. Wu, and Y. H. Wang. DRIVE Net: E-Science Transportation Platform for Data Sharing, Visualization, Modeling, and Analysis. *Transportation Research Record: Journal of the Transportation Research Board*, 2011. 2215: 37–49.

The Standing Committee on Information Systems and Technology (ABJ50) peer-reviewed this paper (18-03940).

Article

TRR
JOURNAL OF THE TRANSPORTATION RESEARCH BOARD

Transportation Research Record
2018, Vol. 2672(45) 76–86

DOI: 10.1177/0361198118795013
journals.sagepub.com/home/trr
SAGE

Satellite-Based InSAR Monitoring of Highway Bridges: Validation Case Study on the North Channel Bridge in Ontario, Canada

Daniel Cusson[1], Ken Trischuk[1], Daniel Hébert[2], Glenn Hewus[3], Matthew Gara[4], and Parwant Ghuman[4]

Abstract

A space-borne bridge displacement monitoring technology was demonstrated on a major highway bridge in Cornwall (Ontario, Canada) over a 15-month period. Two major challenges had to be overcome and solutions were identified and implemented. The first challenge related to the proper selection and analysis of satellite imagery to optimize the number and signal quality of point targets on the bridge for adequate movement detection and analysis. The second challenge was the comparison and validation of independent sets of data having different measurement methods, reference baselines, and distributions in space and time. The key findings include: (i) the regression analysis of interferometric synthetic aperture radar (InSAR) data from the satellite produced the best coherence when fitted against the following three independent variables: height, ambient temperature, and time; (ii) the bridge railings appeared to be excellent natural reflectors with their sharp edges and regular spacing along the bridge, showing a return of over 3,000 clear point targets to the satellite; (iii) the InSAR displacement thermal sensitivity data was found to compare very well to numerical modeling thermal data. The results show great promise and value in applying satellite-based technology for the remote monitoring of highway and railway bridges to alert engineers of excessive movement and, in turn, will help optimize preventive maintenance management, extend structural lifespan, minimize traffic disruptions due to late repair, and ensure structural integrity.

Highway and railway bridges are critical to the nation's road transportation network, especially bridges along the national transportation corridors. Many are in urgent need of rehabilitation due to aging, limited capital investment, and insufficient inspection and maintenance. Bridge owners in North America have been struggling for decades with how to cost-effectively rehabilitate their large portfolios of bridges (*1*).

As a complement to current inspection and maintenance protocols, the judicious use of structural health monitoring (SHM) of highway bridges can contribute to addressing many of these challenges and slowing down the aging of such critical structures. Earlier work has shown that remote monitoring with radar satellites can provide ground motion measurements with sub-centimeter accuracy (*2*). Radar satellites, for example, are capable of monitoring large structures by detecting and measuring movement at hundreds of point targets (PTs) on a single structure, and hundreds of these structures can be monitored and analyzed in a single scene when acquired over major urban centers (*3*).

Previous Work on Interferometric Synthetic Aperture Radar Remote Sensing

Two key limitations of in-situ monitoring—the lack of spatial detail on a given bridge and the inability to instrument all bridges of a major urban center—can be supplemented by remote satellite-based monitoring through repeated radar imaging as a first step before deploying in-situ approaches. The RADARSAT-2 satellite sensor has several advanced beam modes that are suitable for displacement monitoring of specific targets on bridges (e.g., slabs, trusses, towers, abutments, joints, barriers, posts). Although radar-monitoring technology is mature and capable of measuring millimeter-level displacements

[1]National Research Council Canada, Ottawa, ON, Canada
[2]Transport Canada, Ottawa, ON, Canada
[3]Federal Bridge Corporation Limited, Ottawa, ON, Canada
[4]3v Geomatics, Vancouver, BC, Canada

Corresponding Author:
Address correspondence to Daniel Cusson: daniel.cusson@nrc.gc.ca

(*2*), its temporal resolution is limited by the repeat pass period, which is 24 days. Consequently, there is a natural synergy between the high temporal resolution of in-situ sensors and high spatial nominal resolution (1.6 m × 0.8 m for the Spotlight beam mode) and nominal coverage (approximately 18 km × 8 km for the Spotlight beam mode) of RADARSAT-2 radar imagery.

The creation of interferograms from pairs of finely co-registered synthetic aperture radar (SAR) images acquired using nearly identical viewing geometry spawned a host of applications characterizing Earth's surface and its atmosphere. SAR is fundamentally a ranging measurement that generates images by bouncing microwaves from Earth's surface. Each pixel in an interferogram comprises a phase difference (0° to 360°) between two distinct SAR snapshots of a given resolution cell. The interferogram phase is cumulatively sensitive to all geometric and physical variables that affect the return path length of the microwaves between the satellite and Earth's surface. Hence, the phase is proportional to surface topography, ground displacement along satellite line-of-sight, atmospheric pressure, water vapor, and soil moisture.

Interferometric synthetic aperture radar (InSAR) used to measure ground motion was first demonstrated in 1989 (*4*), and since then, InSAR has been successfully applied to measure displacements from landslides (*5*), earthquakes (*6*), volcanoes (*7*), glaciers (*8*), metal mining (*9*), oil/gas fields (*10*), groundwater extraction (*11*), and urban infrastructure (*3, 12*).

In the previously cited publications (*4–12*), SAR interferometry has been applied to diverse applications such as generating elevation maps of Earth, mapping of expansive and localized motion phenomena, and characterization of atmospheric water vapor. This is because SAR sensors capture path images that can be co-registered and interfered with to yield a path difference that is sensitive to the imaging geometry and variations in physical parameters such as displacement, elevation, and atmospheric and soil moisture. Each pixel of an interferogram contains phase contributions from these physical variables, and the contrasting spatio-temporal statistical properties of the constituent signals can be utilized to isolate the desired component. Phase decontamination is the process of segregating the target signal by modeling and removing all other phase contributions.

For the purpose of bridge monitoring, displacement is the target signal and the contaminants are errors due to atmospheric moisture and elevation. Local and expansive variations in atmospheric water vapor get imprinted on interferograms during the travel of microwaves through the atmosphere. Elevation errors occur because accurate vertical positions of bridge targets are unknown during the processing of SAR images.

Figure 1 illustrates PTs identified on the Ironworkers Memorial Bridge in Vancouver, Canada. The higher point density in the image on the right is generated from 3-m resolution SAR imagery (Ultrafine-18 [U18] beam mode) whereas the lower point density on the left is generated by using 8-m resolution images (Fine-Quad-2

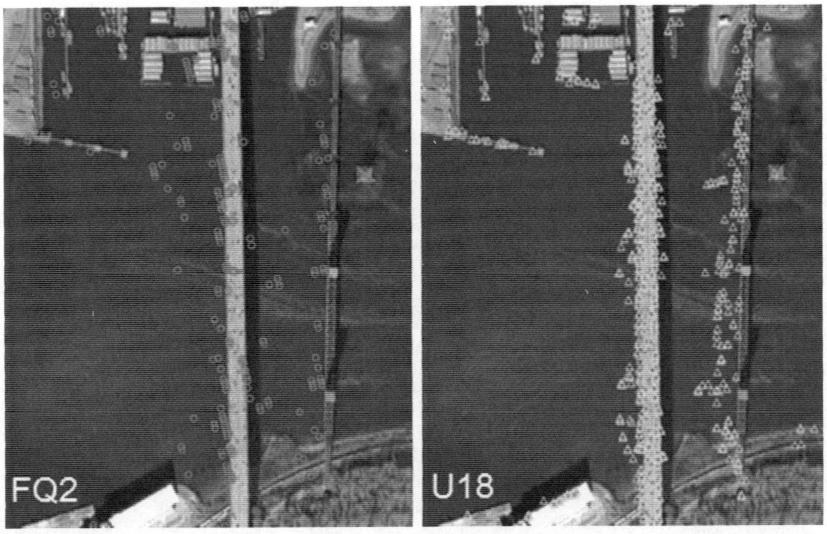

Figure 1. Point targets on the Ironworkers Memorial Bridge, Vancouver, BC, Canada.
Note: FQ2 = Fine-Quad mode with 8-m resolution; U18 = Ultrafine mode with 3-m resolution.

[FQ2] beam mode). The advantage of using lower resolution images in some applications is that they can cover larger ground areas than higher resolution images.

Some PTs—especially from the FQ2 data set—appear to lie in water because of contamination from the aforementioned elevation error manifesting as horizontal positioning inaccuracies. The displacement history at each point can be measured by modeling and removing atmospheric and elevation phase contributions through the process of point target interferometry (PTI), also commonly known as persistent scatterer interferometry. In order to identify PTs and accurately apply PTI, a data stack comprising more than 15 overlapping SAR images is required. Phase denoizing techniques are also applied, whereby multiple redundant measurements are generated, each with independent noise characteristics, and this over-defined set of observations is mathematically solved to suppress noise and emphasize the target displacement signal.

In the United States, University of Virginia has demonstrated the practical use and benefits of InSAR remote sensing for performance monitoring of transportation infrastructure at the network level, including subsidence detection, bridge motion, pavement smoothness, and pavement motion (*13, 14*). This work, which was recognized and supported by the U.S. Department of Transportation, included the development of computer tools to address the above issues and a decision-support system for transportation engineers and decision makers.

Despite the growing effort in applying InSAR remote sensing to transportation infrastructure monitoring, little has been done to compare and validate the displacement results obtained from InSAR remote sensing to known displacement measurement methods on highway bridges.

Project Objectives and Scope of Work

The main objectives of the work described in this paper were (i) to deploy a remote bridge displacement monitoring technology based on satellite imagery, and (ii) to validate it against numerical bridge model predictions of thermal displacements and measurements from displacement/temperature sensors installed on a highway bridge.

To achieve the above objectives, the work described here was conducted in four steps, including:

- InSAR analysis of RADARSAT-2 satellite imagery to obtain displacement measurements
- Sensor-based bridge monitoring of load response (to calibrate a numerical bridge model)
- Numerical bridge modeling (for direct comparison to InSAR results)
- Validation on a single highway bridge by comparing the above sets of results

Acquisition Plan and InSar Analysis of Satellite Imagery

Figure 2 presents the new 335-m long, 4-span, low-level North Channel Bridge (being the subject of this study) and the old high-level North Channel Bridge (under demolition at the time), that cross the Saint Lawrence River between the city of Cornwall (north-left) to

Figure 2. View of new low-level bridge and old high-level bridge in Cornwall, ON, Canada.

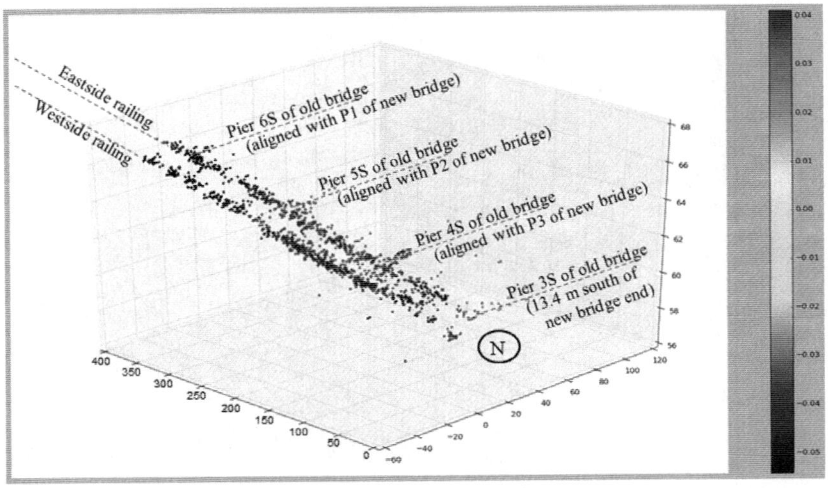

Figure 3. 3D plot (axes in m) of InSAR displacement thermal sensitivity data (color scale in cm/°C).

Cornwall Island (south-right), located in Ontario, Canada. The new steel box girder bridge was completed and opened to traffic in January 2014; the demolition of the old steel truss bridge started in September 2014 and ended in November 2015.

Among different available beam modes on the RADARSAT-2 satellite, the Spotlight mode is best-suited for bridge monitoring applications, which provides the highest accuracy (but the smallest nominal coverage ~18 km × ~8 km). Anticipating possible image viewing geometry issues given the complex configuration and timing of the construction/demolition work, four stacks of images from the Canadian Space Agency had been requested, each with a repeat cycle of 24 days. For each ascending pass and descending pass of the right-looking satellite, two stacks of images with low (26°) and high (47°) incidence angles were selected. Only one stack was to be retained for the final InSAR analysis.

In an initial analysis, challenges related to the proper selection of satellite imagery were addressed to optimize the number and stability of PTs for bridge movement detection. These challenges included: shadow and layover effects (from the adjacent high-level bridge), target stability (affected by construction vehicles and cranes), multi-bounce effects (from the river), height fit (affected by the quality of the available terrain digital elevation model), and the number of undisturbed scenes acquired.

The preliminary analysis revealed that the SLA74-asc stack (ascending pass with a 26° incidence angle) was the most suitable stack for the subsequent detailed analysis.

For the full analysis, SAR data were processed with a multi-model linear regression analysis to extract different sets of data from the SLA74-asc imagery for over 3,000 PTs registered on the new bridge deck. Key extracted parameters included:

- Height (m)—fitted against perpendicular baseline (stereo sensitivity) data (*15*).
- Displacement thermal sensitivity (cm/°C)—fitted against measured bridge temperature data.
- Displacement time rate (cm/year)—fitted against time from the first image acquisition.
- Coherence (ranging from 0 to 1)—confirming the quality of fit and the reliability of processed displacement data, where only PTs with coherence ≥ 0.8 should be trusted for InSAR analysis (*15*). The coherence is computed following a multi-model regression fit, yielding a statistic on both the quality of fit as well as the noise level of the PT.

A Python script was used to extract, de-noise, and visualize selected post-processed data by displaying displacement thermal sensitivity data collected over the bridge deck in a 3D plot, as shown in Figure 3 (X axis = west–east distance [reversed]; Y axis = north–south distance; and Z axis = height). The figure reveals very useful details:

- The displacement thermal sensitivity data show gradual variation along the bridge axis.
- The height and steady slope of the bridge deck along the length of the bridge (1.2° over 335 m) are very well captured by the point cloud.
- The two steel railings of the bridge returned a large amount of clear signals to the satellite. The sharp angles of the railings appear to be excellent natural corner reflectors. Another advantage of

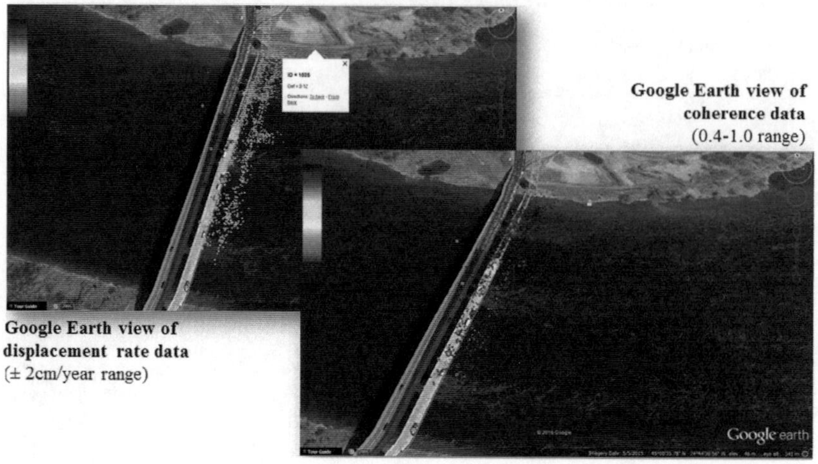

Figure 4. Processed displacement time rate data (left) and corresponding coherence data (right).

railings for InSAR remote sensing is that snow and ice do not accumulate easily on them due to their direct exposure to wind and sun.

- Four groups of outliers leaning toward the east appear to match exactly the locations of the four piers of the old bridge (east–west reversed). Despite being on another bridge, these points are very useful to confirm the exact positioning of the points of interest.

Figure 4 (left) presents the resulting displacement time rate data superimposed onto a Google Earth optical image. With all points being either yellow or green (close to zero), one can conclude that very small movements have occurred over the relatively short monitoring period (15 months). Similarly, Figure 4 (right) illustrates the coherence of the measured displacement data, confirming the high quality and reliability of the processed data, with most points being dark blue (well above 0.8), in which a coherence of 0.8 or higher on directional data is comparable to a standard deviation lower than 0.45 radians.

Sensor-Based Bridge Monitoring

To validate the satellite-based monitoring technology on the North Channel Bridge, the satellite measurements needed to be compared with measurements from other known ground-based technologies, including (i) temperature and displacement sensors installed on the bridge, and (ii) a finite-element (FE) numerical model of the bridge calibrated on the sensors' measurements from truck load test.

Ambient temperature plays a crucial role in this study, as our adopted approach is to fit the satellite displacement data with ambient temperature data to remove the thermal movement and focus on the mechanical movement of the bridge that could be due to settlement, problematic joints or bearings, excessive loading, and so forth.

A first set of temperature data was obtained from temperature sensors installed inside the twin steel box girders, as shown in Figure 5 (left). Being in direct contact with the steel girders, these sensors are expected to provide values similar to ambient air temperature, due to the high thermal conductivity of steel. A second set of temperature data was obtained from the nearby Massena International Airport's weather station (New York State, United States; 23 km away from the bridge), as it was the closest station that could provide hourly values of air temperature. Using the bridge temperature data over the Massena International Airport temperature data to fit SAR displacement data resulted in a small improvement in measured coherence.

Figure 5 (right) illustrates the correspondence between the bridge sensor temperatures and the Massena International Airport temperatures at times corresponding to satellite passes. The strong correlation between the two sets of data and the small improvement in the InSAR coherence data suggest that if sensor measurements are not available at the site, local airport temperatures are suitable for the proposed InSAR analysis approach.

Bridge Numerical Modeling

As radar satellite sensing provides line-of-sight displacement measurements without distinguishing thermal effects from mechanical effects, or between the X, Y, and Z components of displacement, another tool is required

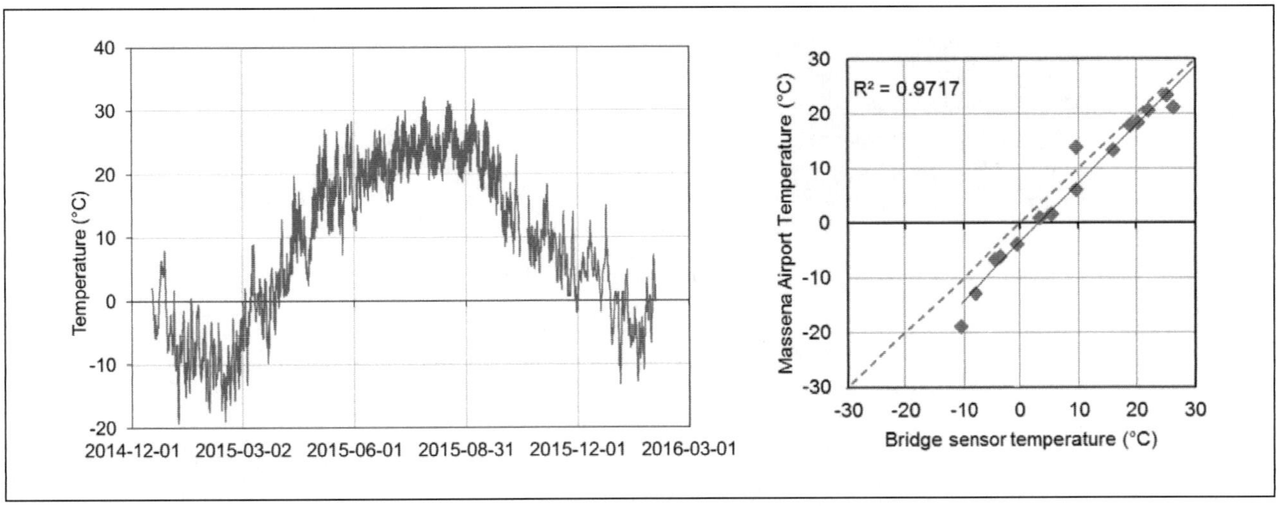

Figure 5. Temperature measured inside steel box girder of bridge (left) and relationship between bridge sensor temperature and local airport temperature (right).

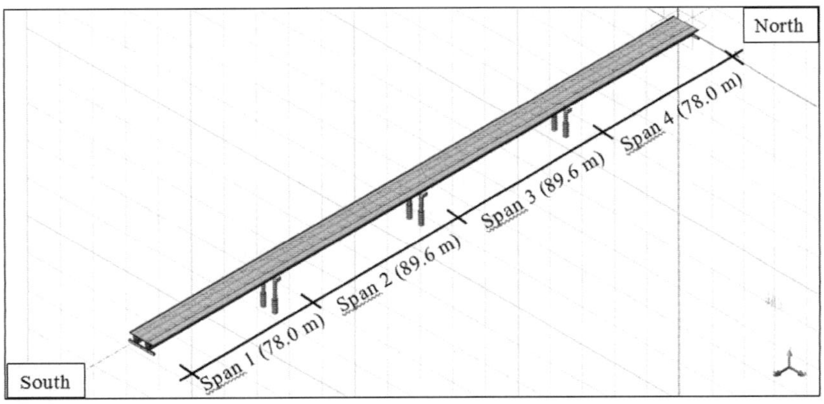

Figure 6. Overall view of complete 3D FE bridge numerical model.

to help determine the orthogonal components of mechanical effects on the bridge that could lead to structural issues if left undetected. This tool could be numerical modeling to determine the thermal movement of the bridge.

The commercial TNO-Diana FE software package was used to build a detailed 3D numerical model of the North Channel Bridge (steel-reinforced concrete deck on twin steel box girders). The construction drawings and specifications were used to obtain detailed information on bridge geometry, boundary conditions, and structural material properties. For a bridge under service loads, a structural static analysis and isotropic linear material models were deemed to be more than adequate for the determination of bridge displacements as a function of service loads and temperature effects.

Figure 6 illustrates the bridge model of the superstructure, three piers and two abutment walls (including 10,000 nodes, 11,154 elements, 6 materials, and 61 properties). The piers were modeled from their fixed connection to the superstructure down to their fixed connection to bedrock. The load-bearing portion of each abutment wall was modeled assuming uniaxial bearing connection (i.e., free movement in longitudinal direction, and blocked in transverse and vertical directions).

Numerical Model Calibration with Load Test Results

Truck load tests were conducted on the bridge on February 2, 2016 and were later simulated using a structural linear analysis of the bridge model for two truck passes in the southbound (SB) and northbound (NB)

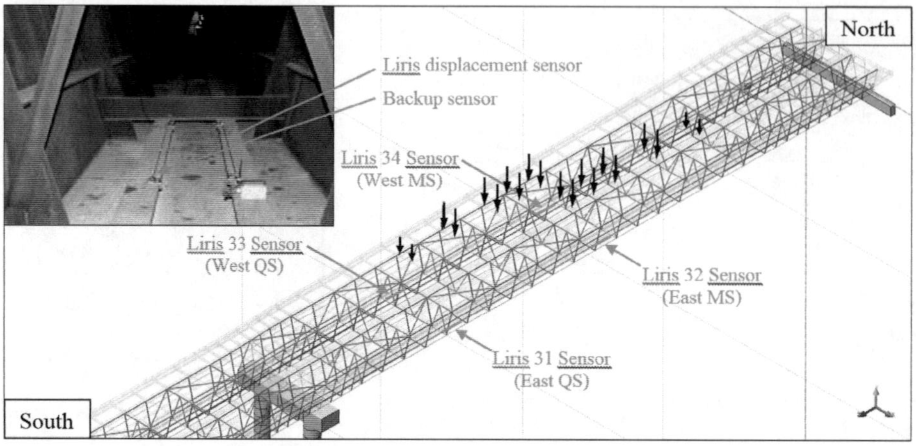

Figure 7. View of FE model showing locations of sensors (green arrows) and truck axle loads (black arrows).

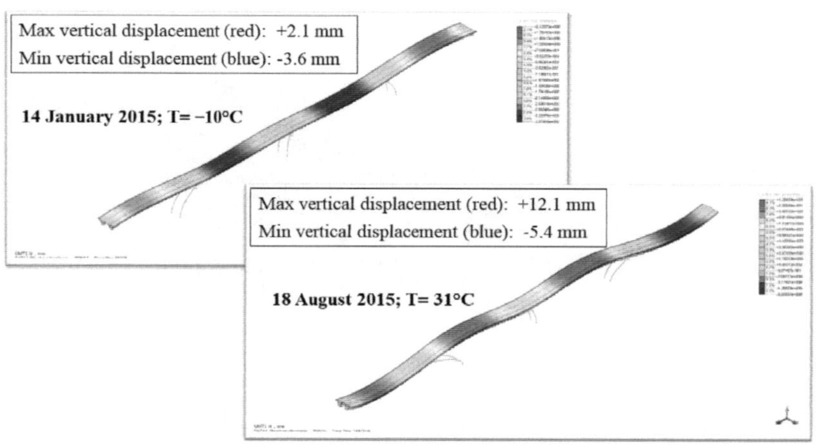

Figure 8. Simulated thermal deformations for a cold winter day (left) and a warm summer day (right).

lanes. Figure 7 indicates the locations of the applied axle loads with the vertical black arrows, and those of the four wireless sensors installed on the steel box girders. After small adjustments to Young's modulus of the deck's reinforced concrete (finally set to 35 GPa), the longitudinal deformations of the bottom layer of steel girders at mid-span (MS) and quarter-span (QS) compared well to sensor measurements.

For example, at mid-span on the southbound lane (1st test), the predicted vs. measured downward displacement values were 0.10 mm vs. 0.10 mm in the west box girder and 0.08 mm vs. 0.07 mm in the east box girder; and at mid-span on the northbound lane (2nd test), the predicted vs. measured values were 0.09 mm vs. 0.07 mm in the west box girder, and 0.11 mm vs. 0.09 mm in the east box girder.

Simulation of Thermal Deformation

Measured bridge temperatures were used as input data in a structural nonlinear analysis of the bridge model for a period of 412 days (from December 17, 2014 to February 2, 2016). For example, Figure 8 (left) presents the thermal deformation (vertical displacement) of the bridge computed for January 14, 2015 corresponding to the satellite pass at the coldest temperature of the year. One can see the deck contraction toward the middle pier, producing inward bending of the two other piers, and related

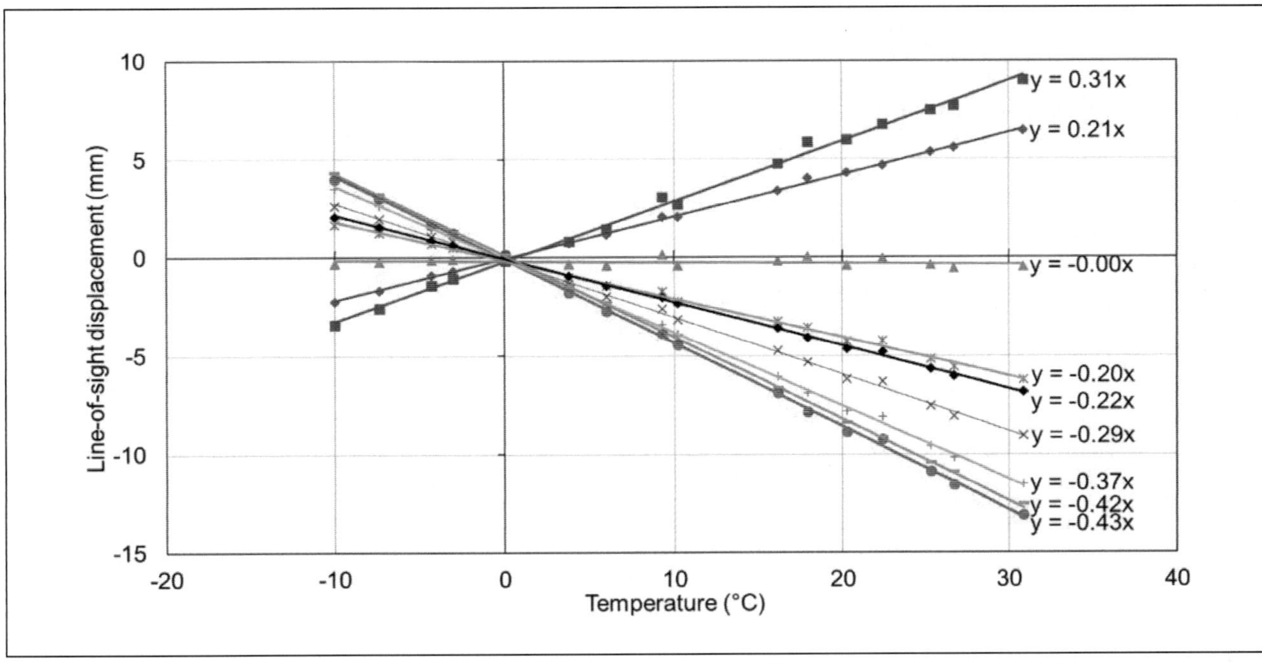

Figure 9. FEM line-of-sight displacement correlated with temperature for 9 locations across the bridge deck.

bending of the concrete deck. Figure 8 (right) presents the thermal deformation (vertical displacement) of the bridge computed for August 18, 2015, corresponding to the satellite pass at the warmest temperature of the year. One can observe the elongation of the bridge deck away from the middle pier, producing an outward bending of the two other piers, and related bending of the concrete deck.

Validation Case Study of Satellite-Based Monitoring Technology

Movement obtained from InSAR analysis of satellite data represents the line-of-sight (LoS) movement, which is the change in the distance between the satellite and the PT being monitored on the ground (here on a bridge). Since the satellite incidence angle is 26.5° (from the vertical), detected movement includes horizontal and vertical components of unknown proportion to each other. For example, a positive LoS displacement (i.e., movement away from satellite) may be the result of a downward vertical displacement, a horizontal displacement away from the satellite, or an unknown combination of both.

To allow comparison of SAR data with finite element model (FEM) data, it was decided to convert the FEM displacement data into LoS displacement with the following equation:

$$\begin{aligned}\Delta \mathrm{LoS} = &-\Delta \mathrm{V}\ (\cos\alpha\cos\gamma - \sin\alpha\sin\beta\sin\gamma)\\ &-\Delta \mathrm{H}\ (\sin\alpha\sin\beta\cos\gamma - \cos\alpha\sin\gamma)\end{aligned} \tag{1}$$

where

ΔLoS is line-of-sight displacement

ΔV is vertical displacement

ΔH is horizontal displacement

α is incidence angle of satellite LoS direction (26.5° for SLA74-asc)

β is the angle between azimuth of satellite track on Earth and azimuth of bridge (26° between the SLA74-asc satellite track and North Channel Bridge)

γ is inclination angle of bridge deck from north to south (1.2° for North Channel Bridge)

Using Equation 1 above, the SAR displacement temperature sensitivity data can be compared directly with FEM thermal data expressed in terms of LoS displacement as a function of change in temperature from a reference temperature (set here as T = 0°C). To that effect, these data were plotted in Figure 9 for nine different positions along the longitudinal axis of the bridge deck (over vertical supports and mid-span locations). Two observations can be made:

- A strong correlation exists between LoS displacement and bridge temperature at any given location on the bridge.

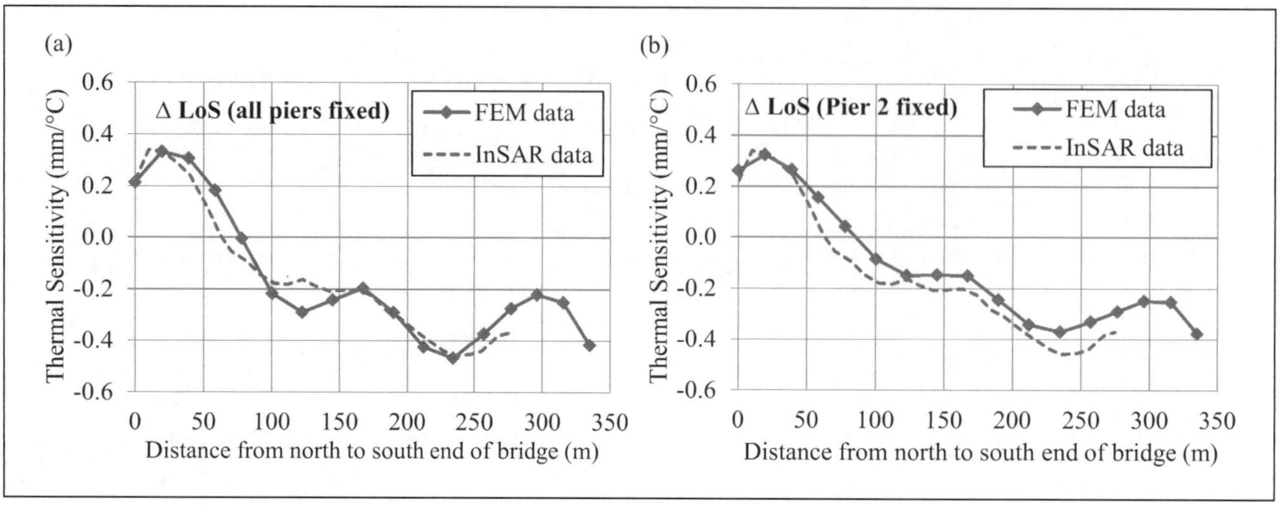

Figure 10. Comparison of FEM and InSAR data sets for thermal sensitivity of line-of-sight displacement: (a) for real bridge support condition (all piers fixed); (b) fictitious bridge support condition (only P2 fixed).

- The slope of the linear regression analysis for each curve is a parameter that can be directly compared with the InSAR LoS displacement temperature sensitivity.

The slopes of the linear regressions presented above were then expressed as the FEM thermal sensitivity along the longitudinal distance on the bridge, and illustrated in Figure 10 (left), which also compares the two sets of independent InSAR and FEM data, where a very good fit can be observed. Note that there were not enough valid InSAR PTs near the south end of the bridge to generate points for $x > 275$ m. A longer period of monitoring, adding more valid PTs to the InSAR analysis, would eliminate this issue.

It is noted that the two regions with lower thermal sensitivity values at approximately $x = 125$ m and 225 m appear more accentuated for the FEM data than for the InSAR data. It was therefore suspected that the support conditions at the three piers could play a role and affect the thermal sensitivity of the FEM displacement data.

To investigate this further, the bridge model was modified to change the support conditions, that is, to release the longitudinal movement of the connections of piers 1 and 3 to the deck structure. In other words, unidirectional bearings were simulated in piers 1 and 3, as opposed to the as-built fixed connection. As shown in Figure 10 (right), the two low regions of the FEM data appear to be flatter than with the previous support condition, and the correlation with the InSAR data seems to have reduced slightly. More monitoring and inspection are required before one can conclude on the status of the support conditions.

Nevertheless, an important point is made from the above discussion and the data in Figure 10 that the InSAR/FEM comparison of the LoS displacement thermal sensitivity data can be used to identify the following issues:

- Frozen or damaged bearings
- Pier flexibility issues
- A numerical modeling issue

As mentioned above, our adopted approach is to use multi-model regression analysis to fit InSAR satellite data against height and temperature data to properly position PTs on the bridge, identify valid data points, and remove thermal movement to obtain the remaining residual data set of displacement time rate, which represents the mechanical movement of the bridge that could be due to unexpected settlement, problematic joints or bearings, excessive loading, and so on.

Discussion

Benefits of Remote InSAR Monitoring

As shown above, remote InSAR technology is capable of monitoring many points per bridge on many transportation structures at once with sub-centimeter accuracy (*2*) for vertical movement detection and measurement both on the bridge superstructure and foundation that could be due to excessive loads and movements, thermal effects, or ground settlement. Satellite-based SHM at the network level can therefore cost-effectively screen many bridges at once for stability issues and flag them for in-situ inspection (*13, 14*). As conventional bridge inspections are typically labor-intensive, InSAR remote monitoring can help with allocating bridge inspections where they are critically needed over a bridge network. A cost–benefit analysis conducted by University of Virginia (*13, 14*) clearly demonstrated that the annual worth savings to the transportation agency can easily exceed the annual worth costs of implementing and using InSAR remote sensing for infrastructure performance monitoring in most situations, depending on a number of key factors (coverage area, satellite image resolution, acquisition rate and cost, incident detection rate, etc.).

The expected outcomes of satellite-based InSAR monitoring are:

- Determination of bridge deformation profiles at different times, also including foundation mapping, resulting from extreme loads and movements, temperature changes, soil settlement, and so forth
- Provision of monthly sets of data to complement biannual visual inspection results
- Rapid identification of bridges with abnormal deformations within a network of bridges
- Cost-effective prioritization and optimization of bridge inspection and maintenance

It is predicted that satellite-based InSAR monitoring will improve the condition assessment of major transportation infrastructure assets including new and aging transportation infrastructure to permit timely alerts on asset condition change, due to incidents or deterioration, and for better asset management of maintenance, repair, and upgrade activities. Such interventions will extend the lifespan of important transportation infrastructure without any compromise to operational safety for the movement of goods and passengers.

Future Work

The InSAR remote sensing technology is ready for validation and application to other bridges across the country and to other types of critical civil engineering infrastructure (e.g., dams, power plants, airports, maritime ports, locks, pipelines, highways, railways, etc.). The technology is expected to provide the nation with a cost-effective method, in combination with other existing methods, for the improved condition assessment of critical infrastructure.

Bridge authorities need tools that can provide them with more accurate and timely information on the real-time (or near real-time) condition of highway bridges. With such readily available information at hand, bridge owners will make more accurate and timely performance predictions resulting in improved bridge maintenance management. Consequently, safety, structural performance, and durability of highway bridges throughout their entire life cycle can be maximized, and their risks of failure and life-cycle costs minimized.

To address the above needs, new and existing decision-support tools (*14*) and guidelines should be developed, adapted and validated against known methods to accelerate acceptance among bridge authorities. These tools should be able to integrate data from different sources (e.g., satellites, bridge sensors, numerical models, and visual inspection reports), provide performance indicators on bridge stability and safety, and assist bridge owners in the decision-making process.

Conclusion

The key findings of this study are as follows:

- Regression analysis of InSAR data from the satellite produced the best coherence when fitted against the following three independent variables: height, ambient temperature, and time. For example, when temperature was excluded from the fit, the coherence of InSAR data reduced significantly.
- The bridge railings appeared to be excellent natural reflectors with their sharp edges and regular spacing along the bridge, returning over 3,000 clear signals to the satellite.
- The InSAR displacement thermal sensitivity data was found to compare very well to FEM thermal data when computed along the longitudinal axis of the bridge. Such data was found to be promising for detecting problematic bearing conditions.
- The results show great promise and value in applying satellite-based technology for the remote monitoring of highway bridges to help optimize preventive maintenance management, extend structural lifespan, minimize traffic disruptions due to late repairs, and ensure structural integrity after extreme events such as major earthquakes and flooding.

Acknowledgments

Direct financial contribution to this project was provided by Transport Canada. Technical assistance and logistics were provided by the National Research Council Canada, Federal Bridge Corporation, 3v Geomatics, Osmos Canada, Seaway International Bridge Corporation, Morrison Hershfield, and the Canadian Space Agency.

Author Contributions

The authors confirm contribution to the paper as follows: study conception and design: DC, DH, GH; data collection: MG, PG, KT; analysis and interpretation of results: DC, MG, PG; draft manuscript preparation: DC. All authors reviewed the results and approved the final version of the manuscript.

References

1. U.S. Department of Transportation, Federal Highway Administration, Federal Transit Administration. *The 2006 Status of the Nation's Highway Bridges and Transit: Conditions & Performance, Report to Congress.* U.S. Department of Transportation, Federal Highway Administration, Federal Transit Administration, Washington, D.C., 2007.
2. Marinkovic, P., G. Ketelaar, F. Van Leijen, and R. Hanssen. In-SAR Quality Control: Analysis of Five Years of Corner Reflector Time Series. Presented at 5th International Workshop on ERS/Envisat SAR Interferometry, 'FRINGE07', Frascati, Italy, 2007.
3. Cusson, D., P. Ghuman, M. Gara, and A. McCardle. Remote Monitoring of Bridges from Space. Presented at 54th Annual Brazilian Congress on Concrete, IBRACON, Maceió, Brazil, 2012.
4. Gabriel, A. K., R. M. Goldstein, and H. Zebker. Mapping Small Elevation Changes Over Large Areas: Differential Radar Interferometry. *Journal of Geophysical Research*, Vol. 94, No. B7, 1989, pp. 9183–9191.
5. Fruneau, B., J. Achache, and C. Delacourt. Observation and Modelling of the Saint-Etienne-de-Tinée Landslide Using SAR Interferometry. *Tectonophysics*, Vol. 265, No. 3–4, 1996, pp. 181–190.
6. Massonnet, D., M. Rossi, C. Carmona, F. Adragna, G. Peltzer, K. Feigl, and T. Rabaute. The Displacement Field of the Landers Earthquake Mapped by Radar Interferometry. *Nature*, Vol. 364, 1993, pp. 138–142.
7. Briole, P., D. Massonnet, and C. Delacourt. Post-Eruptive Deformation Associated With the 1986-87 and 1989 Lava Flows of Etna Detected by Radar Interferometry. *Geophysical Research Letters*, Vol. 24, No. 1, 1997, pp. 37–40.
8. Goldstein, R., H. Engelhardt, B. Kamb, and R. Frolich. Satellite Radar Interferometry for Monitoring Ice Sheet Motion: Application to an Antarctic Ice Stream. *Science*, Vol. 262, No. 5139, 1993, pp. 1525–1530.
9. Rabus, B., E. Eberhardt, D. Stead, P. Ghuman, C. Nadeau, K. Woo, J. Severin, T. Styles, and F. Gao. Application of InSAR to Constrain 3-D Numerical Modelling of Complex Discontinuous Pit Slope Deformations. Presented at International Symposium on Rock Slope Stability in Open Pit Mining, Santiago, Chile, 2009.
10. Fielding, E., R. Blom, and R. Goldstein. Rapid Subsidence Over Oil Fields Measured by SAR Interferometry. *Geophysical Research Letters*, Vol. 25, No. 17, 1998, pp. 3215–3218.
11. McCardle, A., J. McCardle, and F. Ramos. Large Scale Deformation Monitoring and Atmospheric Removal in Mexico City. Presented at Fringe 2009 Workshop, Frascati, Italy, 2009.
12. Rabus, B., and P. Ghuman. A Simple Robust Two-Scale Phase Component Inversion Scheme for Persistent Scatterer Interferometry (Dual-Scale PSI). *Canadian Journal of Remote Sensing*, Vol. 35, No. 4, 2009, pp. 399–410.
13. Acton, S. *InSAR Remote Sensing for Performance Monitoring of Transportation Infrastructure at the Network Level.* Final Report No. RITARS-14-H-UVA. University of Virginia, Charlottesville, VA, 2016.
14. Acton, S. *Sinkhole Detection, Landslide and Bridge Monitoring for Transportation Infrastructure by Automated Analysis of Interferometric Synthetic Aperture Radar Imagery.* Final Report No. RITARS-11-H-UVA. University of Virginia, Charlottesville, VA, 2013.
15. Ferretti, A., A. Monti-Guarnieri, C. Prati, F. Rocca, and D. Massonnet. *InSAR Principles: Guidelines for SAR Interferometry Processing and Interpretation.* ESA Publication TM-19. European Space Agency, Noordwijk, The Netherlands, 2007.

The Standing Committee on Information Systems and Technology (ABJ50) peer-reviewed this paper (18-04394).

Article

TRR
JOURNAL OF THE TRANSPORTATION RESEARCH BOARD

Transportation Research Record
2018, Vol. 2672(45) 87–105

DOI: 10.1177/0361198118798737
journals.sagepub.com/home/trr
SAGE

A Deep Generative Adversarial Architecture for Network-Wide Spatial-Temporal Traffic-State Estimation

Yunyi Liang[1], Zhiyong Cui[2], Yu Tian[3], Huimiao Chen[4], and Yinhai Wang[5]

Abstract

This study proposes a deep generative adversarial architecture (GAA) for network-wide spatial-temporal traffic-state estimation. The GAA is able to combine traffic-flow theory with neural networks and thus improve the accuracy of traffic-state estimation. It consists of two Long Short-Term Memory Neural Networks (LSTM NNs) which capture correlation in time and space among traffic flow and traffic density. One of the LSTM NNs, called a discriminative network, aims to maximize the probability of assigning correct labels to both true traffic-state matrices (i.e., traffic flow and traffic density within a given spatial-temporal area) and the traffic-state matrices generated from the other neural network. The other LSTM NN, called a generative network, aims to generate traffic-state matrices which maximize the probability that the discriminative network assigns true labels to them. The two LSTM NNs are trained simultaneously such that the trained generative network can generate traffic matrices similar to those in the training data set. Given a traffic-state matrix with missing values, we use back-propagation on three defined loss functions to map the corrupted matrix to a latent space. The mapping vector is then passed through the pre-trained generative network to estimate the missing values of the corrupted matrix. The proposed GAA is compared with the existing Bayesian network approach on loop detector data collected from Seattle, Washington and that collected from San Diego, California. Experimental results indicate that the GAA can achieve higher accuracy in traffic-state estimation than the Bayesian network approach.

Traffic state usually refers to traffic flow, traffic density, traffic speed and travel time in the existing literature (*1*). Traffic-state information plays an important role in relieving traffic congestion. With the aid of accurate traffic-state information, administrators can effectively manage and control the transportation network, and travelers can better decide on their departure times and travel routes to make their trips more efficient.

With the rapid development of Intelligent Transportation Systems (ITS), enriched real-time traffic data can be collected by various types of sensors, such as loop detectors, Global Positioning Systems (GPS) and Remote Traffic Microwave Sensors (RTMS). Strategies for traffic management and control, developed based on the application of such data, are often not as effective as expected due to lag time (*2*). In addition, traffic data at some spatial and temporal positions may be missing due to sensor disruption. Therefore, traffic-state estimation has become an important research topic in the transportation area and attracted much attention. Existing studies for traffic-state estimation can be divided into parametric approaches, nonparametric approaches, and hybrid integration methods (*3*). In parametric approaches, model structure is predetermined based on certain theoretical assumptions. Model parameters are calibrated from empirical data. Typical methods of this type

[1]Key Laboratory of Road and Traffic Engineering of the Ministry of Education, Tongji University, Shanghai, P. R. China
[2]Department of Civil and Environmental Engineering, University of Washington, Seattle, WA
[3]School of Economics and Management, Tongji University, Shanghai, P. R. China
[4]Department of Electrical Engineering, Tsinghua University, Beijing, P. R. China
[5]Department of Civil and Environmental Engineering, University of Washington, Visiting Director of Tongji University Center for Transportation Data Science, Seattle, WA

Corresponding Author:
Address correspondence to Yinhai Wang: yinhai@uw.edu

include cell-transmission-model-based methods (4) and dynamic-traffic-assignment-based methods (5). In nonparametric approaches, neither model structure or parameters are fixed. This type of approach can be further divided into classical statistical models and Computational Intelligence (CI) models (6). Typical models of the former type include the Kalman Filter (7, 8) and support vector machine (9, 10). The artificial neural network (ANN) (11–13) is a typical model of the latter type. Deep learning technology refers to the neural networks with more than three layers. It is one of the most promising Artificial Intelligence (AI) technologies. Compared with traditional neural networks, they are able to effectively extract features from training data and play an important role in many areas, such as image recognition, speech recognition and image blending. In the field of transportation, deep learning models are mainly applied for prediction problems, such as traffic-flow prediction (14) and speed prediction (15); to date, they have achieved promising results. Hybrid integration approaches combine the parametric approaches and nonparametric approaches. As pointed out by Vlahogianni et al. (6), CI models outperform classical statistical models in traffic-state estimation because they have less a priori assumptions for input variables and are more capable of processing disrupted data.

Traffic-flow theory, such as the cell transmission model (4, 16, 17), indicates that traffic-flow parameters in a transportation network have correlation over both time and space. Incorporating traffic-flow theory is beneficial for improving the accuracy of traffic-state estimation. However, most of the CI models in the literature fail to do so due to their inflexiable model structures.

Generative adversarial networks (GANs) are the frontier of deep learning technology. ANNs can be categorized into discriminative networks and generative networks. A trained discriminative network is able to predict output given input by learning the conditional probability distribution of output given input. While a trained generative network is able to generate samples having a similar distribution to the training samples by learning the joint distribution of the input and the output. Since Goodfellow et al. proposed them (18), GANs have become a very hot topic in the AI area. They have been applied in image inpainting, image blending, image translation and super-resolution. So far, the most striking successes in the application of ANNs for traffic-state estimation is based on discriminative networks. Having a flexible framework, GANs are able to incorporate the spatial-temporal relationship between traffic-flow parameters in traffic-state estimation and improve the accuracy of traffic-state estimation.

The methodology presented this paper is inspired by GANs used for image inpainting (19). The principle of image inpainting using GANs mainly depends on two types of information: 1) the prior information in the corrupted image, and 2) the experience from the training data set. The former makes the recovered image logical while the latter makes the recovered image realistic. The methodology proposed by this study is analogous to the principle of image inpainting. We regard traffic state within the time and space plane as an image. Then we estimate traffic state using three types of information. The first type of information is initial spatial-temporal traffic state; the second is traffic-state correlation in time and space learned from training data; and the last type of information is the conservation law of traffic flow which in theory describes the spatial-temporal relationship between traffic flow and traffic density. By combining these pieces of information, we are able to estimate spatial-temporal traffic state with a high level of accuracy.

The remainder of this article is organized as follows. The next section briefly introduces GANs and LSTM NNs. The generative adversarial architecture (GAA) for spatial-temporal traffic-state estimation is proposed in the subsequent section. Thereafter, the proposed GAA is compared with the existing Bayesian network approach on two data sets: loop detector data collected from Seattle, Washington and that collected from San Diego, California. The final section concludes this study and discusses future work.

GANs and LSTM NNs

GANs

GANs are frameworks for training generative models. GANs consist of two neural networks, a generative neural network G, and a discriminative neural network D. G maps a random vector $\mathbf{z}$, sampled from a prior distribution $p_{\mathbf{z}}$, to the sample space. D maps an input sample to a likelihood. G aims to generate realistic samples to get true labels from D. While D plays an adversarial role to discriminate between the sample generated from G and the sample from the training data. The training procedure for GANs corresponds to a minimax two-player game. It converges when Nash equilibrium is achieved. Thus, the two neural networks are simultaneously trained by optimizing the following loss function:

$$\min_G \max_D V(G, D) = E_{\mathbf{h} \sim p_{data}(\mathbf{h})}[\log(D(\mathbf{h}))] + E_{\mathbf{z} \sim p_z(\mathbf{z})}[\log(1 - D(G(\mathbf{z}))] \tag{1}$$

where

h = a vector sampled from the training data distribution p_{data};

z = a random vector sampled from a prior distribution p_z.

The convergence to the global optimality of the training procedure has been proved in (*18*). After convergence, *G*can produce a high-quality sample similar to the training data distribution from the latent space and *D* is not able to tell generated samples from training data. That said, *D* is able to reject samples that are too fake or dissimilar. The principle of GANs is illustrated in Figure 1*a*. Note that the two neural networks can be of any type. In this study, we use two long short-term memory neural networks (LSTM NNs) to capture the spatial-temporal correlation of traffic state in a transportation network.

LSTM NNs

Recurrent neural networks (RNNs) are a class of ANNs. They have the capability of modeling nonlinear time series problems by feeding back the current state of each hidden layer to itself along with inputs at the next time step.

The principle of a basic RNN is illustrated in Figure 1*b*. Where $\boldsymbol{x}_t$, $\boldsymbol{o}_t$ and $\boldsymbol{h}_t$ are the input to the RNN at time step *t*, the output of the RNN at time step *t* and the hidden layer state of the RNN at time step *t*, respectively. For further details about RNNs, readers are referred to (*15*). Despite the superior capability of RNNs in modeling nonlinear time series problems, when modeling time series problems with long time lags, the traditional RNNs encounter the vanishing gradient problem. The vanishing gradient problem in the context of training RNNs means that the update of weight matrix *W* is nearly unaffected by the hidden layer state *h* in time steps long before the current one.

To overcome the vanishing gradient problem of traditional RNNs, LSTM NNs were initially introduced by (*20*). The primary objectives of LSTM NNs are to model long-term dependencies and determine the optimal time lag for time series problems. As pointed out in (*15*), these two features are desirable for traffic-state estimation. A basic LSTM NN is composed of one input layer, one hidden layer and one output layer. The hidden layer is the core of an LSTM NN. Compared with that of a traditional RNN, the hidden layer of an LSTM NN has one more state, the cell state **c**, for long-term memory storage. In an LSTM NN, the cell state**c**is controlled by two gates, a forget gate and an input gate. The former determines how much of the cell state c_{t-1} at the last time step is kept at the current cell state c_t. The latter determines how much of the current input x_t of the LSTM NN is kept at the current cell state c_t. The output of the forget gate $\mathbf{f_t}$ and the input gate $\mathbf{i_t}$ can be calculated using

$$\mathbf{f_t} = \sigma(\mathbf{W_f} \cdot [\mathbf{h_{t-1}}, \mathbf{x_t}] + \mathbf{b_f}) \tag{2}$$

$$\mathbf{i_t} = \sigma(\mathbf{W_i} \cdot [\mathbf{h_{t-1}}, \mathbf{x_t}] + \mathbf{b_i}) \tag{3}$$

where

$\mathbf{W_f}$ = the weight matrices of forget gate;

$\mathbf{W_i}$ = the weight matrices of input gate;

$[\mathbf{h_{t-1}}, \mathbf{x_t}]$ = the new vector formed by connecting $\mathbf{h_{t-1}}$ and $\mathbf{x_t}$;

$\mathbf{b_f}$ = the bias vectors of forget gate;

$\mathbf{b_i}$ = the bias vectors of input gate;

$\sigma(\cdot)$ = the standard logistic sigmoid function defined in Equation 4.

$$\sigma(x) = \frac{1}{1 + e^{-x}} \tag{4}$$

The current memory $\mathbf{Q_t}$ can be expressed by

$$\mathbf{Q_t} = \tanh(\mathbf{W}_c \cdot [\mathbf{h_{t-1}}, \mathbf{x_t}] + \mathbf{b_c}) \tag{5}$$

where

$\mathbf{W}_c$ = the corresponding weight matrix;

$\mathbf{b_c}$ = the corresponding bias vector.

Function $\tanh(\cdot)$ is defined as

$$\tanh(x) = \frac{e^x - e^{-x}}{e^x + e^{-x}} \tag{6}$$

Then the current cell state $\mathbf{c_t}$ can be calculated by

$$\mathbf{c_t} = \mathbf{f_t} \odot \mathbf{c_{t-1}} + \mathbf{i_t} \odot \mathbf{Q_t} \tag{7}$$

where

$\odot$ = the scalar product of two vectors.

Using Equation 7, we combine long-term memory $\mathbf{c_{t-1}}$and current memory $\mathbf{Q_t}$ to determine the current cell state$\mathbf{c_t}$in an LSTM NN. Therefore, the LSTM NN is able to store the information long before it is due for processing at the forget gate, while it is also able to incorporate the current information into memory.

The output gate calculated by Equation 8 describes the effect of long-term memory on the current output:

$$\mathbf{o_t} = \sigma(\mathbf{W_o} \cdot [\mathbf{h_t}, \mathbf{x_t}] + \mathbf{b_o}) \tag{8}$$

where

$\mathbf{W}_c$ = the weight matrix of the output gate;

$\mathbf{b_c}$ = the bias vector of the output gate.

The final output of an LSTM cell can be determined by the output gate $\mathbf{o_t}$ and the current cell state $\mathbf{c_t}$

$$\mathbf{h_t} = \mathbf{o_t} \odot \tanh(\mathbf{c_t}) \tag{9}$$

The principle of a basic LSTM NN is illustrated in Figure 1*c*.

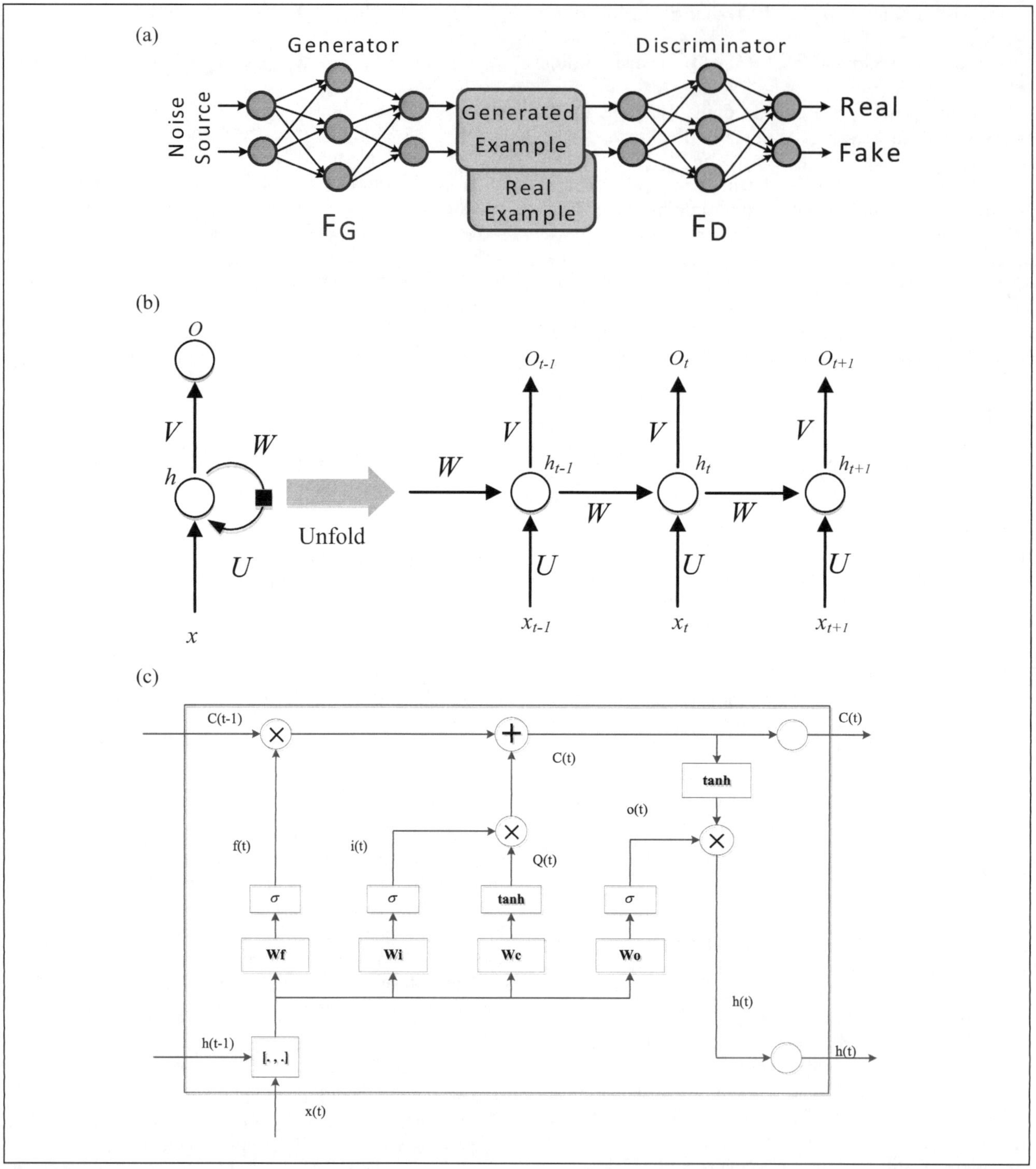

Figure 1. The principle of generative adversarial networks (GANs), a basic recurrent neural network (RNN), and a basic long short-term memory neural network (LSTM NN).

(*a*) The principle of GANs. (*b*) the principle of a basic RNN. (*c*) the principle of a basic LSTM NN.

GAA for Spatial-Temporal Traffic-State Estimation

Motivation and Problem Description

Traffic-flow theory, such as the cell transmission model (*4*), indicates that traffic flow, traffic density and traffic velocity in a transportation network are correlated. Futher, these correlations exist in both time and space. Incorporating these spatial-temporal correlations can improve the accuracy of traffic-state estimation. For example, Ma et al. (*15*) chose traffic flow and traffic velocity as inputs for LSTM NNs to improve traffic velocity prediction. Spatial-temporal correlations among traffic-flow parameters are incorporated in traffic-state estimation in (*2, 9, 21*). However, in those studies, spatial-temporal correlations among traffic-flow parameters are only mined from data and no specific theoretical correlations are defined due to their relatively fixed model structures. As shown below, GANs are able to incorporate correlations among traffic-flow parameters from both the data and theory due to their flexible framework.

The traffic-state estimation problem addressed in this study can be described as follows. We choose traffic flow and traffic density as inputs to our proposed model. As traffic flow and traffic density evolve with time and space, both traffic flow and traffic density within a given spatial-temporal area can be represented as a two-dimensional matrix. We assume the number of time intervals for the study period is n, and that the road segment can be divided into m discrete cells. The $n*(m+1)$ traffic-flow matrix is denoted as F with each element $q(t, l)$ representing the traffic flow at location l at time t. The $n*m$ traffic-density matrix is denoted as K with each element $k(t, s)$ representing the traffic density in cell$s$at time t. The problem addressed in this study is how to estimate or predict the missing values in F and K given known elements in each matrix. Note that the proposed model can be viewed as a deep-learning-based and data-driven cell transmission model if the known elements are boundary conditions. The problem description can be explained by Figure 2. Note that we do not add traffic velocity as an input in this study. This is mainly because traffic velocity$v$can be determined by traffic flow and traffic density using $v = q/k$ and thus it makes little contribution to traffic-state estimation.

Network Architecture

The GAA proposed in this study consists of two LSTM NNs. In our GAA, the generative LSTM NN randomly generates traffic-state matrices. Another LSTM NN, the discriminative network, estimates the probability that a

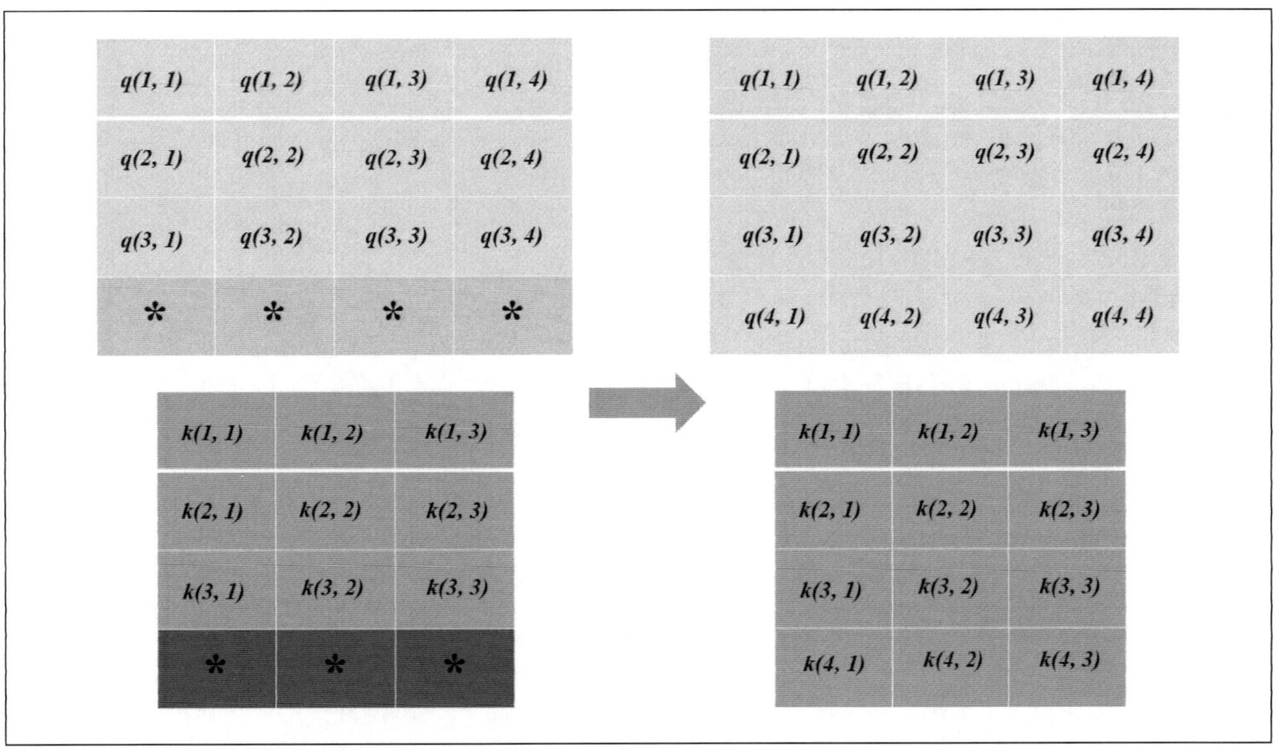

Figure 2. Description of the traffic-state estimation problem addressed in this study.

traffic-state matrix comes from training data rather than the generative LSTM NN. We train the discriminative LSTM NN to maximize the probability of assigning correct labels to both traffic-state matrices from training data and traffic-state matrices from the generative LSTM NN. The generative LSTM NN is trained to generate realistic traffic-state matrices which maximize the probability that the discriminative LSTM NN assigns true labels to them. The discriminative LSTM NN has four layers: one input layer, two hidden layers and one output layer (see Figure 3*a*). The input of the discriminative LSTM NN is a sequence with time steps. At each time step, the inputs are the traffic flow at each location and the traffic density at each cell. The first hidden layer contains an LSTM cell as described above. The output layer maps the output sequence of the second hidden layer into a value between 0 and 1 such that it is able to calculate the probability that the input traffic-state matrix is from training data. The sigmoid function defined by Equation 4 is used in the output layer.

The input of the generative LSTM NN is a sequence randomly sampled from a uniform distribution following (*18*). Note that other kinds of distribution can be used to replace the uniform distribution. Its output is a traffic-state matrix having the same dimension as the output of the second hidden layer of the discriminative LSTM NN. The model structure of the generative LSTM NN is shown in Figure 3*b*.

The two LSTM NNs are trained simulatenously using the GANs framework described above. The training procedure using minibatch stochastic gradient descent is illustrated in Figure 3*c*.

Within the GANs framework, each of the two LSTM NNs are trained by the algorithm proposed in (*22*). The steps of the training algorithm for each LSTM NN are described as follows.

Step 1: forward calculate the values of $\mathbf{f}_t$, $\mathbf{i}_t$, $\mathbf{c}_t$, $\mathbf{o}_t$ and $\mathbf{h}_t$ using Equations 2–9.

Step 2: Backward calculate the error in LSTM NN training. The backward calculation of the error includes the error that propagates backward over time (i.e., calculate the error at each previous time step from the current time step) and the error that propagates backward to previous layers from the current layer. The error at time step$p$backward propagating from time step t can be calculated by

$$\delta_p^T = \prod_{j=p}^{t-1} \delta_{o,t}^T \mathbf{W}_{oh} + \delta_{f,t}^T \mathbf{W}_{fh} + \delta_{i,t}^T \mathbf{W}_{ih} + \delta_{Q,t}^T \mathbf{W}_{ch} \quad (10)$$

where

$$\delta_{o,t}^T = \delta_t^T \circ \tanh(\mathbf{c}_t) \circ \mathbf{o}_t \circ (1 - \mathbf{o}_t) \quad (11)$$

$$\delta_{f,t}^T = \delta_t^T \circ \mathbf{o}_t \circ (1 - \tanh(\mathbf{c}_t)^2) \circ \mathbf{c}_{t-1} \circ \mathbf{f}_t \circ (1 - \mathbf{f}_t) \quad (12)$$

$$\delta_{i,t}^T = \delta_t^T \circ \mathbf{o}_t \circ (1 - \tanh(\mathbf{c}_t)^2) \circ \mathbf{Q}_{t-1} \circ \mathbf{i}_t \circ (1 - \mathbf{i}_t) \quad (13)$$

$$\delta_{i,t}^T = \delta_t^T \circ \mathbf{o}_t \circ (1 - \tanh(\mathbf{c}_t)^2) \circ \mathbf{i}_t \circ (1 - \mathbf{Q}_t^2) \quad (14)$$

The error at layer $(l-1)$ backward propagating from layer l can be calculated by

$$\frac{\partial E}{\partial \mathbf{net}_t^{l-1}} = (\delta_{o,t}^T \mathbf{W}_{oh} + \delta_{f,t}^T \mathbf{W}_{fh} + \delta_{i,t}^T \mathbf{W}_{ih} + \delta_{Q,t}^T \mathbf{W}_{ch}) \circ z'(\mathbf{net}_t^{l-1}) \quad (15)$$

where

$\mathbf{net}_t^{l-1}$ = the input at layer $(l-1)$ at time step$t$;

z = the activation function at layer $(l-1)$.

Step 3: calculate the gradient of weight matrices and the bias vectors according to their corresponding errors using

$$\frac{\partial E}{\partial \mathbf{W}_{oh}} = \sum_{j=1}^{t} \delta_{o,j}^T \mathbf{h}_{j-1}^T \quad (16)$$

$$\frac{\partial E}{\partial \mathbf{W}_{fh}} = \sum_{j=1}^{t} \delta_{f,j}^T \mathbf{h}_{j-1}^T \quad (17)$$

$$\frac{\partial E}{\partial \mathbf{W}_{ih}} = \sum_{j=1}^{t} \delta_{i,j}^T \mathbf{h}_{j-1}^T \quad (18)$$

$$\frac{\partial E}{\partial \mathbf{W}_{ch}} = \sum_{j=1}^{t} \delta_{Q,j}^T \mathbf{h}_{j-1}^T \quad (19)$$

$$\frac{\partial E}{\partial \mathbf{b}_o} = \sum_{j=1}^{t} \delta_{o,j} \quad (20)$$

$$\frac{\partial E}{\partial \mathbf{b}_i} = \sum_{j=1}^{t} \delta_{i,j} \quad (21)$$

$$\frac{\partial E}{\partial \mathbf{b}_f} = \sum_{j=1}^{t} \delta_{f,j} \quad (22)$$

$$\frac{\partial E}{\partial \mathbf{b}_c} = \sum_{j=1}^{t} \delta_{Q,j} \quad (23)$$

$$\frac{\partial E}{\partial \mathbf{W}_{ox}} = \delta_{o,t} \mathbf{x}_t^T \quad (24)$$

$$\frac{\partial E}{\partial \mathbf{W}_{fx}} = \delta_{f,t} \mathbf{x}_t^T \quad (25)$$

$$\frac{\partial E}{\partial \mathbf{W}_{ix}} = \delta_{i,t} \mathbf{x}_t^T \quad (26)$$

$$\frac{\partial E}{\partial \mathbf{W}_{cx}} = \delta_{Q,t}\mathbf{x}_t^T \quad (27)$$

where $\mathbf{W}_{oh}$, $\mathbf{W}_{fh}$, $\mathbf{W}_{ih}$, $\mathbf{W}_{ch}$, $\mathbf{W}_{ox}$, $\mathbf{W}_{fx}$, $\mathbf{W}_{ix}$ and $\mathbf{W}_{cx}$ satisfy

$$[\mathbf{W}_o]\begin{bmatrix}\mathbf{h}_{t-1}\\ \mathbf{x}_t\end{bmatrix} = [\mathbf{W}_{oh}\ \mathbf{W}_{ox}]\begin{bmatrix}\mathbf{h}_{t-1}\\ \mathbf{x}_t\end{bmatrix} = \mathbf{W}_{oh}\mathbf{h}_{t-1} + \mathbf{W}_{ox}\mathbf{x}_t \quad (28)$$

$$[\mathbf{W}_f]\begin{bmatrix}\mathbf{h}_{t-1}\\ \mathbf{x}_t\end{bmatrix} = [\mathbf{W}_{fh}\ \mathbf{W}_{fx}]\begin{bmatrix}\mathbf{h}_{t-1}\\ \mathbf{x}_t\end{bmatrix} = \mathbf{W}_{fh}\mathbf{h}_{t-1} + \mathbf{W}_{fx}\mathbf{x}_t \quad (29)$$

$$[\mathbf{W}_i]\begin{bmatrix}\mathbf{h}_{t-1}\\ \mathbf{x}_t\end{bmatrix} = [\mathbf{W}_{ih}\ \mathbf{W}_{ix}]\begin{bmatrix}\mathbf{h}_{t-1}\\ \mathbf{x}_t\end{bmatrix} = \mathbf{W}_{ih}\mathbf{h}_{t-1} + \mathbf{W}_{ix}\mathbf{x}_t \quad (30)$$

$$[\mathbf{W}_c]\begin{bmatrix}\mathbf{h}_{t-1}\\ \mathbf{x}_t\end{bmatrix} = [\mathbf{W}_{ch}\ \mathbf{W}_{cx}]\begin{bmatrix}\mathbf{h}_{t-1}\\ \mathbf{x}_t\end{bmatrix} = \mathbf{W}_{ch}\mathbf{h}_{t-1} + \mathbf{W}_{cx}\mathbf{x}_t \quad (31)$$

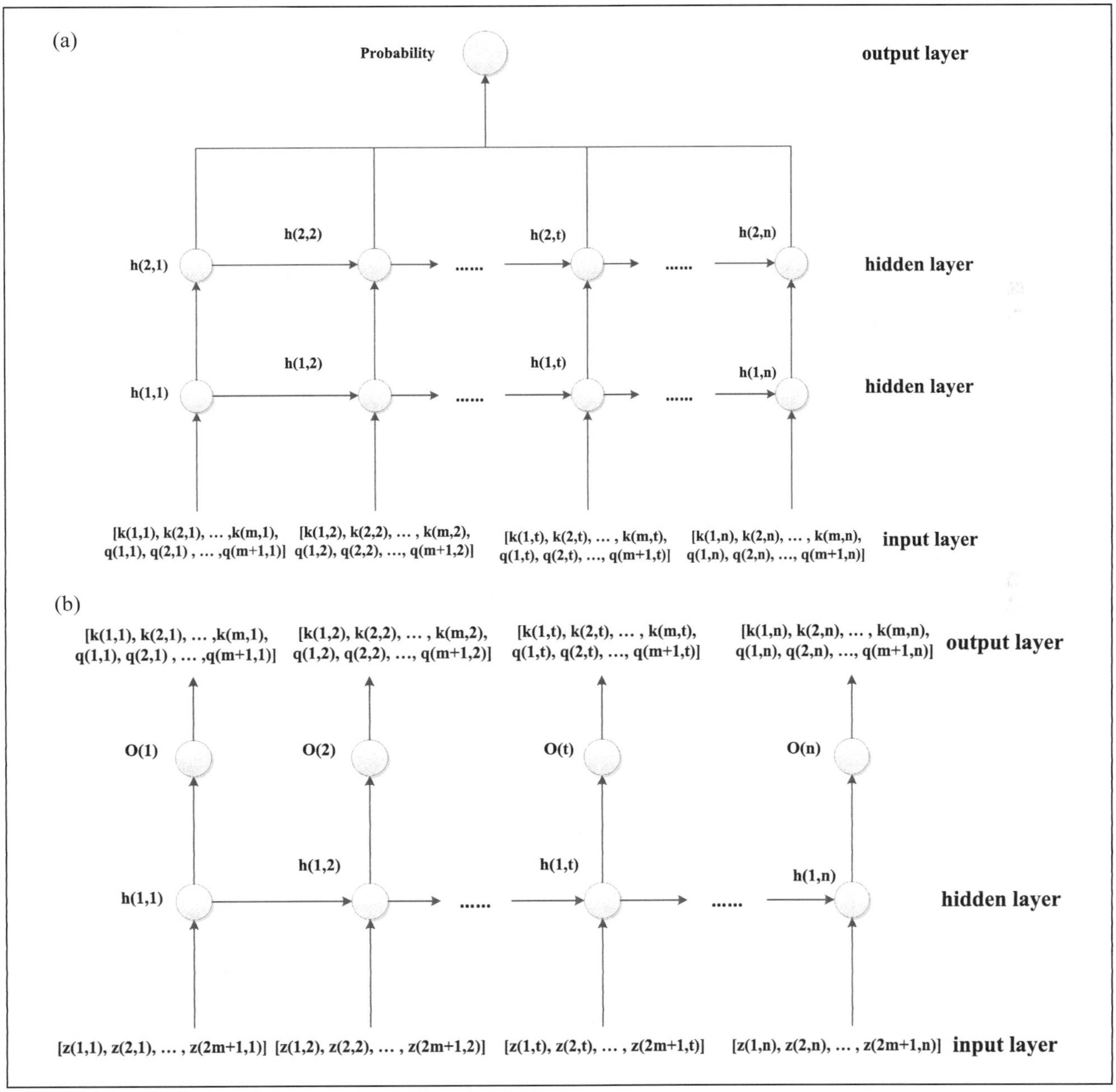

Figure 3. Model structure of the proposed GAA.
(*a*) Model structure of the discriminative LSTM NN. (*b*) model structure of the generative LSTM NN. (*c*) minibatch stochastic gradient descent training of GANs.

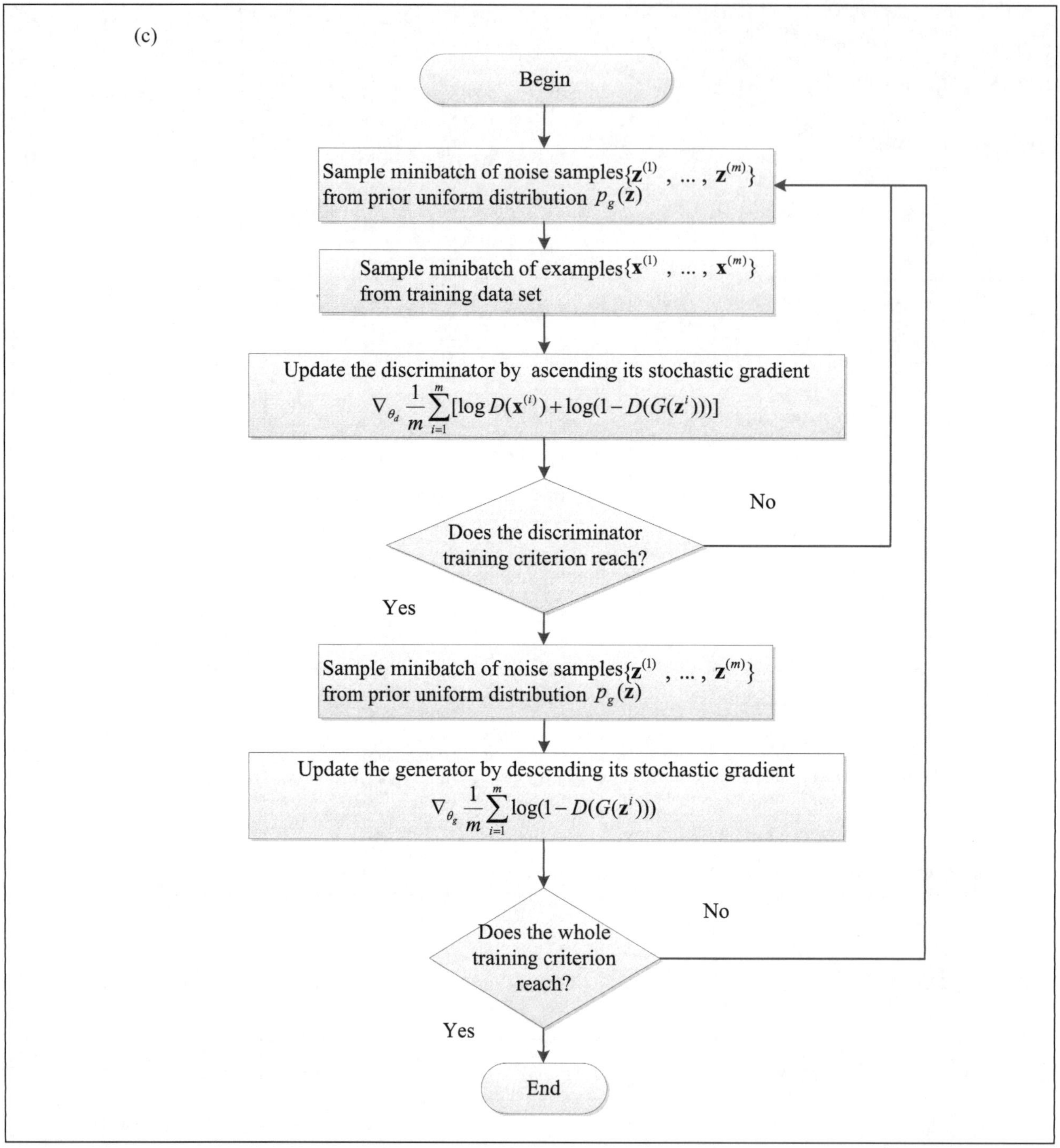

Figure 3. (continued)

Traffic-State Estimation with Pre-Trained GANs

Our methodology for traffic-state estimation utilizes the G and D networks, pre-trained with uncorrupted data, to reconstruct traffic-state. Denote the corrupted traffic-state matrix as $\mathbf{y}$. We do not use D to update $\mathbf{y}$ by maximizing $D(\mathbf{y})$. Similar to the image inpainting in (*19*), maximizing $D(\mathbf{y})$ does not lead to the desired reconstruction. This is mainly becauseyis neither on the p_{data} manifold, nor on the G manifold and the corrupted data is not drawn from those distributions. Therefore, we consider using both G and D for reconstruction. To quantify the 'closest' mapping from $\mathbf{y}$ to the reconstruction, we define three loss functions: a contextual loss, a perceptual loss and a conservative loss.

Contextual Loss. We need to incorporate the information from the known information of the given corrupted traffic-state matrix into the traffic-state estimation. The contextual loss is used to measure the context similarity between the reconstructed traffic-state matrix and the uncorrupted portion. The contextual loss is defined as:

$$L_{contextual}(\mathbf{z}) = \|M \odot G(\mathbf{z}) - M \odot \mathbf{y}\|_1 \tag{32}$$

where

M = a binary matrix having the same dimension as the output traffic matrices of G, denoting the mask of the corrupted traffic matrix.

Elements equal to one in M indicate the corresponding element in the corrupted traffic matrix is uncorrupted. Elements equal to zero in M indicate the corresponding element in the corrupted traffic matrix is missing. $\odot$denotes the element-wise product operation. The corrupted portion, i.e., $(1 - M) \odot \mathbf{y}$ is not used in the loss. The choice of the ℓ_1-norm is empirical. As shown in (*19*), in image inpainting, images recovered with the ℓ_1-norm tend to be sharper and of higher quality compared with the ones reconstructed with the ℓ_2-norm.

Perceptual Loss. The perceptual loss encourages the reconstructed traffic-state matrix to be similar to the samples drawn from the training set. This is achieved by updating $\mathbf{z}$ to fool D. As a result, D will predict $G(\mathbf{z})$ to be from the training data with a high probability. We use the same loss for fooling D as in GANs:

$$L_{perceptual}(\mathbf{z}) = \log(1 - D(G(\mathbf{z}))) \tag{33}$$

Without $L_{perceptual}$, some reconstructed traffic-state matrices tend to be unrealistic. We illustrate this by showing the examples where we optimized with and without $L_{perceptual}$ below.

Conservative Loss. According to the conservation law of traffic flow (*4*), the spatial-temporal relationship between traffic volume and traffic density can be described as:

$$k(t+1, s) = k(t, s) + \frac{\Delta t}{\Delta x_s}(q_{in}(t, s) - q_{out}(t, s)) \tag{34}$$

where

$k(t, s)$ = the vehicle density of cell s at time index t;

$q_{in}(t, s)$ = the total flows (in vehicles per unit time) entering cell s during the time interval $[t \cdot \Delta t, (t+1) \cdot \Delta t)$;

$q_{out}(t, s)$ = the total flows (in vehicles per unit time) leaving cell s during the time interval $[t \cdot \Delta t, (t+1) \cdot \Delta t)$;

Δt = the sampling duration;

Δx_s = the length of cell$s$.

In the context of traffic-state estimation using GAA, the conservative loss is defined as:

$$K_{G(\mathbf{z})}(t+1, s) = K_{G(\mathbf{z})}(t, s) + \frac{\Delta t}{\Delta x_s}(F_{G(\mathbf{z})}(t, s) - F_{G(\mathbf{z})}(t, s+1)) \tag{35}$$

where

$K_{G(\mathbf{z})}$ = the traffic-density matrix generated by the pre-trained$G$;

$F_{G(\mathbf{z})}$ = the traffic-flow matrix generated by the pre-trained$G$.

Equation 35 describes the theoretical spatial-temporal correlation among traffic flow and traffic density. Thus, it helps to improve the accuracy of traffic-state estimation.

Traffic-State Estimation

With the defined loss functions, the traffic-state matrix with missing values can be mapped to the closest $\mathbf{z}$ in the latent representation space. $\mathbf{z}$ is updated using back-propagation with the total loss:

$$\hat{\mathbf{z}} = \arg\min_{\mathbf{z}} (L_{contextual}(\mathbf{z}) + \lambda_p L_{perceptual}(\mathbf{z}) + \lambda_c L_{conservative}(\mathbf{z})) \tag{36}$$

where

λ_p = the weighting parameter of the perceptual loss;

λ_c = the weighting parameter of the conservative loss.

In practice, the weighting parameters have to be relatively small to ensure the traffic-state estimation accuracy. After finding $\hat{\mathbf{z}}$, the estimated traffic state can be obtained by:

$$\mathbf{x}_{reconstructed} = \mathbf{M} \odot \mathbf{y} + (1 - \mathbf{M}) \odot G(\hat{\mathbf{z}}) \tag{37}$$

Experiments

The proposed GAA for spatial-temporal traffic-state estimation is evaluated on two data sets: loop data collected from a segment of the I-5 freeway in Seattle, Washington (see Figure 4*a*) and loop data from a segment of the CA-52 state highway in San Diego, California (see Figure 4*b*). The segment of interest of the I-5 freeway has six loop detectors installed along it. The data is collected every 20 seconds from January 1st, 2015 to December 31st, 2015 and is stored in the Digital Roadway Interactive Visualization and Evaulation Network (DRIVE Net) system operated by the Smart Transportation Application and Research Laboratory (STAR Lab) at the University of Washington. The segement of interest along the CA-52 state highway has four loop detectors installed along it. Here, the data is collected every 30 seconds from January 1st, 2016 to December 31st, 2016 and is stored in the Performance Measurement System operated by the California Department of Transportation.

Each training data record in the two experiments consists of data across one hour from the six loop detectors and the four loop detectors, respectively. This indicates that the dimension of each training traffic-state matrix is 12*11 and 12*7 in the two experiments, respectively. In the two evaluations, we randomly select two-thirds of the data for training and the rest is used for validation.

Four types of GAAs are evaluated in the two experiments. These types include a GAA without perceptual loss and conservative loss (GAA model 1), a GAA without perceptual loss (GAA model 2), a GAA without conservative loss (GAA model 3) and a GAA with perceptual loss and conservative loss (GAA model 4). The four types of GAAs are compared with the existing Bayesian network approach proposed by (*21*) in each experiment. Similar to our proposed GAA models, the Bayesian network approach considers spatial-temporal correlation among traffic-flow parameters in traffic-state estimation. Therefore, the Bayesian network approach is chosen for comparison in this study.

Discriminative LSTM NNs and generative LSTM NNs in the four GAAs respectively have the same topology as shown in Figure 3*a* and *b* and they have 16 hidden neurons in the hidden layer.

Each record in the two data sets is a pair of average traffic flow and traffic-density measurements computed over one hour. We used the data in the previous half hour (i.e., 6*5 minutes) to predict traffic flow and traffic density in the next half hour (i.e., the next 6*5 minutes). Note that this time lag is not limited to one hour as the LSTM NN is capable of capturing long time dependency among traffic-flow parameters. Figures 5 and 6 compare the traffic state estimated by the five models on the two data sets, respectively. Based on the data collected for the I-5 freeway segment, subfigures 5*a–f* compare the traffic density estimated by the five models from 7:00 a.m to 7:00 p.m. on October 10th between each pair of loop detectors, respectively. Subfigures 5*g–k* compare the traffic flow estimated by the five models in the same way. As opposed to the traffic density defined between each pair of loop detectors, traffic flow is defined on each loop detector. Figure 6 presents results in the analogous and aforementioned manner, but based on the data collected along the CA-52 state highway.

To measure the effectiveness of the five models, the Mean Absolute Percentage Error (MAPE) and Mean Squared Error (MSE) are calculated. The MAPE and MSE obtained by the five models for traffic density estimation and traffic-flow estimation are shown in Table 1.

Experiment on the Data Collected from Seattle

For traffic-density estimation, the GAA without conservative loss performs almost as well as the GAA with conservative loss and perceptual loss, with only 0.08% larger MAPE. They both outperform the other two GAA models in terms of MAPE. While in terms of MSE, the GAA with perceptual loss and conservative loss outperforms the other three GAA models. In this case, the GAA with perceptual loss and conservative loss receive 2.84% lower MSE than GAA without conservative loss.

For traffic-flow estimation, the GAA conservative loss and perceptual loss performs best among all GAA models in terms of both MAPE and MSE.

For traffic-density estimation, the proposed GAA with conservative loss and perceptual loss outperforms the Bayesian network approach with 1.00% lower MAPE and 1.66 lower MSE. For traffic-flow estimation, the GAA with conservative loss and perceptual loss performs nearly as well as the Bayesian network approach with only 0.07% larger MAPE. While the GAA with conservative loss and perceptual loss outperforms the Bayesian network approach with 0.81 lower MSE.

For traffic-density estimation, the GAA without conservative loss outperforms the Bayesian network approach in terms of MAPE with 0.92% lower MAPE. In all other cases, the Bayesian network approach outperforms the other three GAA models.

Experiment on the Data Collected from San Diego

For traffic-density estimation, the GAA with conservative loss and perceptual loss outperforms the other three GAA models in terms of both MAPE and MSE. For traffic-flow estimation, the GAA without conservative loss performs almost as well as the GAA with conservative loss and perceptual loss, with only 0.07% larger MAPE. They both outperform the other two GAA

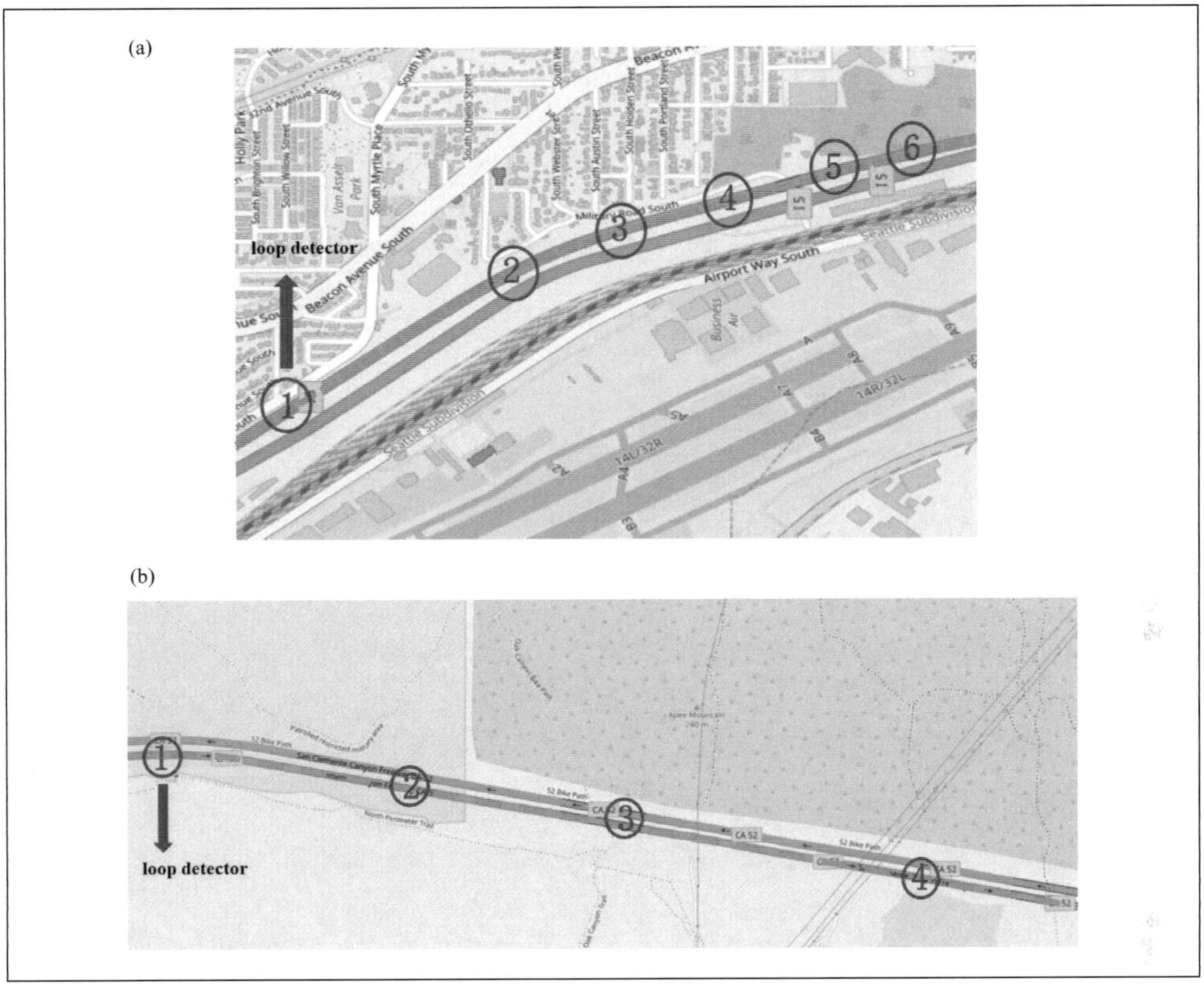

Figure 4. The sites for data collection.
(*a*) The I-5 freeway segment in Seattle, Washington. (*b*) the CA-52 state highway segment in San Diego, California.

models in terms of MAPE. While in terms of MSE, the GAA with perceptual loss and conservative loss outperforms the other three GAA models. In this case, the GAA with perceptual loss and conservative loss receives 3.54 lower MSE than the GAA without conservative loss.

For traffic-density estimation, the proposed GAA with conservative loss and perceptual loss outperforms the Bayesian network approach with 0.95% lower MAPE and 1.04 lower MSE. The Bayesian network approach outperforms the GAA without conservative loss with 0.89% lower MAPE and 3.09 lower MSE. For traffic-flow estimation, the performance of the Bayesian network approach, the GAA without conservative loss and the GAA with conservative loss and perceptual loss are close in terms of MAPE.The difference between the MAPE of the Bayesian network approach and the MAPE of the GAA without conservative loss is only 0.05%. The difference between the MAPE of the Bayesian network approach and the MAPE of the GAA with conservative loss and perceptual loss are only 0.02%. While the GAA with conservative loss and perceptual loss outperforms the Bayesian network approach with 0.85 lower MSE. In this case, the Bayesian network approach outperforms the GAA without conservative loss with 2.79 lower MSE.

In the first experiment, for traffic-flow estimation, the difference between the GAA without conservative loss and the GAA without perceptual loss is marginal in terms of MAPE with only 0.08% MAPE difference. While in other cases, the GAA without conservative loss

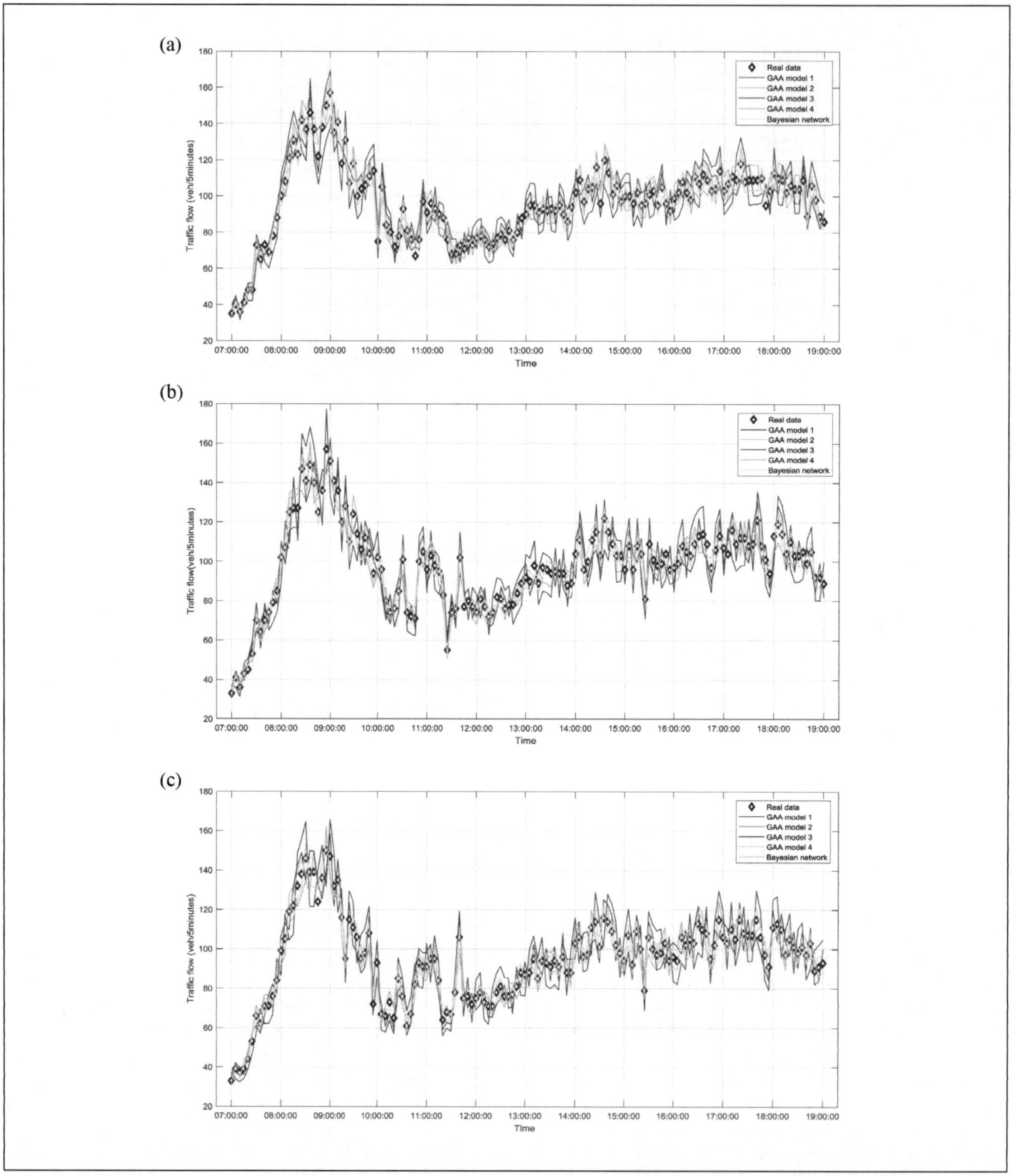

Figure 5. Model comparison based on the data collected for the I-5 freeway segment between 7:00 a.m. and 7:00 p.m. on October 10th. (*a*) Predicted traffic flow for loop detector 1; (b) predicted traffic flow for loop detector 2; (c) predicted traffic flow for loop detector 3; (d) predicted traffic flow for loop detector 4; (e) predicted traffic flow for loop detector 5; (f) predicted traffic flow for loop detector 6; (g) predicted traffic density between loop detector 1 and 2; (h) predicted traffic density between loop detector 2 and 3; (i) predicted traffic density between loop detector 3 and 4; (j) predicted traffic density between loop detector 4 and 5; (k) predicted traffic density between loop detector 5 and 6.

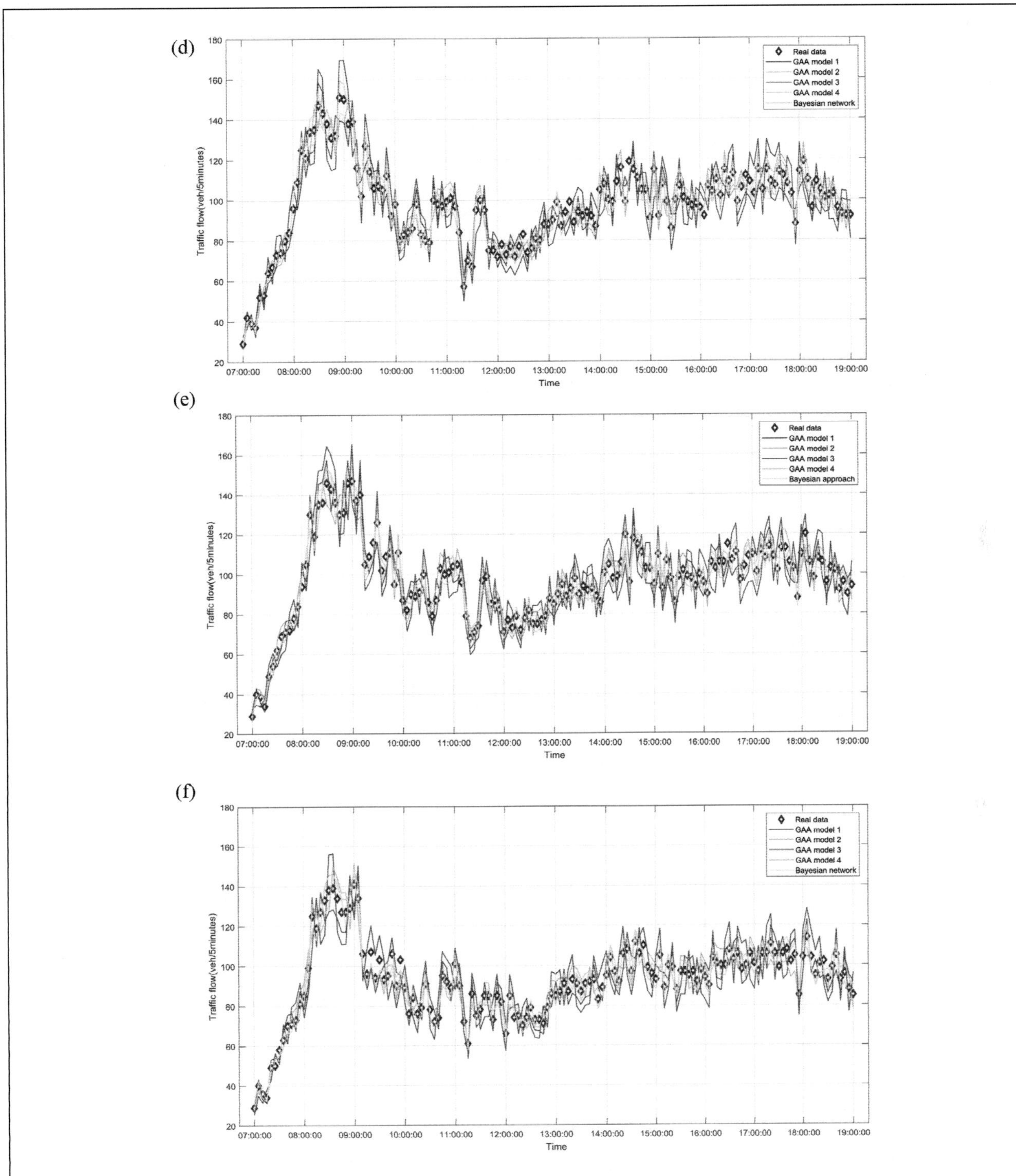

Figure 5. (continued)

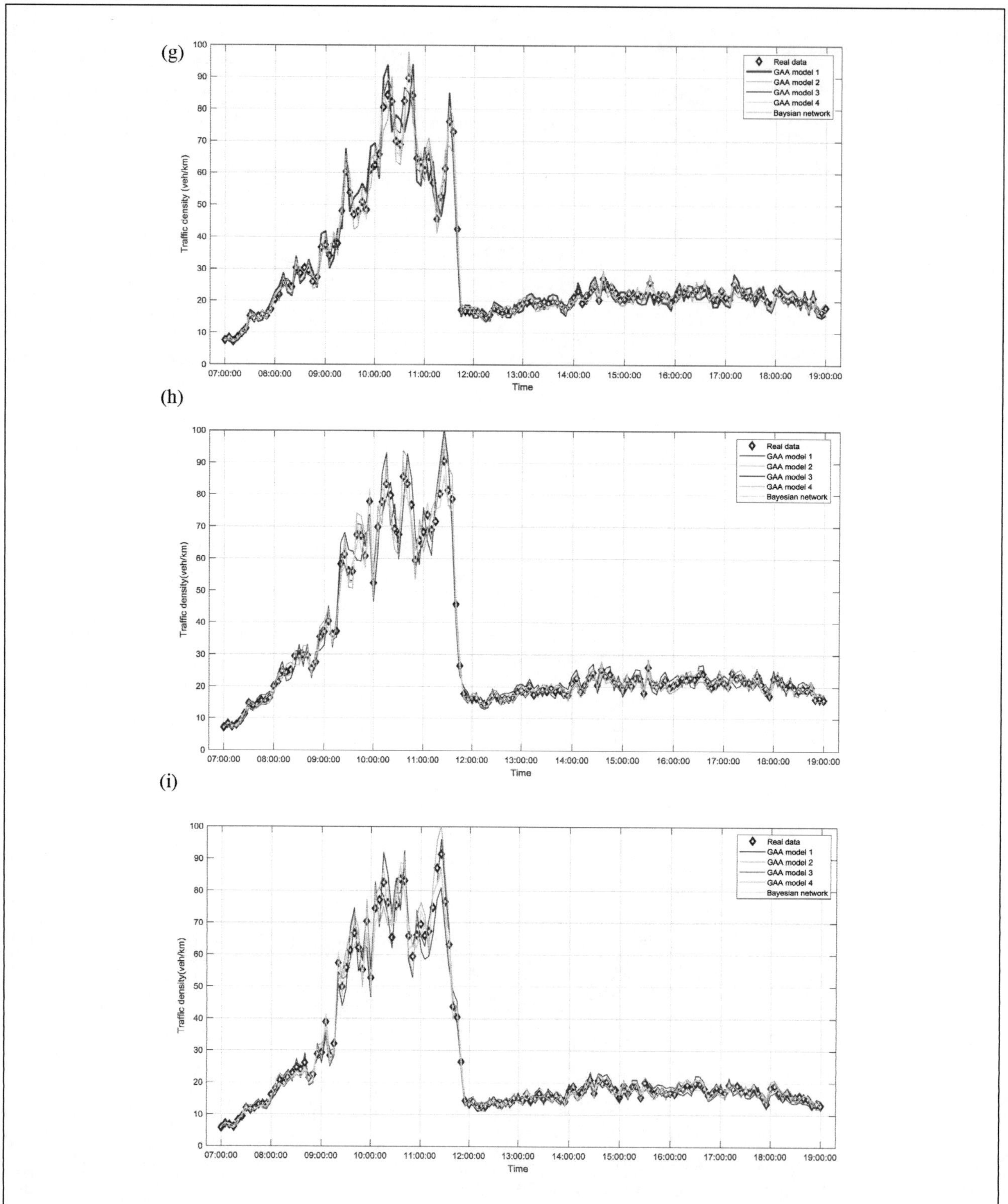

Figure 5. (continued)

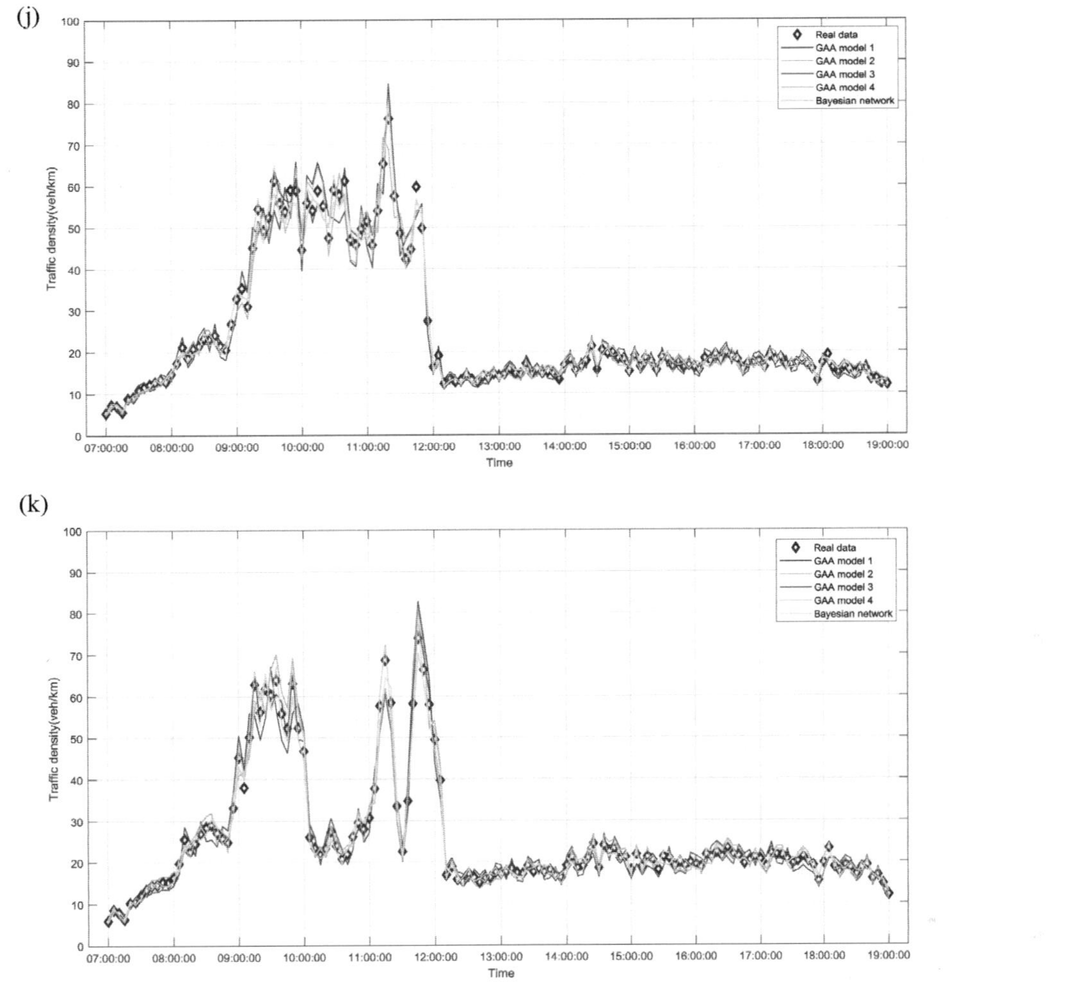

Figure 5. (continued)

outperforms the GAA without perceptual loss. In the second experiment, the GAA without conservative loss outperforms the GAA without perceptual loss in traffic-density estimation and traffic-flow estimation in terms of both MAPE and MSE. This indicates that learning from the training dataset brings greater improvements in traffic-state estimation accuracy than learning from the traffic-flow conservation law.

In both experiments, for both traffic-density estimation and traffic-flow estimation, the GAA without conservative loss and perceptual loss performs worst in terms of both MAPE and MSE.

Note that in both experiments, in traffic-flow estimation, the difference between the Bayesian network approach and the GAA with conservative loss and perceptual loss is marginal in terms of MAPE. It seems that the GAA with conservative loss and perceptual loss has more advantages over the Bayesian network approach in traffic-density estimation compared with that in traffic-flow estimation.

Discussions and Conclusions

This study proposes a GAA for spatial-temporal traffic-state estimation. The proposed GAA is compared with the existing Bayesian network approach on loop detector data collected from Seattle, Washington and that collected from San Diego, California. Experimental results indicate that as the GAA incorporates traffic-flow theory, it can achieve higher accuracy in traffic-state estimation than the Bayesian network approach. In addition, two useful findings can be summarized from the experiments: 1) adding

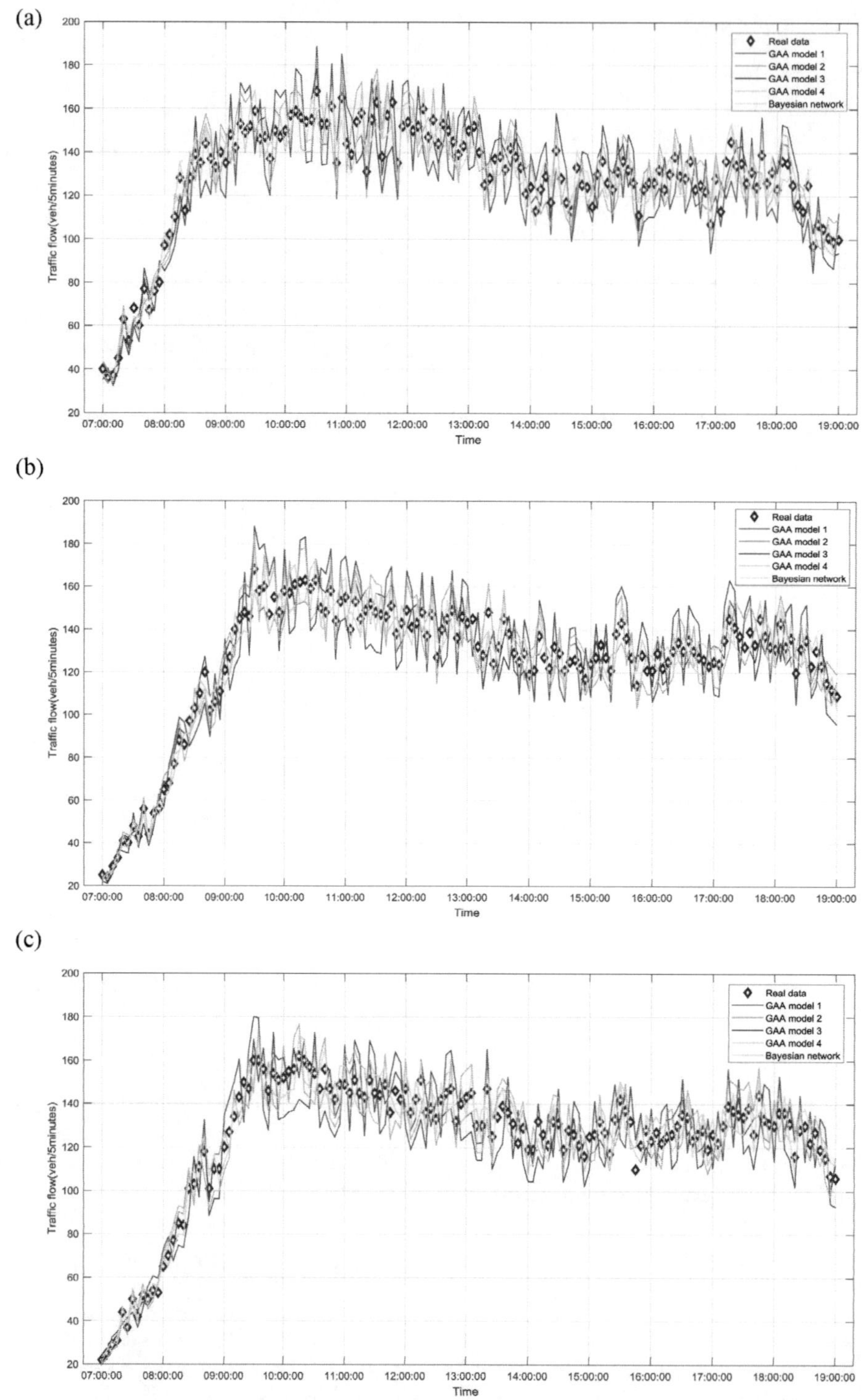

Figure 6. Model comparison based on the data collected for the CA-52 state highway segment between 7:00 a.m. and 7:00 p.m. on October 10th.

(*a*) Predicted traffic flow for loop detector 1. (*b*) predicted traffic flow for loop detector 2. (*c*) predicted traffic flow for loop detector 3. (*d*) predicted traffic flow for loop detector 4. (e) predicted traffic density between loop detector 1 and 2. (*f*) predicted traffic density between loop detector 2 and 3. (*g*) predicted traffic density between loop detector 3 and 4.

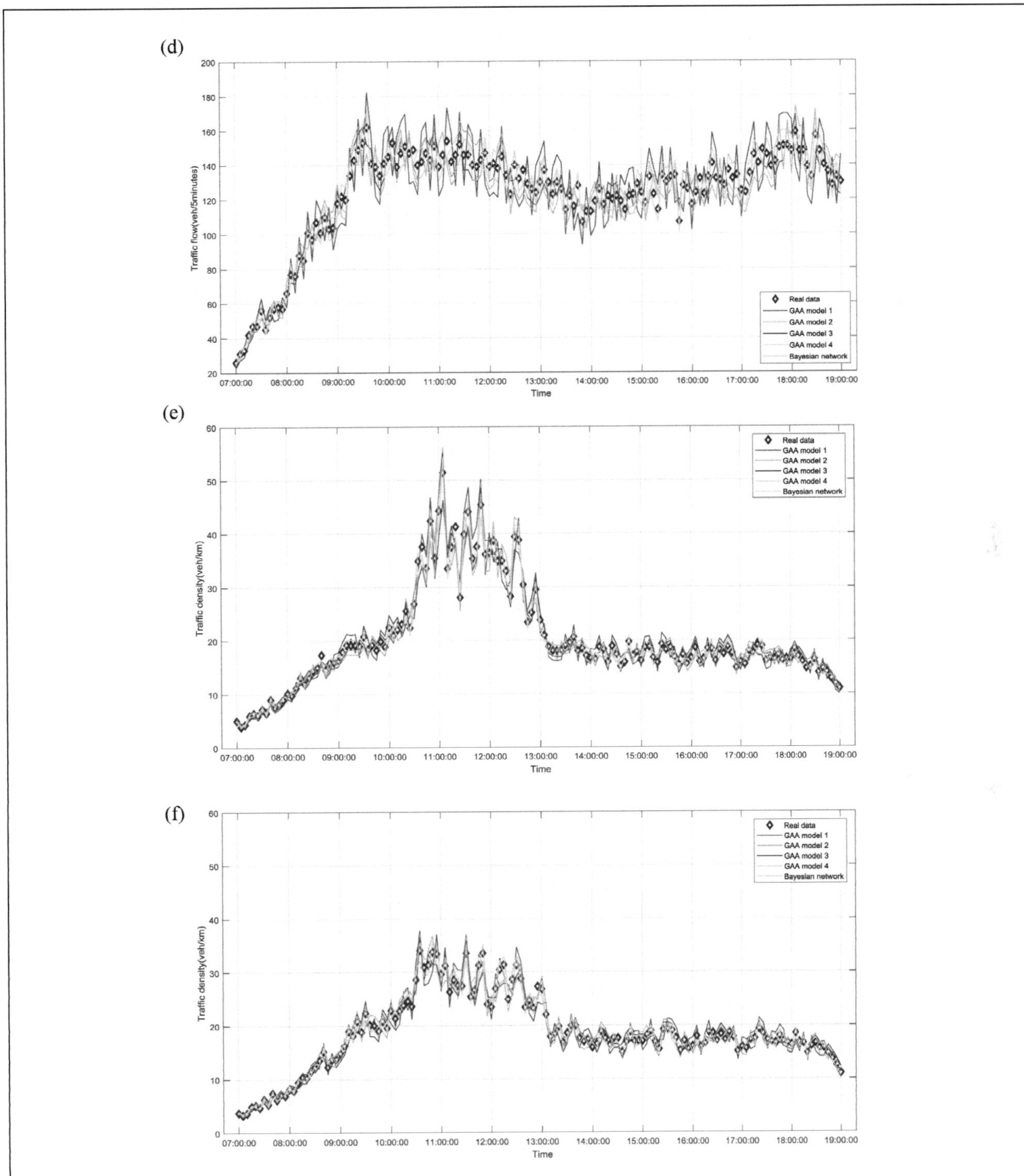

Figure 6. (continued)

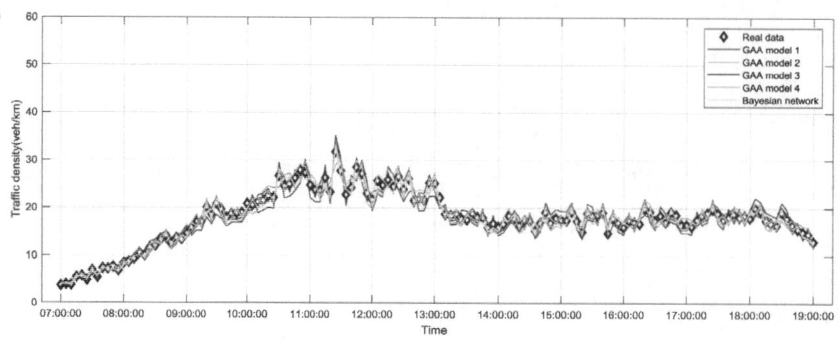

Figure 6. (continued)

Table 1. Performance of Different Models in Traffic-State Estimation

	Traffic-density estimation		Traffic-flow estimation	
	MAPE	MSE	MAPE	MSE
Experiment 1				
GAA without percetual loss and conservative loss	11.47%	125.62	12.36%	110.55
GAA without perceptual loss	9.80%	20.55	7.91%	19.30
GAA without conservative loss	5.12%	9.11	7.83%	8.42
GAA with perceptual loss and conservative loss	5.04%	6.27	5.26%	5.95
Bayesian network	6.04%	7.93	5.19%	6.76
Experiment 2				
GAA without percetual loss and conservative loss	10.76%	119.59	12.08%	108.13
GAA without perceptual loss	9.07%	19.88	9.16%	25.64
GAA without conservative loss	6.42%	10.23	6.02%	7.92
GAA with perceptual loss and conservative loss	4.68%	6.10	5.95%	4.28
Bayesian network	5.53%	7.14	5.97%	5.13

the conservative loss can improve the accuracy of traffic-state estimation, and adding the perceptual loss can lead to further improvement; 2) the GAA with conservative loss and perceptual loss is better at density and traffic-flow estimation than the Bayesian network approach, and the magnitude of the benefit is stronger for density estimation.

The proposed GAA can be applied in real-time traffic prediction tasks, as well as in historical traffic-state reconstruction. Further, artificical intelligent technology is shown to be promising in the transportation area. Even though a GAA requires a large amount of training traffic data to be trained and quite a long time to achieve Nash equilibrium in the training procedure, these two shortcomings can be easily tackled with the aid of enriched data collected by various advanced sensors and advanced computation technologies.

Despite the promising results obtained from this study, traffic-flow phenomena, such as breakdown and capacity drop, cannot be caught by the proposed GAA in our experiments. The difficulty of using artificial intelligent technology for predicting complicated traffic-flow phenonmena lies in two aspects. The first aspect is that it is difficult to understand how they learn to predict breakdown and capacity drop from training data. Moreover, if there are no traffic-flow breakdown events described in training dataset, it is almost impossible for the proposed GAA to create knowledge about breakdown. Future research directions can work to modify the GAA and make it able to learn and predict more complicated traffic-flow phenomena.

Acknowledgments

This research was supported in part by the National Natural Science Foundation of China (Grant No. 51329801) and the Shenzhen Science and Technology Planning Project (Grant No. GJHZ20150316154158400).

Author Contributions

The authors confirm contribution to the paper as follows: study conception and design: Yunyi Liang, Zhiyong Cui, Yinhai Wang; data collection: Zhiyong Cui; analysis and interpretation of results: Yunyi Liang, Yinhai Wang; draft manuscript preparation: Yunyi Liang, Yu Tian, Huimiao Chen. All authors reviewed the results and approved the final version of the manuscript.

References

1. Zhang, X., E. Onieva, A. Perallos, E. Osaba, and V. C. Lee. Hierarchical Fuzzy Rule-Based System Optimized with Genetic Algorithms for Short Term Traffic Congestion Prediction. *Transportation Research Part C: Emerging Technologies*, Vol. 43, 2014, pp. 127–142.
2. Cai, P., Y. Wang, G. Lu, P. Chen, C. Ding, and J. Sun. A Spatiotemporal Correlative k-nearest Neighbor Model for Short-term Traffic Multistep Forecasting. *Transportation Research Part C: Emerging Technologies*, Vol. 62, 2016, pp. 21–34.
3. Pan, T. L., A. Sumalee, R.-X. Zhong, and N. Indra-Payoong. Short-term Traffic State Prediction Based on Temporal–Spatial Correlation. *IEEE Transactions on Intelligent Transportation Systems*, Vol. 14, No. 3, 2013, pp. 1242–1254.
4. Daganzo, C. F. The Cell Transmission Model: A Dynamic Representation of Highway Traffic Consistent with the Hydrodynamic Theory. *Transportation Research Part B: Methodological*, Vol. 28, No. 4, 1994, pp. 269–287.
5. Peeta, S., and A. K. Ziliaskopoulos. Foundations of Dynamic Traffic Assignment: The Past, the Present and the Future. *Networks and Spatial Economics*, Vol. 1, No. 3, 2001, pp. 233–265.
6. Vlahogianni, E. I., M. G. Karlaftis, and J. C. Golias. Short-term Traffic Forecasting: Where We are and Where We're Going. *Transportation Research Part C: Emerging Technologies*, Vol. 43, 2014, pp. 3–19.
7. Wang, Y., M. Papageorgiou, and A. Messmer. RENAISSANCE–A Unified Macroscopic Model-based Approach to Real-time Freeway Network Traffic Surveillance. *Transportation Research Part C: Emerging Technologies*, Vol. 14, No. 3, 2006, pp. 190–212.
8. Van Lint, J. W. C. Online Learning Solutions for Freeway Travel Time Prediction. *IEEE Transactions on Intelligent Transportation Systems*, Vol. 9, No. 1, 2008, pp. 38–47.
9. Asif, M. T., J. Dauwels, C. Y. Goh, A. Oran, E. Fathi, M. Xu, M. M. Dhanya, N. Mitrovic, and P. Jaillet. Spatiotemporal Patterns in Large-scale Traffic Speed Prediction. *IEEE Transactions on Intelligent Transportation Systems*, Vol. 15, No. 2, 2014, pp. 794–804.
10. Zhang, Y., and Y. Liu. Traffic Forecasting Using Least Squares Support Vector Machines. *Transportmetrica*, Vol. 5, No. 3, 2009, pp. 193–213.
11. Zhang, X., E. Onieva, A. Perallos, E. Osaba, and V. C. S. Lee. Hierarchical Fuzzy Rule-based System Optimized with Genetic Algorithms for Short Term Traffic Congestion Prediction. *Transportation Research Part C: Emerging Technologies*, Vol. 43, 2014, pp. 127–142.
12. Li, L., Y. Li, and Z. Li. Efficient Missing Data Imputing for Traffic Flow by Considering Temporal and Spatial Dependence. *Transportation Research Part C: Emerging Technologies*, Vol. 34, 2013, pp. 108–120.
13. Abdi, J., B. Moshiri, B. Abdulhai, and A. K. Sedigh. Forecasting of Short-term Traffic-flow Based on Improved Neurofuzzy Models via Emotional Temporal Difference Learning Algorithm. *Engineering Applications of Artificial Intelligence*, Vol. 25, No. 5, 2012, pp. 1022–1042.
14. Polson, N. G., and V. O. Sokolov. Deep Learning for Short-term Traffic Flow Prediction. *Transportation Research Part C: Emerging Technologies*, Vol. 79, 2017, pp. 1–17.
15. Ma, X., Z. Tao, Y. Wang, H. Yu, and Y. Wang. Long Short-term Memory Neural Network for Traffic Speed Prediction Using Remote Microwave Sensor Data. *Transportation Research Part C: Emerging Technologies*, Vol. 54, 2015, pp. 187–197.
16. Ma, D., X. Luo, S. Jin, D. Wang, W. Guo, and F. Wang. Lane-based Saturation Degree Estimation for Signalized Intersections Using Travel Time Data. *IEEE Intelligent Transportation Systems Magazine*, Vol. 9, No. 3, 2017, pp. 136–148.
17. Ma, D., X. Luo, W. Li, S. Jin, W. Guo, and D. Wang. Traffic Demand Estimation for Lane Groups at Signal-controlled Intersections Using Travel Times from Video-imaging Detectors. *IET Intelligent Transport Systems*, Vol. 11, No. 4, 2017, pp. 222–229.
18. Goodfellow, I., J. Pouget-Abadie, M. Mirza, B. Xu, D. Warde-Farley, S. Ozair, A. Courville, and Y. Bengio. Generative Adversarial Nets. *Advances in Neural Information Processing Systems*, 2014, pp. 2672–2680.
19. Yeh, R., C. Chen, T. Y. Lim, A. G. Schwing, M. Hasegawa-Johnson, and M. N. Do. Semantic Image Inpainting with Perceptual and Contextual Losses. *arXiv preprint arXiv*, 2016, 1607.07539.
20. Hochreiter, S., and J. Schmidhuber. Long Short-term Memory. *Neural Comput*, Vol. 9, No. 8, 1997, pp. 1735–1780.
21. Sun, S., C. Zhang, and G. Yu. A Bayesian Network Approach to Traffic Flow Forecasting. *IEEE Transactions on Intelligent Transportation Systems*, Vol. 7, No. 1, 2006, pp. 124–132.
22. Gers, F. Long Short-term Memory in Recurrent Neural Networks. PhD dissertation. école Polytechnique Fédérale De Lausanne, 2001.

The Standing Committee on Information Systems and Technology (ABJ50) peer-reviewed this paper (18-06235).

Article

TRR
JOURNAL OF THE TRANSPORTATION RESEARCH BOARD

Transportation Research Record
2018, Vol. 2672(45) 106–114

DOI: 10.1177/0361198118775841
journals.sagepub.com/home/trr
SAGE

Automatic Background Filtering Method for Roadside LiDAR Data

Jianqing Wu[1], Hao Xu[1], Yuan Sun[1], Jianying Zheng[2], and Rui Yue[1]

Abstract

The high-resolution micro traffic data (HRMTD) of all roadway users is important for serving the connected-vehicle system in mixed traffic situations. The roadside LiDAR sensor gives a solution to providing HRMTD from real-time 3D point clouds of its scanned objects. Background filtering is the preprocessing step to obtain the HRMTD of different roadway users from roadside LiDAR data. It can significantly reduce the data processing time and improve the vehicle/pedestrian identification accuracy. An algorithm is proposed in this paper, based on the spatial distribution of laser points, which filters both static and moving background efficiently. Various thresholds of point density are applied in this algorithm to exclude background at different distances from the roadside sensor. The case study shows that the algorithm can filter background LiDAR points in different situations (different road geometries, different traffic demands, day/night time, different speed limits). Vehicle and pedestrian shape can be retained well after background filtering. The low computational load guarantees this method can be applied for real-time data processing such as vehicle monitoring and pedestrian tracking.

Connected-vehicle technologies and applications are developed to help avoid crashes, decrease travel times, and reduce emissions and fuel consumption using the wireless communication of v2v, v2i, and v2x (*1*). The full benefits of connected-vehicle technologies rely on all road users being connected and sharing their real-time information about location, speed, and direction. Though connected-vehicle technologies have been promoted by U.S. Department of Transportation (U.S. DOT) since 2003, there are still limited numbers of connected vehicles on the roads, and most of these are mainly used for pilot programs (*2*). It will take time to put connected-vehicle devices into all vehicles, especially old ones. That means mixed traffic with connected vehicles and unconnected vehicles will last for the coming decades (*3*). In the mixed traffic situation, connected vehicles can not obtain the full benefits as only part of the vehicle movement information can be obtained. New methodologies need to be considered to collect high-resolution micro traffic data (HRMTD) and fill the data gap caused by the mixed traffic. The HRMTD should provide the location, speed, and trajectories of all roadway users with high frequency and accuracy. The three-dimensional (3D) LiDAR (light detection and ranging) sensors deployed on the roadside provide a solution to the need to obtain HRMTD to serve the mixed traffic situation.

The 3D LiDAR technology can give a 360° 3D scan of surrounding objects with high accuracy, which gives an effective method to fill the data gap in mixed traffic flow. A LiDAR sensor can track and report the precise location and speed of each object in its scanned range. LiDAR sensors have been employed in advanced autonomous vehicles for several years to detect road boundaries and lane marks (*4*, *5*). Point cloud outputs from the LiDAR sensor provide the necessary data for the onboard system to determine where potential obstacles exist in the environment and where the vehicle is in relation to those potential obstacles. In the onboard detection systems, LiDAR sensors often work with other sensors such as onboard cameras to detect and track other vehicles and pedestrians. Different from traditional onboard LiDAR sensors, roadside LiDAR sensors are deployed at a static location around intersections or along roads. Roadside LiDAR sensors are expected to independently offer HRMTD without high-resolution camera sensors, considering the cost and limitation of infrastructures and right of way. As roadside deployment is a new application of LiDAR technology, how to

[1]Department of Civil and Environmental Engineering, University of Nevada, Reno, NV
[2]School of Urban Rail Transportation, Soochow University, Suzhou City, China

Corresponding Author:
Address correspondence to Hao Xu: haox@unr.edu

process the data collected by LiDAR sensors is the first thing to be solved to obtain HRMTD. The data processing of roadside LiDAR sensors needs to accurately detect and track all vehicles, pedestrians, and other road users, with cost-effective LiDAR sensors and over a wide area. Therefore, the data processing procedure needs to offer reliable output with limited object points when the objects are some way away from the sensors. With LiDAR sensors deployed at intersections and along roads, the HRMTD of each road user can be extracted and shared with traffic facilities and other roadway users through wireless communication technologies such as dedicated short range communications (DSRC) or cellular communications, as illustrated in Figure 1.

To obtain the HRMTD of all roadway users, background filtering serves as the first step and the basis for enhancing the accuracy and computational efficiency of the subsequent lane identification, object clustering, and vehicle tracking process. Many methods have been developed for background filtering of different data types in previous studies. Zheng et al. developed an algorithm to extract background images from traffic video streams using the pixel information in each frame (*6*). That algorithm analyzes each pixel's color values in a series of frames and then uses the mode of the series as the correct color value for the background image. Zhang et al. used the median value of each color channel to calculate the difference between background and non-background (*7*). The results showed that this method worked reasonably well for freeway applications under free-flow to moderately congested conditions. These algorithms developed for vision-based data processing could not be used for

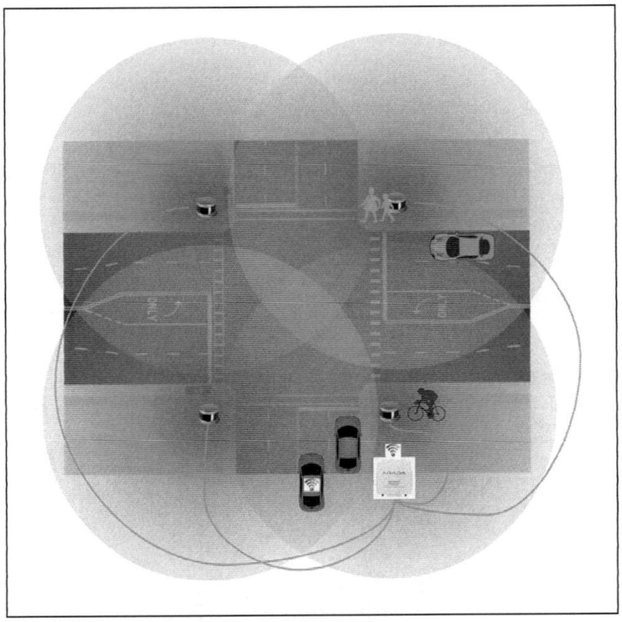

Figure 1. LiDAR sensors for a connected-vehicle system.

LiDAR data directly because the roadside LiDAR data are a series of points instead of raster pixel information. Some methods have been developed to exclude background for LiDAR sensors installed on autonomous vehicles. For the detection systems of autonomous vehicles, the background points are usually not identified directly, being determined by the background changes with the movement of the vehicle (*8*). The tracking algorithms typically use patch segmentation and classification to extract objects of interest from the raw sensor (*9*). Then the remaining area is considered as background and excluded from the database. Though some LiDAR sensors report the intensity of the laser return, these intensity values are usually not very discriminative in general and may be unreliable, as they could vary significantly because of incidence angle or between laser sensors (*10*). So no algorithms can be directly used to exclude backgrounds for HRMTD extraction from roadside LiDAR data. An effective background filtering method is then needed to serve the new-generation traffic monitoring system with roadside LiDAR sensors, a method which excludes background points and keeps vehicles and pedestrians records as much as possible.

This research paper presents a process to exclude the background automatically from the 3D points collected by a roadside LiDAR. An innovative background filtering method was developed to automatically exclude background points from LiDAR data based on the spatial distribution of laser points. The case study showed that this method excludes background points but at the same time keeps most vehicle and pedestrian points. The average time used to filter the background of one frame was 100 ms, which can serve the real-time data processing task well. This paper is organized as follows. The second part introduces the proposed background filtering methodology and procedure. The third part documents a case study to evaluate the effectiveness and accuracy of the algorithm. The last part summarizes the research conclusions and discusses future extended research.

Methodology for Background Filtering

An automatic background filtering method named 3D-density-statistic-filtering (3D-DSF) was developed in this study to filter both static and dynamic background. The 3D-DSF was developed with consideration of the density distribution of LiDAR cloud points in the 3D space. Though, in theory, the algorithm could serve any LiDAR sensors regardless of brand, this paper used the VLP-16 LiDAR sensor manufactured by Velodyne LiDAR™ for the case study. The VLP-16 LiDAR can generate a 360° 3D point cloud by using 16 laser/detector pairs mounted in a compact housing. The LiDAR has a rotational speed of 5–20 rotations per second, which can generate 600,000

3D points per second. It can cover a 360° horizontal field of view and a 30° vertical field of view with ± 15° up and down. The VLP-16 sensor reports spherical coordinates (r, ω, α), which can be converted into Cartesian coordinates (x, y, z). Based on the GPS location of the LiDAR and a reference point, all the laser points can be projected to a position expressed by longitude, latitude, and elevation. Though the manufacturer claimed that the VLP-16 could detect objects within a radius of 100 m, the practice shows that vehicles can be effectively detected within 60 m from the LiDAR sensor. Out of the 60 m range, the background can still be partly scanned, but the points of vehicles may be lost. In this study, points within 60 m distance from the sensor were processed. Points out of the range were excluded from the data frames.

The 3D-DSF method can be elaborated into four major parts: frame aggregation, points statistics, threshold (TD) learning, and real-time filtering. A 3D matrix representing the locations of background areas is generated in the third step. In real-time background filtering, the LiDAR points in the identified background spaces are excluded from the LiDAR data. Figure 2 shows the flowchart of the 3D-DSF method.

Frame Aggregation

The algorithm initially collected raw data in a period as initial input for the background learning. The data frames in a period were aggregated based on LiDAR point coordinates. Theoretically, the more frames used for the background identification, the better the accuracy which can be achieved. However, more frames require more computer memory and increase the time for the background identification. In contrast, the algorithm is unable to identify vibrating background points if the number of frames for background identification is too low. This research compared the performance of the method with different numbers of frames to find the optimum number of frames for aggregation. The other factors were controlled as the same. Table 1 summarizes the results of one frame before and after background filtering.

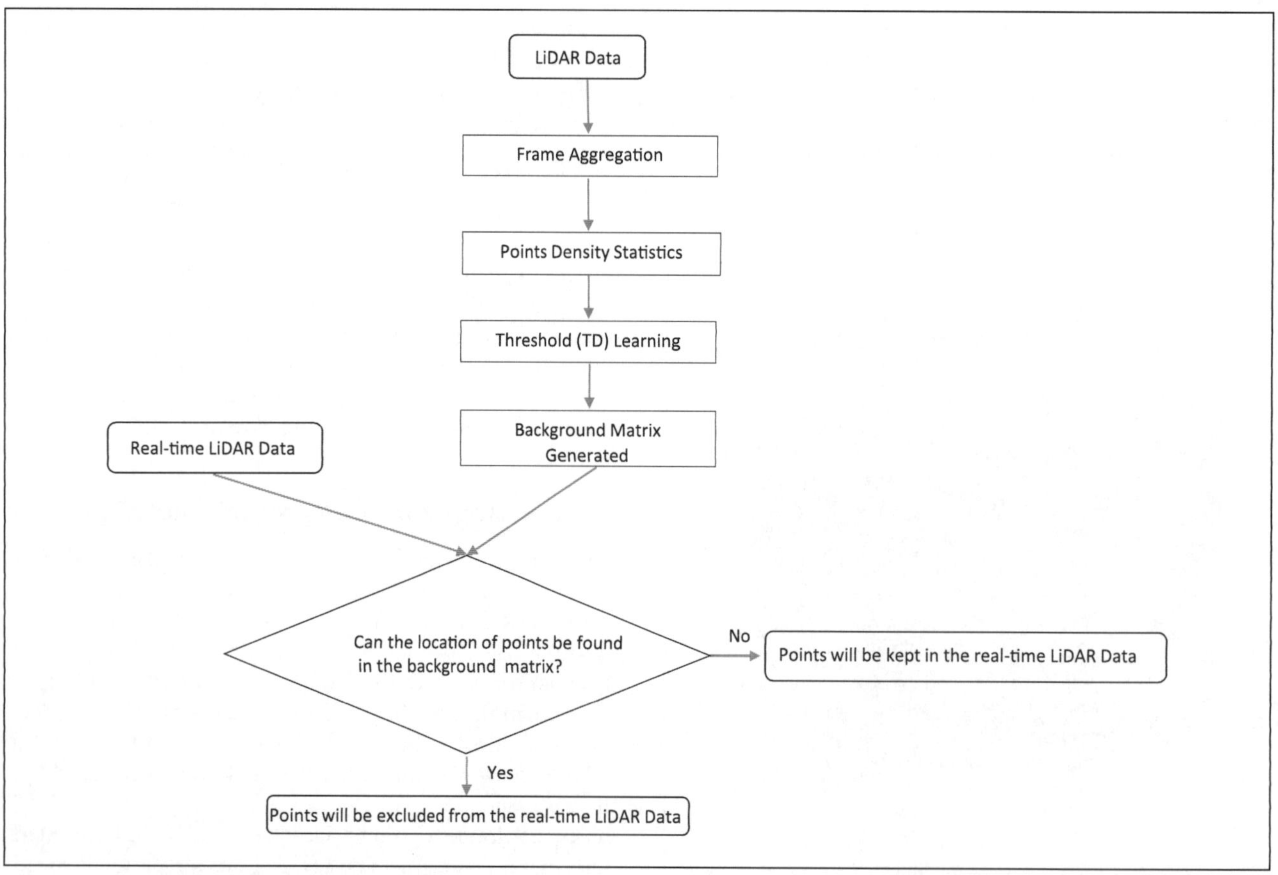

Figure 2. Flow chart of 3D-DSF algorithm.

Table 1. Background Filtering Results using Different Number of Frames

	Number of frames for aggregation								
	100	500	1000	1500	2000	2500	3000	3500	4000
Number of points in a frame before background filtering (NB)	17947								
Number of points in the frame after background filtering (NA)	3220	452	142	123	117	108	72	69	69
Percentage of background left (NA/NB)	17.94%	2.52%	0.79%	0.69%	0.65%	0.60%	0.40%	0.38%	0.38%
Time used to generate the 3D matrix (min)	8	9	10	12	14	17	19	24	30

Table 2. Background Filtering using Different Side Lengths of the Cube

	Side length of the cube			
	0.05 m	0.1 m	0.5 m	1 m
Number of background points before background filtering (NB)	17944			
Number of vehicle points before background filtering (NVB)	36			
Number of background points after background filtering (NA)	103	42	1	1
Percentage of background left (NA/NB)	0.57%	0.23%	0.005%	0.005%
Number of vehicle points after background filtering (NVA)	34	32	0	0
Percentage of vehicle points excluded (NVA-NVB)/NVB	5.6%	11.1%	100%	100%
Time used to generate the 3D matrix (min)	46	6.2	2.2	2.2

As shown in Table 1, with the number of aggregated frames higher than 3000, the missed background points did not change much but the time used for calculation dramatically increased. So the recommended number of aggregated frames is between 1500 and 3000. If the background analysis time is not critical for an application, 3500 aggregation frames are suggested.

Points Statistics

After frame aggregation, the 3D space was then chopped into small continuous cubes. Each cube can be a background space or not. A corresponding 3D matrix was built to represent the whole space so that each matrix element represented a cube to record the number of aggregated points in the cube. The critical parameter for cube division is the side length of the cube, which influences the number of rows and columns of the 3D matrix. A short side length can increase the time used to build the 3D matrix whereas a long side length may report non-background points as background by mistake. The influence of side length on the performance of the algorithm was then explored. The results of filtering a sample frame are illustrated in Table 2.

As shown in Table 2, the processing time for generating the 3D matrix increased with the shorter side length. When the side length was long (0.5 m or 1 m), though almost all background can be filtered, all vehicle points were also excluded from the space. When the side length was changed from 0.1 m to 0.05 m, the number of vehicle points did not change too much but the number of background points left increased a lot (NA rose from 42 to 103) and the time cost increased by almost 40 min. Therefore, the recommended side length of the cube is 0.1 m.

Threshold (TD) Learning

The density of aggregated 3D points in each cube is calculated in this step. In general, the density of background space should be higher compared with the density of space with moving vehicles or pedestrians. A threshold of point density of the cubes (TD) then needs to be determined to distinguish background and non-background cubes. A low TD may take slow-moving vehicles or pedestrians as background whereas a high TD may not exclude the background far away from the LiDAR. A method to learn TD automatically is introduced as follows.

The point density of the same object scanned by the LiDAR varies with its distance from the LiDAR. In general, the number of points decreases if the distance between the object and the LiDAR sensor increases.

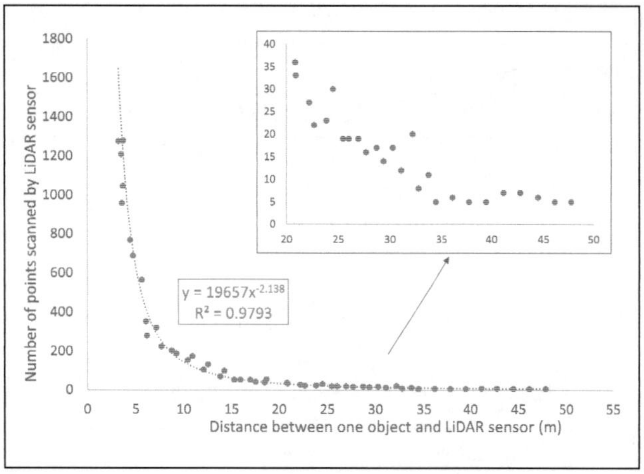

Figure 3. Distribution of vehicle points in LiDAR data.

Figure 3 shows an example of the change of the number of one (vehicle) object's points when it is within different distances from the LiDAR. The distance in Figure 3 is from the center of vehicle points to the LiDAR sensor.

As can be seen in Figure 3, the number of points decreases dramatically with the distance between vehicles and LiDAR sensor increasing. The relation between range and the number of points followed a power function (as shown in Figure 3), with an R^2 of 0.9793. So when background objects are at different distance from the LiDAR sensor, the background point densities are also different, which means the TD should vary in different detection ranges. The whole scan range was then divided into six parts for background filtering based on distance to the LiDAR sensor: 0–5 m, 5–10 m, 10–20 m, 20–30 m, 30–40 m, longer than 40 m. This study analyzed the point-density distribution in each cube in two scenarios: 1. there were pedestrians and vehicles in the LiDAR data; 2. there were no moving objects (pedestrians and vehicles) in the LiDAR data. Figure 4 shows the frequency of point density that is 5 m away from the LiDAR in different scenarios with 2000 frames aggregated.

The data in Figure 4*a* were collected during daytime, and the data in Figure 4*b* were collected at midnight. With pedestrians and vehicles in a frame (Figure 4*a*), the low-density cubes are greater in number than the low-density cubes in the frame without pedestrians and vehicles. The moving vehicles and pedestrians generate low-density cubes. In the background frame, the numbers of cubes with different densities are consistent whereas the numbers of low-density cubes of pedestrians and vehicles change consistently. Therefore the threshold of identifying background cubes is determined by

$$\text{Slope} = \frac{F_i - F_{i-1}}{N_i - N_{i-1}} \quad (1)$$

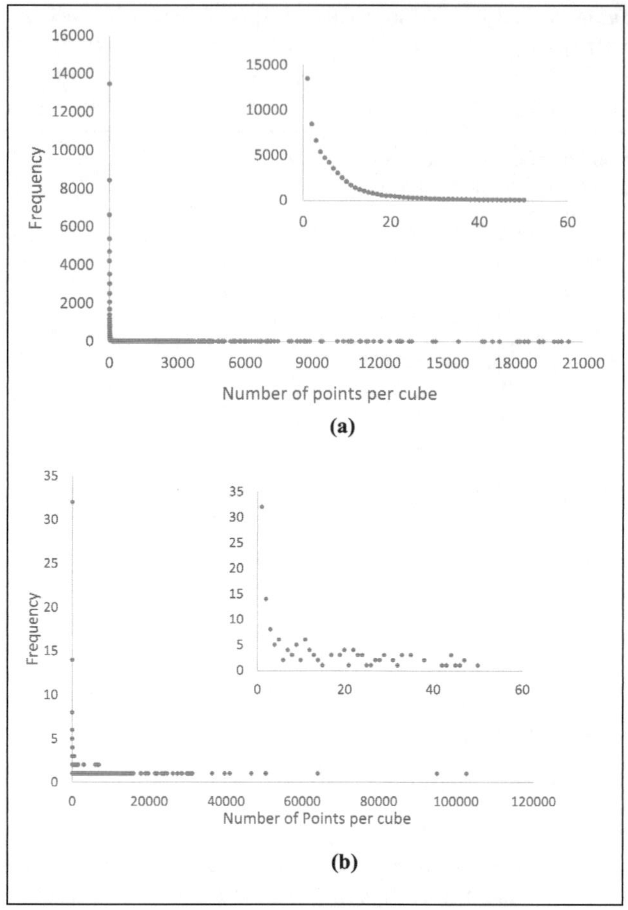

Figure 4. Frequency distribution of cubes with different numbers of points: (*a*) 0–5 m (with pedestrians or vehicles) and (*b*) 0–5 m (no pedestrians or vehicles).

where

N_i is the ith number of points per cube (from lowest to highest), and

F is the frequency of ith number of points per cube.

When the slope first becomes 0 or positive, the frequency of number of points per cube in Equation 1 (F) is used as TD.

After identifying the background cubes with the determined threshold, the background positions are saved in the 3D matrix. For real-time data processing, any LiDAR point in the background cubes is identified as a background point and excluded.

Case Study

Data Collection Sites

The LiDAR data were collected at six different sites in Reno, Nevada. The first place is the intersection of North Virginia Street and 15th Street, and the data were

Table 3. Information from Testing Sites

Location	Frames aggregated for background filtering	Hz	AADT	Speed limit (mph)	Data collection time
North Virginia Street at 15th Street	3000	10	11000	25	Night time: no pedestrians or vehicles
North Virginia Street at 10th Street	1500	5	11000	25	Peak hour: high volume of pedestrians and vehicles
Evans Avenue at Enterprise Road	2500	10	3500	25	Non-peak hour
Parking plot	3000	10	NA	5	Non-peak hour
Kietzke Lane	3000	10	22000	45	Peak hour: high vehicle volume
I-80 in Elko, Nevada	3500	10	11800	80	Non-peak hour

Note: Hz = hertz; AADT = annual average daily traffic; NA = not applicable.

(a) (b)

Figure 5. Different installation methods of roadside LiDAR sensors, showing (*a*) permanent installation and (*b*) temporary installation.

collected during night time without vehicles and pedestrians. The second site is North Virginia Street at 10th Street, and the data were collected during the afternoon peak with high vehicle/pedestrian volume. The third place is Evans Avenue at Enterprise Road, and the data were collected during a non-peak hour in daytime. The fourth site is a parking plot at the University of Nevada, Reno, and the data were collected during a non-peak hour in daytime. The fifth location is near the T intersection of Kietzke Lane and Miami Way. Kietzke Lane is a two-way arterial road with a 45 mph speed limit, serving the major north-south urban traffic in the east area of Reno. The last site is near a wildlife overpass on the I-80 freeway in the rural area of Elko, Nevada. The speed limit of this I-80 segment was 80 mph. Table 3 summarizes the data collection information at different sites.

The LiDAR sensor at the intersection of North Virginia Street and 15th Street was installed on the top of a pedestrian signal, as shown in Figure 5*a*. Data at other sites were collected using a VLP-16 LiDAR mounted on a tripod for short-term data collection, as shown in Figure 5*b*.

Before and After Study

Table 4 shows the TD of different distances to the sensor at all the sites. The TDs were determined by the algorithm given above.

Two examples of LiDAR data in the parking plot and at North Virginia Street and10th Street were processed with the developed background filtering method. The before–after comparison is presented in Figure 6.

Table 4. TD for Different Sites

Distance to LiDAR sensor	North Virginia Street at 15th Street	North Virginia Street at 10th Street	Evans Avenue at Enterprise Road	Parking plot	Kietzke Lane	I-80
0–5 m	4	29	24	20	19	18
5–10 m	3	26	17	11	10	11
10–20 m	3	17	14	10	7	9
20–30 m	3	10	8	9	7	5
30–40 m	3	13	8	9	7	5
40+ m	3	12	4	9	7	5

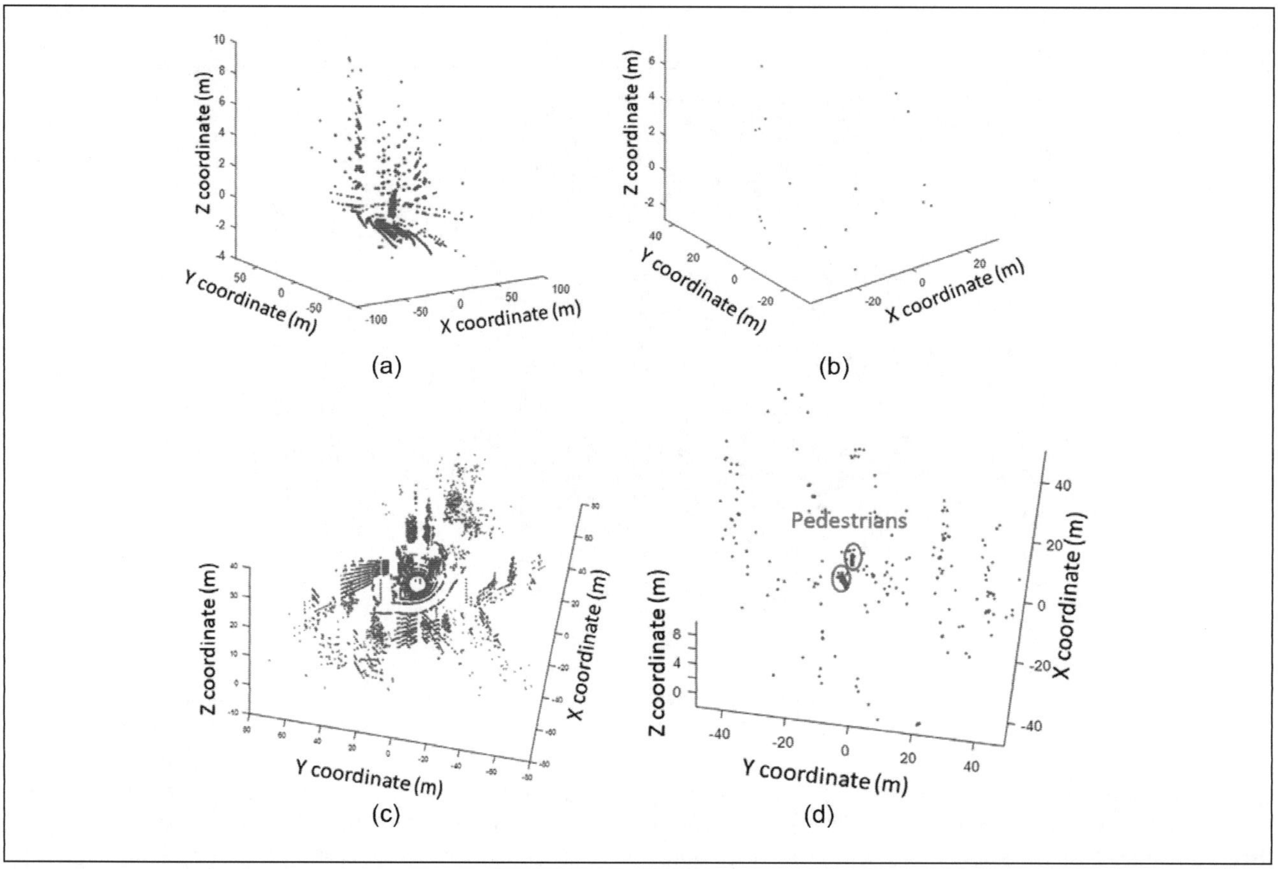

Figure 6. Before and after background filtering: (*a*) before: parking plot; (*b*) after: parking plot; (*c*) before: North Virginia Street at 10th Street; (*d*) after: North Virginia Street at 10th Street.

Figure 6 shows the performance of the algorithm. The algorithm filtered the background well (Figure 6, *a* and *b*). When the vehicle and pedestrian volume was high, a little bit more noise was left after background filtering (Figure 6, *c* and *d*). The practice also indicated that the accuracy of background filtering was reduced if the aggregated data frames for background identification were collected at peak hours. Therefore, LiDAR data with low/no traffic are recommended for background identification. Table 5 presents a detailed evaluation of the background filtering method with numbers of the background points, vehicle points, and pedestrian points before and after background filtering.

As shown in Table 5, the proposed method filtered more than 99% of background when there were no pedestrians or vehicles (North Virginia Street at 15th Street). Even at the site with high pedestrian and vehicle volume (North Virginia Street at 10th Street), the method

Table 5. Evaluation of Algorithm Performance at Different Sites

		Parking plot	North Virginia Street at 10th Street	Evans Avenue at Enterprise Road	North Virginia Street at 15th Street	Kietzke Lane	I-80
Before	Background points	18316	45132	18621	18035	18217	17162
After	Background points	21	1362	176	25	46	20
Filtering percentage		99.8%	97.0%	99.1%	99.9%	99.7%	99.9%
Before	Vehicle points	NA	507	451	NA	531	651
After	Vehicle points	NA	493	411	NA	506	606
Vehicle excluded percentage		NA	2.8%	8.9%	NA	4.7%	6.9%
Before	Detected vehicle length (ft)	NA	46.1	44.8	NA	43.2	51.6
After	Detected vehicle length (ft)	NA	40.1	42.5	NA	40.0	49.1
Before	Pedestrian points	NA	209	553	NA	NA	NA
After	Pedestrian points	NA	196	532	NA	NA	NA
Pedestrian excluded percentage		NA	6.2%	3.8%	NA	NA	NA

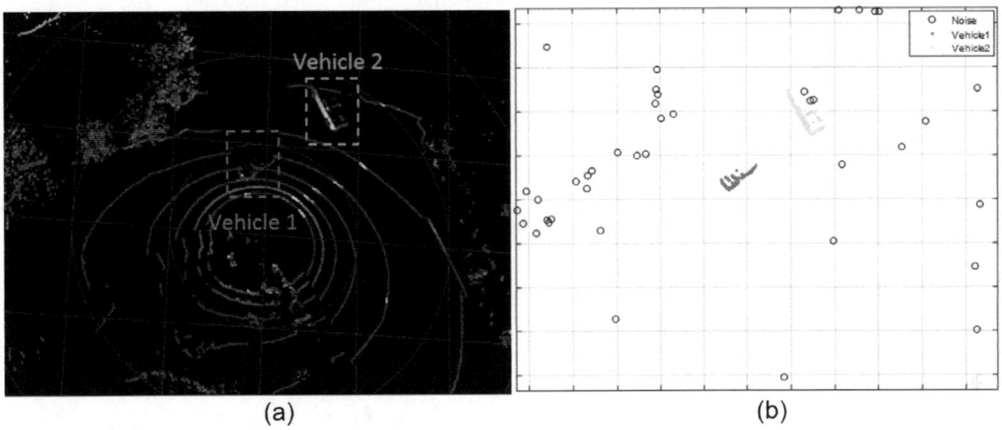

Figure 7. DBSCAN clustering results using LiDAR data after background filtering, showing (*a*) raw LiDAR data and (*b*) clustering result.

still excluded more than 97% of background points. For the data frames with pedestrians and vehicles, background filtering needs to keep the pedestrian and vehicle shapes. The vehicle length was calculated by manually selecting the front point and back point of the vehicle. The results in Table 5 show that the vehicle shapes were kept well after background filtering. The sites included in this case study have different function classes, geometries, and speed limits (from 5 mph to 80 mph). The results show that the 3D-DSF can filter the background very well at all those sites. Though there were still some noise points left after the background filtering, the following data processing proved that the remaining noise points do not affect the performance of clustering methods such as k-means clustering or density-based spatial clustering of applications with noise (DBSCAN). An example of the follow-up vehicle clustering is shown in Figure 7.

To get the real-time HRMTD, the time cost of background filtering needs to be as low as possible. The algorithm has been implemented in Matlab and was deployed on a Dell desktop equipped with an Intel Core i7-4790 CPU (3.60GHz) and 16 GB of RAM. After obtaining the 3D matrix, the time used to exclude background is about 100 ms, which can meet the requirement of real-time data processing.

Conclusion

This paper introduces a process to extract background of roadside LiDAR data automatically and accurately. This research is an essential component of the system applying roadside LiDAR sensors to serve connected vehicles and other traffic engineering practices. The 3D-DSF method developed in this study can filter both static background and vibrating background efficiently. This algorithm can learn the threshold of cube-point-density (TD) automatically. Only two parameters are needed: the number of aggregated frames and the side length of cubes. This paper also provided the recommended values for these two parameters based on the sensitivity analysis. The case study results proved that the 3D-DSF method serves well for background filtering in different road situations (high traffic volume; night time; different speed limits). After obtaining the 3D matrix representing the location of background points, only the third step is needed for real-time data processing, which can significantly reduce the time cost.

The obtained 4,400 naturalistic driving study (NDS) trips provide rich and detailed data for driving cycle development, but they could not meet the data requirement of all combinations of road properties and levels of service (LOS). Therefore, driving cycle gaps still exist. The 4,400 trips are only part of the Strategic Highway Research Program (SHRP) 2 NDS dataset, which was limited by the budget for the data service fee. The whole SHRP 2 NDS data can be used to generate an actual comprehensive database of driving cycles.

It should be noted that it is difficult to exclude all background points without reducing vehicle and pedestrian points. Though there was still some noise left after applying 3D-DSF, the clustering results showed that the remaining noise did not affect the object identification. In future research, it is expected that other machine learning or artificial intelligence methods will be incorporated into the 3D-DSF to further improve the accuracy. For redressing the background, it is recommended to run the algorithm backstage every several hours to update any changes of environment, which can further improve the vehicle clustering accuracy. This paper did not evaluate the effectiveness of the algorithm for processing the data under extreme weather conditions, such as heavy snow, which will also be analyzed in further studies.

Acknowledgments

This research was funded by the SOLARIS Institute, a Tier 1 University Transportation Center (UTC) under Grant No. DTRT13-G-UTC55 and matching funds by the Nevada Department of Transportation (NDOT) under Grant No. P224-14-803/TO #13. The authors gratefully acknowledge this financial support. This research was also supported by engineers with the Nevada Department of Transportation, the Regional Transportation Commission of Washoe County, Nevada, and the City of Reno.

Author Contributions

The authors confirm contribution to the paper as follows: study conception and design: Jianqing Wu, Hao Xu, Jianying Zheng; data collection: Yuan Sun, Rui Yue; analysis and interpretation of results: Jianqing Wu, Yuan Sun, Jianying Zheng; draft manuscript preparation: Hao Xu. All authors reviewed the results and approved the final version of the manuscript.

References

1. Lu, N., N. Cheng, N. Zhang, X. Shen, and J. W. Mark. Connected Vehicles: Solutions and Challenges. *IEEE Internet of Things Journal*, Vol. 1, No. 4, 2014, pp. 289–299.
2. Harding, J., G. Powell, R. Yoon, J. Fikentscher, C. Doyle, D. Sade, M. Lukuc, J. Simons, and J. Wang. *Vehicle-to-Vehicle Communications: Readiness of V2V Technology for Application.* Report No. DOT HS 812 014. U.S. Department of Transportation, National Highway Traffic Safety Administration, Washington, D.C., 2014.
3. Wu, J., H. Xu, and J. Zheng. Automatic Background Filtering and Lane Identification with Roadside LiDAR Data. *Proc., IEEE 20th International Conference on Intelligent Transportation Systems (ITSC)*, Yokohama, Japan, IEEE, New York, 2017, pp. 1–6.
4. Reina, G., J. Underwood, G. Brooker, and H. Durrant-Whyte. Radar-Based Perception for Autonomous Outdoor Vehicles. *Journal of Field Robotics*, Vol. 28, No. 6, 2011, pp. 894–913.
5. Levinson, J., J. Askeland, J. Becker, J. Dolson, D. Held, S. Kammel, J. Z. Kolter, D. Langer, O. Pink, V. Pratt, and M. Sokolsky. Towards Fully Autonomous Driving: Systems and Algorithms. *Proc., Intelligent Vehicles Symposium (IV)*, Baden-Baden, Germany, IEEE, New York, 2011, pp. 163–168.
6. Zheng, J., Y. Wang, N. Nihan, and M. Hallenbeck. Extracting Roadway Background Image: Mode-Based Approach. *Transportation Research Record: Journal of the Transportation Research Board*, 2006. 1944: 82–88.
7. Zhang, G., R. Avery, and Y. Wang. Video-Based Vehicle Detection and Classification System for Real-Time Traffic Data Collection Using Uncalibrated Video Cameras. *Transportation Research Record: Journal of the Transportation Research Board*, 2007. 1993: 138–147.
8. Brackstone, M., M. McDonald, and B. Sultan. Dynamic Behavioral Data Collection Using an Instrumented Vehicle. *Transportation Research Record: Journal of the Transportation Research Board*, 1999. 1689: 9–16.
9. Wang, D. Z., I. Posner, and P. Newman. What Could Move? Finding Cars, Pedestrians and Bicyclists in 3d Laser Data. *Proc., IEEE International Conference on Robotics and Automation (ICRA)*, IEEE, New York, 2012, pp. 4038–4044.
10. Stauffer, C., and W. E. L. Grimson. Learning Patterns of Activity Using Real-Time Tracking. *IEEE Transactions on Pattern Analysis and Machine Intelligence*, Vol. 22, No. 8, 2000, pp. 747–757.

The Standing Committee on Artificial Intelligence and Advanced Computing Applications (ABJ70) peer-reviewed this paper (18-00286).

TRR
JOURNAL OF THE TRANSPORTATION RESEARCH BOARD

Article

Transportation Research Record
2018, Vol. 2672(45) 115–123

DOI: 10.1177/0361198118776139
journals.sagepub.com/home/trr
SAGE

Network Scale Travel Time Prediction using Deep Learning

Yi Hou[1] and Praveen Edara[2]

Abstract

In recent years, deep learning models have been receiving increased attention within the artificial intelligence (AI) community because of their high prediction accuracy. In this paper, two deep learning models, long short-term memory (LSTM) and convolutional neural network (CNN), are proposed to predict travel time in a road network. One major advantage of using deep learning for travel time prediction is that it can make accurate predictions for all the segments in the transportation network with a single model structure, instead of building customized models for each segment separately. The proposed models were evaluated on a transportation network in the City of Saint Louis, Missouri. The prediction results show that deep learning can provide accurate prediction for both congested and uncongested traffic conditions, and can successfully capture the traffic dynamics of unexpected incidents or special events. The study findings show that deep learning offers a promising approach to real-time prediction of travel times on a network scale.

Travel time is an important performance measure used by travelers and traffic management agencies alike. Because of its importance, accurate travel time prediction has been a popular research topic in transportation for many years. A wide range of approaches have been proposed in the literature, including time series models (*1–4*), Kalman filter (*5–7*), historical average and smoothing (*8–10*), statistical regression (*11–13*), and traffic flow theory based methods (*14–16*). Recently, machine learning algorithms became a promising approach in the field of traffic prediction. Successful applications include artificial neural network (ANN) (*17–22*), support vector machine (SVM) (*23–25*), Bayesian (*26*, *27*), and k-nearest neighbors (KNN) (*28*, *29*). As compared with traditional statistical modeling methods, machine learning techniques do not require detailed mathematical forms and assumptions on error distributions; as a result, machine learning is capable of capturing the underlying relationships among different variables in an environment of uncertainty even when they are not easily apparent.

Recently, one of the branches of machine learning, deep learning, has achieved great success in the domain of computer vision, natural language processing, and speech recognition. Traffic data has grown exponentially for the last decade because of the fast development of GPS, telecommunication, and sensor technology. The increasing availability of GPS and sensor data has allowed researchers to examine traffic conditions in a road network as a whole. However, it is still challenging for traditional data-driven methods to predict travel time of all the segments in the network simultaneously by taking all the traffic data of the network as input. This is because of the curse of dimensionality: the required training data increase exponentially as input dimensions grow. The emergence of deep learning offered a remedy. Deep learning is especially suitable for dealing with large amounts of data in a high-dimensional space.

However, applications of deep learning approaches in traffic prediction are limited. Polson and Sokolov introduced a deep learning structure combining a linear model and a sequence of tanh layers to traffic flow prediction (*30*). Huang et al. used deep belief networks (DBN) to predict traffic flow (*31*). Ma et al. developed an RBM–RNN model that combines deep restricted Boltzmann machines (RBM) with a recurrent neural network (RNN) for congestion prediction (*32*). Lv et al. proposed a deep learning–based traffic flow prediction model with traffic features learned by stacked autoencoder model (*33*). Duan et al. applied deep learning to impute the missing traffic data collected by loop detectors (*34*). Ma et al. employed a long short-term memory (LSTM) for traffic speed prediction using remote

[1]National Renewable Energy Laboratory, Golden, CO
[2]University of Missouri-Columbia, Columbia, MO

Corresponding Author:
Address correspondence to Yi Hou: yi.hou@nrel.gov

microwave sensor data (35). Ma et al. proposed a convolutional neural network (CNN) for network-wide speed prediction (36). Spatial-temporal traffic data was converted to images, and then CNN was used to learn traffic as images.

In previous studies, deep learning has demonstrated its superiority over other methods in traffic prediction. To the best of the authors' knowledge, no study applied deep learning methods to predict travel time over several segments in a network simultaneously. This paper proposes two deep learning algorithms, LSTM and CNN, for network scale travel time prediction. These two algorithms have the capability to deal with input data with high dimensionality, making them suitable for network scale travel time prediction. LSTM is a special RNN architecture designed to learn time series data with long time spans and high dimensions. It is able to overcome the issue facing a traditional RNN algorithm, that is, to capture the long temporal dependency for input sequences. CNN is an effective image recognition algorithm that has been successfully applied in computer vision. The convolutional layers of a CNN not only enable CNN to capture the local feature of the image data, but also reduce model parameters that need to be estimated. A time–space travel time matrix can be constructed by spatial–temporal travel time data. Then, CNN is able to learn traffic as images and make predictions.

The remainder of the paper is organized as follows. It begins with a detailed description of the fundamental theory of LSTM and CNN, followed by a section describing the application of LSTM and CNN to travel time prediction. The performance of the deep learning algorithms over a case study of a portion of a downtown road network in the Saint Louis region of Missouri is then presented. The results section reports the prediction accuracy for the two algorithms and compares it with other machine learning algorithms from literature. The paper concludes with a discussion of the advantages of deep learning algorithms and directions for future research.

Methodology

Long Short-Term Memory (LSTM)

LSTM is a special RNN architecture that was first introduced by Hochreiter and Schmidhuber (37). A typical RNN structure is displayed in Figure 1. A neural network layer A looks at the input at time t, x_t, and outputs a value h_t, with a loop allowing temporal sequence information to be passed down from one step to the next. The internal loop of RNN can also be seen as multiple copies of the same network, each passing some information to its successor with output value h_T in the end. In the context of network scale travel time prediction, x_t can be considered as an array of travel times on a total of N segments in the network at time t, $\{TT_{1t}, TT_{2t}, \ldots, TT_{Nt}\}$, and output h_t is an array of predicted travel times on all segments at time $t+1$, $\left\{\widehat{TT}_{1(t+1)}, \widehat{TT}_{2(t+1)}, \ldots, \widehat{TT}_{N(t+1)}\right\}$. Then, the final output h_T is an array of predicted travel times on all segments at time $T+1$, $\left\{\widehat{TT}_{1(T+1)}, \widehat{TT}_{2(T+1)}, \ldots, \widehat{TT}_{N(T+1)}\right\}$.

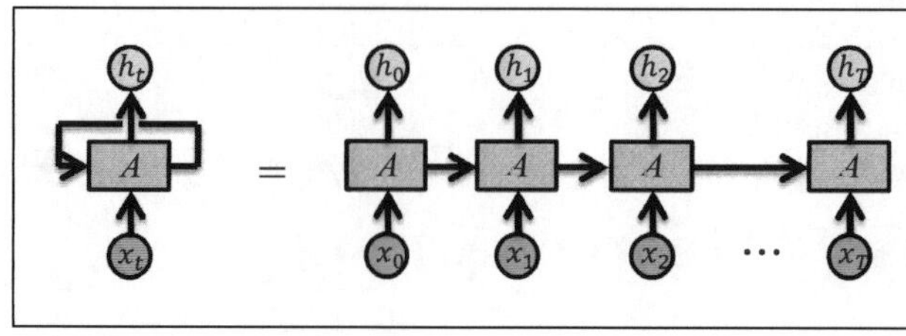

Figure 1. Structure of RNN.

LSTM has the same chainlike structure as RNN, but the repeating module has a different structure. Instead of a single neural network layer, LSTM has four layers interacting in a special manner. The repeating module of LSTM is presented in Figure 2. The calculation steps in an LSTM repeating module are performed as follows.

1. A sigmoid layer σ called the "forget gate layer" makes a decision on what information to throw away from the cell state. It looks at the output of previous step $h_{t-1}$and current input x_t and outputs a binary decision for each number in the previous cell state C_{t-1}. Let $[W_f, b_f]$ denote the weights of the "forget gate layer". The output f_t is calculated as

$$f_t = \sigma(W_f.[h_{t-1}, x_t] + b_f) \tag{1}$$

2. LSTM decides what new information to store in the cell state. A sigmoid layer σ called the "input gate layer" decides which values to update. Then, a tanh layer outputs an array of new candidate values, $\tilde{C}_t$ to add to the previous cell state C_{t-1}. The tanh function is a rescaled version of the sigmoid function with output range of $[-1,1]$ instead of $[0,1]$. Let $[W_i, b_i]$ and i_t denote the weights and output of the "input gate layer". The process in step 2 is calculated as

$$i_t = \sigma(W_i.[h_{t-1}, x_t] + b_i) \tag{2}$$

$$\tilde{C}_t = \tanh(W_C.[h_{t-1}, x_t] + b_C) \tag{3}$$

3. The previous cell state C_{t-1} is updated by

$$C_t = f_t{*}C_{t-1} + i_t{*}\tilde{C}_t \tag{4}$$

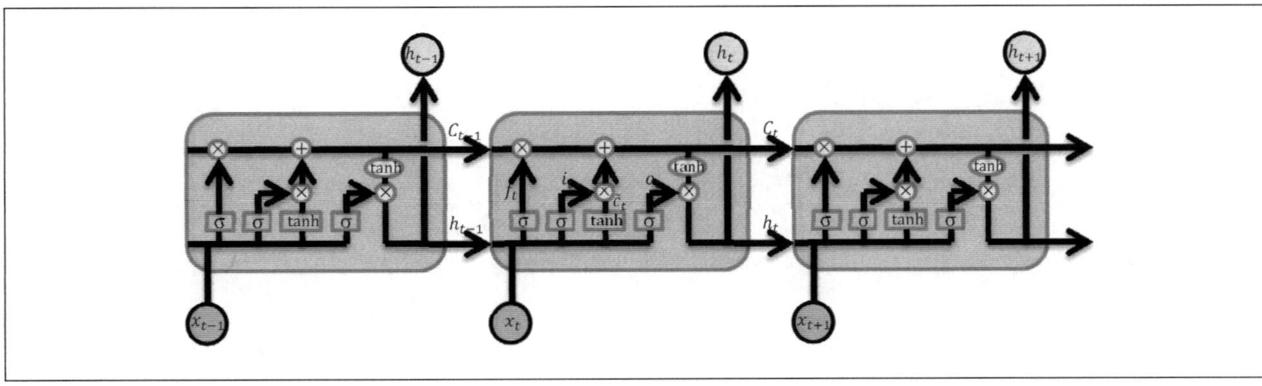

Figure 2. Repeating module in an LSTM.

4. Finally, LSTM computes the output. First, a sigmoid layer σ called the "output gate layer" decides what part of the cell state to output. Next, the cell state is passed through a tanh layer and multiplied by the output of the "output gate layer", o_t. The mathematical form of the calculations is

$$o_t = \sigma(W_o.[h_{t-1}, x_t] + b_o) \quad (5)$$

$$h_t = o_t * \tanh(C_t) \quad (6)$$

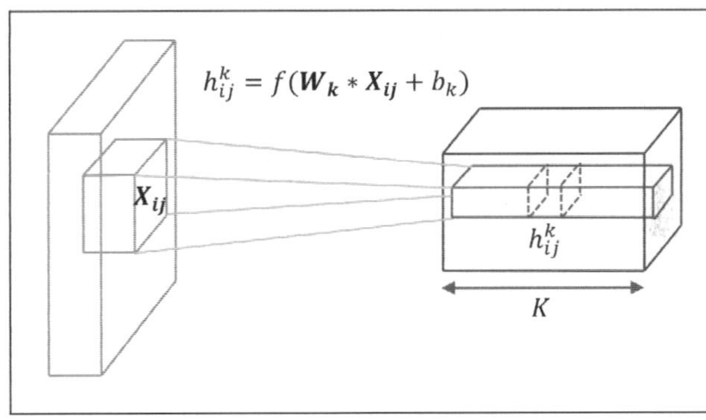

Figure 3. Convolutional layer.

Convolutional Neural Network (CNN)

Lecun and Bengio (*38*) used CNN architecture to design LeNet -5 model for character recognition. Modern convolutional neural networks such as AlexNet (*39*) achieved astounding performance in image recognition and won the ImageNet object recognition challenge in 2012. CNN is very similar to ordinary neural networks except for the different hidden layer structures. Other than fully-connected layers in ordinary neural networks, CNN contains two more main types of layers: convolutional layers and pooling layers. The convolutional layer is the core building block of a CNN and does most of the computational analysis. For each convolutional layer, there is a filter or kernel which is an array of weights or parameters. The filter slides across the width and height of an image and computes dot products between the weights in the filter and a small patch of the image locally. After sliding the filter over all locations, it outputs a new image, or feature maps with a new width, height, and depth. An example of a convolutional layer is shown in Figure 3. At a given layer with depth of K, the kth feature map is denoted by h^k. The filter is determined by the weights $\{\boldsymbol{W_k}, k = 1, 2, \ldots, K\}$ and biases $\{b_k, k = 1, 2, \ldots, K\}$, where $\boldsymbol{W_k}$ is a three dimensional matrix and b_k is a single numeric value. Let $\boldsymbol{X_{ij}}$, a three dimensional matrix, denote the corresponding patch when the filter slides to the location of (i, j) on the input image, and $f(.)$ denote the activation function. The feature map h^k is obtained by

$$h_{ij}^k = f\left(\boldsymbol{W_k} * \boldsymbol{X_{ij}} + b_k\right) \quad (7)$$

It is common to insert a pooling layer between successive convolutional layers in CNN architecture. The main purpose of the pooling layer is to reduce the spatial size of the image and the number of parameters in the network. The most common pooling layer is max pooling. For every point on the image, it computes the maximum value of a small neighborhood around that point.

In this study, the network travel time can be learned as an image by CNN. The spatiotemporal travel time information of a network can be represented by this time–space matrix

$$\begin{bmatrix} TT_{11} & TT_{12} & \ldots & TT_{1N} \\ TT_{21} & TT_{22} & \ldots & TT_{2N} \\ \vdots & \vdots & \ldots & \vdots \\ TT_{T1} & TT_{T2} & \ldots & TT_{TN} \end{bmatrix} \quad (8)$$

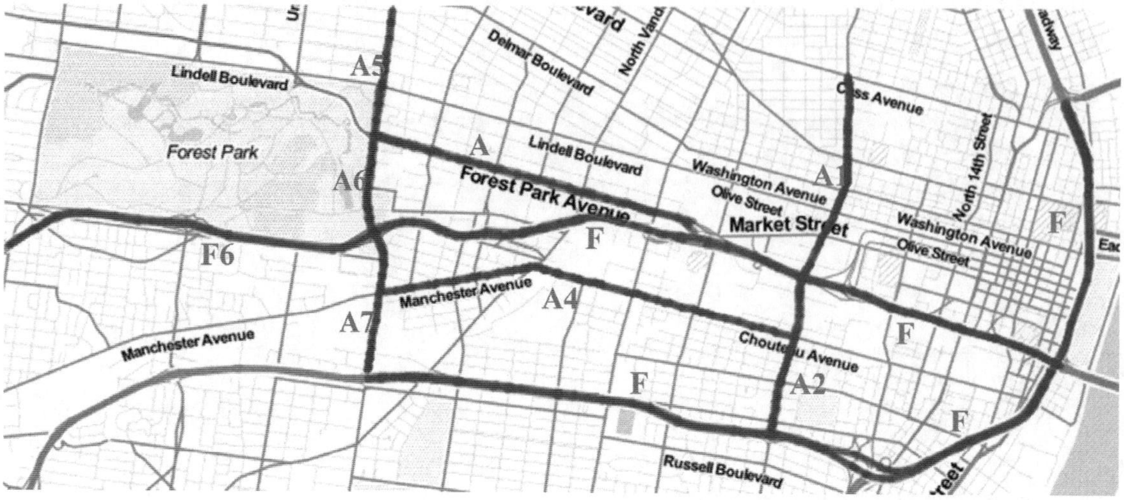

Figure 4. Traffic network of study.

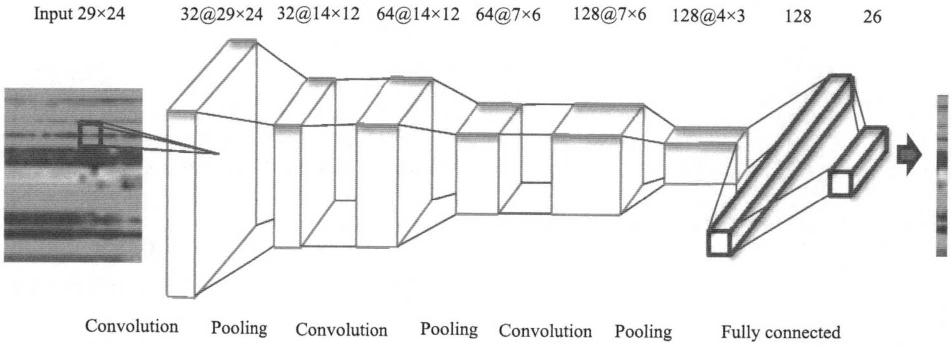

Figure 5. Structure of CNN.

where T is the length of time intervals, N is the number of segments or links in the network, and TT_{ij} is the travel time on segment j at the i th time interval. The time–space matrix can be seen as an image of size $T \times N$ with one color channel.

Model Development

Data and Case Study

The probe-based travel time data were collected in the Saint Louis region and are accessible via RITIS (Regional Integrated Transportation Information System) and Nokia HERE. Two years of travel time data with 2-min resolution in the Saint Louis region from June 2014 to May 2016 were retrieved for analysis. The data size is about 80GB and the dataset contains about 1.5 billion rows of travel time records.

A network of 13 road segments in downtown Saint Louis was selected for this study. It includes six freeway segments and seven signalized arterial segments. These segments were found to experience a large variation in travel times during the day. As shown in Figure 4, the selected traffic network is composed of the highlighted segments. The freeway segments are denoted by "F" followed by a number. The signalized arterial segments are denoted by "A" followed by a number. The segment length ranges from 0.8 mi to 3.0 mi. The travel time on each segment averages 3.0 min with a standard deviation of 1.5 min. In this study, the travel times of the past twenty-four 15-min intervals (i.e., 6 h) on all road segments in the network are considered as input variables.

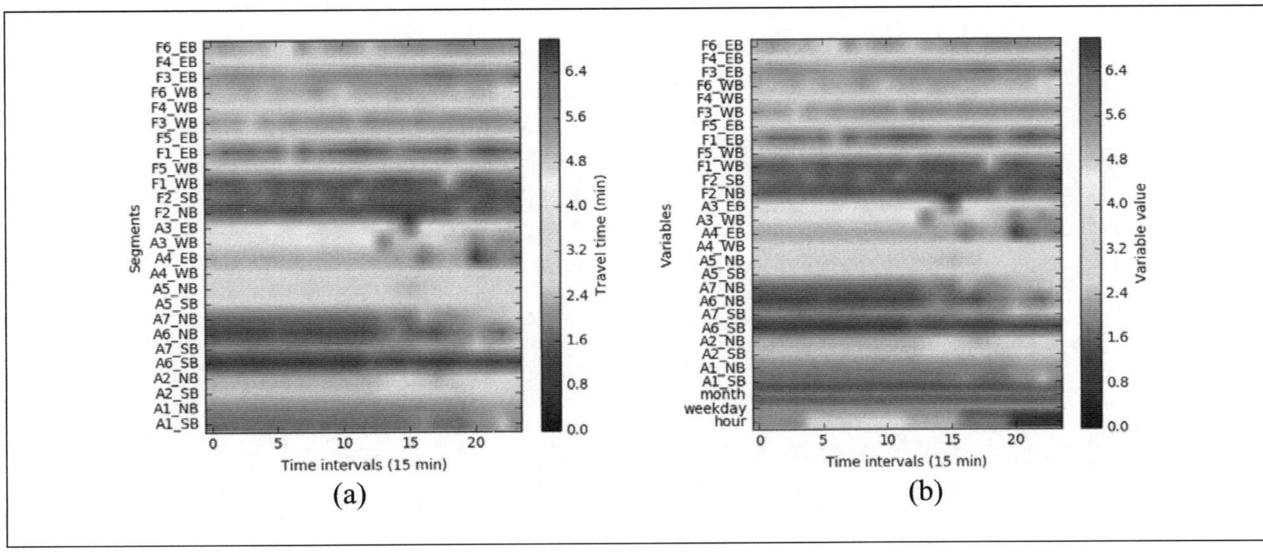

Figure 6. (*a*) Sample image representing spatiotemporal travel time, and (*b*) sample of final input image for CNN.

In addition, three other factors—month of the year, day of the week, and time of day—were also used in the model. The proposed models provide prediction of travel times on all segments in the network 60 min into the future. For example, if the current time is 3:00 p.m., the travel times on all segments in the network recorded in every 15 min from 9:00 a.m. to 3:00 p.m. are considered as input variables to predict the travel times on all segments in the network at 4:00 p.m. After data reduction, a total number of 69,415 observations were extracted from the large dataset; 60,266 of them from June 2014 to February 2016 were used for training and 9,149 of them from March to May 2016 were used to test the models.

There are four hyperparameters that need to be considered when implementing LSTM. They are the number of LSTM layers, the number of hidden neurons in each layer, the number of epochs, and the batch size. For CNN, more hyperparameters need to be considered, including the depth of model structure; hyperparameters concerned with convolutional layers such as kernel size, strides, and activation function type; number of epochs; and batch size. For both LSTM and CNN, 10% of the training data was used for validation in every epoch of the training process. An exhaustive trial-and-error process was adopted for tuning hyperparameters for both models. In this study, LSTM with one LSTM layer of 128 hidden neurons, 20 epochs, and batch size of 128 yielded the highest model accuracy. The structure of CNN for this study is shown in Figure 5. It consists of three convolutional layers with filters of size 3 $\times$ 3. For each convolutional layer, a pooling layer with pooling region size of 2 $\times$ 2 follows. At the final stage, two fully-connected layers with 128 and 26 neurons transform the feature maps to travel time predictions. The convolutional layers consecutively transfer the number of feature map depth to 32, 64, and 128. The CNN structure with 30 epochs and batch size of 128 produced the best model performance.

A sample image generated by time–space matrix is shown in Figure 6*a*. The *x*-axis is a time range of the previous 6 h (twenty-four 15-min intervals). The *y*-axis represents the segments of the network. Figure 6*b* shows the final input image by adding more relevant factors such as month, day, and hour as input variables. Such an image contains rich travel time information for a whole network that can be learned by CNN. Ten different alignments of spatial position of the segments in the time–space matrix were tried for CNN training. The results showed that the variation of segment spatial layout does not affect the model accuracy.

Instantaneous travel time was used as a baseline predictor for comparison. Instantaneous travel time uses the most recent travel time as a proxy for the travel time of the future intervals. In addition, two other machine learning models, random forests (RF) and gradient boosting machines (GBM), were also developed for comparison with the deep learning models. Hou et al. provide additional descriptions of the RF and GBM model applications to travel time prediction (*40*). Both spatial and temporal correlations of travel time were considered in RF and GBM. RF and GBM can only output the travel time prediction for each segmentone at a time. Therefore, a group of individual models need to be built for network scale travel time prediction. Feature selection was performed for each RF and GBM model by

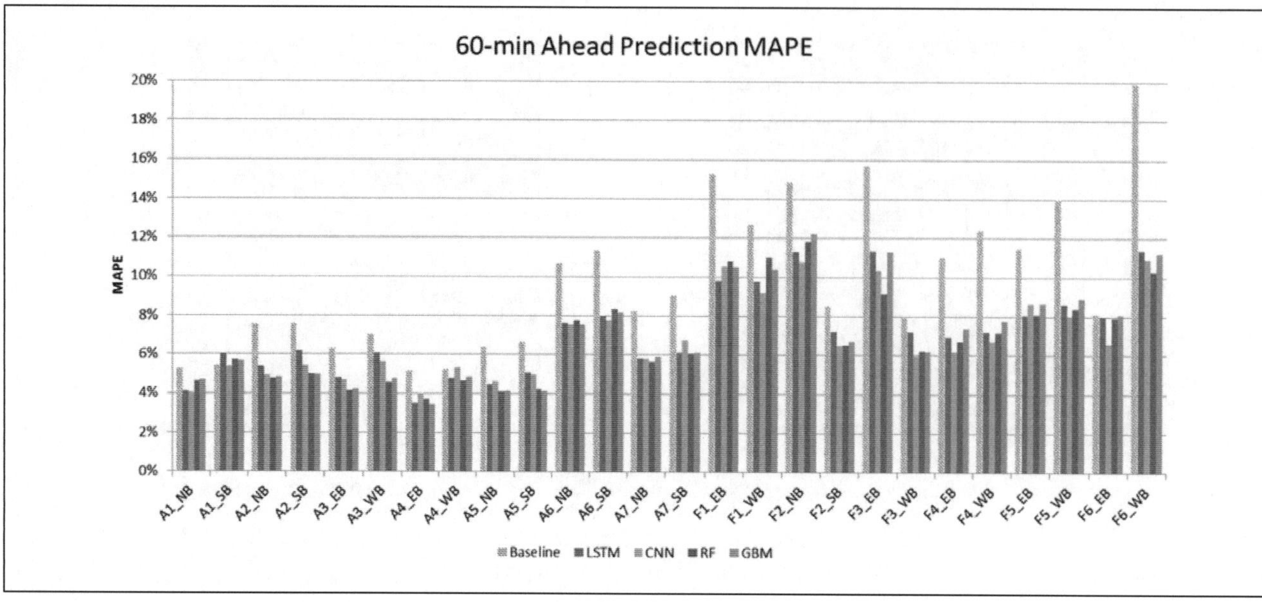

Figure 7. Network scale travel time prediction results for 60-min ahead.

Table 1. 60-min Ahead Travel Time Prediction Results

	MAPE		
Model	Minimum	Mean	SD
Baseline	9.8%	9.80%	0.00%
LSTM	7.0%	7.25%	0.15%
CNN	6.8%	7.09%	0.19%
RF	6.8%	6.83%	0.00%
GBM	7.0%	7.03%	0.00%

Note: SD = Standard deviation.

Table 2. Model Performances for Peak and Off-Peak Hours

Model	Peak	Off-peak
LSTM	10.5%	6.3%
CNN	9.6%	6.3%

calculating correlation. In contrast, LSTM and CNN are able to output network scale travel time prediction in a single model. All models were evaluated using mean absolute percentage error (MAPE), computed as

$$\text{MAPE} = \frac{1}{N}\sum_{i=1}^{N}\left|\frac{t_i - \widehat{t_i}}{t_i}\right| \quad (9)$$

where t_i is the observed travel time, $\widehat{t_i}$ is the predicted or forecasted travel time, and N is the total number of observations.

Results and Comparison

To test both model prediction accuracy and reliability, all methods were trained for 10 times using the already fine-tuned optimal hyperparameters. The prediction results for test data are displayed in Table 1. It shows the minimum, mean, and standard deviation of MAPE across the 10 trained models for each method. It indicates that the minimum MAPE measures of 60-min ahead travel time prediction for LSTM and CNN are 7.0% and 6.8%, as opposed to baseline MAPE of 9.8%. The prediction results for LSTM and CNN are very close to RF and GBM, the MAPE measures for which are 6.8% and 7.0%. However, both LSTM and CNN have higher standard deviation of MAPE than RF and GBM. The larger randomness in deep learning results is expected, because there are many more parameters to estimate in the deep learning training process than in RF and GBM and the values of initial parameters were randomly assigned. The model performance of LSTM and CNN during different traffic conditions, such as peak and off-peak hours, is displayed in Table 2. Peak hour is defined as periods from 7:00 to 10:00 a.m. and from 4:00 to 7:00 p.m. Table 2 indicates that the prediction accuracy in off-peak hour is higher than during the peak hour for both models. This is expected because of the larger variation in travel time during peak hour than during off-peak hour.

The prediction MAPE for each segment is displayed in Figure 7. Compared with baseline, the improvement in prediction accuracy made by LSTM and CNN is significant. The maximum decrease of MAPE reaches about

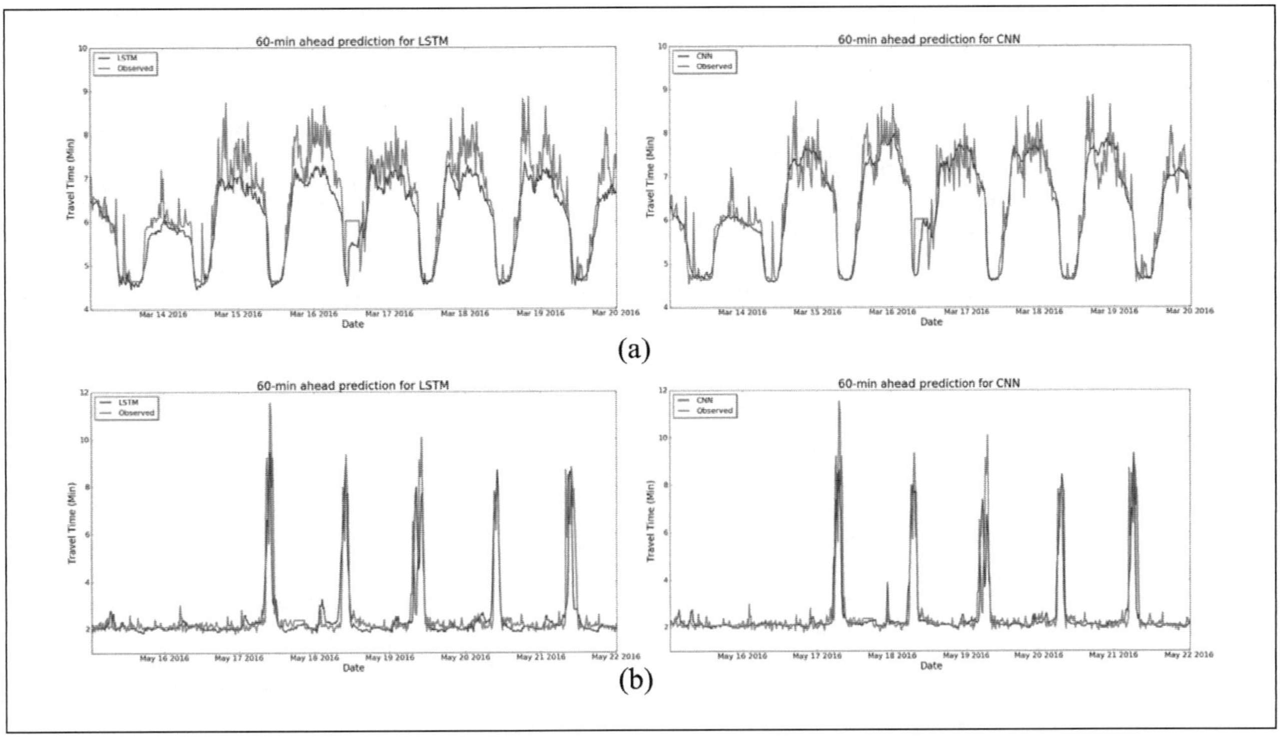

Figure 8. One-week temporal variation of predicted and observed travel time on (*a*) westbound of segment A3 and (*b*) eastbound of segment F3.

10%. For all segments, LSTM and CNN have very similar prediction accuracy to RF and GBM. In general, the difference in MAPE between baseline and deep learning models for freeways was found to be larger than the difference for signalized arterials.

A robust travel time prediction model must perform well under all traffic conditions, congested and uncongested. Thus, the developed deep learning models were tested by predicting the temporal variation of travel time for a whole week. A segment was randomly selected from both freeways and signalized arterials for test. Then, a 1-week time window from Sunday to Saturday, not included in the model training process, was also randomly selected for each test segment. Figure 8 shows the prediction results for LSTM and CNN at 60-min ahead for signalized arterial A3 for the 1-week period from March 13 to March 19, 2016, and freeway segment F3 for the 1-week period from May 15 to May 21, 2016. The blue line indicates predicted travel time. The red line represents observed travel time. Figure 8 shows that LSTM and CNN were effective in predicting travel time accurately under both congested and uncongested traffic conditions. The plotted travel times clearly demonstrate that there are five distinct peak hours during weekdays. LSTM and CNN can satisfactorily capture the variations during peak hours. It also appears that the LSTM and CNN respond to the effects of incidents or other nonrecurring events. For example, as shown in Figure 8*b*, there is a travel time spike on segment F3 on May 18, 2016 (possibly caused by an incident). LSTM and CNN are able to capture the travel time effect.

Table 3. Computation Time

Model	Training	Prediction
LSTM	5.2 min	1.1 s
CNN	29.7 min	4.0 s
RF	26.3 min	10.4 s
GBM	27.1 min	125.3 s

Table 3 presents the computation time for the four models on an Intel Core i7 central processing unit with 32-GB random access memory. LSTM has the lowest training time of 5.2 min as opposed to 29.7 min for CNN, 26.3 min for RF, and 27.1 min for GBM. LSTM also outperformed other models in that its prediction time was 1.1 s.

Conclusion

In this study, deep learning algorithms, LSTM and CNN, were used to predict travel times in a road network. The algorithms were demonstrated using a case study of a 13-segment road network in the Saint Louis downtown area. The models predict travel times up to 60 min into the future on all segments simultaneously. The prediction results showed that LSTM and CNN significantly outperformed the naïve baseline estimates. The results were also compared with two other machine learning models from literature. However, the RF and GBM models had to be trained separately for each segment, that is, 26 separate models had to be developed. The results of LSTM and CNN were similar to RF and GBM models. Both LSTM and CNN learned the temporal variation of travel time during the day and made accurate predictions for both congested (e.g., peak hours) and uncongested conditions. As for computational efficiency, LSTM consumes significantly lower computation time than CNN, RF, and GBM.

There are several advantages that make deep learning a promising method for network scale travel time prediction. Deep learning is especially suitable for modeling with large amounts of data in a high-dimensional space. For example, travel times in a network are time series data with long time spans and high number of dimensions (i.e., large number of segments). As was shown, LSTM needs small computation times to generate the same level of accuracies as other machine learning approaches. Deep learning approaches are scalable. When dealing with a large transportation network with hundreds or thousands of segments, deep learning can save huge amounts of work for training and updating models. With more advanced computation technology such as Graphic Processing Units (GPU), deep learning models of network travel times can be trained even faster. Future research should test the deep learning approaches, especially LSTM, against a larger road network with over 100 segments. Such a network size would require the use of GPUs.

Author Contributions

The authors confirm contribution to the paper as follows: study conception and design: Yi Hou, Praveen Edara; data collection: Praveen Edara; analysis and interpretation of results: Yi Hou; draft manuscript preparation: Yi Hou, Praveen Edara. All authors reviewed the results and approved the final version of the manuscript.

References

1. Lee, S., and D. B. Fambro. Application of Subset Autoregressive Integrated Moving Average Model for Short-Term Freeway Traffic Volume Forecasting. *Transportation Research Record: Journal of the Transportation Research Board*, 1999, 1678: 179–188.
2. Billings, D., and J.-S. Yang. Application of the ARIMA Models to Urban Roadway Travel Time Prediction: A Case Study. *Proc., IEEE International Conference on System, Man and Cybernetics*, Taipei, Taiwan, Vol. 3, IEEE, 2006, pp. 2529–2534.
3. Yang, J.-S. A Study of Travel Time Modeling via Time Series Analysis. *Proc., IEEE Conference on Control Applications*, Toronto, Canada, 2005.
4. Nakata, T., and J. Takeuchi. Mining Traffic Data from Probecar System for Travel Time Prediction. *Proc., International Conference on Knowledge Discovery and Data Mining*, Seattle, WA, 2004.
5. Harvey, A. C. *Forecasting, Structural Time Series Models and the Kalman Filter*. Cambridge University Press, Cambridge, 1990.
6. Chen, M., and I. J. Chien. Dynamic Freeway Travel Time Prediction Using Probe Vehicle Data: Link-Based vs. Path-Based. *Transportation Research Record: Journal of the Transportation Research Board*, 2001. 1768: 157–161.
7. van Lint, J. W. C. Online Learning Solutions for Freeway Travel Time Prediction. *IEEE Transactions on Intelligent Transportation Systems*, Vol. 9, No. 1, 2008, pp. 276–281.
8. Williams, B., P. Durvasula, and D. Brown. Urban Freeway Traffic Flow Prediction: Application of Seasonal Autoregressive Integrated Moving Average and Exponential Smoothing Models. *Transportation Research Record: Journal of the Transportation Research Board*, 1998. 1644: 132–141.
9. Sadabadi, K. F., M. Hamedi, and A. Haghani. Evaluating Moving Average Techniques in Short-Term Travel Time Prediction Using an AVI Data Set. Presented at 89th Annual Meeting Transportation Research Board, Washington, D.C., 2010.
10. Smith, B., and M. Demetsky. Traffic Flow Forecasting: Comparison of Modeling Approaches. *Journal of Transportation Engineering*, Vol. 123, No. 4, 1997, pp. 261–266.
11. Min, W., and L. Wynter. Real-Time Road Traffic Prediction with Spatio-Temporal Correlations. *Transportation Research Part C*, Vol. 19, No. 4, 2011, pp. 606–616.
12. Fei, X., C. C. Lu, and K. Liu. A Bayesian Dynamic Linear Model Approach for Real-Time Short-Term Freeway Travel Time Prediction. *Transportation Research Part C*, Vol. 19, No. 6, 2011, pp. 1306–1318.
13. Li, L., Y. Li, and Z. Li. Efficient Missing Data Imputing for Traffic Flow by Considering Temporal and Spatial Dependence. *Transportation Research Part C*, Vol. 34, 2013, pp. 108–120.
14. Li, L., X. Chen, Z. Li, and L. Zhang. Freeway Travel-Time Estimation Based on Temporal–Spatial Queueing Model. *IEEE Transactions on Intelligent Transportation Systems*, Vol. 14, 2013, pp. 1536–1541.
15. Li, L., X. Chen, and L. Zhang. Multimodel Ensemble for Freeway Traffic State Estimations. *IEEE Transactions on Intelligent Transportation Systems*, Vol. 15, No. 3, 2014, pp. 1323–1336.
16. Yildirimoglu, M., and N. Geroliminis. Experienced Travel Time Prediction for Freeway Systems. *Proc., 15th International IEEE Conference on Intelligent Transportation Systems*, Anchorage, Alaska, 2012.
17. van Lint, J. W. C., S. P. Hoogendoorn, and H. J. van Zuylen. Robust Freeway Travel time Prediction with State-space Neural Networks. Presented at the 13th EWGT

Mini Euro Conference, Polytechnic of Bari, Bari, Italy, 2003.
18. van Lint, J. W. C., S. P. Hoogendoorn, and H. J. van Zuylen. Accurate Travel Time Prediction with State-Space Neural Networks under Missing Data. *Transportation Research Part C*, Vol. 13, No. 5, 2005, pp. 347–369.
19. Krikke, R. Short-Range Travel Time Prediction Using an Artificial Neural Network. *Proc., 9th World Congress on Intelligent Transport Systems*, Chicago, Illinois, 2002.
20. Wei, Y., and M.-C. Chen. Forecasting the Short-Term Metro Passenger Flow with Empirical Mode Decomposition and Neural Networks. *Transportation Research Part C*, Vol. 21, No. 1, 2012, pp. 148–162.
21. van Hinsbergen, C., J. van Lintvan, and H. Zuylen. Bayesian Committee of Neural Networks to Predict Travel Times with Confidence Intervals. *Transportation Research Part C*, Vol. 17, No. 5, 2009, pp. 498–509.
22. Sun, S., R. Huang, and Y. Gao. Network-Scale Traffic Modeling and Forecasting with Graphic Lasso and Neural Networks. *Journal of Transportation Engineering*, Vol. 138, No. 11, 2012, pp. 1358–1367.
23. Wu, C. H., J. M. Ho, and D. T. Lee. Travel-Time Prediction with Support Vector Regression. *IEEE Transactions on Intelligent Transportation Systems*, Vol. 5, No. 4, 2004, pp. 276–281.
24. Wang, J., and Q. Shi. Short-Term Traffic Speed Forecasting Hybrid Model Based on Chaos-Wavelet Analysis-Support Vector Machine Theory. *Transportation Research Part C*, Vol. 27, 2012, pp. 219–232.
25. Hong, W.-C. Traffic Flow Forecasting by Seasonal SVR with Chaotic Simulated Annealing Algorithm. *Neurocomputing*, Vol. 74, No. 12–13, 2011, pp. 2096–2107.
26. Sun, S., C. Zhang, and G. Yu. A Bayesian Network Approach to Traffic Flow Forecasting. *IEEE Transportation Intelligent Transportation System*, Vol. 7, No. 1, 2006, pp. 124–132.
27. van Hinsbergen, C. P. I., and J. W. C. van Lint. Bayesian Combination of Travel Time Prediction Models. Presented at 87th Annual Meeting of the Transportation Research Board, Washington DC, USA, 2008.
28. Clark, S. Traffic Prediction Using Multivariate Nonparametric Regression. *Journal of Transportation Engineering*, Vol. 129, No. 2, 2003, pp. 161–168.
29. Tak, S., S. Kim, and H. Yeo. Travel Time Prediction for Origin-Destination Pairs without Route Specification in Urban Network. *Proc., 17th International IEEE Conference on Intelligent Transportation Systems*, Qingdao, China, 2014.
30. Polson, N., and V. Sokolov. Deep Learning Predictors for Traffic Flows. arXiv, arXiv:1604.04527, 2016.
31. Huang, W., G. Song, H. Hong, and K. Xie. Deep Architecture for Traffic Flow Prediction: Deep Belief Networks with Multitask Learning. *IEEE Transaction Intelligent Transportation System*, Vol. 15, No. 5, 2014, pp. 2191–2201.
32. Ma, X., H. Yu, Y. Wang, and Y. Wang. Large-Scale Transportation Network Congestion Evolution Prediction Using Deep Learning Theory. *PLoS One*, Vol. 10, No. 3, 2015, e0119044.
33. Lv, Y., Y. Duan, W. Kang, Z. Li, and F. Y. Wang. Traffic Flow Prediction with Big Data: A Deep Learning Approach. *IEEE Transaction Intelligent Transportation System*, Vol. 16, No. 2, 2014, pp. 865–873.
34. Duan, Y., Y. Lv, W. Kang, and Y. Zhao. A Deep Learning Based Approach for Traffic Data Imputation. *Proc., 17th International IEEE Conference on Intelligent Transportation Systems*, Qingdao, China, 2014.
35. Ma, X., Z. Tao, Y. Wang, H. Yu, and Y. Wang. Long Short-Term Memory Neural Network for Traffic Speed Prediction Using Remote Microwave Sensor Data. *Transportation Research Part C*, Vol. 54, 2015, pp. 187–197.
36. Ma, X., Z. Dai, Z. He, J. Ma, Y. Wang, and Y. Wang. Learning Traffic as Images: A Deep Convolutional Neural Network for Large-Scale Transportation Network Speed Prediction. *Sensors*, Vol. 17, No. 4, 2017, 818.
37. Hochreiter, S., and J. Schmidhuber. Long Short-Term Memory. *Neural Computation*, Vol. 9, No. 8, 1997, pp.1735–1780.
38. LeCun, Y., and Y. Bengio. Convolutional Networks for Images, Speech, and Time Series. In *The Handbook of Brain Theory and Neural Networks* (M. A. Arbib, ed.), Cambridge, MA, MIT Press, 1998, pp. 255–258.
39. Krizhevsky, A., I. Sutskever, and G. E. Hinton. ImageNet Classification with Deep Convolutional Neural Networks. *Proc., 25th Anual Conference on Neural Information Processing System*, 2012.
40. Hou, Y., P. Edara, and Y. Chang. Road Network State Estimation using Random Forest Ensemble Learning. *Proc., 20th IEEE Intelligent Transportation Systems Conference*, Yokohama, Japan, 2017.

The Standing Committee on Artificial Intelligence and Advanced Computing Applications (ABJ70) peer-reviewed this paper (18-00942).

Article

TRR
JOURNAL OF THE TRANSPORTATION RESEARCH BOARD

Transportation Research Record
2018, Vol. 2672(45) 124–135

DOI: 10.1177/0361198118786824
journals.sagepub.com/home/trr
SAGE

Bayesian Network for Freeway Traffic State Prediction

Ho-Chul Park[1], Dong-Kyu Kim[2], and Seung-Young Kho[3]

Abstract
Traffic state prediction is an important issue in traffic operations. One of the main purposes of traffic operations is to prevent a flow breakdown. Therefore, it is necessary to predict the traffic state in such a way as to reflect the stochastic process of traffic flow. To predict accurately the traffic state, machine learning-based models have been widely adopted, but they have difficulty in obtaining insights for traffic state prediction due to black-box procedures of the models. A Bayesian network (BN) is a methodology that is suitable for dealing with problems that involve uncertainty, and it can also improve the understanding of such problems. In this study, we develop a traffic state prediction model using a BN to reflect the dynamic and stochastic characteristics of traffic flow. To improve the BN, which has been used with a simple structure in transportation problems, we propose a modeling procedure using a mixture of Gaussians (MoGs). In the performance evaluation, the BN has better performance than a logistic regression, and it has the same level of performance as an artificial neural network based on machine learning. Also, by performing sensitivity analyses, we provide the understanding of traffic state prediction and the guidelines for improving the model. The BN developed in this study can be considered as a traffic state prediction model with good prediction accuracy and interpretability.

Traffic state prediction is an important issue in traffic information systems. As the result of extensive research, a model for predicting the traffic state has been developed and used to obtain specific values of traffic variables, such as flow, speed, and density, which can describe the traffic state. These values are useful for drivers and operators of a freeway in terms of traffic information. Traffic state prediction also is important in terms of traffic operations. One of the main purposes of traffic operations is to prevent or delay flow breakdown (transition from the free-flow traffic state to the congested traffic state) because it takes a long time to recover when flow breakdown occurs. It is, therefore, necessary to predict whether the future traffic state will be congested or not and to respond in advance to that predicted future traffic state. However, the predicted specific values of the traffic variables are limited in their ability to clearly predict the traffic state (in particular, when it is near capacity) because traffic flow has intrinsic stochastic and dynamic characteristics. A model that directly predicts the traffic state is, therefore, needed for actual traffic operations. Consequently, the research into traffic state estimation and prediction can be divided into two parts: estimating or predicting traffic variables; and directly estimating or predicting traffic states (e.g., free-flow or congested).

Various modeling approaches have been used to estimate or predict the traffic state. The modeling approaches are distinguished by their parametric approaches, which have a finite number of parameters, and nonparametric approaches. But the number of parameters is potentially infinite. In parametric approaches, the autoregressive integrated moving average (ARIMA) family of models (*1, 2*) and the Kalman filtering models (*3, 4*) have been used to predict traffic variables, and logistic regressions (*5, 6*) were used to predict traffic states directly. The models that use the parametric approach generally have well-established theoretical backgrounds, and transportation researchers have validated these models. Also, in these models, we can interpret the coefficients of variables and obtain insights into a problem based on the functional relationship between

[1]Institute of Construction and Environmental Engineering, Seoul National University, Seoul, South Korea
[2]Department of Civil and Environmental Engineering, Seoul National University, Seoul, South Korea
[3]Department of Civil and Environmental Engineering and Institute of Construction and Environmental Engineering, Seoul National University, Seoul, South Korea

Corresponding Author:
Address correspondence to Dong-Kyu Kim: dongkyukim@snu.ac.kr

the explanatory variable and the dependent variable. However, the models that use the parametric approach have difficulty obtaining good prediction performance when dealing with complex and nonlinear data (e.g., traffic state prediction). It is also difficult to implement large-scale networks or large datasets.

Models that use the nonparametric approach can deal effectively with relatively large-scale networks and large datasets. They can also handle problems with complex and non-linear properties, and many studies have validated their performance. Neural networks (*7*, *8*), nearest neighbors (*9*, *10*), and support vector regression (*11*) have been used in predicting traffic variables. Clustering-based models (*12*, *13*), Markov models (*14*, *15*), neural networks (*7*), and support vector machines (*16*) have also been used in the direct prediction of traffic states. The models that use nonparametric approaches, however, have a limitation on their interpretation as a result of the difficulty in the analysis of the relationships between variables. In neural networks, in particular, black-box procedures are the main disadvantage. Since most of the parameters in the model have no physical meaning related to the targeted problem, it is difficult to obtain insights concerning the traffic mechanisms through the interpretation. As commented by previous literature, data-driven based models, such as nearest neighbors, support vector regression, support vector machines, and other nonparametric models have lower interpretability compared with parametric approach models which are rules-based (*17*). Therefore, we need a traffic state prediction model that overcomes the disadvantages of both approaches.

The Bayesian network (BN), as a probabilistic graphical model, can offer a solution to the issues mentioned above. The BN is suitable for uncertain knowledge or stochastic process problems (e.g., traffic state prediction). The BN combines expert knowledge and a data-driven approach, which allows the prediction of the traffic state with good accuracy. Also, the model's analyses are based on a graphical representation, which provides valuable insights concerning the traffic state prediction, and these insights can be used to establish the guidelines for the development of new models. However, the BN has not been used extensively for transportation problems. A few studies have used the BN for traffic state prediction (*18*), traffic incident detection (*19*), and crash prediction (*20*, *21*). However, most studies only considered a simple cause–effect relationship between variables to construct the structure of the BN. Such structures may guarantee the model's performance, but the interpretation of the model's parameters is limited, and this limits the insights that BNs can provide concerning transportation problems. In previous studies, BNs were developed and evaluated, but there was almost no discussion through model analysis.

The purpose of this study is to develop a probabilistic prediction model that reflects the stochastic process of the traffic state using a BN. To improve the naïve BN structures used in previous studies, we propose a BN structure for traffic state prediction. We also propose analysis measurements to interpret the model quantitatively. The new structure and the analysis measurements provide an interpretable BN model for insights that would help in understanding the traffic state transition mechanism. By comparing the BN with other methodologies, we evaluate the performance of the developed BN. Therefore, the BN developed in this study can be considered as a traffic state prediction model with good prediction accuracy and interpretability.

Data Collection and Preparation

Study Site

The study site was determined by considering the quality of the detector data, the geometry of the freeway, including the on and off ramps, and the congestion patterns. The site of the study is Interstate highway I80-W in the city of Berkeley, Alameda County, California, and the selected area had both spontaneous breakdown and congestion due to a downstream queue. The main data sources were aggregated stationary detectors on the freeway. The data were obtained from the Caltrans Performance Measurement System (PeMS) provided by the California Department of Transportation. The data were aggregated at 30-second intervals, and 57,600 samples were collected over a period of eight weeks, between 1/2/2017 and 2/24/2017, on non-holiday weekdays only. We collected data for traffic flow, speed, and occupancy in the target area as well as upstream and downstream. We also collected data on flow just on the ramps. We removed the data taken during adverse weather conditions. We used 70% of the data for learning and 30% for testing.

Traffic State Identification

Traffic state identification is needed to construct the dataset for supervised learning of the traffic state prediction model. However, there is no consistent method or guidelines for detector location and type of performance measures to identify the traffic state (*22*). Therefore, it was necessary to select a measure based on the fixed detector data to perform reliable traffic state identification.

Most studies have used speed as a primary performance measure to identify the traffic state. In general, speed responds instantaneously to changes in traffic state relative to other indicators (*22*). In previous research, the basic logic of the identification algorithm was to set the speed threshold at which the traffic state changes from

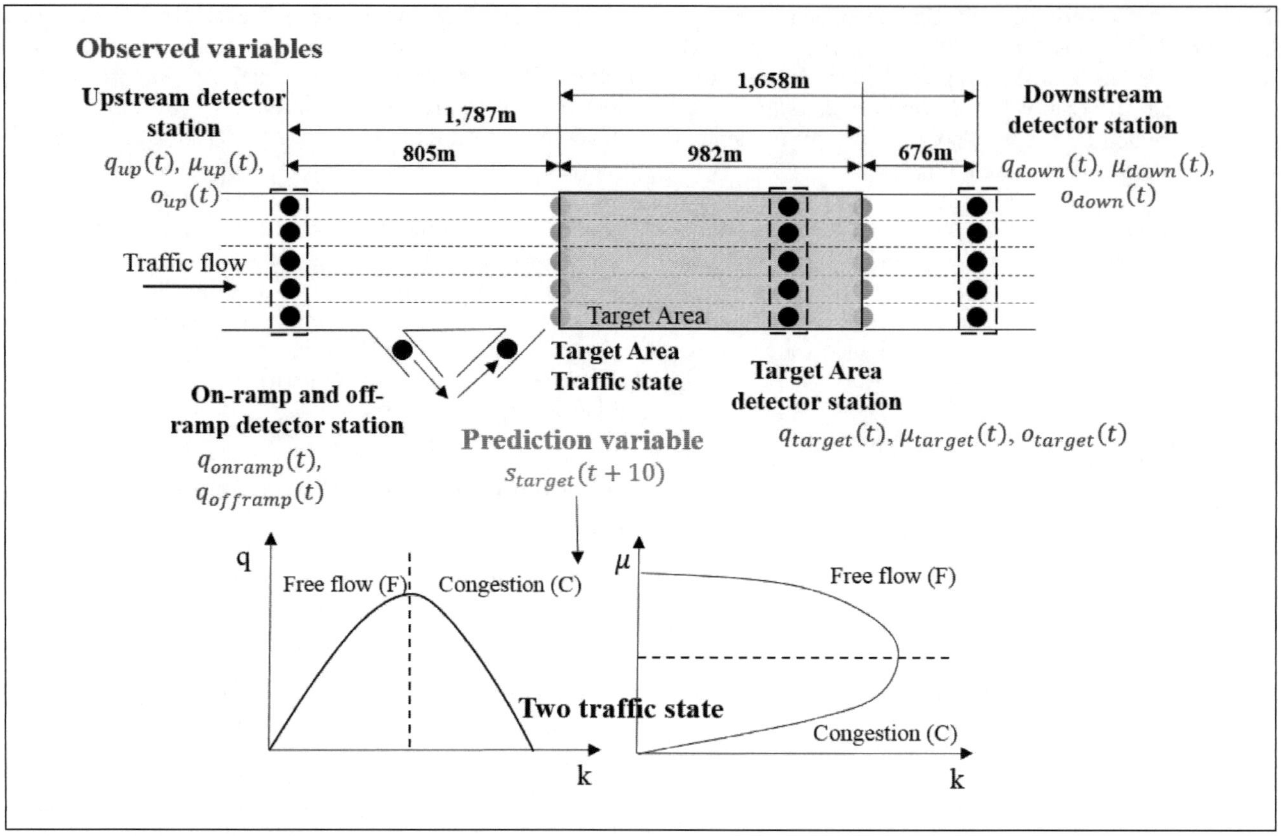

Figure 1. Description of the data.

the free-flow traffic state to the congested traffic state or vice versa (*23*, *24*). In addition, some researchers considered a time duration (*23–25*). If the time duration of the speed below the threshold was greater than a certain time, it was determined that the traffic state changed from the free-flow state to the congested state or vice versa. In this study, we used the traffic state identification algorithm, which includes the speed threshold and the time duration based on link speed. Using the fundamental diagram, the speed threshold was set to 80 km/h, and the time duration was set to 10 time steps (5 minutes).

Description of the Data

In this section, the data are described to model BN for traffic state prediction. As shown in Figure 1, the data consisted of 12 variables: one prediction variable, which is the result of traffic state identification in the target area; and 11 observed variables in each area. In this study, the aggregate interval time of the data was 30 seconds, and we predicted the traffic state after 5 minutes. Also, the prediction variable, as the target of this study, was assumed to be a binary traffic state, that is, a free-flow state or a congested state. We performed BN modeling using this dataset.

Methodology

Bayesian Network (BN)

In this section, we briefly explain the specification of BN following Jenson and Nielsen's convention of notation (*26*). BN is classified as a probabilistic graphical model and is well established for providing practical representations of knowledge for reasoning under uncertainty (*27*). Also, this model can be used to understand complex and stochastic problems and to predict probabilistic events.

The BN consists of two main components. One component is a structure that consists of nodes that represent the variables and directed links that represent the relationships between the variables. The nodes and links form a directed acyclic graph (DAG). The direction of the links can be determined in a cause–effect relationship, but it is not a strict requirement for modeling the structure of the BN. The other component is a parameter, the conditional probability distribution (or table) of each node for a given structure. To calculate the parameter, we define BN over a universe of variables $U = \{A_1, \ldots, A_n\}$. Then we specify a unique joint probability distribution, $P(U)$ (*26*):

$$P(U) = \prod_{i=1}^{n} P(A_i | pa(A_i)) \quad (1)$$

where $pa(A_i)$ is a set of parent variables of variable A_i. If $P(U)$ is known and new multiple findings $e = \{e_1, \ldots, e_m\}$ are provided, we can calculate the joint probability distribution given evidence e :

$$P(U,e) = \prod_{i=1}^{n} P(A_i|pa(A_i)) \prod_{j=1}^{m} e_j \tag{2}$$

By marginalizing $P(U,e)$, the probability of any variable A within U can be calculated as follows:

$$P(A|e) = \frac{\sum_{U/A} P(U,e)}{P(e)} = \frac{\sum_{U/A} P(U,e)}{\sum_{A} P(A,e)} \tag{3}$$

Using this type of two-way evidential reasoning, all of the probabilistic queries can be calculated (*19*).

Modeling of BN

In previous studies, as noted above, transportation researchers constructed a naïve structure that had a patent node as an effect variable to all the other nodes or a structure that considered a simple cause–effect relationship between the variables. In this study, we proposed a new structure of BN for traffic state prediction. The modeling procedure consists of constructing the structure and learning the parameters. First, we constructed a structure and then estimated the conditional probability distribution of each node through parameter learning. In this study, without variable selection procedures, we included all observed variables to analyze the effects of all variables on future traffic state in the BN model.

To construct the structure based on expert knowledge, we used an object-oriented modeling approach (*28*). This approach can be used when the structure of BN includes copies of almost-identical network fragments. Instead of repeatedly constructing the same network fragment, a generic network fragment was constructed once and it was instantiated to other objects. In this approach, a generic network fragment is called a module, and the approach has a top-down process for the construction.

First, we constructed the interface of modules (abstraction level) without specifying their internal details. The interface was constructed by considering the spatiotemporal relationship of the modules based on the information for the prediction of the traffic state. The modules were divided into five areas: target area; upstream; downstream; on-ramp; and off-ramp. It was assumed that all modules at time t are mutually independent and have causal impacts on the future traffic state of the target area.

After setting the interface of the modules, we designed the modules. We observed data from a fixed-loop detector and used those data as sample or test results that represented the traffic state of a limited area and cannot have direct causality to other variables. Therefore, we added the current traffic state node, which represents the entire area, to each module (background variables). The links between symptom variables and background variables were defined by Neil et al.'s structure idioms, as shown in Figure 2*a* (*29*). Then, we used a mixture of Gaussians (MoGs) distribution because the modules included both the continuous variables and discrete variables. It is well known that any probability distribution can be approximated by using MoGs (*28*). An MoGs is the sum of finite number of Gaussian distributions with weights of the components. If we assume that X is a continuous variable and $S = \{s_1, \ldots, s_n\}$ is a parent node of X, the MoGs of X is as follows (*28*):

$$f(X) = \sum_{i=1}^{n} p_i N(m_i, \sigma_i^2) \tag{4}$$

where m_i, σ_i^2 and p_i are the mean, variance, and weight of the ith component in the mixture ($m_i, \sigma_i^2 \in R$ and $0 \le p_i \le 1$, such that $\sum_i p_i = 1$). By this formula, the continuous variable X is represented with a conditional linear Gaussian distribution.

The traffic states (parent node) for MoGs modeling were classified by K-mean cluster analysis. The optimal number of traffic states was selected based on the evaluated results. As a result, four traffic states with the highest overall accuracy and small computation times were selected. The fundamental diagrams of the four traffic states are shown in Figure 2*b*. The traffic states are classified into one four states: free flow (state 1); two transitional states (states 2 and 3); and one congested (state 4). By performing the design of the interface and modules, we derived the structure of the BN, as shown in Figure 2*c*.

After determining the structure of BN, we conducted verification analysis based on the dependency and independency relationships of the nodes. Then, parameter learning was performed to estimate the conditional probability distribution of the BN. We used maximum likelihood estimation (MLE) to learn the parameters. The completed BN, which includes the structure and parameters, enables probabilistic inference. In this study, we used the Bayes Net Toolbox (BNT) MATLAB package to construct the BN and to infer this model given evidence.

Performance Measurements

In this study, to evaluate the prediction results for binary data, five performance measurements were calculated with a probability cut point of 0.5 as follows:

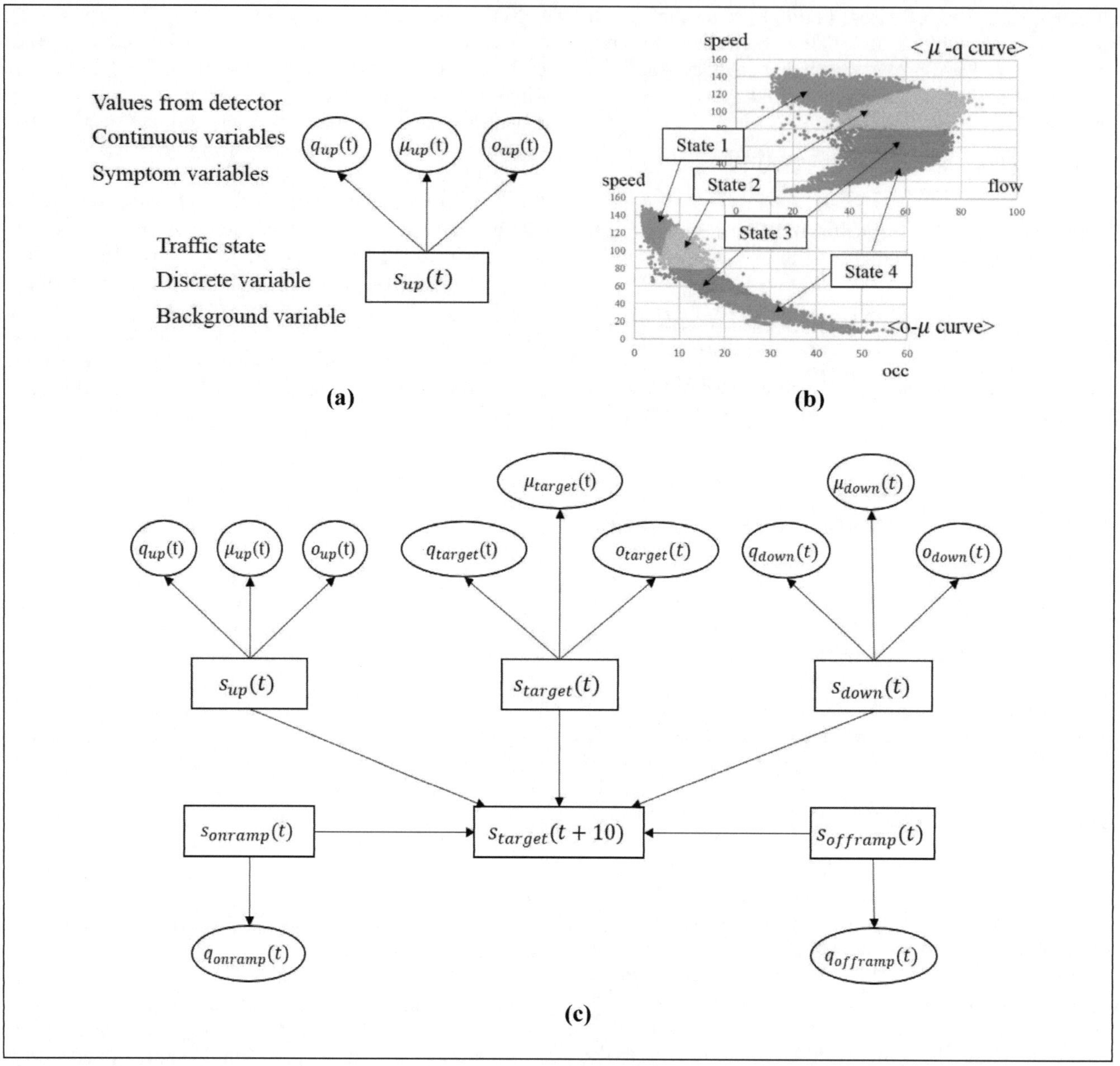

Figure 2. Structure of the Bayesian network: (*a*) MoGs in each module, (*b*) Fundamental diagrams of MoGs, and (*c*) Structure of the BN for traffic state prediction.

$$
\begin{aligned}
&\text{Specificity} = \frac{\text{Correctly predicted counts of free} - \text{flow traffic states}}{\text{Counts of free} - \text{flow traffic states}} \times 100\% \\
&\text{Sensitivity} = \frac{\text{Correctly predicted counts of congested traffic states}}{\text{Counts of congested traffic states}} \times 100\% \\
&\text{False positive rate} = 1 - \text{specificity} \\
&\text{False negative rate} = 1 - \text{sensitivity} \\
&\text{Overall accuracy} = \frac{\text{Correctly predicted counts of both traffic states}}{\text{Counts of both traffic states}} \times 100\%
\end{aligned}
\tag{5}
$$

In addition to these measurements, we used the receiver operating characteristic (ROC) curves to evaluate the predictive accuracy. The ROC curve is a very widely used measurement for classification and provides a comprehensive and visually attractive way to summarize the predictive accuracy. Given the ROC curve, the area under the ROC (AUC) is the most commonly used performance measurement to evaluate the predictive accuracy. This ranges from 0.5 for a perfectly random classifier to 1.0 for a perfect classifier (*30*).

Analysis Measurements for Evidence Sensitivity Analysis

To interpret the model quantitatively, we proposed three analysis measurements, such as cost-of-omission (CoO), minimum and maximum beliefs (MMB), and normalized likelihood (NL), suitable for the traffic state prediction model. Based on the structure and conditional probability distributions of BN, an unbiased probability can be derived using incomplete evidence. This function enables us to perform analyses of various scenarios and provides valuable insights on the modeled problem. In this study, we analyzed the probability changes in traffic state prediction according to the given evidence, and we identified the effect of each variable (evidence sensitivity analysis). The background information consisted of the variables that represented the traffic state of each area, and it had discrete forms, but the symptom information had continuous forms as measured values from the detectors. Therefore, considering these two types of data, the performance measurements were selected.

The CoO, which is a performance measurement based on the Kullback-Leibler distance, was used to analyze the background information. The Kullback-Leibler distance comes from the field of information theory and is given as (*28*):

$$K(P,Q) = H(P,Q) - H(P) = -\sum_{i=1}^{n} p_i \log_2 q_i + \sum_{i=1}^{n} p_i \log_2 p_i = \sum_{i=1}^{n} p_i \log_2 \frac{p_i}{q_i}, \tag{6}$$

where $H(P,Q)$ is the cross-entropy of P and Q which expresses the overall difference between two distributions, and $H(P)$ is the entropy of P, which is a measure of how much information P carries. The value of this measure ranges from 0 to ∞. Here, the CoO is the measurement that replaced P with $P(X|\varepsilon)$ and Q with $P(X|\varepsilon\backslash\{\varepsilon_i\})$. Using this measurement, we can analyze the impact of the omitted evidence, ε_i, on the problem variable X. Let $X = \{x_1, \ldots x_n\}$ be a problem variable and let $\varepsilon = \{\varepsilon_1, \ldots, \varepsilon_n\}$ be a set of evidence. Given that, the CoO of ε_i is follows (*28*):

$$c(\varepsilon_i) = c(P(X|\varepsilon), P(X|\varepsilon\backslash\{\varepsilon_i\})) = \sum_{x\in dom(X)} P(x|\varepsilon) log\left(\frac{P(x|\varepsilon)}{P(X|\varepsilon\backslash\{\varepsilon_i\})}\right) \tag{7}$$

An evidence is considered to be unimportant if the calculated probability is the same with and without the evidence. Conversely, if the calculated probability is affected significantly by the absence of that evidence, it is considered to be important. In the CoO, the problem variable X is the future traffic state at time $t + 10$, and evidence, ε_i, means the traffic state of each area at time t. The problem variable X has two alternatives (the free-flow and the congested traffic state), so the probability changes that occur in the two alternatives are symmetric and opposite when evidence is given. However, CoO shows an asymmetrical result. For example, when evidence ε_i is omitted and the probability of the free-flow traffic state is then increased, the CoO of ε_i in the congested traffic state is larger than that in the free-flow traffic state. This means that a higher CoO is obtained in the traffic state in which the probability is increased by inputting omitted evidence. Therefore, it is possible to analyze how each piece of evidence increases or decreases the probability of the traffic state.

Next, to perform evidence sensitivity analysis on symptom information, MMB and NL were calculated. MMB is a performance measure to find the maximum and minimum probability of the problem variable X given all possible observations of a variable Y. Here, Y is the variable of interest, and it is not included in evidence ε (*28*).

$$\min_{y\in Y} P(x_i|\varepsilon, y), \ \max_{y\in Y} P(x_i|\varepsilon, y) \tag{8}$$

This measurement identifies the range of the posterior belief in a problem variable X according to the variable Y, which indicates the impact of Y on the probability of X. In the study, the measurement of Y was calculated given no evidence ε.

NL is a useful performance measure that can be used to analyze the impact of different subsets of the evidence on the problem variable x_i. The magnitude and direction of the probability change of problem variable x_i are analyzed according to evidence. The NL of a problem variable x_i given an evidence ε' is as follows (*28*):

$$NL = \frac{P(\varepsilon'|x_i)}{P(\varepsilon')} = \frac{P(\varepsilon', x_i)/P(x)}{P(\varepsilon')} = \frac{P(x_i|\varepsilon')P(\varepsilon')/P(x)}{P(\varepsilon')} = \frac{P(x_i|\varepsilon')}{P(x_i)}, \tag{9}$$

where it is assumed that $P(\varepsilon')>0$ and $P(x)>0$. If an NL is 1, it can be concluded that the evidence had no impact on the change of the probability where no evidence is

Table 1. Performance Evaluation Results

Model	Specificity, %	Sensitivity, %	False positive rate, %	False negative rate, %	Overall accuracy, %	AUC
BN	93.3	96.8	6.7	3.2	95.1	0.9929
Logistic regression	80.7	89.6	19.3	10.4	85.1	0.9469
ANN	94.2	96.3	5.8	3.7	95.2	0.9931

given. If an NL is greater than 1, the evidence, as a specific value of the symptom information variable, has an impact on increasing the probability and vice versa.

Results and Discussion

Evaluation Results

To evaluate the performance of the traffic state prediction model using the BN, we compared the BN with a logistic regression model, as a parametric approach, and an artificial neural network (ANN) as a nonparametric approach.

Logistic regression is a regression model with categorical dependent variables. In this study, the binary logistic regression was used to predict the future traffic state. The logistic regression for traffic state prediction includes traffic flow, whose unit is vehicles per 30 seconds, and speed in the target area, whose unit is km/h, as input variables (predictors). The occupancy in the target area and the traffic variables in other areas were excluded from the model due to the multicollinearity problem. The constructed logistic regression model is as follows:

$$\left(s_{target}(t+10) = con\right) = \frac{exp\left(11.864 - 0.047q_{target}(t) - 0.110\mu_{target}(t)\right)}{1 + exp\left(11.864 - 0.047q_{target}(t) - 0.110\mu_{target}(t)\right)} \tag{10}$$

ANN, which is a very popular machine learning method, has a good predictive ability and modeling flexibility with large, multidimensional data. In this study, we used three-layer, feed-forward ANN as one of type of ANN. The input layer had 11 input (variables) neurons and one bias neuron. The hidden layer was composed of 10 hidden neurons and one bias neuron, and the output layer had two output neurons. The input variables were values that were observed only from stationary loop detectors without current traffic state variables as mediating nodes in the BN. Also, the number of hidden neurons was determined by evaluating the performance of the model, and these neurons had no physical meaning. The output neurons were free-flow and congested traffic state as binary data. Supervised training of the ANN was performed using the Levenberg-Marquardt backpropagation algorithm of MATLAB's NN toolbox.

The evaluation results of which BN, logistic regression, and ANN were modeled using the same training and testing datasets are shown in Table 1. The overall accuracies of the ANN and the BN were 95.2% and 95.1%, respectively, which are significantly higher than the accuracy of the logistic regression. In comparing the BN and the ANN, we found that both had a high level of performance with overall accuracies greater than 95%. Also, the AUCs of the ANN and the BN were 0.9931 and 0.9929, respectively. Consequently, the BN constructed in this study had similar performance to the methodology based on machine learning.

Evidence Sensitivity Analysis

Background Information Variables. Using the CoO, we analyzed the impact of the current traffic state of each area at time t on the future traffic state of the target area at time $t + 10$. The CoO has asymmetrical results in the probability of the binary alternative. Therefore, it is possible to determine which variables increase the probability of a certain traffic state. We divided the CoO results into the free-flow traffic state, the congested traffic state, and the sum of the two states.

As shown in Figure 3*a*, the CoO was divided by the target area, upstream, downstream, on-ramp, and off-ramp. In case of the sum, it was determined that each current traffic state had a great impact on the future traffic state in the following order: target area; downstream; upstream; on-ramp; and off-ramp. Also, the effects of the current traffic states acted in favor of the congested traffic state in the future (except for on-ramp). In the case of upstream, the CoO of the congested traffic state was more than twice as much as that of the free-flow traffic state, whereas these values were similar in the case of downstream. This shows that the current traffic state of the upstream acts more in the direction of increasing the probability of the congested traffic state in the future.

As shown in Figure 3*b*, to conduct a detailed analysis, the CoO in each area was divided according to the current traffic state of the target area, $s_{target}(t)$, which has the greatest influence on the prediction of the traffic state. Figure 3*b* has four graphs of the CoO in each area. Also, the CoO of the future traffic state in each area was

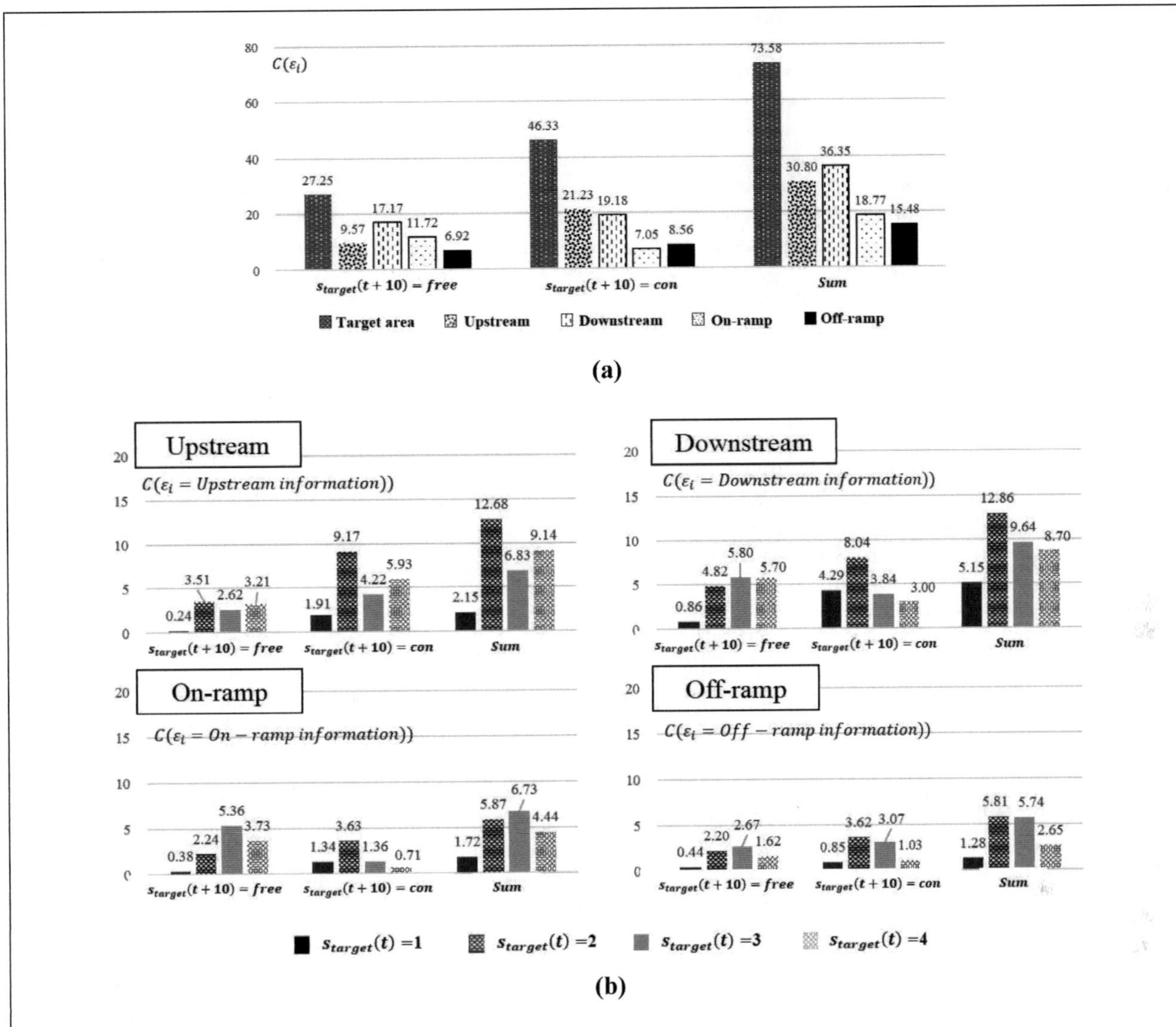

Figure 3. Results of CoO: (*a*) Results of CoO in each area; and (*b*) Results of CoO in each area according to the current traffic state of the target area.

divided into the four traffic states of the target area, and we could analyze the impact of the information of each area depending on the current traffic state of the target area. In the sum of each area, the highest CoO was in transitional state 2 or 3 of the target area. This is because there is a lot of uncertainty in the transitional states about a future traffic state, which can be lowered by knowing the information of each area. That is, in the transitional state, all information of the areas is important in the traffic state prediction. It should be noted that the patterns of CoO of upstream and downstream according to $s_{target}(t)$ are different. The upstream information has a greater impact on the probability of congested traffic state in all current traffic states of the target area. However, when the current traffic state of the target area is congested ($s_{target}(t) = 3$ or $s_{target}(t) = 4$), the downstream information has a greater impact on increasing the probability of a free-flow traffic state.

Symptom Information Variables. To analyze the impact of the traffic variables on the traffic state prediction, we calculated the MMB and NL. Figure 4*a* shows the results of the MMB of the traffic variables. The probability $P(s_{target}(t+10) = congested|no\ evidence)$ is 44.3%. In the terms of the area, the traffic variables in the target area have a greater impact on $P(s_{target}(t+10) = congested|y)$ than the upstream or downstream variables. Also, upstream and downstream information has different directions of the impacts. Upstream information can increase $P(s_{target}(t+10) = congested|y)$ to a larger extent than decrease the probability. This is, upstream information mainly contributes to increasing the probability of the congested traffic state. However, downstream information can decrease $P\left(s_{target}(t+10) = congested|y\right)$ to a larger extent than increase the probability. This is, the downstream information contributes mainly to decreasing the probability of the congested traffic state. The impacts of ramp information on $P(s_{target}(t+10) = congested|y)$ were relatively small. The impacts on $P(s_{target}(t+10) = congested|y)$ were different, according to the traffic variables. The impact of traffic flow on $P(s_{target}(t+10) = congested|y)$ was relatively small compared with other variables. In particular, the variation of the probability based on the upstream flow was very small. The traffic state prediction with only flows may have limitations. Speed and occupancy had similar results. Therefore, speed and occupancy in the same area may be regarded as substitution variables in traffic state prediction.

Next, NL was calculated to analyze how the probability of future traffic state changes as the traffic variables change. Figure 4*b* shows the results of the NL of the congested traffic state according to the traffic variables in each area. In the results of flow, it has a significant impact on lowering $P(s_{target}(t+10) = congested|e')$ when the flows in the downstream and the target area are smaller than about 40 vehicles per 30 seconds. Consistent with previous results, the upstream flow has little impact on $P(s_{target}(t+10) = congested|e')$. In the results of speed and occupancy, the NL of speed usually decreased as the speed increased, and conversely, the NL of occupancy usually increased as the occupancy increased. Also, there were points at which abrupt changes occurred in the NLs of speed and occupancy. The flows of ramps had less impact than other traffic variables of the mainline.

It has been confirmed that traffic flow, speed, and occupancy have different impacts on predicting future traffic state. Traffic flow is a good contributor to predict the free-flow traffic state, which was consistent with a previous study (*13*). The results also showed that the speed and occupancy in the same area had almost similar effects. These two variables may have a relationship as substation variables for traffic state prediction. This suggests that the impact of different traffic variables and different areas on predicting future traffic state should be analyzed carefully when selecting traffic variables for a traffic state prediction model.

Conclusion

The purpose of this study was to develop a probabilistic prediction model for traffic state with good prediction accuracy and interpretability by using BN. Despite its advantages, BN has not been used extensively for transportation problems. A few studies have used BN with the naïve structures that did not consider the relationship between variables and structures that considered a simple cause-effect relationship between the variables. The proposed structure, which was based on theoretical background, improved the previous simple structures. This structure was not only logical but also provided the correct interpretation of the problem. To interpret the model quantitatively, we proposed three analysis measurements suitable for the traffic state prediction model, such as cost-of-omission (CoO), minimum and maximum beliefs (MMB), and normalized likelihood (NL). Through the measurements in evidence sensitivity analysis, we obtained the insights on the impact of the information in each area on traffic state prediction and the impact of traffic variables as input data. In the performance evaluation, we showed that the BN had the same level of performance as existing models with good prediction accuracy.

We also provided an understanding of traffic state prediction and the guidelines for model improvement by performing sensitivity analyses. The impacts on the future traffic state of the target area were different depending on the location of the information and traffic variables. Therefore, traffic state prediction modeling that considers the impacts, that is, structure design and variable selection, is needed. A consideration is that speed and occupancy are substitution variables. If necessary (e.g., for simplification of the model), only one of the two—speed and occupancy—may be used to reflect the effect. The traffic state prediction with only traffic flows may have limitations. In addition, we found that the current traffic state of the target area is most important. To improve the performance of traffic state prediction, it is necessary to set a measure that represents the target area accurately (e.g., density).

Future research is required to improve our BN, which included only traffic variables at time t and assumed that the current traffic states of different areas at the same time were independent. Future research should consider adding nodes for historical traffic variables and links for dependency between all nodes. The new model may have high complexity but could be a useful improvement. When the length of study site becomes longer and the number of variables is larger, variable selection procedure should be considered. We can identify the influential variables through the variable selection, which makes the prediction model accurate and efficient. The MoGs used in this study assumed that the distribution of each state of variables was Gaussian, but actual distributions may not satisfy the Gaussian distribution. Thus, future research to examine the effect of the distribution on the

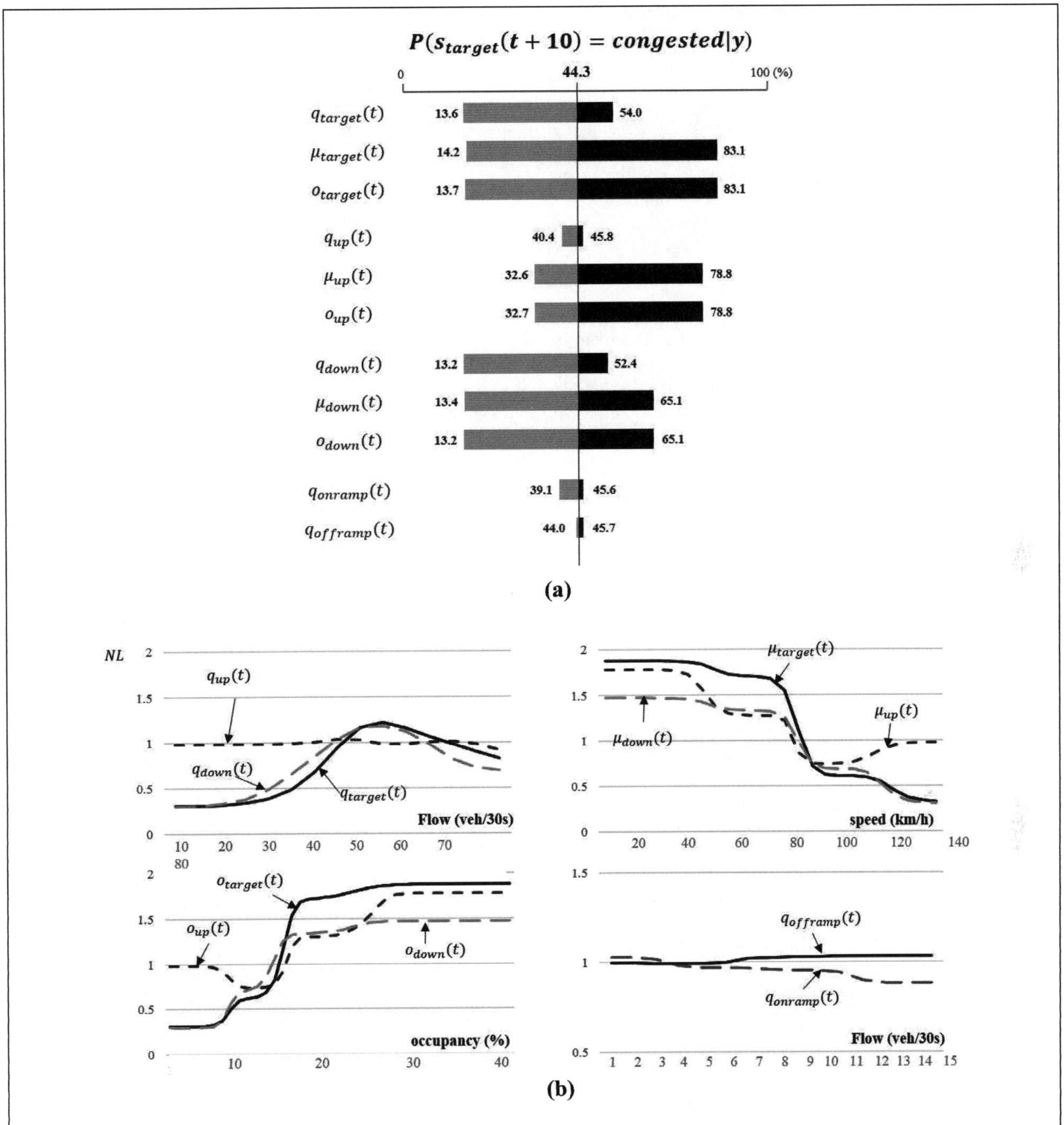

Figure 4. Results of MMB and NL: (*a*) Results of MMB and (*b*) Results of NL of the congested traffic state.

performance of the model is necessary. Finally, transferability of the developed BN should be examined for future research. In this study, we analyzed the five-lane freeway with an on- and an off-ramp. If the geometry of the study site is changed, for example, the number of lanes and ramps, lane reduction, slope, and other geometries, the transferability may be reduced. Also, traffic and weather conditions can affect the transferability. With those influencing factors, BN, which has a structure based on general expert knowledge, can maintain its transferability. The structure based on general knowledge can be transferable, and the parameters based on data can be partially transferable. Confirming the transferability of the BN will increase its applicability to predict traffic states.

Acknowledgments

This research was supported by the Basic Science Research Program through the National Research Foundation of Korea (NRF) funded by the Ministry of Education, South Korea (2016R1C1B1008492).

Author Contributions

The authors confirm contributions to the paper as follows: study conception and design: Park, Kim, and Kho; data collection: Park; analysis and interpretation of results: Park, Kim and Kho; draft manuscript preparation: Park. All authors reviewed the results and approved the final version of the manuscript.

References

1. Ahmed, M. S., and A. R. Cook. Analysis of Freeway Traffic Time-Series Data by Using Box-Jenkins Techniques. *Transportation Research Record: Journal of the Transportation Research Board*, 1979. 722: 1–9.
2. Williams, B., P. Durvasula, and D. Brown. Urban Freeway Traffic Flow Prediction: Application of Seasonal Autoregressive Integrated Moving Average and Exponential Smoothing Models. *Transportation Research Record: Journal of the Transportation Research Board*, 1998. 1644: 132–141.
3. Okutani, I., and Y. J. Stephanedes. Dynamic Prediction of Traffic Volume through Kalman Filtering Theory. *Transportation Research Part B: Methodological*, Vol. 18, No. 1, 1984, pp. 1–11.
4. Wang, Y., and M. Papageorgiou. Real-Time Freeway Traffic State Estimation Based on Extended Kalman Filter: A General Approach. *Transportation Research Part B: Methodological*, Vol. 39, No. 2, 2005, pp. 141–167.
5. Li, W., C. Li, X. Du, K. Qian, H. Zhang, and D. Hou. A Traffic Flow Prediction Model Based on Ordered Logistic Regression. *Proc., 6th International Conference on the Digital Content, Multimedia Technology and its Applications (IDC)*, Seoul, South Korea, IEEE, New York, 2010, pp. 213–216.
6. Huili, D., W. Guohua, and G. Min. Traffic Guidance Oriented Model of Traffic State Probability Forecast. *Journal of Transportation Systems Engineering and Information Technology*, Vol. 11, No. 2, 2011, pp. 27–32.
7. Florio, L., and L. Mussone. Neural-Network Models for Classification and Forecasting of Freeway Traffic Flow Stability. *Control Engineering Practice*, Vol. 4, No. 2, 1996, pp. 153–164.
8. Vlahogianni, E. I., M. G. Karlaftis, and J. C. Golias. Optimized and Meta-Optimized Neural Networks for Short-Term Traffic Flow Prediction: A Genetic Approach. *Transportation Research Part C: Emerging Technologies*, Vol. 13, No. 3, 2005, pp. 211–234.
9. Davis, G. A., and N. L. Nihan. Nonparametric Regression and Short-Term Freeway Traffic Forecasting. *Journal of Transportation Engineering*, Vol. 117, No. 2, 1991, pp. 178–188.
10. Myung, J., D. Kim, S. Kho, and C. Park. Travel Time Prediction Using k Nearest Neighbor Method with Combined Data from Vehicle Detector System and Automatic Toll Collection System. *Transportation Research Record: Journal of the Transportation Research Board*, 2012. 2256: 51–59.
11. Wang, J., and Q. Shi. Short-Term Traffic Speed Forecasting Hybrid Model Based on Chaos–Wavelet Analysis-Support Vector Machine Theory. *Transportation Research Part C: Emerging Technologies*, Vol. 27, 2013, pp. 219–232.
12. Xia, J., and M. Chen. A Nested Clustering Technique for Freeway Operating Condition Classification. *Computer-Aided Civil and Infrastructure Engineering*, Vol. 22, No. 6, 2007, pp. 430–437.
13. Xia, J., W. Huang, and J. Guo. A Clustering Approach to Online Freeway Traffic State Identification Using Its Data. *KSCE Journal of Civil Engineering*, Vol. 16, No. 3, 2012, pp. 426–432.
14. Antoniou, C., H. N. Koutsopoulos, and G. Yannis. Dynamic Data-Driven Local Traffic State Estimation and Prediction. *Transportation Research Part C: Emerging Technologies*, Vol. 34, 2013, pp. 89–107.
15. Noroozi, R., and B. Hellinga. Real-Time Prediction of Near-Future Traffic States on Freeways Using a Markov Model. *Transportation Research Record: Journal of the Transportation Research Board*, 2014. 2421: 115–124.
16. Sun, Z. Q., J. Q. Feng, W. Liu, and X. M. Zhu. Traffic Congestion Identification Based on Parallel SVM. *Proc., 2012 8th International Conference on the Natural Computation*, Chongqing, China, IEEE, New York, 2012, pp. 286–289.
17. Karlaftis, M. G., and E. I. Vlahogianni. Statistical Methods versus Neural Networks in Transportation Research: Differences, Similarities and Some Insights. *Transportation Research Part C: Emerging Technologies*, Vol. 19, No. 3, 2011, pp. 387–399.
18. Sun, S., C. Zhang, and G. Yu. A Bayesian Network Approach to Traffic Flow Forecasting. *IEEE Transactions on Intelligent Transportation Systems*, Vol. 7, No. 1, 2006, pp. 124–132.
19. Zhang, K., and M. A. Taylor. Toward Universal Freeway Incident Detection Algorithms. *Transportation Research Part C: Emerging Technologies*, Vol. 14, No. 2, 2006, pp. 68–80.
20. Hossain, M., and Y. Muromachi. A Bayesian Network Based Framework for Real-Time Crash Prediction on the Basic Freeway Segments of Urban Expressways. *Accident Analysis and Prevention*, Vol. 45, 2012, pp. 373–381.
21. Lin, L., Q. Wang, and A. W. Sadek. A Novel Variable Selection Method Based on Frequent Pattern Tree for Real-Time Traffic Accident Risk Prediction. *Transportation Research Part C: Emerging Technologies*, Vol. 55, 2015, pp. 444–459.
22. Kondyli, A., L. Elefteriadou, W. Brilon, F. L. Hall, B. Persaud, and S. Washburn. Development and Evaluation of Methods for Constructing Breakdown Probability Models. *Journal of Transportation Engineering*, Vol. 139, No. 9, 2013, pp. 931–940.

23. Lorenz, M., and L. Elefteriadou. Defining Freeway Capacity as Function of Breakdown Probability. *Transportation Research Record: Journal of the Transportation Research Board*, 2001. 1776: 43–51.
24. Dong, J., and H. S. Mahmassani. Flow Breakdown and Travel Time Reliability. *Transportation Research Record: Journal of the Transportation Research Board*, 2009. 2124: 203–212.
25. Kim, J., H. Mahmassani, and J. Dong. Likelihood and Duration of Flow Breakdown: Modeling the Effect of Weather. *Transportation Research Record: Journal of the Transportation Research Board*, 2010. 2188: 19–28.
26. Jensen, F. V., and T. D. Nielsen. *Bayesian Networks and Decision Graphs*. Springer Science + Business Media, New York, 2009.
27. Druzdel, M., and L. C. Van Der Gaag. Building Probabilistic Networks: Where Do the Numbers Come From? *IEEE Transactions on Knowledge and Data Engineering*, Vol. 12, No. 4, 2000, pp. 481–486.
28. Kjaerulff, U. B., and A. L. Madsen. *Bayesian Networks and Influence Diagrams*. Springer Science + Business Media, New York, 2008.
29. Neil, M., N. Fenton, and L. Nielson. Building Large-Scale Bayesian Networks. *The Knowledge Engineering Review*, Vol. 15, No. 3, 2000, pp. 257–284.
30. Hand, D. J. Measuring Classifier Performance: A Coherent Alternative to the Area under the ROC Curve. *Machine Learning*, Vol. 77, No. 1, 2009, pp. 103–123.

The Standing Committee on Artificial Intelligence and Advanced Computing Applications (ABJ70) peer-reviewed this paper (18-01043).

Article

TRR
JOURNAL OF THE TRANSPORTATION RESEARCH BOARD

Transportation Research Record
2018, Vol. 2672(45) 136–146

DOI: 10.1177/0361198118772689
journals.sagepub.com/home/trr
SAGE

Evaluation of the Gradient Boosting of Regression Trees Method on Estimating Car-Following Behavior

Sina Dabiri[1] and Montasir Abbas[1]

Abstract

Car-following models, as the essential part of traffic microscopic simulations, have been utilized to analyze and estimate longitudinal drivers' behavior for sixty years. The conventional car-following models use mathematical formulas to replicate human behavior in car-following phenomenon. The incapability of these approaches to capture the complex interactions between vehicles calls for deploying advanced learning frameworks to consider more detailed behavior of drivers. In this study, we apply the gradient boosting of regression tree (GBRT) algorithm to vehicle trajectory data sets, which have been collected through the Next Generation Simulation (NGSIM) program, to develop a new car-following model. First, the regularization parameters of the proposed method are tuned using cross-validation technique and sensitivity analysis. Second, prediction performance of the GBRT is compared to the world-famous Gazis-Herman-Rothery (GHR) model, when both models have been trained on the same data sets. The estimation results of the models on unseen records indicate the superiority of the GBRT algorithm in capturing the motion characteristics of two successive vehicles.

Drivers' stochastic behavior renders more complexity, uncertainty, and non-linearity in transportation systems. Microscopic simulation architectures are key tools in replicating the stochastic nature of traffic conditions (*1*, *2*). These frameworks allow traffic engineers to evaluate the performance of real-world traffic networks while considering the interaction between individual vehicles and various sections of roads such as signalized intersections. This results in reducing high expenditure on collecting field data and monitoring real transportation systems. In microscopic models, details of each vehicle's motion characteristics such as speed, acceleration/deceleration, time and space headway are described with the help of behavioral models, including gap-acceptance, speed adaption, lane-changing, ramp metering, overtaking, and car-following models (*3*). Lateral and longitudinal interactions between vehicles are the two principal components for understanding the dynamic behavior of travelers in traffic microscopic simulators (*4*). Lateral behaviors, described by lane-changing models, comprise the decision to change lane, choice of target lane, and gap-acceptance (*5*). Car-following models reproduce the longitudinal behavior of vehicles, where the movement features of the following vehicle are restricted by the leading vehicle in the same lane. The focus of this study is on car-following models as the principal type of microscopic simulation model.

A growing trend towards developing car-following models began sixty years ago (*6*). Many classical car-following models have been constructed according to traffic flow theories and explicit assumptions. Examples of conventional car-following models range from the Gazis-Herman-Rothery (GHR) model (*7*), and safety distance or collision avoidance models (*8*), to desired measures (*9*, *10*), optimal velocity models (*11*), linear models (*12*), cellular automaton models (*13*), and psycho-physical or action points models (*14*). The relative distance and speed between following and leading vehicles, and the speed of the following vehicle, as well as the driver's perception/reaction time are the standard inputs, while the following vehicle's acceleration/deceleration or speed are the main outputs in the car-following models. On the other hand, the rapid growth

[1]Charles Edward Via, Jr., Department of Civil and Environmental Engineering, Virginia Polytechnic Institute and State University, Blacksburg, VA

Corresponding Author:
Address correspondence to Sina Dabiri: sina@vt.edu

of data-driven approaches in addition to the availability of field data have led to a proliferation of deploying data mining techniques in transportation fields (*15*). In contrast to classical models, closed-form mathematical functions, calibrating, and manual errors are not concerns in data-driven techniques (*16*). The capability to extract more details on drivers' responses as well as the flexibility to involve additional information (e.g., environmental effects and roadway geometry) are further benefits to using data-driven techniques in car-following models.

This study begins by reviewing car-following literature with a focus on models developed with the aid of data-driven approaches. It will then present the gradient boosting of regression trees (GBRT) algorithm and the GHR model in the Methodology section. Trajectory data sets and preprocessing procedures are described in the Data Description section, and in the Results and Discussion section, the optimal parameters of the proposed model are determined using sensitivity analysis and the cross-validation technique. The prediction performance of the proposed model is also compared with the GHR model in this section. The Conclusion summarizes the study.

Literature Review

As mentioned in the previous section, car-following models can be categorized into two main groups: conventional and data-driven models (*4*). A systematic review on the classical car-following models, their calibration, and evaluation have been reported by Brackstone et al. (*6*). Studies by Olstam et al. (*3*) and Panwai et al. (*17*) highlighted and compared the performance of classical car-following models in commercial traffic microsimulation packages, including AIMSUM with the safety distance model, MITSIM with the GHR model, and VISSIM and Paramics with the psycho-physical models. A group of conventional car-following models, such as GHR, Gipps, Cellular Automaton, S-K, Intelligent Driver, and Wiedemann, was assessed in terms of fundamental diagrams and vehicle trajectory data sets by Hamdar (*18*). A qualitative study by Saifuzzaman (*19*) described extension and improved versions of the classical car-following models from both engineering and human factor points of view. In accordance with the scope of the current study and the existence of holistic survey papers on classical car-following models, only the cutting-edge car-following models developed by data mining techniques are investigated.

Fuzzy logic, artificial neural network (ANN), and combinations of the two approaches, (e.g., locally linear neural-fuzzy models (*20*)), have widely been utilized to simulate the future states of a vehicle in a car-following process. A number of studies have also tested the performance of other machine learning algorithms, including k-nearest neighbor (*16*), locally weighted regression (*21*), and support vector regression (*22*), for better understanding drivers' dynamic behavior.

The reaction of a driver to a given stimulus stems from a set of ambiguous and vague driving rules rather than a deterministic relationship, which is typically enforced in the classic car-following models. To overcome such an issue, Kikuchi et al. (*23*), for the first time, presented a fuzzy inference system to simulate the driver's action as a result of a fuzzy reasoning process determined by natural-language-based driving rules. In their study, the relative distance and speed between the two vehicles as well as acceleration/deceleration of the leading vehicle were fed as inputs into the GHR model. Each of the inputs was grouped into six fuzzy sets. A triangular membership function was assigned to each fuzzy set. Specifying a set of rules generates a set of outputs in fuzzy numbers. Then, averaging weighted outputs over the time, as the final output, leads to estimating the following-vehicle movement relative to the leading vehicle. More work in this context can be found in Wu et al. and Gao et al. (*24, 25*). The main challenge in developing such models was the determination of fuzzy sets and the corresponding membership functions, which are basic and essential concepts in fuzzy logic.

The stochastic and non-linearity nature of human behavior and traffic streams call for more sophisticated models. ANNs have received the most attention in modeling car-following behavior in comparison with other data-driven approaches due to their remarkable ability to derive meaningful concepts from complicated and incomplete data sets. Hongfei et al. (*26*) introduced the application of ANNs in the car-following models for the first time. A two-hidden-layer back propagation ANN was applied to develop a car-following model, where the training and test data sets were collected by means of a five-wheel system in an open roadway condition. Moreover, the performance of other types of ANN architectures, including fuzzy adaptive resonance theory (ART) neural networks, radial basis function networks, and swarm algorithm, have been investigated. However, the results implied the superiority of the regular neural network compared to almost all other ANN training algorithms (*27, 28*). As ANNs require no pre-settings for considering the various combination of input variables or adding new features, the contribution of other factors, including the motion features of the preceding vehicle in front of the leading vehicle, can also be examined using ANNs (*29*). Unlike many studies that put the reaction time of the driver-vehicle unit as a constant value, a number of modified ANNs have been built to predict car-following behavior based on the instantaneous reaction delay (*30, 31*). Several studies have attempted to fuse

ANNs with other approaches such as fuzzy logic and local linear regression models (*4, 32*).

Notwithstanding that several data-driven approaches for modeling car-following behavior have been postulated in literature, a lack of examining ensemble algorithms in this context has encouraged us to evaluate the capability of this class of machine learning algorithms. Since all variables in the context of car-following models are continuous, regression models are more reliable algorithms to build the relationship between the variables. Accordingly, the GBRT, as an effective type of ensemble algorithms, is deployed that takes advantage of predictive power, robustness to the noise by combining a set of base learners and minimizing the prediction error as learning proceeds.

Methodology

GHR Models

The general GHR model, proposed by Gazis, Herman, and Rothery at General Motors laboratory (*7*), is used as the basis of comparison for examining the performance of the GBRT-based car-following model. All GHR models express a general idea that a driver responds to a stimulus according to the following relationship: Response = function (sensitivity, stimulus), which is also known as the follow-the-leader theory of traffic. The response is always considered as the acceleration/deceleration of the following vehicle, which is a factor that can be easily controlled by drivers through accelerator or brake pedals. According to the follow-the-leader theory, it has been shown that there is a high correlation between the following driver's response and the relative speed difference of the follower and leader vehicles. This results in considering relative speed as the stimulus in all versions of the GHR models (*7*). Therefore, the variations among GHR models is contingent on how to formulate the sensitivity term. While a variety of functional forms were proposed and evaluated as the sensitivity term, the general GHR model was introduced to cover all other types of GHR models:

$$acc_{n+1}(t+\tau) = \alpha v_{n+1}^{m}(t+\tau)\frac{\Delta v_{n,n+1}(t)}{\Delta x_{n,n+1}{}^{l}(t)} \tag{1}$$

Denoting the current time as t, $acc_{n+1}(t+\tau)$ and $v_{n+1}(t+\tau)$ are the acceleration/deceleration and speed of the following vehicle at the instant $(t+\tau)$, respectively. τ is the time lag of the driver's response to observed stimulus predictors. $n+1$ and n are subscripts for the following and leading vehicles. $\Delta v_{n,n+1}(t)$ and $\Delta x_{n,n+1}(t)$ respectively represent the relative speed and the relative distance between the two vehicles at the time instant t. The model contains three parameters α, m, and l that need to be calibrated through the problem data set. Although a vast majority of studies have investigated the GHR model in an attempt to determine the best combination of the parameters, we define the optimal values by solving the optimization problem formulated in Equations 2 and 3. Optimizing the parameters for each data set separately, instead of using pre-defined values, leads to capturing the car-following behavior that is compatible with the traffic conditions that exist in the data set.

$$\min_{\alpha,m,l}\sum_{t=1}^{T}\left[\alpha v_{n+1}^{m}(t+\tau)\frac{\Delta v_{n,n+1}(t)}{\Delta x_{n,n+1}{}^{l}(t)} - acc_{n+1}(t+\tau)\right]^{2} \tag{2}$$

$$\text{S.t: } \mathbf{LB} \leq \alpha, m, l \leq \mathbf{UP} \tag{3}$$

where T is the number of training instances. Equation 2 measures the sum of the squares of the errors between the estimated and true (observed) values of acceleration/deceleration. Equation 3 constrains the model parameters between the lower bound vector, $\mathbf{LB} = [0,0,0]$, and upper bound vector, $\mathbf{LB} = [3,3,3]$. The vectors are designed in such a way as to cover all values used in the literature for GHR models, as summarized in the References (*6*).

Gradient Boosting of Regression Trees

In this study, the response and explanatory variables of the proposed car-following model are considered similar to the GHR model. Accordingly, the acceleration/deceleration of the following vehicle at the time instant $(t+\tau)$, $acc_{n+1}(t+\tau)$ is the model's response variable predicted as a function of the explanatory variables $[v_{n+1}(t+\tau), \Delta v_{n+1}(t), \Delta x_{n+1}(t)]$. Since all variables, particularly the response variable, are continuous, regression models are suitable approaches to predict car-following behavior. In this paper, we assess the performance of an ensemble regression algorithm in estimating car-following models.

The core idea in ensemble methods is to combine multiple base learners, which are usually called weak learners, to improve the accuracy and robustness of the final model. Two major questions need to be answered in ensemble methods: what type of weak learner should be used for the training process? How are weak learners trained and fused to yield the final model? Accordingly, ensemble algorithms fall into two main families: averaging and boosting methods. Each family follows a distinct procedure towards training and combining individual base learners (*33*). In the averaging methods, base learners are first built independently on randomly selected training instances, and then are averaged to generate the final model. Examples of this category are bagging and random forest methods. Naïve Bayes classifiers, decision trees, and neural networks are typically utilized as the machine learning techniques for training weak learners. In the boosting methods, weak learners are trained iteratively and in a stage-wise fashion to find a model that reduces the bias and variance of prediction. In each iteration, a strength weight is assigned to the weak learner that implies its prediction error rate; meanwhile, each training instance is reweighted by how incorrectly it was classified. The method of weighting training data and base learners distinguishes between boosting algorithms such as adaptive boosting and gradient boosting. The final model is the sum of the weighted weak learners (*33*).

In this study, we envision utilizing the GBRT algorithm, which was initially developed by Friedman (*34*), for estimating the car-following model. The key goal in the GBRT algorithm is to fit a regression tree at each step to the difference between the observed response and the aggregated prediction of all learners grown previously. As indicated by the name of the algorithm, the decision trees algorithm with a fixed size is chosen as the weak learner.

Suppose the training data set contains N samples of (y_i, $\mathbf{x}_i$), in which y_i and $\mathbf{x}_i$ are the response variable and explanatory vector for the sample i, respectively. The goal is to learn a model $F^*(\mathbf{x})$, the best approximation of the real function $F(\mathbf{x})$ that maps $\mathbf{x}$ to y, while the expected value of the loss function $L(y, F(\mathbf{x}))$ is minimized as shown in Equation 4:

$$F^* = \arg\min_F E_{y,\mathbf{x}} L(y, F(\mathbf{x})) = \arg\min_F E_{y,\mathbf{x}}[(y - F(\mathbf{x})]^2 \quad (4)$$

Notwithstanding that many forms of loss functions exist in the literature, the squared-error loss function is chosen in Equation 4 due to the superior computational properties of the least-square algorithm for minimizing the squared-error loss function (*34*). The GBRT algorithm seeks to find the function $F(\mathbf{x})$ in the form of the sum of the weighted weak learners, $h(\mathbf{x}; \mathbf{a}_m)$:

$$F(\mathbf{x}; \{\beta_m, \mathbf{a}_m\}_1^M) = \sum_{m=1}^{M} \beta_m h(\mathbf{x}; \mathbf{a}_m) \quad (5)$$

where M is the number of weak learners, and β is the corresponding weight for each learner. $h(\mathbf{x}; \mathbf{a}_m)$is a small regression tree defined by a finite set of parameters $\mathbf{a}_m$, including splitting variables, split locations, and leaf nodes. Analogous to other boosting algorithms, the GBRT builds an additive model in the form of Equation 6, when $F_m(\mathbf{x})$ in the last iteration equals $F(\mathbf{x})$ in Equation 5:

$$F_m(\mathbf{x}) = F_{m-1}(\mathbf{x}) + \beta_m h(\mathbf{x}; \mathbf{a}_m) \quad (6)$$

Thus, at each step m, the estimator $h(\mathbf{x}; \mathbf{a}_m)$ is added to the imperfect model $F_{m-1}(\mathbf{x})$ in the previous step to build a better additive model in a forward stage-wise fashion presented in Equation 6. The decision tree $h(\mathbf{x}; \mathbf{a}_m)$ and its associated weight β_m are trained to minimize the loss function of observed responses and $F_m(\mathbf{x})$ as shown in Equation 7:

$$[h(\mathbf{x}; \mathbf{a}_m), \beta_m] = \arg\min_{h,\beta} \sum_{i=1}^{N} L(y_i, F_{m-1}(\mathbf{x}_i) + \beta h(\mathbf{x}_i; \mathbf{a}_m)) \quad (7)$$

where the loss functions in Equations 4 and 7 are the same. As Equation 7 is a hard optimization problem, one way to more easily resolve the problem is to apply the gradient descent algorithm. Considering Equation 6 and given the estimator $F_{m-1}(\mathbf{x})$, the function $\beta_m h(\mathbf{x}; \mathbf{a}_m)$ can be viewed as the best greedy step towards the most accurate approximation of $F_m(\mathbf{x})$. To apply the gradient descent algorithm to Equations 6 and 7, $h(\mathbf{x}; \mathbf{a}_m)$ is trained using the training set $\{(\mathbf{x}_i, g_m(\mathbf{x}_i)\}_i^N$, where $g_m(\mathbf{x}_i)$ is the negative gradient of loss function computed by Equation (8):

$$g_m(\mathbf{x}_i) = -\left[\frac{\partial L(y_i, F(\mathbf{x}_i))}{\partial F(\mathbf{x}_i)}\right]_{F(x) = F_{m-1}(\mathbf{x})} = [y_i - F_{m-1}(\mathbf{x}_i)] \quad (8)$$

where the right-hand side is computed by substituting the squared-error loss function. Hence, at each step m, the regression tree $h(\mathbf{x}; \mathbf{a}_m)$ is fitted to the difference between the observed response and the aggregated prediction of all learners grown so far (i.e., $h(\mathbf{x}; \mathbf{a}_m)$ is trained using the training set$\{(\mathbf{x}_i, g_m(\mathbf{x}_i)\}_{i=1}^N$).

The step size, β_m, is calculated by solving the one-dimensional optimization problem using Equation 9:

$$\beta_m = \arg\min_{\beta} \sum_{i=1}^{N} L(y_i, F_{m-1}(\mathbf{x}_i) + \beta h(\mathbf{x}_i; \mathbf{a}_m)) \quad (9)$$

It should be noted that solving Equation 9 can be thought of as finding the optimal step size in the gradient descent algorithm.

The GBRT algorithm steps are summarized as follows:

1. Initialize the approximation function$F(\mathbf{x})$:

$$F_0(\mathbf{x}) = \arg\min_{\beta} \sum_{i=1}^{N} L(y_i, \beta) \text{ or } F_0(\mathbf{x}) = \frac{1}{N}\sum_{i=1}^{N} y_i$$

2. For $m = 1$ to M do:
   - Calculate the pseudo-responses: $g_m(\mathbf{x}_i) = [y_i - F_{m-1}(\mathbf{x}_i)]$ for $i = 1, \ldots, N$

- Fit the regression tree $h(\mathbf{x}; \mathbf{a}_m)$using the training set: $\{(\mathbf{x}_i, g_m(\mathbf{x}_i)\}_{i=1}^{N}$
- Calculate the step size β_m using the line search:

$$\beta_m = \arg\min_{\beta} \sum_{i=1}^{N} L(y_i, F_{m-1}(\mathbf{x}_i) + \beta h(\mathbf{x}_i; \mathbf{a}_m))$$

- Update the model: $F_m(\mathbf{x}) = F_{m-1}(\mathbf{x}) + \beta_m h(\mathbf{x}; \mathbf{a}_m)$

3. End the algorithm. $F_M(\mathbf{x}) = F^*(\mathbf{x})$ is the final output of the algorithm.

Regularization Parameters

Regularization is a process used in an attempt to prevent the overfitting problem in statistical models by explicitly controlling the model complexity and constraining the fitting procedure (*34*). Overfitting occurs when the model is extremely complex with too many parameters. A large number of parameters engenders the learning process to be performed through memorization of the training data set rather than understanding the concepts and attributes inside the data. An overfitted model perfectly predicts only on training examples due to the memorization but has a poor predictive performance on the unseen data. For the GBRT model, adding too many weak learners with over-complex trees as well as the high speed of learning are the main sources of overfitting, which call for introducing regularization parameters to control the issue.

The value of M, the number of weak learners, is one of the regularization parameters that regulates the expected loss reduction over the training data. Moreover, handling the speed of learning through a shrinkage process has been found to be an efficient approach for avoiding complexity and overfitting, particularly in additive models (*34*). The shrinkage strategy can be obtained by incorporating the learning rate ν, which scales the contribution of each weak learner, as shown in Equation 10. Thus, the additive model in the second step of the algorithm is replaced by:

$$F_m(\mathbf{x}) = F_{m-1}(\mathbf{x}) + \nu.\beta_m.h(\mathbf{x}; \mathbf{a}_m) \quad 0 < \nu \leq 1 \qquad (10)$$

The parameters M and ν interact strongly with each other. Thus, the value of each affects the other. Accordingly, a sensitive analysis that considers both parameters simultaneously needs to be performed for determining their optimal values. Model evaluation techniques such as cross-validation is deployed to compute and compare the generalization error under each combination of M and ν.

The size of regression trees, denoted as D, is another regularization parameter that controls the overfitting problem. The tree depth is defined by tuning parameters like the maximal number of decision splits and/or the minimum number of observations required to be at a leaf node. The decision tree is usually overfitted on the data set with a few samples but a large number of features. Fortunately, the current data set contains a large number of samples with only three features, which reduces the possibility of generating overfitted trees. Even though the size of trees has a slight impact on the overfitting problem due to the mentioned characteristics of the car-following data set, we compute its optimal value by performing a sensitivity analysis. So we not only obtain the highest decrease in the test error but ensure that the important structures of the ultimate model are captured (*35*).

It is worth mentioning that, in this paper, the driver's reaction time (τ) is not an explanatory variable in both GHR and GBRT models; instead, it is used as a constant value in the models. Indeed, its value affects the order of predictor variables with their corresponding response variable. For example, the triple $[v_{n+1}(t+\tau), \Delta v_{n+1}(t), \Delta x_{n+1}(t)]$ is used as input variables for predicting the following car acceleration at the time instant $(t+\tau)$, $acc_{n+1}(t+\tau)$. As a consequence, each data instance is defined as $\{acc_{n+1}(t+\tau), [v_{n+1}(t+\tau), \Delta v_{n+1}(t), \Delta x_{n+1}(t)]\}$, which is contingent on the value of τ. Considering the reaction time range between 0.1–3.0 seconds as the interval that covers all ranges offered by a variety of studies (*21*), the optimal value of τ is determined through a sensitivity analysis for both models. It is worth mentioning that the optimal value of τ might be different for each model.

Data Description

The proposed car-following model is developed and tested based on vehicle trajectory data sets collected by the Federal Highway Administration (FHWA) under the NGSIM program. The NGSIM program aims to collect high-quality data on vehicle motion parameters over a time window for which driver behavior and interactions with traffic objects are inferred through microscopic modeling. The data in this study are selected from the northbound traffic on I-80 in Emeryville, California, recorded on April 2005. The NGSIM I-80 data set contains three subsets recorded in three time periods: 4:00 to 4:15 p.m., 5:00 to 5:15 p.m., and 5:15 to 5:30 p.m. Deploying seven video cameras installed on a tall building, the trajectory data was recorded at the one-tenth-of-second resolution. A variety of vehicle trajectory features have been extracted and provided in each data set, including acceleration, speed, location coordinates, ID of the preceding and the following vehicle of the subject vehicle, type and length of the vehicle, and space and time headway with respect to the leading vehicle.

However, the NGSIM data include measurement errors in computing location coordinates, which in turn produce the unrealistic large speed and acceleration

Table 1. Basic Statistics for Acceleration, Speed, and Relative Distance of All Data Series

		Statistics for acceleration (m/s^2)			Statistics for speed (m/s)			Statistics for relative distance (m)		
Data series	No. of samples	min	max	mean	min	max	mean	min	max	mean
Auto-1	845	−3.41	3.41	−0.01	0.57	9.16	5.26	8.30	28.22	14.39
Truck-1	473	−3.41	3.41	0.03	4.56	9.23	6.85	3.41	38.24	25.76
Auto-2	1444	−3.41	3.41	0.02	0.00	3.97	1.31	6.43	13.62	9.93
Truck-2	1130	−3.41	3.41	0.00	1.36	6.71	3.42	25.08	41.58	32.91
Auto-3	1412	−3.41	3.41	0.00	0.00	3.07	1.80	5.68	14.70	10.08
Truck-3	1415	−3.41	3.41	0.04	0.67	7.25	2.74	6.12	15.14	10.16

values. Raw data has also been affected by high-to-medium-frequency disturbances due to speed transitions, which need to be modified. Several studies in literature have proposed techniques to rectify errors in the raw data before being used in traffic flow theory problems (*36, 37*). In this study, the proposed multistep filtering procedure in the References (*36*) is applied to the NGSIM data to create reconstructed reliable trajectory data. First, the data points out that their absolute acceleration/deceleration exceeds a certain threshold value, which is 3 m/s^2 within this scope, and are detected as unrealistic values. Using a filtering technique such as natural cubic spline interpolation, new longitudinal locations of outliers are interpolated between a sequence of non-outlier points before and after the outliers. The outlier speed and acceleration are then calculated from the new distance locations through the basic equations of motion. Secondly, a low-pass filter is utilized to remove the random error component from speed and acceleration profiles. Further details on the vehicle-trajectory-reconstruction procedure can be found in the References (*36*).

In this study, a total of six data series, two from each subset, are utilized for the analysis. In each of the three subsets, two pairs of following and leading vehicles that have the highest number of data points were selected. The following vehicle in one pair is an Auto, and in the other is a Truck vehicle. Each pair builds one data series that ends up with the total six data series, named Auto-1, Truck-1, Auto-2, Truck-2, Auto-3, and Truck-3. Summary statistics of the acceleration, speed, and relative distance of the following vehicle in the original data series are provided in Table 1. Using the above-mentioned filtering procedure, the reconstructed data of speed and acceleration profiles for Auto-2 data series, as the data series with the maximum number of samples, are plotted against the original NGSIM data in Figure 1. The proposed car-following algorithm and the corresponding analysis are applied to all data series.

Results and Discussion

The model evaluation process for the proposed GBRT-based car-following model is implemented in two steps. First, the regularization parameters (v, M, D, τ) are optimized using 5-fold cross-validation to ensure a good fit of data. In the second stage, using the optimal values found in the first stage, the performance of the GBRT model is compared with the optimum GHR model that has been optimized based on Equations 2 and 3. It should be noted the optimal reaction time is also determined for the GHR model to have a fair comparison. As mentioned in the Methodology section, the data instances that are fed into both algorithms vary according to the reaction time (τ) value. Such a time dependency between consequent data instances hinders splitting them randomly into the training and test sets. Consequently, in this study, the first 80% of each data series is selected as the training data set while holding on to the remaining 20% of the data as the unseen records to be used only in the second step of the model evaluation process. Both steps are consecutively implemented in each of the six data series separately and their results are reported.

First Step: Regularization Parameters

In the sensitivity analysis for computing regularization parameters, 5-fold cross-validation is utilized to obtain the average prediction error. Due to the inherent time dependency between the explanatory and response variables of the consequent data samples, unlike general cross-validation, each fold is not created randomly. While the test set is still held on to for the second step of the evaluation, the training set is subdivided into five subsets in such a way that the first 20% of the training portion of each data series constitutes the first fold, the second 20% forms the second fold, and so forth.

A matrix with different combinations of v and M is designed for simultaneously optimizing the learning rate and the number of weak learners. The vectors $\mathbf{V} = [0.1, 0.3, 0.5, 0.8, 1]$ and $\mathbf{M} = [1:1:9, 10:10:90, 100:100:1000]$ constitute the rows (learning rate) and columns (the number of weak learners) of the matrix, respectively. For each element of the designed VM matrix (i.e., a pair of (v, M)), 5-fold cross-validation is applied to the training set to measure the effectiveness of

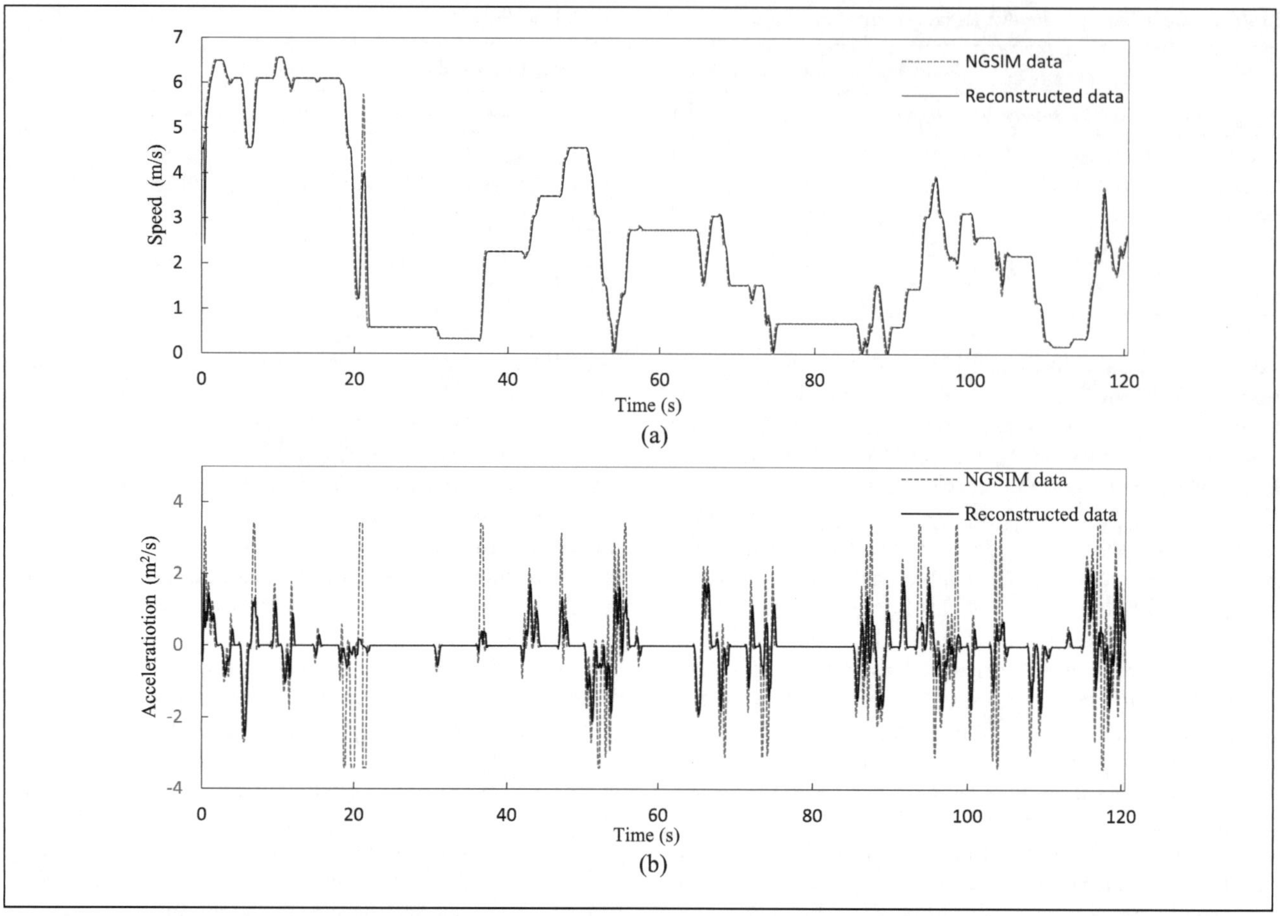

Figure 1. Comparison between the NGSIM and reconstructed data of speed and acceleration profiles for the Auto-2 data series: (*a*) speed profile and (*b*) acceleration profile.

the model. At each iteration, the GBRT is fitted to the four folds and validated on the remaining fold by computing the mean square error (MSE) as the performance metric. The final performance measure pertaining to each pair (v, M) is the average of the MSE values computed in the loop.

Figure 2 illustrates how the GBRT prediction error is influenced by various combinations of (v, M) for the data series Auto-2 and Truck-2, as the data sets with the highest number of samples. Each line represents the variation of the average MSE for a particular value of v over the full range of the vector **M**. Initially, after a fluctuation for the values of M less than 10, the average MSE increases gradually when the number of weak learners also rise. However, the rate of average MSE levels out after M increases to hundreds of weak learners. Intuitively, it was expected that prediction performance is enhanced by raising the number of base learners. Nonetheless, the results indicate that augmenting the M renders the model over-complex and overfitted, which in turn leads to a deteriorating accuracy of the GBRT model. Thus, the lower values of M are more efficient in generalizing the GBRT model to unseen instances. According to the sensitivity analysis of (v, M) for all data series, $M = 8$ is chosen as the optimum value for further analysis. Furthermore, as shown in the Figure 2, the smaller values of learning rate give a more improved predictive model. Although setting the learning rate to values lower than 0.1 will almost always ameliorate prediction performance, we choose $v = 0.1$ as the optimal value to maintain the computational cost at a reasonable level (*38*). It is worth mentioning that similar trends in Figure 2 have been obtained for other data series as well.

Figure 3*a, b* delineates how accuracy of the GBRT model is influenced by changing the size of the decision trees (D) and the reaction time (τ), respectively. The average MSE is computed for each value of D and τ with

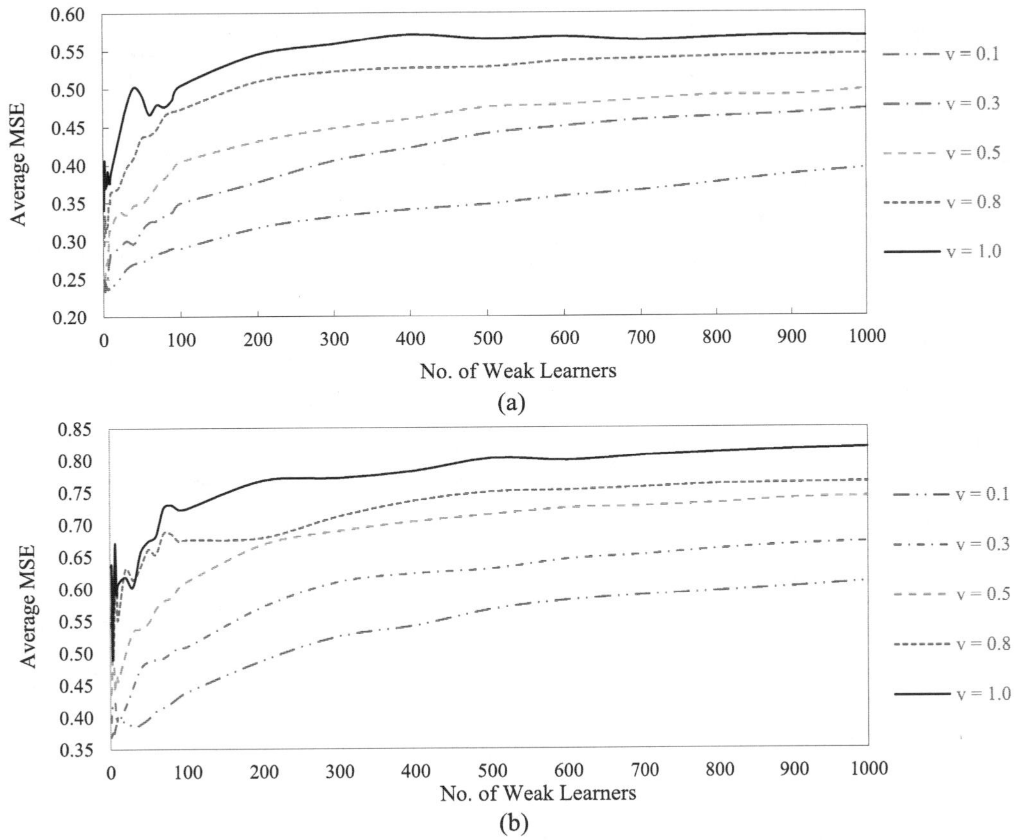

Figure 2. Average MSE for the GBRT model upon various combinations of the number of weak learners (*M*) and the learning rate (*v*) for two data series: (*a*) Auto-2, (*b*) Truck-2.

the analogous approach to the v and M. As depicted in Figure 3*a*, the average MSE for almost all data series declines slightly by setting $D = 3$. However, the variation in MSE remains constant for values of $D > 3$. Accordingly, the $D = 3$ is selected as the optimum value of the decision tree depth. Such a plateau stems from that fact that the car-following model contains only a few features. As a consequence, the possibility of facing the overfitting problem decreases even if the decision trees become over-complex by increasing the value of D. Furthermore, as depicted in Figure 3*b*, the GBRT prediction performance remained approximately constant over the reaction time interval although the lower values of the reaction time work better for some of the data series such as Auto-1 and Auto-2. It is clear that the optimal value of the reaction time is an essential characteristic of the driver in a given data set rather than the learning algorithm, which may be distinct for other data sets. Since the scope of this study is to tune the GBRT parameters instead of determining a reliable reaction time interval for all possible data sets, we use the exact optimal value of the reaction time obtained for each data series. In summary, as mentioned in the Methodology section, the most effective regularization parameters for the GBRT-based car-following models are (v, M) while the (D, τ) do not significantly impact on controlling the degree of fit to the data.

Step 2: Comparison between GBRT and GHR Models

First, we optimize the parameters of the GBRT and GHR models for establishing a fair comparison between the two predictive models, as one of the salient objectives in this study. Then, for having a robust comparison, the fitted models are examined on the test samples of each data series, which have not been deployed in the training process. For the test portion of each data series, the acceleration of the following vehicle is estimated with both techniques and the corresponding MSEs are reported as shown in Figure 4.

The comparison between the two models proves the superiority of the GBRT model in all data series,

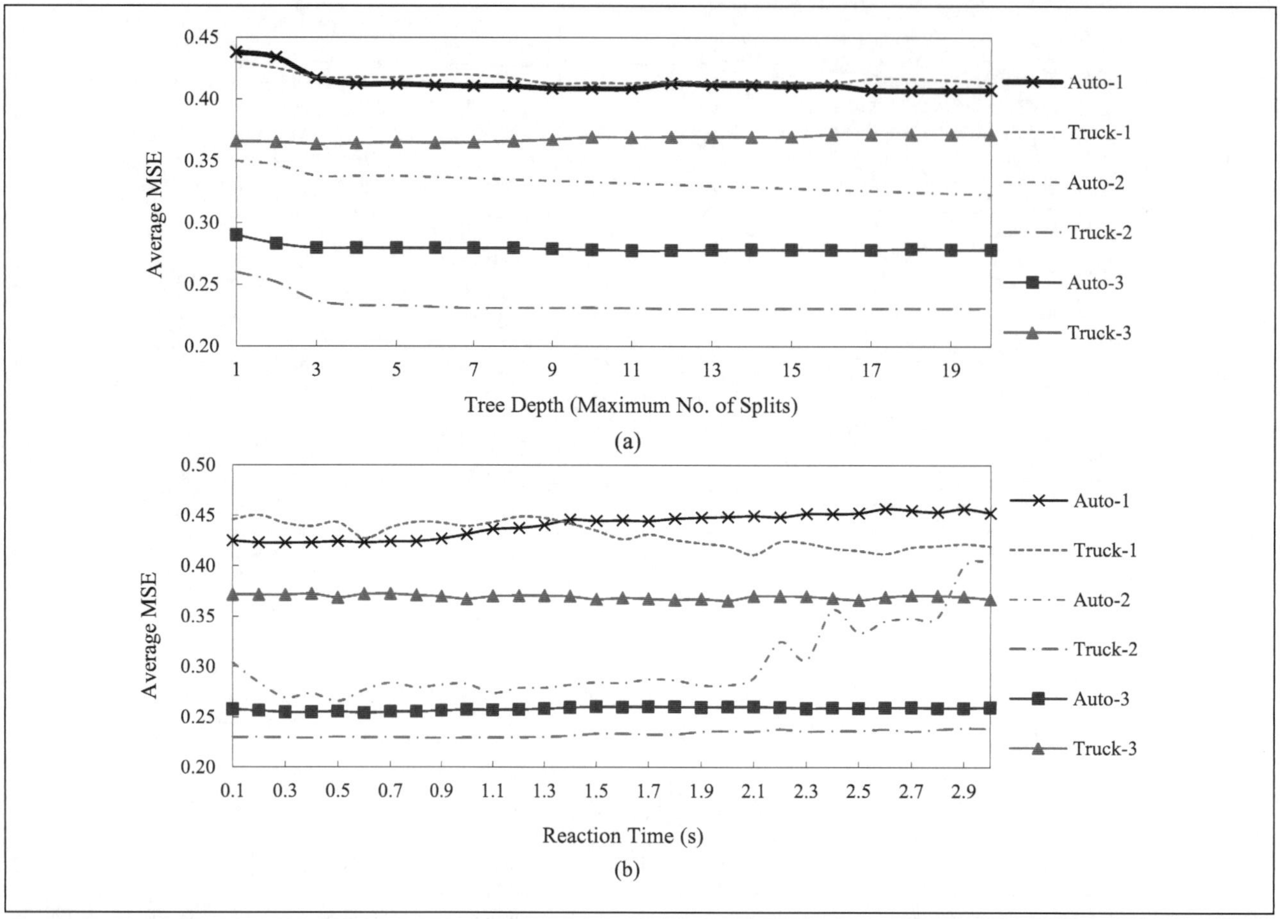

Figure 3. Average MSE of the GBRT model upon different values of: (*a*) Tree Depth (Maximum Number of Splits), (*b*) Reaction Time(s).

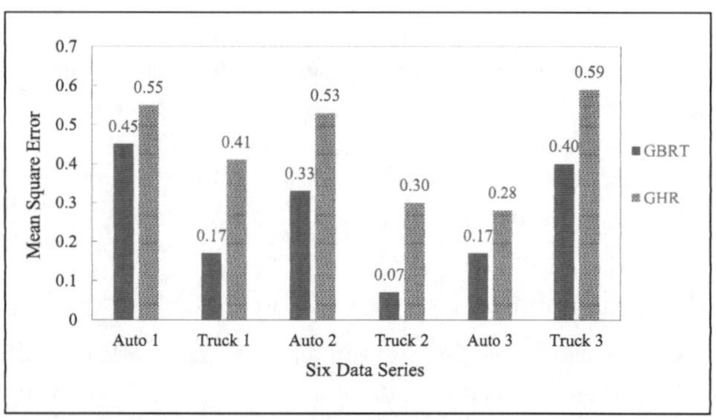

Figure 4. Performance comparison between the GBRT and GHR models based on MSE, as the goodness-of-fit measure, for six data series.

regardless of whether the auto or truck is the follower vehicle. The primary reason for the superiority of GBRT originates from the fact that such models are unrestricted in growth according to the relationship among the variables. However, constraining the interdependency between features of the car-following behavior to a predefined structure lessens the GHR power to capture more aspects of driver behavior.

Conclusion

With the emergence of advanced data-driven approaches and technologies for collating the massive vehicle trajectory data sets, the sophisticated car-following models can be developed so as to alleviate the shortcomings of the classical car-following models, such as restriction to mathematical equations and inability to incorporate additional effective factors. With the help of data mining techniques, there is no need to define the functionality form of the model. This is a practical benefit particularly in the complex traffic flow condition where building an appropriate model is almost impossible. Furthermore, data-driven approaches give room to researchers to investigate the effect of new factors on car-following behavior, including other kinds of the microscopic and macroscopic parameters, weather conditions, leading vehicle types, time of day or week, road geometry, and the motion inputs related to the vehicles other than the leading vehicle.

In this study, the GBRT algorithm was presented to reproduce the response action of the following vehicle in accordance with the motion characteristic of the leading vehicle. Data sets for the analysis were captured from the NGSIM program. Data preprocessing techniques were applied to the raw data to remove random error measurements and obtain a reliable data source. To ensure a promising model, first, the regularization parameters of the model were determined. Second, the optimal GBRT and GHR models were trained on the same training set. The prediction results of applying both models to the unseen test data demonstrated the superiority of the proposed GBRT algorithm compared to the well-known GHR model.

As a future research direction, it would be interesting to simultaneously capture other drivers' stochastic behavior. Using advanced machine learning algorithms, lateral and longitudinal behaviors of a vehicle can be simulated by integrating the lane-changing, car-following, and gap-acceptance concepts. Furthermore, outstanding frameworks for replacing the traditional models with the data-driven approaches in the commercial microsimulation packages such as VISSIM and CORSIM should be provided.

Author Contributions

The authors confirm contribution to the paper as follows: Research Supervision: Dr. Montasir Abbas; study conception and design: Sina Dabiri, Dr. Montasir Abbas; data collection: Publically Available; analysis and interpretation of results: Sina Dabiri; draft manuscript preparation: Sina Dabiri. All authors reviewed the results and approved the final version of the manuscript.

References

1. Dabiri, S., and M. Abbas (eds.). *Arterial traffic signal optimization using Particle Swarm Optimization in an integrated VISSIM-MATLAB simulation environment*. Presented at 19th International Conference on Intelligent Transportation Systems (ITSC), IEEE, 2016.
2. Dabiri, S., K. Kompany, and M. Abbas. Introducing a Cost-Effective Approach for Improving the Arterial Traffic Performance Operating under the Semi-Actuated Coordinated Signal Control. *Transportation Research Record: Journal of the Transportation Research Board*, 2018. In press.
3. Janson Olstam, J, and A. Tapani. *Comparison of Car-following Models*. Statens väg-och transportforskningsinstitut, 2004.
4. Aghabayk, K., N. Forouzideh, and W. Young. Exploring a Local Linear Model Tree Approach to Car-Following. *Computer-Aided Civil and Infrastructure Engineering*, Vol. 28, No. 8, 2013, pp. 581–593.
5. Ahmed, K. I. *Modeling Drivers' Acceleration and Lane Changing Behavior*. Massachusetts Institute of Technology, 1999.
6. Brackstone, M, and M. McDonald. Car-following: A Historical Review. *Transportation Research Part F: Traffic Psychology and Behaviour*, Vol. 2, No. 4, 1999, pp. 181–196.
7. Gazis, D. C., R. Herman, and R. W. Rothery. Nonlinear Follow-the-leader Models of Traffic Flow. *Operations Research*, Vol. 9, No. 4, 1961, pp. 545–567.
8. Gipps, P. G. A. Behavioural Car-following Model for Computer Simulation. *Transportation Research Part B: Methodological*, Vol. 15, No. 2, 1981, pp. 105–111.
9. Koshi, M., M. Kuwahara, and H. Akahane. Capacity of Sags and Tunnels on Japanese Motorways. *ITE Journal*, Vol. 62, No. 5, 1992, pp. 17–22.
10. Treiber, M., A. Hennecke, and D. Helbing. Congested Traffic States in Empirical Observations and Microscopic Simulations. *Physical Review E*, Vol. 62, No. 2, 2000, p. 1805.
11. Bando, M., K. Hasebe, A. Nakayama, A. Shibata, and Y. Sugiyama. Dynamical Model of Traffic Congestion and Numerical Simulation. *Physical Review E*, Vol. 51, No. 2, 1995, p. 1035.
12. Helly, W. (1959) Simulation of Bottlenecks in Single Lane Traffic Flow. Proceedings of the Symposium on Theory of Traffic Flow, Research Laboratories, General Motors, 207–238.
13. Nagel, K, and M. Schreckenberg. A Cellular Automaton Model for Freeway Traffic. *Journal de Physique I*, Vol. 2, No. 12, 1992, pp. 2221–2229.
14. Wiedemann, R, and U. Reiter. Microscopic Traffic Simulation: The Simulation System MISSION, Background and Actual State. *Project ICARUS (V1052) Final Report*, Vol. 2, 1992, pp. 1–53.
15. Dabiri, S., and K. Heaslip. Transport-domain applications of widely used data sources in the smart transportation: A survey. arXiv preprint arXiv:1803.10902 (2018).
16. He, Z., L. Zheng, and W. Guan. A Simple Nonparametric Car-following Model Driven by Field Data. *Transportation Research Part B: Methodological*, Vol. 80, 2015, pp. 185–201.
17. Panwai, S, and H. Dia. Comparative Evaluation of Microscopic Car-following Behavior. *IEEE Transactions on Intelligent Transportation Systems*, Vol. 6, No. 3, 2005, pp. 314–325.
18. Hamdar, S. *Driver Behavior Modeling*. Handbook of Intelligent Vehicles, Springer, 2012. pp. 537–558.
19. Saifuzzaman, M. and Z. Zheng. Incorporating Human-factors in Car-following Models: A Review of Recent Developments and Research Needs. *Transportation Research Part C: Emerging Technologies*, Vol. 48, 2014, pp. 379–403.
20. Khodayari, A., A. Ghaffari, R. Kazemi, and R. Braunstingl (eds). Modify Car-Following Model by Human Effects Based on Locally Linear Neuro Fuzzy. *Intelligent Vehicles Symposium (IV)*, 2011, pp. 661–666.

21. Papathanasopoulou, V, and C. Antoniou. Towards Data-driven Car-following Models. *Transportation Research Part C: Emerging Technologies*, Vol. 55, 2015, pp. 496–509.
22. Wei, D, and H. Liu. Analysis of Asymmetric Driving Behavior Using a Self-learning Approach. *Transportation Research Part B: Methodological*, Vol. 47, 2013, pp. 1–14.
23. Kikuchi, S, and P. Chakroborty. Car-following Model Based on Fuzzy Inference System. *Transportation Research Record: Journal of the Transportation Research Record*, 1992. 1365: 82–91.
24. Wu, J., M. Brackstone, and M. McDonald. Fuzzy Sets and Systems for a Motorway Microscopic Simulation Model. *Fuzzy Sets and Systems*, Vol. 116, No. 1, 2000, pp. 65–76.
25. Gao, Q., S. Hu, and C. Dong (eds). *The Modeling and Simulation of the Car-following Behavior Based on Fuzzy Inference*. Presented at WMSO'08 International Workshop on Modelling, Simulation and Optimization, IEEE, 2008.
26. Hongfei, J., J. Zhicai, and N. Anning (eds). Develop a Car-following Model using Data Collected by "Five-wheel System". *Proc., Intelligent Transportation Systems*, 2003.
27. Panwai, S, and H. Dia. Neural Agent Car-following Models. *IEEE Transactions on Intelligent Transportation Systems*, Vol. 8, No. 1, 2007, pp. 60–70.
28. Colombaroni, C, and G. Fusco. Artificial Neural Network Models for Car Following: Experimental Analysis and Calibration Issues. *Journal of Intelligent Transportation Systems*, Vol. 18, No. 1, 2014, pp. 5–16.
29. Tanaka, M, and N. Uno. Evaluation of Car-Following Input Variables and Development of Three-Vehicle Car-Following Models with Artificial Neural Networks. *Journal of the Eastern Asia Society for Transportation Studies*, Vol. 11, 2015, pp. 1826–1841.
30. Khodayari, A., A. Ghaffari, R. Kazemi, and R. Braunstingl. A Modified Car-following Model Based on a Neural Network Model of the Human Driver Effects. *IEEE Transactions on Systems, Man, and Cybernetics-Part A: Systems and Humans*, Vol. 42, No. 6, 2012, pp. 1440–1449.
31. Zheng, J., K. Suzuki, and M. Fujita. Car-following Behavior with Instantaneous Driver–Vehicle Reaction Delay: A Neural-network-based Methodology. *Transportation Research Part C: Emerging Technologies*, Vol. 36, 2013, pp. 339–351.
32. Chong, L., M. M. Abbas, A. M. Flintsch, and B. Higgs. A Rule-based Neural Network Approach to Model Driver Naturalistic Behavior in Traffic. *Transportation Research Part C: Emerging Technologies*, Vol. 32, 2013, pp. 207–223.
33. scikit-learn. *Ensemble Methods*. http://scikit-learn.org/stable/modules/ensemble.html. Accessed July, 2017.
34. Friedman, J. H. Greedy Function Approximation: A Gradient Boosting Machine. *Annals of Statistics*, 2001, pp. 1189–1232.
35. Friedman, J., T. Hastie, and R. Tibshirani. *The Elements of Statistical Learning*. Springer Series in Statistics Springer, Berlin, 2001.
36. Montanino, M, and V. Punzo. Making NGSIM Data Usable for Studies on Traffic Flow Theory: Multistep Method for Vehicle Trajectory Reconstruction. *Transportation Research Record: Journal of the Transportation Research Board*, 2013. 2390: 99–111.
37. Punzo, V., M. T. Borzacchiello, and B. Ciuffo. On the Assessment of Vehicle Trajectory Data Accuracy and Application to the Next Generation SIMulation (NGSIM) Program Data. *Transportation Research Part C: Emerging Technologies*, Vol. 19, No. 6, 2011, pp. 1243–1262.
38. Ridgeway, G. Generalized Boosted Models: A Guide to the GBM Package. *Update*, Vol. 1, No. 1, 2007, p. 2007.

The Standing Committee on Artificial Intelligence and Advanced Computing Applications (ABJ70) peer-reviewed this paper (18-01813).

Article

TRR
JOURNAL OF THE TRANSPORTATION RESEARCH BOARD

Transportation Research Record
2018, Vol. 2672(45) 147–159

DOI: 10.1177/0361198118788427
journals.sagepub.com/home/trr
SAGE

Simultaneous Estimation of Flexible Models and Associated Hyperparameters: An Application to Travel Activity-Duration Modeling

Daniel Emaasit[1] and Alexander Paz[1]

Abstract

Estimation of flexible-statistical models of travel demand involves tuning varying parameters, hyperparameters, manually and iteratively. Proper tuning of hyperparameters results in superior models. However, considerable expertise, including technical knowledge of statistics, data mining or machine learning, and experience are required to tune hyperparameters and consequently generate appropriate models. Moreover, tuning hyperparameters is prone to subjective error and consequently produces travel demand models that are difficult to reproduce and extend, and makes the development more an art than a science. There is a need for methods to reduce or eliminate subjectivity during the tuning process. This study proposed a framework to reduce subjectivity during the tuning of hyperparameters required for the estimation of nonparametric models of activity-duration. That is, a flexible-statistical framework, which leverages state-of-the-art innovations in Bayesian optimization (BO), was proposed to estimate Gaussian process models of activity duration and associated hyperparameters. The framework was applied to estimate duration models for five types of out-of-home non-mandatory activity episodes for household individuals in the greater Los Angeles area. Experiments demonstrate that the accuracy of results from the proposed framework are superior to those from the current tuning process, and are obtained in a fraction of the time. The proposed framework could potentially increase the productivity of modelers by reducing time required to tune hyperparameters.

The life cycle of a typical travel demand modeling process requires upgrades to multiple statistical analyses to capture changes in travel behavior patterns. These statistical analyses are tedious and involve an iterative procedure that requires expert knowledge and time to perform several tasks including but not limited to preprocessing data, selecting appropriate attributes, selecting appropriate models, finding model parameters, and critically analyzing results (*1*). The procedure is governed by multiple assumptions such as functional forms. In general, models that require weak assumptions about the phenomena under study are preferred. Models estimated under weak assumptions are termed in the literature as flexible and the data drives their estimation (*2*). Flexible models are also known as nonparametric and have been demonstrated to have superior predictive performance compared with traditional econometric methods (*3*) (Emaasit and Paz, unpublished).

Estimation of flexible-statistical models of travel demand (*3*) involves tuning varying parameters, hyperparameters, manually and iteratively. Hyperparameters enable the estimation of different models for special-purpose applications and datasets. Typically, the values for these hyperparameters are specified by an analyst. An example includes a penalty hyperparameter to control the size of coefficients in a linear ridge regression model. This hyperparameter is different from model parameters, the coefficients in the linear regression model that are estimated by a search algorithm. Considering that there is no general optimizer, all optimization models use hyperparameters that require tuning. Often, optimization algorithms are selected based on how involved the tuning of their hyperparameters is for the solution of a particular problem (*4*).

[1]Transportation Research Center, Howard R. Hughes College of Engineering, Department of Civil & Environmental Engineering, University of Nevada, Las Vegas, NV

Corresponding Author:
Address correspondence to Daniel Emaasit: emaasit@unlv.nevada.edu

Values for the hyperparameters are set by a modeler to initialize and estimate a flexible-statistical model of travel demand (*3*). Proper tuning of hyperparameters results in superior models (*5*). However, considerable expertise, including technical knowledge of statistics, data mining or machine learning, and experience are required to tune hyperparameters and consequently select appropriate models. Tuning hyperparameters is prone to subjective error and consequently produces travel demand models (TDMs) that are difficult to reproduce and extend, and makes the development more an art than a science (*5*). Hence, there is a need for methods to reduce or eliminate subjectivity during the tuning process.

Estimation of TDMs results in costly functions that require significant amounts of computation resources and time. Involving the modeler in the tuning of hyperparameters is computationally too expensive because even a single run may take days or weeks to complete. This is a major limitation in cases with large amounts of data. Using a subset of a large dataset to produce models having superior performance is more computation- and time-efficient than using the entire large dataset to produce models having similar performance. Therefore, there is a need for methods that are data efficient such that only required data points are used to obtain adequate performing models.

Currently, there is increasing use of emerging methods for simultaneous hyperparameter tuning and model selection in scientific fields, including biology (*6*), natural language processing (*7*), robotics (*8*), autonomous vehicles (*9*) and medicine (*10*). However, there is very limited research and application in transportation planning. Algorithms that take the role of the modeler, or assist in the task of tuning hyperparameters and selecting appropriate models, are important for transportation planning in at least three ways:

- To reduce subjective error in model estimation.
- To increase productivity of modelers by reducing amount of time spent on tuning and model selection. Modelers can focus their time on other tasks, such as model interpretation and reporting as well as running more experiments.
- To enable modelers with minimal technical knowledge to estimate models.

This study proposed an approach to reduce subjectivity during the tuning of hyperparameters required for the estimation of nonparametric models. That is, a flexible-statistical framework was proposed to estimate a model and associated hyperparameters. The framework was applied, as an example, to estimate flexible activity-duration models for out-of-home non-mandatory activity episodes for household individuals in the greater Los Angeles area (similar to Ferdous et al. [*11*]).

Activity-duration modeling has received a considerable amount of research attention (*3*, *12–19*). This is because activity-based modeling (ABM) places emphasis on understanding participation characteristics of individuals including the duration spent performing an activity (*1*). The amount of time spent at activity locations contributes to dynamic changes in the built environment (*18*, *19*). For instance, locations that experience longer durations of out-of-home non-mandatory activities, such as shopping and recreation, will attract leisure services (such as shopping malls, markets) thereby affecting travel behavior. Several operational frameworks of ABM, implemented in many metropolitan areas in the U.S., rely on estimates of durations for out-of-home work activities for use in generating and scheduling out-of-home non-mandatory activities (*14*). Some studies have investigated the factors that influence durations of activity episodes. For instance, Habib (*16*) studied factors affecting out-of-home physical activities by children in the city of Dhaka, Bangladesh. Car ownership, tenure type, activity location, and parent-escort arrangements were found to be the most notable predictors of the duration of physical activity. Insights from this study provided important policy implications. The reader is referred to Pinjari et al. (*14*) for a more detailed exposition of other benefits of activity-duration modeling.

The rest of this paper is organized as follows. Related Work reviews literature in the transportation domain that have proposed some comparative methods for hyperparameter tuning and model selection. Methodology presents the details of the proposed methodology including: a methodological context that highlights the current emerging methods for automated hyperparameter tuning and model selection; and methodology formulation which develops two important components of the proposed framework. Experiments presents the experiments performed including a description of the data used, the sample formation process, and empirical analysis for the five different activity episodes. Discussion of Results, presents a general discussion of the results as it relates to activity-duration modeling and implications for travel planning. Finally, a conclusion summarizes the important findings from this study, including potential directions for future research.

Related Work

The current state of the art in transportation has proposed several methods for the estimation of statistical models while reducing the need for inputs from the modeler. For instance, the selection of variables to be considered has received some attention. Recently, Habibi

(*20*) used a non-random hold-out validation framework to select variables to be considered for predicting car type demand. Another study by Ahmed et al. (*21*) used different search techniques including stepwise, forward, and backward search to identify significant variables associated with crash occurrence. All three search techniques identified the same variables to be significant in their multivariate analysis. Both proposed frameworks only addressed variable selection but did not consider other key aspects such as model selection, parameter estimation, and hyperparameter tuning.

Other studies have focused on search techniques to obtain the best model specification. For instance, Washington and Wolf (*22*) compared Ordinary Least Squares (OLS) regression and Hierarchical tree-based regression (HTBR) methods for estimating trip generation models. They demonstrated that HTBR methods automatically search for statistical relationships between response and predictor variables. Kim et al. (*23*) compared several techniques for calibrating traffic microsimulation flow models including: manual search, gradient-based, simplex-based, and artificial intelligence techniques. Their proposed approach required pre-specifying a statistically based objective function for use in the calibration process. Athanasopoulos et al. (*24*) implemented three time series algorithms, including Forecast Pro, ARIMA, and exponential smoothing based algorithms, for forecasting tourism demand. Their methods followed a step-wise algorithm, proposed by Hyndman and Khandakar (*25*), to traverse the space of models efficiently to obtain a model with the lowest Akaike Information Criterion (AIC) value. Tulic et al. (*26*) developed a procedure for predicting route travel times and mean speeds in urban networks. Their approach considered features of input data, particularly measurement errors, by including a model for heteroscedasticity.

Advanced methods for parameter-tuning have also been proposed. For instance, Abraham and Hunt (*27*) used a heuristic search method, Levenberg-Marquardt, to tune parameters to improve the goodness of fit of their nested logit model for travel destination and mode choice. This heuristic search method was a much faster approach than an initial manual process. Hollander (*28*) proposed a multi-objective approach to predict travel times and modes. The approach involved combining several objective functions, rather than using one as is the case of maximum likelihood estimation. The full set of mode choices made by travelers was estimated. Then, the objective functions were used to test the extent to which the estimated choices met travelers' needs. This was performed repeatedly with different possible sets of parameters. The downhill simplex method was used to search for the combination of parameters that best estimated travel times and modes. The main drawback of this approach is that it is computationally expensive in terms of the number of interactions required to reach convergence, which is not always guaranteed.

To the best of the authors' knowledge, comprehensive estimation including model selection, hyperparameter tuning and parameter estimation using current emerging techniques (described in the methodology section) have not been explored for developing TDMs.

Methodology

Methodology Context

A systematic approach to handle the iterative process of hyperparameter tuning, model selection, and parameter estimation would be to treat the process as an optimization problem (*29*) in order to search for the best fit between the model and observed data (*29*). Generally, such optimization can be categorized into two groups, including methods that:

- search in a finite model space of known standard models, such as logit, probit, and mixed logit; and
- use a nonparametric approach to search in an infinite model space for the model and parameters that best explain the observed data.

The latter category, to a great extent, lets the data dominate the process of model discovery. This category often results in the discovery of new model specifications not included in the existing literature (*30*). The methodology proposed in this study is based on the latter category of methods.

Optimization methods that search in an infinite model space seek to find a model that is fundamentally superior to others in terms of predictive performance (*31*). This includes estimating the corresponding parameters to produce models that best describe the observed data and can generalize well to new observations (*32*). Grid search is an example of such optimization methods (*30*), and requires specifying a set of configurations for the hyperparameters so that a grid of models and parameters is generated. Each of these configurations is fitted, and the one with the best predictive performance is selected. Grid search becomes more computationally intensive as the data or model increases in complexity.

These drawbacks can be addressed by using random search (RS) (*33*). Instead of searching the entire grid of models and hyperparameters, RS takes random walks over this large grid. This method, which has been shown to be superior to grid search (*30*), is considered the de facto method for tuning hyperparameters because of its simplicity and efficiency. However, in both grid and RS, explorations are done randomly and blindly. That is, any set of hyperparameters and models is explored independently of the previous one. This results in loss of valuable information that could help the next iteration.

This drawback can be addressed by using sequential model-based methods, such as Sequential Model-Based Global Optimization (SMBO) (*5, 30, 34–36*). SMBO methods, such as Bayesian optimization (BO) (*30*), use a surrogate function to approximate the unknown function of the "true" model. Given that this surrogate function is simple to evaluate, it can be optimized to select points that yield the best approximation of the true function while leveraging information gathered from previous iterations. BO has demonstrated success in optimizing millions of hyperparameters in neural networks (*9*). The proposed approach in this study leverages BO to formulate a framework that simultaneously seeks adequate hyperparameters and selects a flexible activity-duration model.

Methodology Formulation

Generally, the proposed methodology consists of two main stages:

- Stage 1: A surrogate function for the model that relates activity duration and associated factors is proposed.
- Stage 2: The surrogate is optimized iteratively using a few data points until it approaches the "true" model of activity duration.

To formulate the methodology, consider for each individual traveler, i, that y_i represents the log of activity duration and $\mathbf{x}_i$ is a vector of potential explanatory variables such as trip distance, income levels, and household characteristics. Regression modeling can involve estimating a latent function f, which maps a vector of input data, $\mathbf{x}_i$, to output data y_i for $i = 1, 2,..., N$, where N is the total number of travelers. Each of the input vectors $\mathbf{x}_i$ is of dimension D, and $\mathbf{X}$ is a N x D matrix with rows $\mathbf{x}_i$. The observations are assumed to satisfy:

$$y_i = f(\mathbf{x}_i) + \varepsilon, \quad where\ \varepsilon \sim \mathbf{N}(0, \sigma_\varepsilon^2) \tag{1}$$

The noise term, ε, is assumed to be normally distributed with a zero mean and variance, σ_ε^2. Latent function f represents a hidden underlying travel behavior that produced the observed activity-duration data. The objective of the optimization framework is to find the set of arguments and values, $\mathbf{x}_*$, that maximizes (or minimizes) a continuous function $f(\mathbf{x})$ over a compact space x:

$$\mathbf{x}_* = arg\ max_{x \in x} f(\mathbf{x}) \tag{2}$$

Given the complexity of the functional relationship between activity duration and many associated factors, it is difficult to know a priori the function to use for $f(\mathbf{x})$. That is, it is difficult to know whether $f(\mathbf{x})$ is linear, quadratic, sinusoidal or polynomial to some order. Hence, the first stage of proposed methodology involves using a surrogate function to approximate $f(\mathbf{x})$.

A probabilistic function sampled from a Gaussian process (GP) prior (*2, 37, 38*) was chosen as the surrogate for the functional relationship between activity duration and associated factors. The reason for this choice is because a GP prior provides a probability distribution over an infinite space of possible functions (*37*). This surrogate function is fully parameterized by a mean vector and covariance matrix (kernel), denoted as

$$\mathbf{f} \sim GP(\mathbf{m}, \mathbf{K}_{N,N}), \tag{3}$$

where

$\mathbf{f}$ = a surrogate function that relates activity duration and associated factors,

$\mathbf{m}$ = a mean vector (mean function) that relates activity duration and associated factors, and

$\mathbf{K}_{N,N}$ = a covariance matrix (kernel function) that captures correlation between regressors.

Bayes theorem then was used to estimate the posterior distribution over the unknown function evaluations $\mathbf{f}$ at all data points $\mathbf{x}_i$, as illustrated by:

$$\begin{aligned} p(\mathbf{f}|\mathbf{y}, \mathbf{X}) &= \frac{p(\mathbf{y}|\mathbf{f}, \mathbf{X})p(\mathbf{f})}{p(\mathbf{y}|\mathbf{X})}, \\ &= \frac{p(\mathbf{y}|\mathbf{f}, \mathbf{X})\mathrm{N}(\mathbf{f}|\mathbf{m}, \mathbf{K}_{N,N})}{p(\mathbf{y}|\mathbf{X})}, \end{aligned} \tag{4}$$

where

$p(\mathbf{f}|\mathbf{y}, \mathbf{X})$ = the posterior distribution of functions that best explain log activity duration, given the covariates,

$p(\mathbf{y}|\mathbf{f}, \mathbf{X})$ = the likelihood of log of activity duration, given the functions and covariates,

$p(\mathbf{f})$ = the prior over all possible functions of log of activity duration, and

$p(\mathbf{y}|\mathbf{X})$ = the data (constant).

This posterior is a GP composed of a distribution of possible functions that best explain activity duration patterns. Given that the data are fixed, Equation 4 was reformulated as the unnormalized posterior distribution:

$$p(\mathbf{f}|\mathbf{y}, \mathbf{X}) \propto p(\mathbf{y}|\mathbf{f}, \mathbf{X})\mathrm{N}(\mathbf{f}|\mathbf{m}, \mathbf{K}_{N,N}). \tag{5}$$

The kernel function, $\mathbf{K}_{N,N}$, was used to encode prior information about the smoothness of the function that relates activity duration and associated attributes. In this study, it was assumed that ?travelers with similar personal and household characteristics spend more or less similar time during out-of-home non-mandatory activity episodes.? For instance, travelers with similar characteristics, such as income levels and job types, were assumed to have more similar activity durations than individuals with different characteristics. This smoothness assumption was encoded using the Squared Exponential-Automatic Relevance Determination (SE-ARD) kernel (*39*). Each covariance element in this kernel, $k(x_i, x_j)$, was evaluated as:

$$k_{se-ard}(\mathbf{x}_i, \mathbf{x}_j^{'}) = \boldsymbol{\sigma}_f^2 exp(-\frac{1}{2}\sum_{d=1}^{D}\frac{(x_{i,d} - x_{j,d})^2}{l_d^2}), \quad (6)$$

where σ_f^2 is the signal or function variance to capture the variance in the estimated functions; and l_d is the characteristic length-scale in dimension d, to capture the relevance of each covariate to activity duration. The likelihood function of activity duration was expressed as:

$$p(\mathbf{y}|\mathbf{f}, \mathbf{X}) = \mathbf{N}(\mathbf{f}, \boldsymbol{\sigma}_\varepsilon^2). \quad (7)$$

The hyperparameters in this surrogate model include $\theta = \{\sigma_f^2, l_d, \sigma_\varepsilon\}$. Given the GP prior in Equation 3 and the data likelihood in Equation 7, the posterior of the hyperparameters and parameters of the surrogate were estimated using Bayesian inference as

$$p(\mathbf{f}|\mathbf{y}, \mathbf{X}) \propto p(\mathbf{y}|\mathbf{f}, \mathbf{X})\mathbf{N}(\mathbf{f}|\mathbf{m}, \mathbf{K}_{N,N}). \quad (8)$$

The second stage involves optimizing stage 1 to run efficiently. Given that evaluations of the surrogate model $f(\mathbf{x})$ maybe expensive due to its complexity and/or use of large dataset, an approach is proposed to select a subset of the large dataset to produce a model of comparable performance. The approach selects a few data points at a time to estimate the surrogate function iteratively. That is, as new data points, $D_{1:T} = \{x_{1:T}, y_{1:T}\}$, are accumulated from each iteration t, where T is the total number of iterations, the GP prior is combined with the data likelihood to update the posterior distribution of the surrogate, $p(\mathbf{f}|\mathbf{y}, \mathbf{X})$, using Equation 8. The posterior in iteration t becomes the prior for the next iteration, $t + 1$. To select this subset of data points, a secondary utility function, called an acquisition function, is pre-specified. The purpose of this acquisition function, $a(.)$, is to select the next data point to use for re-estimating the surrogate. The acquisition function was designed to make effective and efficient choices of optimal data points by simultaneously considering the following two requirements (*30*):

- The acquisition function should exploit data points that are close to previous optimal iterations because they may potentially offer improvement.
- The acquisition function should explore data points where the uncertainty is highest because the optimal point may lie in those areas.

The acquisition function was designed to trade-off these two competing requirements to determine points to be explored and exploited with preference given to those points that produce the best improvement of the surrogate function over previous iterations. The acquisition function used in this study is the expected improvement (E.I) (*40*), expressed as follows: Consider that $f(x_t)$ is the current best (extremal) point evaluation of the surrogate model. That is:

$$\mathbf{x}_t = arg\ max_{x \in x} f(\mathbf{x}) \quad (9)$$

and $f(x_{t+1})$ is the evaluation of the surrogate function at the next iteration. The magnitude of improvement, $I(x_{t+1})$, over the current best point is given by:

$$I(x_{t+1}) = max\{0, f(x_{t+1}) - f(x_t)\} \quad (10)$$

where $I(x_{t+1})$ is positive when $f(x_{t+1})$ is higher than the $f(x_t)$ and zero otherwise. The likelihood of improvement, I, on a normal posterior distribution specified by $\mu(x_{t+1}), \sigma(x_{t+1})$ can be computed from the normal density function given by:

$$\frac{1}{\sigma(x_{t+1})\sqrt{2\pi}} exp\left(-\frac{(\mu(x_{t+1}) - f(x_t) - I)^2}{2\sigma^2(x_{t+1})}\right) \quad (11)$$

The expected improvement is the integral over the likelihood of improvement, given by:

$$\mathbb{E}(I) = \int_{I=0}^{I=\infty} \frac{1}{\sigma(x_{t+1})\sqrt{2\pi}} exp\left(-\frac{(\mu(x_{t+1}) - f(x_t) - I)^2}{2\sigma^2(x_{t+1})}\right) dI \quad (12)$$

$$\mathbb{E}(I) = \sigma(x_{t+1}) \left[\frac{\mu(x_{t+1}) - f(x_t)}{\sigma(x_{t+1})} \Phi\left(-\frac{\mu(x_{t+1}) - f(x_t)}{\sigma(x_{t+1})}\right) + \phi\left(-\frac{\mu(x_{t+1}) - f(x_t)}{\sigma(x_{t+1})}\right)\right] \quad (13)$$

where $\Phi(.)$ and $\phi(.)$ are the probability and cumulative distribution functions of the standard normal distribution, respectively. The analytic solution to $E(I)$ (*40*) in Equation 13 is given by:

$$EI(x_{t+1}) = \begin{cases} (\mu(x_{t+1}) - f(x_t) - \xi)\Phi(Z) + \sigma(x_{t+1})\phi(Z), & \text{if } \sigma(x_{t+1}) > 0 \\ 0, & \text{if } \sigma(x_{t+1}) = 0 \end{cases} \quad (14)$$

where $\xi \geq 0$ is a parameter that controls the trade-off between global search and local optimization (i.e., exploration/exploitation) suggested by Lizotte (*41*), and:

$$Z = \begin{cases} \frac{\mu(x_{t+1}) - f(x_t) - \xi}{\sigma(x_{t+1})}, & \text{if } \sigma(x_{t+1}) > 0 \\ 0, & \text{if } \sigma(x_{t+1}) = 0 \end{cases} \quad (15)$$

For each iteration, E.I is maximized to select a new additional data point, x_{t+1}. This data point is then augmented with previous optimal data points and collectively used to update the surrogate. After updating the surrogate function, the cycle is repeated until convergence or a constraint on the total number of iterations, T, is met. Each new evaluation decreases the distance between the true global maximum and the expected maximum given the model.

The two stages described above (i.e., the surrogate from a GP prior and the expected improvement acquisition function) were combined to produce the proposed method for hyperparameter tuning and selection of an activity-duration model. The proposed framework is summarized in the following algorithm:

- Step 1. Initialization:
  - Step 1.1. Select two random data points of activity duration and corresponding attributes $(x_1, y_1), (x_2, y_2)$ from the dataset and define their likelihood, p(y | f, X).
  - Step 1.2. Define a GP prior, p(f | X), that shows that individuals with similar attributes have similar activity durations.
  - Step 1.3. Combine the prior with the likelihood, using Bayes theorem, to produce the GP posterior distribution, p(f | y, X), i.e., the surrogate of the activity-duration model.
- Step 2. Selection of an optimal data point:
  - Step 2.1. Use the surrogate of the activity-duration model to construct an acquisition function $EI(x)$.
  - Step 2.2. Maximize $EI(x)$ to find an optimal data point x_{t+1} that produces the highest expected improvement of the activity-duration model over the previous data point.
- Step 3. Evaluation of objective function at selected point:
  - Step 3.1. Use the optimal data point x_{t+1} to re-evaluate the surrogate of the activity-duration model to produce a new GP posterior.
  - Step 3.2. Query the new GP posterior to obtain the corresponding value of activity duration y_{t+1}.
  - Step 3.3. Augment the new selected data point to the previous points to obtain updated data set of activity durations and corresponding attributes $(x_1, y_1), (x_2, y_2), (x_{t+1}, y_{t+1})$.
- Step 4. Updating probabilistic model of surrogate:
  - Step 4.1. Use the updated data $(x_1, y_1), (x_2, y_2), (x_{t+1}, y_{t+1})$ to update the surrogate of the activity-duration model.
  - Step 4.2. Identify extremal value of the activity-duration model, f_{max}.
- Step 5. Stopping criteria: If the prediction error of the surrogate of the activity-duration model has met the desirable error, then the algorithm is terminated. Or if the algorithm has reached the maximum number of possible iterations (due to time or compute constraints), then the algorithm is terminated. Otherwise go back to step 2. That is, if $error \leq error_{desirable}$ or $T = T_{desirable}$, then stop the optimization algorithm. Otherwise, go to step 2.

Figure 1 illustrates the proposed procedure for performing BO using an integrated acquisition function and a probabilistic model of the objective function.

Experiments

Raw Data and Sample Formation

The data used for empirical analysis in this study was drawn from the 2010–2012 California household travel survey (CHTS) collected by the California Department of Transportation (Caltrans), which is available online at the Transportation Secure Data Center (*42*). This survey consists of socioeconomic and travel behavior information of 42,431 California households.

To obtain the sample for the current analysis, the CHTS data was processed as follows. First, data from the Los Angeles area were filtered. Second, only activities performed on a weekday that was not a holiday were selected, because the focus of the current paper is to study individuals' activity participation patterns on a typical weekday. Third, only five out-of-home non-mandatory activity types were considered including (similar to Ferdous et al. [*11*]):

- personal/family care (including personal care, caring for children in the household, pick-up/drop-off of children/adults, and caring for extended family members)
- shopping (grocery shopping, purchasing gas/food, banking, window shopping, clothes, and electronic shopping)
- meals

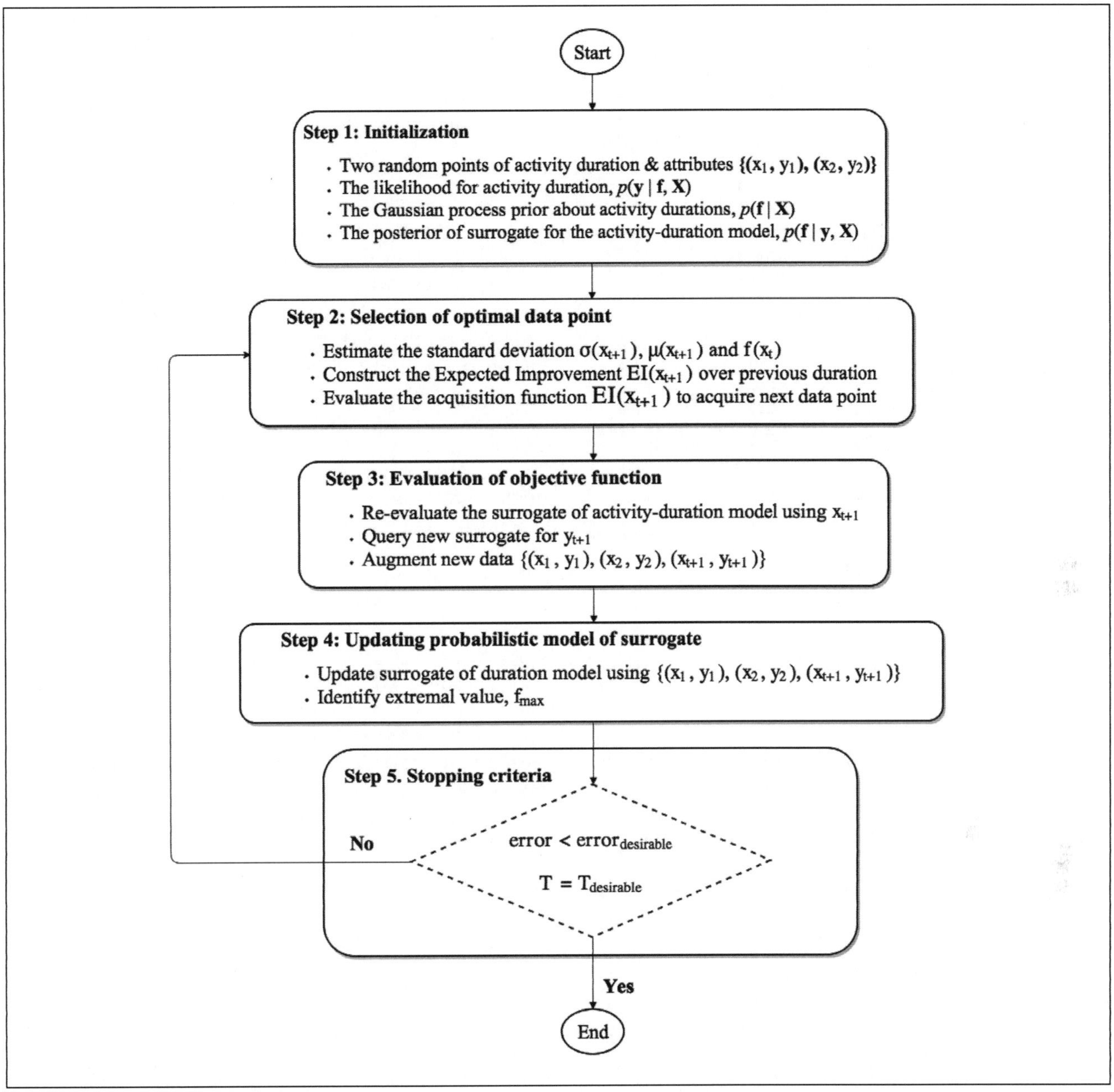

Figure 1. Proposed algorithm to tune hyperparameters and generate an activity-duration model.

- physically active recreation (sports, exercise, walking, bicycling, recreational and volunteer activities)
- physically inactive recreation (social, relaxing, movies, and attending religious/cultural/sports events)

Fourth, data regarding travelers' socioeconomic statuses, trip purposes, household characteristics, and personal characteristics were appended to their activity types. Table 1 provides descriptive statistics of some of the categorical variables. Fifth, the choices for commute mode that were considered include auto driver, auto passenger, non-motorized, bus, and train. Travelers with missing data were removed from the sample, resulting in a total sample size of 23,906 observations.

Table 2 presents the proportion (in minutes) of each activity type in the final sample. Personal activities were the most frequently undertaken activity type with 51.4% proportion followed by meals (20.2%), shopping (15.6%), physically inactive recreation (7.9%) and physically active recreation (4.8%). Figure 2 presents the

Table 1. Summary Statistics of Some Variables in the Sample

Variable	Categories	Frequency	Percentage (%)
Gender	Female	12353	51.7
	Male	11553	48.3
Income	Low	9591	40.1
	Medium	9301	38.9
	High	5014	21.0
Education	College	14569	60.9
	Some college	5633	23.6
	No college	3704	15.5
Mode	Auto-driver	20372	82.7
	Non-motorized	1294	5.5
	Auto-passenger	1054	4.4
	Bus	873	3.6
	Train	313	1.3
Race	White	16627	69.6
	Hispanic	2393	10.0
	Asian	2202	9.2
	Indian	1,382	5.7
	Black	1302	5.4
Occupation	Management	11670	48.8
	Clerical	5607	23.5
	Social	2607	10.9
	Manufacturing	2449	10.2
	Services	1573	6.6
Residence type	House	18920	79.1
	Apartment	4986	20.9
Home ownership	Own	17755	74.3
	Rent	6151	25.7

density distribution of each activity type in the sample. Generally, most of the duration density of all activity types was concentrated within the 300-min mark indicating that most out-of-home non-mandatory activities last at most 300 min (5 h). However, personal activities and meals had two distinct peaks in their duration density. For instance, the data suggests that there were two groups of dining behavior; a larger and a small group that spent on average 60 and 550 min (9 h), respectively. It is typical for dining to last on average one hour. The second, smaller group may constitute long-planned dinning events such as parties or other types of celebrations. Given that the duration density for shopping is more compact and narrow (i.e. has a small variance), it suggests that shopping is more time-consistent than all others. Almost all shoppers took on average the same amount of time of 35 min as confirmed in Table 2. These insights on activity duration could be crucial in planning the built environment which in turn influences travel behavior. For instance, parking duration for shopping centers could be optimized to 30 min per shopper.

Empirical Analysis

To tune carefully the GP model parameters using RS, a rigorous and thorough procedure, adopted from Kuhn and Johnson (*43*), was followed. This procedure, which is considered state of the art in machine learning, was customized in this study as shown in Algorithm 1. For each activity type, the following procedure was followed. A set of hyperparameter values to be evaluated was generated using an RS algorithm (*33*). The set of hyperparameters to be tuned in the fitted GP model included one signal variance parameter σ_f and 33 length-scale parameters l_d for each dimension d of the regressors. One hundred unique combinations of the 34 hyperparameters were generated using RS, resulting in 3,400 tunable hyperparameters. The data was partitioned into two with 70% and 30% for training and testing, respectively. The type of resampling procedure was then specified to involve 10-fold cross validation (CV), each repeated 5 times (i.e. five separate 10-fold CVs). This is known as repeated k-fold CV. This resulted in 50 training instances of the GP model. For each training instance, 9-folds of data partition was used for training and 1-fold used as hold-out partition for testing the fitted GP model. Root mean square error (RMSE) values were estimated for each testing instance as metrics of model performance. The hyperparameter set which produced the lowest RMSE value was chosen as the best signal variance, $\{\sigma_f^{best}, l_d^{best}\}_{RS}$. Given that each of the re-sampled data sets is independent of the others, the training of the models was performed in parallel so as to increase computational efficiency. This study utilized a Linux server with 80 cores and 256GB of RAM to spread the computation across these cores. Table 3 presents only values for σ_f^{best} by activity type. Values of 0.012, 0.013, 0.021, 0.015, and 0.023 were estimated for personal activities, meals, shopping, inactive, and active recreation, respectively. Finally, $\{\sigma_f^{best}, l_d^{best}\}_{RS}$ was used to fit the GP model to all the training data (70% portion). The final estimated model was tested on the test dataset (30% portion) to determine

Table 2. Descriptive Statistics of Durations by Activity Type

Activity type	Mean	Std. dev.	Min	Max	Frequency	Percentage (%)
Personal activities	309.5	269.0	1.0	1,370	12,296	51.4
Meals	342.1	262.1	1.0	1,249	4,837	20.2
Shopping	35.2	44.3	1.0	635	3,735	15.6
Physically inactive recreation	157.3	187.1	1.0	1,069	1,896	7.9
Physically active recreation	186.4	165.9	1.0	1,254	1,142	4.8

Table 3. A Comparison of Duration Prediction Accuracies between RS and BO by Activity Type

Activity type	Procedure	Best parameter σ_f^{best}	RMSE (min)	MAPE (%)	R-squared
Personal activities	RS	0.012	14.26	44.13	0.30
	BO	0.004	4.72	24.39	0.57
Meals	RS	0.013	12.24	25.33	0.33
	BO	0.006	2.18	13.85	0.59
Shopping	RS	0.021	12.16	31.45	0.12
	BO	0.013	2.69	12.35	0.49
Active recreation	RS	0.015	13.11	23.62	0.18
	BO	0.004	2.82	23.31	0.46
Inactive recreation	RS	0.013	14.24	24.68	0.18
	BO	0.024	3.99	16.99	0.47

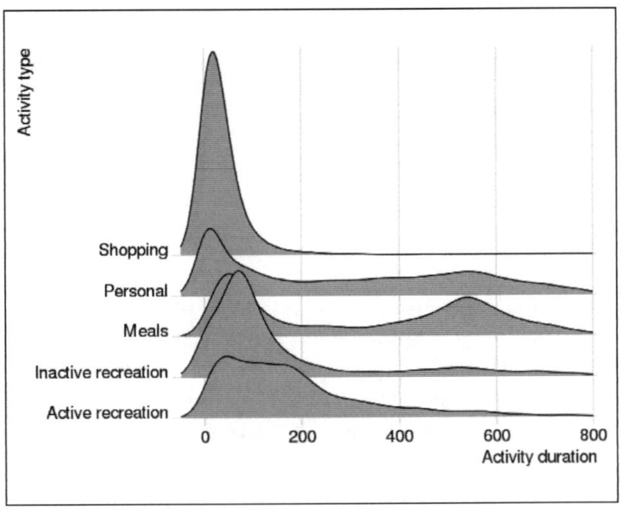

Figure 2. Distribution of activity duration by activity type.

Algorithm 1 Procedure to Carefully Tune Hyperparameters of a GP Model using RS

```
Procedure
  Generate a set of hyperparameters to evaluate, (σ_f, l_d), using RS
  Specify a re-sampling procedure, such as 10-fold CV repeated 5 times
  for each set of(σ_f, l_d) do
    for each of the 5 repeations of the 10 − folds of the datapartition do
      Hold-out 1 random fold of the data
      Fit the GP model on the remainder of the partitions
      Use fitted GP model to predict the hold-out samples
    end for Calculate the RMSE and R-Squared values across hold-out partitions
  end for Determine the optimal parameter set, {σ_f^best, l_d^best}.
  Fit the final model to all the training data using the optimal parameter set
end procedure
```

the RMSE, MAPE and R-squared values, which are presented in Table 3.

The proposed BO procedure was then applied to the same data to search for the best 34 hyperparameters $\{\sigma_f^{best}, l_d^{best}\}_{BO}$. The following two settings were chosen to initialize the BO run (as previously described in the proposed algorithm in Figure 1). First, two random data points of activity duration with their corresponding attributes were chosen. Second, the total number of iterations, $T_{desirable}$, was chosen to be 100. For each iteration, the corresponding $\{\sigma_f, l_d\}$ and RMSE were evaluated.

The hyperparameter set that produced the lowest RMSE was chosen as the optimal/best set, $\{\sigma_f^{best}, l_d^{best}\}_{BO}$. Table 3 only shows the σ_f^{best}. Values of 0.004, 0.006, 0.013, 0.004, and 0.024 were evaluated for personal activities, meals, shopping, active recreation, and inactive recreation, respectively. These values from BO are considerably different from the RS procedure. It would have been difficult for an expert to suggest such values given their small scale and that there is an infinite number of possibilities in the range $\{0, 0.9\}$. Finally, $\{\sigma_f^{best}, l_d^{best}\}_{BO}$ was used to fit the GP model to all the training data. The final estimated model was tested on test data to evaluate the RMSE, MAPE and R-squared values, which are presented in Table 3.

The performance metrics for the proposed BO framework were found to be superior to the RS procedure. The RMSE and MAPE from BO were smaller than those from RS across all activity types. For instance, an error of 24.39% was found from BO for personal activities compared to an error of 44.13% from the RS procedure. The R-squared values from BO were higher than those from RS across all activity types. For instance, an R-squared value of 0.57 was found from BO for personal activities compared to an R-squared value of 0.30 from the RS procedure.

Insights related to transportation planning obtained from the final estimated model for personal activities only are discussed in the next section.

Table 4. Estimates for the Length-Scale Parameters for the Attributes

Covariate	Estimate	Std error	Conf.low	Conf.high	Rhat	N_{eff}
Auto driver	0.98	1.01	0.12	2.08	1.00	375
Auto passenger	0.97	0.90	0.19	2.05	1.00	400
Non-motorized	1.21	1.09	0.16	2.38	1.01	368
Trip distance (miles)	0.88	1.00	0.03	2.14	1.01	223
Vehicle count	1.22	1.04	0.18	2.52	0.99	400
Operational vehicle count	1.21	1.11	0.20	2.35	0.99	400
Low income ($\leq$ \$34, 999)	0.99	0.91	0.10	2.11	0.99	400
Medium income (\$35,000 - \$74,999)	1.22	1.11	0.27	2.67	1.00	400
Home ownership: rent	0.86	0.79	0.11	1.78	1.01	400
Trip count	1.06	1.15	0.14	2.43	1.01	329
Person trips	1.13	0.70	0.35	2.03	1.01	400
Activity count	0.67	0.65	0.11	1.26	1.01	400
No college	1.03	0.93	0.17	2.16	1.02	231
Some college	1.07	1.02	0.18	2.28	1.01	324
Black	0.96	0.92	0.16	1.97	1.00	400
Hispanic	1.23	1.47	0.16	2.64	1.00	188
Indian	0.99	0.91	0.19	2.14	1.00	400
White	1.02	0.90	0.19	2.13	1.01	400
Management occupation	0.95	0.82	0.16	2.15	1.00	324
Manufacturing occupation	1.03	0.97	0.09	2.32	1.00	400
Social occupation	1.06	1.04	0.13	2.11	1.00	310
House residence type	1.08	1.17	0.13	2.24	1.00	400

Discussion of Results

Another set of empirical results included the characteristic length-scale parameters, l_d, for each of the attributes in the final estimated models. For illustration and discussion, only length-scale parameters for personal activities are provided in Table 4. This non-negative parameter is a measure of relevance of an attribute (which could be interpreted as "significance" in frequentist terminology) to predicting activity duration. The larger the l_d, the smaller the covariance of the corresponding attribute. Consequently samples from a distribution with a small covariance are more or less similar. This indicates that such attributes do not vary with activity duration. This study retained attributes with $l_d \leq 5$ as those that were relevant to predicting activity durations. In addition, Table 4 includes statistical properties of their corresponding length-scale parameters such as a potential scale reduction statistic (Rhat) (*44*) to show whether estimation reached global convergence. The Rhat value for all parameters were less than 1.05 indicating that all the MCMC chains in the model converged to the global equilibrium (*44*).

Given that the length-scale parameter is a non-negative value, the sign does not have any meaning. Plots of the marginal posterior mean and covariance distributions of each individual attribute in the estimated GP models were generated to investigate the relationship between the attributes and activity duration. For illustration, Figure 3 presents plots for a few selected variables. The red and blue lines are the posterior mean and covariance functions, respectively. They constitute functions that best explain the observed data. The black dots are samples of test data plotted over the prediction functions. Given that additivity made the offset of each function arbitrary, the critical properties indicated from the graphs were the pattern of variation as well as the relative magnitude of variation across the different covariates.

For instance, the marginal posterior distribution illustrated in Figure 3*a* suggests that individuals who are auto drivers spend longer durations at personal activities than individuals who take the bus (bus was the base category). Figure 3*b* suggests that individuals who are auto passengers spend shorter durations at personal activities than individuals who take the bus. Considering household- and personal-level factors, the findings revealed that individuals who live in house residences spent longer durations at personal activities than those who lived in apartments, the base category (Figure 3*c*). Individuals who work in the manufacturing occupation spend longer durations at personal activities than those in the clerical business (Figure 3*d*). Regarding racial groups, the findings in Figure 3*e* suggest that Black individuals spend shorter durations at personal activities than Asians, the base category. Figure 3*f* suggests that individuals with some college spend shorter durations at personal activities than those without college education.

Other factors, not shown Figure in 3, that were found to positively affect personal activity duration include vehicle count, operational vehicle count, and medium income.

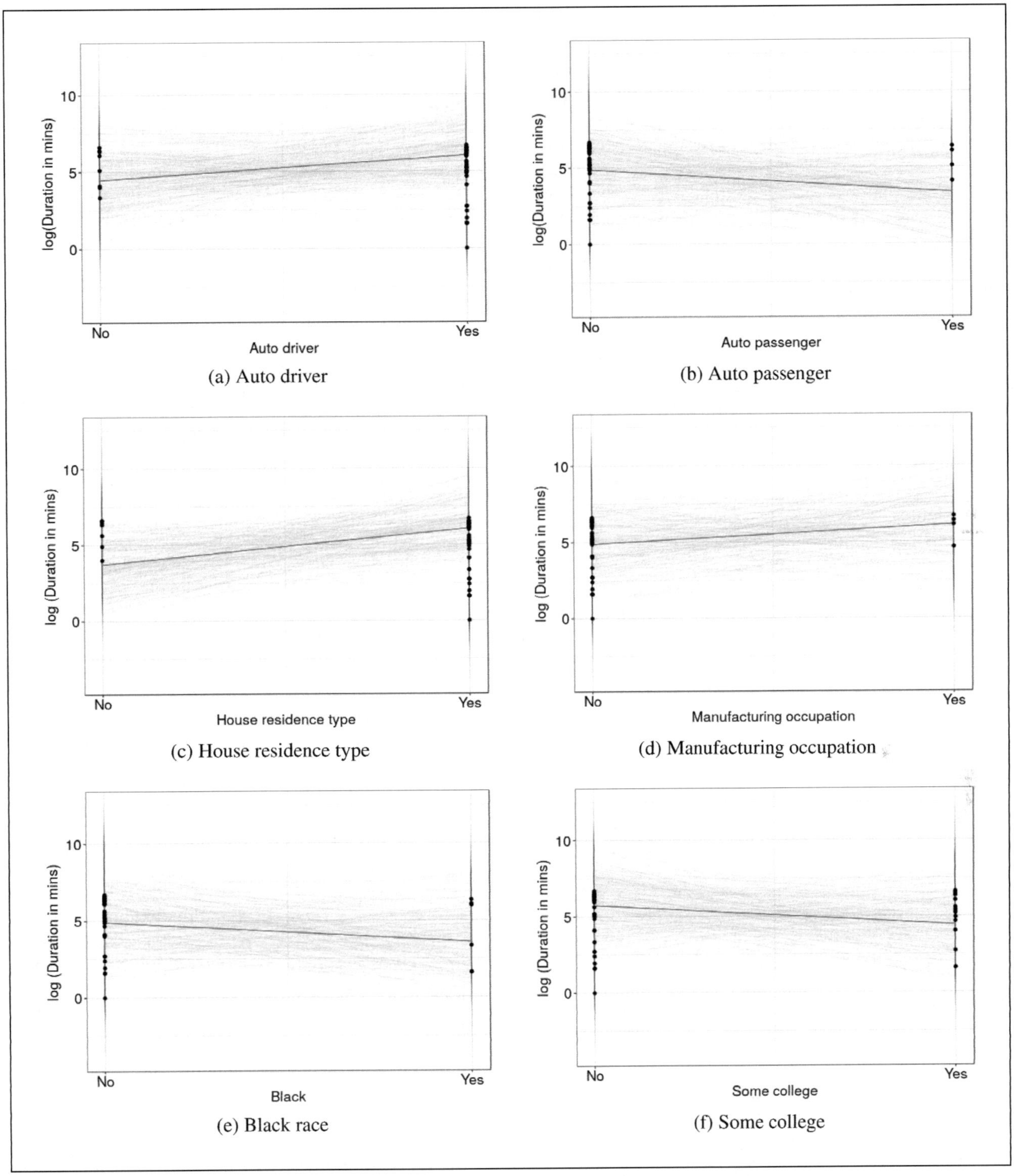

(a) Auto driver (b) Auto passenger (c) House residence type (d) Manufacturing occupation (e) Black race (f) Some college

Figure 3. The relationship between attributes and duration of personal activities.

Factors that negatively affected personal activity duration include trip distance, activity counts, trip counts, and person trips. As the distance to the activity location increases, the duration spent there decreases. The number of trips and activity episodes undertaken decrease the amount of time spent at each individual activity location.

Conclusions and Future Work

This paper proposed a BO framework to simultaneously tune hyperparameters and generate flexible duration models for five types of out-of-home non-mandatory activity episodes. The proposed framework addresses the current challenge in travel demand estimation whereby significant time and expertise are required to adjust hyperparameters. Algorithms that take the role of the modeler or assist his/her task of tuning hyperparameters are important because they reduce subjectivity in model estimation, increase productivity of modelers, and enable modelers with minimal technical background of statistics to select appropriate hyperparameters. Experiments demonstrated that results from this framework outperform expert-level performance and are obtained in a fraction of time compared to traditional approaches used in practice. Future work will involve increasing sample sizes and optimizing the algorithm to be efficient with larger datasets.

Author Contributions

The authors confirm contribution to the paper as follows: study conception and design: Emaasit, Paz; data collection: Emaasit; analysis and interpretation of results: Emaasit; draft manuscript preparation: Emaasit, Paz. All authors reviewed the results and approved the final version of the manuscript.

References

1. Castiglione, J., M. Bradley, and J. Gliebe. *SHRP 2 Report S2-C46-RR-1: Activity-Based Travel Demand Models: A Primer*. Transportation Research Board of the National Academies, Washington, D.C., 2015.
2. Ghahramani, Z. Bayesian Non-parametrics and the Probabilistic Approach to Modelling. *Philosophical Transactions of the Royal Society A*, Vol. 371, No. 1984, 2013, p. 20110553.
3. Emaasit, D., and A. Paz, An Integrated Travel Demand Model of Work Duration and Commute-mode Choice Decisions: A Flexible Modeling Approach. *Presented at International Scientific Conference on Mobility and Transport, mobil.TUM,* 2017.
4. Paz, A., V. Molano, E. Martinez, C. Gaviria, and C. Arteaga, Calibration of Traffic Flow Models Using a Memetic Algorithm. *Transportation Research Part C: Emerging Technologies*, Vol. 55, 2015, pp. 432–443.
5. Bergstra, J. S., R. Bardenet, Y. Bengio, and B. Kégl, Kégl, Algorithms for hyper-parameter optimization. *Advances in Neural Information Processing Systems*, 2011, pp. 2546–2554.
6. Ulmasov, D., C. Baroukh, B. Chachuat, M. P. Deisenroth, and R. Misener, Bayesian Optimization with Dimension Scheduling: Application to Biological Systems. *arXiv preprint arXiv:1511.05385*, 2015.
7. Yogatama, D., and N. A. Smith, Bayesian Optimization of Text Representations. *arXiv preprint arXiv:1503.00693*, 2015.
8. Wang, Z., B. Zhou, and S. Jegelka, Optimization as Estimation with Gaussian Processes in Bandit Settings. *Proc., 19th International Conference on Artificial Intelligence and Statistics*, 2016, pp. 1022–1031.
9. Schneider, J., Bayesian Optimization and Embedded Learning Systems. *Proc., 22nd ACM SIGKDD International Conference on Knowledge Discovery and Data Mining*, ACM, 2016, pp. 413–413.
10. Snoek, J., *Bayesian Optimization and Semiparametric Models with Applications to Assistive Technology*. PhD thesis. Citeseer, 2013.
11. Ferdous, N., N. Eluru, C. R. Bhat, and I. Meloni, A Multivariate Ordered-response Model System for Adults' Weekday Activity Episode Generation by Activity Purpose and Social Context. *Transportation Research Part B: Methodological*, Vol. 44, No. 8, 2010, pp. 922–943.
12. Bhat, C. R., and J. L. Steed. A Continuous-time Model of Departure Time Choice for Urban Shopping Trips. *Transportation Research Part B: Methodological*, Vol. 36, No. 3, 2002, pp. 207–224.
13. Vovsha, P., and M. Bradley. Hybrid Discrete Choice Departure-time and Duration Model for Scheduling Travel Tours. *Transportation Research Record: Journal of the Transportation Research Board*, 2004. 1894: 46–56.
14. Pinjari, A. R., R. M. Pendyala, C. R. Bhat, and P. A. Waddell. Modeling the Choice Continuum: An Integrated Model of Residential Location, Auto Ownership, Bicycle Ownership, and Commute Tour Mode Choice Decisions. *Transportation*, Vol. 38, No. 6, 2011, p. 933.
15. Habib, K. M. N. Modeling Commuting Mode Choice Jointly with Work Start Time and Work Duration. *Transportation Research Part A: Policy and Practice*, Vol. 46, No. 1, 2012, pp. 33–47.
16. Habib, M., and N. Daisy. Modeling Frequency and Duration of Out-of-Home Participation in Physical Activity by School-Age Children. *Transportation Research Record: Journal of the Transportation Research Board*, 2013. 2357: 116–124.
17. Srinivasan, S., and C. R. Bhat. Modeling Household Interactions in Daily In-home and Out-of-home Maintenance Activity Participation. *Transportation*, Vol. 32, No. 5, 2005, pp. 523–544.
18. Sener, I. N., and C. R. Bhat. Modeling the Spatial and Temporal Dimensions of Recreational Activity Participation with a Focus on Physical Activities. *Transportation*, Vol. 39, No. 3, 2012, pp. 627–656.
19. Handy, S., X. Cao, and P. Mokhtarian. Correlation or Causality Between the Built Environment and Travel Behavior? Evidence from Northern California. *Transportation Research Part D: Transport and Environment*, Vol. 10, No. 6, 2005, pp. 427–444.
20. Habibi, S. *Prediction-driven Approaches to Discrete Choice-models with Application to Forecasting Car Typedemand*, 2016.

21. Ahmed, M., M. Abdel-Aty, and R. Yu. Bayesian Updating Approach for Real-time Safety Evaluation with Automatic Vehicle Identification Data. *Transportation Research Record: Journal of the Transportation Research Board*, 2012. 2280: 60–67.
22. Washington, S., and J. Wolf. Hierarchical Tree-based Versus Ordinary Least Squares Linear Regression Models: Theory and Example Applied to Trip Generation. *Transportation Research Record: Journal of the Transportation Research Board*, 1997. 1581: 82–88.
23. Kim, S.-J., W. Kim, and L. Rilett. Calibration of Microsimulation Models Using Nonparametric Statistical Techniques. *Transportation Research Record: Journal of the Transportation Research Board*, 2005. 1935: 111–119.
24. Athanasopoulos, G., R. J. Hyndman, H. Song, and D. C. Wu. The Tourism Forecasting Competition. *International Journal of Forecasting*, Vol. 27, No. 3, 2011, pp. 822–844.
25. Hyndman, R., and Y. Khandakar. *Automatic Time Series Forecasting: The Forecast Package for R 7*. 2008. https://www.jstatsoft.org/article/view/v027i03 [WebCite Cache], 2007. Accessed February 24, 2016.
26. Tulic, M., D. Bauer, and W. Scherrer. Link and Route Travel Time Prediction Including the Corresponding Reliability in An Urban Network Based on Taxi Floating Car Data. *Transportation Research Record: Journal of the Transportation Research Board*, 2014. 2442: 140–149.
27. Abraham, J., and J. Hunt. Parameter Estimation Strategies for Large-scale Urban Models. *Transportation Research Record: Journal of the Transportation Research Board*, 2000. 1722: 9–16.
28. Hollander, Y. Multiple Objectives in Travel Demand Modeling. *Transportation Research Record: Journal of the Transportation Research Board*, 2010. 2175: 120–129.
29. Liu, H., Q. Yu, W. Ding, D. Ni, H. Wang, and S. Shannon. Feasibility Study for Automatic Calibration of Transportation Simulation Models. *Proc., 44th Annual Simulation Symposium, Society for Computer Simulation International*, 2011, pp. 87–94.
30. Snoek, J., H. Larochelle, and R. P. Adams. Practical Bayesian Optimization of Machine Learning Algorithms. *Advances in Neural Information Processing Systems*, 2012, pp. 2951–2959.
31. Luo, G. A Review of Automatic Selection Methods for Machine Learning Algorithms and Hyper-parameter Values. *Network Modeling Analysis in Health Informatics and Bioinformatics*, Vol. 5, No. 1, 2016, pp. 1–16.
32. Hutter, F., H. H. Hoos, K. Leyton-Brown, and T. Stützle. ParamILS: An Automatic Algorithm Configuration Framework. *Journal of Artificial Intelligence Research*, Vol. 36, No. 1, 2009, pp. 267–306.
33. Bergstra, J., and Y. Bengio. Random Search for Hyper-parameter Optimization. *Journal of Machine Learning Research*, Vol. 13, No. Feb, 2012, pp. 281–305.
34. Hutter, F., H. H. Hoos, and K. Leyton-Brown. Sequential Model-based Optimization for General Algorithm Configuration. *International Conference on Learning and Intelligent Optimization*, 2011, pp. 507–523.
35. Swersky, K., J. Snoek, and R. P. Adams. Multi-task Bayesian Optimization. *Advances in Neural Information Processing Systems*, 2013, pp. 2004–2012.
36. Shahriari, B., K. Swersky, Z. Wang, R. P. Adams, and N. de Freitas, Taking the Human Out of the Loop: A Review of Bayesian Optimization. *Proceedings of the IEEE*, Vol. 104, No. 1, 2016, pp. 148–175.
37. Williams, C. K., and C. E. Rasmussen. Gaussian Processes for Machine Learning. *The MIT Press*, Vol. 2, No. 3, 2006, p. 4.
38. Gelman, A., J. B. Carlin, H. S. Stern, and D. B. Rubin. *Bayesian Data Analysis*. Vol. 2. Chapman & Hall/CRC Boca Raton, FL, 2014.
39. Neal, R. M. *Bayesian Learning for Neural Networks*. Vol. 118, Springer Science & Business Media, 2012.
40. Mockus, J., V. Tiesis, and A. Zilinskas. Toward Global Optimization, Vol. 2, In *Chapter Bayesian Methods for Seeking the Extremum*, 1978.
41. Lizotte, D. J. *Practical Bayesian Optimization*. University of Alberta, Edmonton, AB, Canada, 2008.
42. Transportation Secure Data Center. *Transportation Secure Data Center*, 2016. Accessed October 26, 2016.
43. Kuhn, M., and K. Johnson. *Applied Predictive Modeling*, Vol. 810. Springer, New York, NY, 2013.
44. Gelman, A., and D. B. Rubin. Inference from Iterative Simulation Using Multiple Sequences. *Statistical Science*, 1992, pp. 457–472.

The Standing Committee on Artificial Intelligence and Advanced Computing Applications (ABJ70) peer-reviewed this paper (18-02780).

Article

TRR
JOURNAL OF THE TRANSPORTATION RESEARCH BOARD

Transportation Research Record
2018, Vol. 2672(45) 160–172

DOI: 10.1177/0361198118774691
journals.sagepub.com/home/trr
SAGE

Vehicle Reidentification in a Connected Vehicle Environment using Machine Learning Algorithms

Zuoyu Miao[1], K. Larry Head[1], and Byungho Beak[1]

Abstract

Deployment of connected vehicles will become possible for most American cities in the next 10 to 20 years. Connected vehicle (CV) applications (e.g., mobility, safety, environment) are constantly receiving vehicle data. The current ID protection mechanism assumes a vehicle's ID changes every 5 minutes, so the topic of rematching vehicles is of interest in privacy protection and performance measure research. This paper explores the possibility of rematching connected vehicles' IDs using popular machine learning techniques, including logistic regression (LR), linear discriminant analysis (LDA), quadratic discriminant analysis (QDA), linear and nonlinear support vector machine (SVM) and nearest neighbor algorithms. An experiment is conducted using a microscopic traffic simulation model through a software-in-the-loop technique. The best average mismatching rate is 14%. To assess potential factors' effects on matching accuracy, a Poisson mixed regression model is analyzed under the Bayesian inference framework. Findings are: different matching algorithms vary in matching performance and the linear SVM, the QDA and the LDA have the best accuracy results; traffic volume and market penetration rate have little impact on matching results; location and number of vehicles to be matched are considered significant. The results make the performance measurement of future CV applications feasible and also suggest that more secure mechanisms are needed to protect the public.

Connected vehicle (CV) technology has made the realization of vehicle to vehicle (V2V) and vehicle to infrastructure (V2I) communication possible. New applications are emerging under the V2I framework that enable the road side unit (RSU) to collect real-time vehicle data to improve traffic mobility, as well as safety. According to the FHWA V2I Deployment Guidance, limited deployment of RSUs and equipped vehicles has begun and is expected to become widespread over the next 10–20 years (*1*). With the rapid development of CV technologies, more and more users' data will be collected.

Although the enriched data facilitates CV applications, openly publishing data could potentially pose significant privacy and security concerns. The Federal Trade Commission (FTC) and the NHTSA held an open workshop on June 28, 2017 in Washington, D.C to discuss such privacy and security issues (*2*). The spatial and temporal attributes of trajectories can be used as powerful quasi-identifiers linking personal identification information (PII). Movements of connected vehicles can be observed by chance or on purpose, even with few observations. By matching trajectories having the same spatial and temporal attributes, a user's trip information may be inferred, including home and work addresses. Therefore, third party access to such trajectory data may cause serious privacy and security threats. In the CV environment, the equipped vehicles generate random identities every 300 s to break the trajectory into pieces and mix it with other vehicles. However, even with such a pseudonym protecting mechanism, exposure of the de-identified vehicle trajectory data to adversaries is still considered to be risky. This paper serves as evidence to show that strict CV protection strategies are essential to secure consumer privacy.

When new Intelligent Transportation System (ITS) applications are implemented in the CV environment, the individual connected vehicle's travel time and delay are important measures for performance analysis. Without a vehicle ID matching process, the real-world system-level performance measures cannot be fully implemented,

[1]System and Industrial Engineering Department, University of Arizona, Tucson, AZ

Corresponding Author:
Address correspondence to Zuoyu Miao: zymiao@email.arizona.edu

because one vehicle may have different IDs as it moves among intersections. Therefore, matching connected vehicles' IDs becomes desirable. More research work is needed in vehicle ID matching in the quickly maturing CV environment as the newly emerged technologies (V2I and V2V communications) are scheduled to be deployed nationwide in the next few years.

This paper explores trajectory matching possibilities by implementing several machine learning classification techniques and systematically assessing factors potentially affecting the matching accuracy. The organization of this paper is as follows. The Literature Review provides a review on vehicle matching and experimental analysis of random factors; the Description of Connected Vehicle Reidentification and Machine Learning Methods sections introduce algorithms for CV matching; the Results section consists of model specification, validation, and factor effect analysis; and the Conclusion provides a summary and suggestions for future work.

Literature Review

There appears to be no literature on vehicle reidentification using CV data. Significant research has been focused on maintaining the vehicle ID as it travels through multiple video camera sites. Zeng and Crisman proposed matching vehicles using body colors and the matching accuracy was 16.42% with an 8-vehicle test sample (*3*). Kogut and Trivedi investigated color features and the spatial organization of vehicle platoons together for vehicle matching (*4*). They achieved an accuracy rate of 45%. But their research work was based on a small number of samples and did not consider the site location impact on matching accuracy. Later, additional vehicle features were considered, such as external dimensions, points of optical demarcation, and so on (*5*). Another approach was based on a sparse representation algorithm which was originally proposed as a facial recognition method (*6*). The authors claimed an accuracy of 57.84%. The limitations of video-based vehicle matching methods include illumination variations, occlusion phenomena, and object overlapping. Such difficulties, and the cost of video detector deployment and maintenance, make this vehicle matching method unfavorable for transportation engineering practitioners. Some research has used unique vehicle IDs for identification. Examples include matching unique Bluetooth MAC addresses (*7–13*), GPS device addresses (*14*), and magnetic signatures (*15–18*). However, not all these ID matching facilities are widely deployed in the U.S. Another approach is vehicle trajectory tracking. The general routine is to detect vehicles in a picture by feature classification and then to track the object using Kalman filtering and its variants. A thorough literature review was conducted in on-road vision-based vehicle detection, tracking, and behavior understanding by Sivaraman et al. (*19*). Their review found that the tracked object is always confined in one camera's view angle and only suitable for continuous vehicle trajectory estimation. In the CV environment, vehicle features are automatically and precisely recorded in basic safety messages (BSMs) and extracting them is effortless compared with image processing. This advantage of structured data provides the possibility of matching vehicle IDs in a more complex real-life environment with better accuracy compared with the unstructured image data.

To systematically assess the performance of matching methods and other factors, it is helpful to use designed experiments. The traditional experiment analysis is the analysis of variance (ANOVA) approach. Even though this classical framework has served the simulation and field data analytics for decades, it also has two limitations. First, this approach requires well balanced sample sizes with the number of observations for each set of conditions being almost equal. In a real data collection process, observations are often missing because of human error, database malfunction, and condition limitations. Some modifications to the ANOVA were proposed by using approximate methods (*20*, *21*), but they are only appropriate for situations in which the dataset is mildly unbalanced (*22*). Second, non-normal data are usually transformed to make them appear normal (*23*). The transformation should be conducted very carefully, because it can fail if count data contain many zero values, model assumptions are violated, or the scope of inference is limited. To account for these disadvantages, the residual maximum likelihood (REML) method is often employed for parameter estimation as an alternative to the maximum likelihood (ML) method when both random and fixed factors are considered in the general linear mixed model (GLMM) (*24*). Because the likelihood equations are highly nonlinear, closed analytical solutions are not available. Approximation methods include the quasi-likelihood (QL) approach which uses a Taylor series expansion (*25–27*) and the Gauss–Hermite quadrature (GHQ) method which uses optimal subdivisions to numerically evaluate the integrand (*28*). The QL approach induces bias for inference purposes and the GHQ approaches work poorly with more than two random factors (*29*). Instead of trying to solving the likelihood functions, the Markov chain Monte Carlo (MCMC) algorithm (*30*) generates random samples from the distributions of parameter values of factor effects and is free of these drawbacks.

In this paper, with the help of software-in-the-loop (SIL) simulation, CV trajectory data is collected through a VISSIM simulation. Seven supervised machine learning classification algorithms are applied with considerations of different scenario settings. To explore the factors affecting the matching results, a designed experiment and its corresponding statistical analysis are presented.

Description of CV Reidentification

Typically, RSUs are mounted on top of traffic light signal poles of equipped intersections. Using dedicated short range communication (DSRC), approaching vehicles send BSMs up to 10 times per second to an RSU. According to the Multi-Modal Intelligent Traffic Signal System (MMITTS) project (*31*), for purposes of traffic signal control and performance measures, BSMs are collected and stored every second when vehicles are in the range of DSRC which is between 300 and 600 m. More details of this database design can be found on page 61 of the final report (*31*).

According to the Society of Automotive Engineers (SAE) J2735 standard (*32*), the BSM data contains vehicle position, time, speed, heading acceleration, brake system status, phases, and so on. In this research, only the vehicle position and the timestamp are used to match CV IDs and other vehicle properties are not included in the matching algorithms. This is because of two considerations. First, vehicle physical properties can serve as the quasi-ID to match vehicles before the machine learning techniques are applied. This pre-process can reduce the number of vehicle to be matched. In the situation in which each vehicle has its unique properties, the reidentification can be done without further matching algorithms. But the methods proposed in this paper can deal with more complex scenarios in which many vehicles share common physical properties. Second, incorporating the physical properties as covariates will cause computation difficulties. For example, the constant values will make covariance matrices singular for the LDA and QDA methods.

Machine Learning Methods

Trajectory attributes used to train classifiers are a vehicle's location coordinates and corresponding time stamps. For a pair of intersections (A and B in Figure 1), the training dataset is from one intersection RSU's BSM data and the test dataset is from the other intersection. Let Y represent the ID variable and a 3×1 vector variable $\mathbf{X} = [X_1 \quad X_2 \quad X_3]^T$ represent the trajectory attributes where X_1 and X_2 are vehicles' local coordinates with the origin defined by the intersection center's coordinates (*33*), and X_3 represents the trajectory's time stamp. With a trained classifier, the identity of a vehicle is determined by its trajectory points' most classified ID. In this research, an odd number of trajectory points

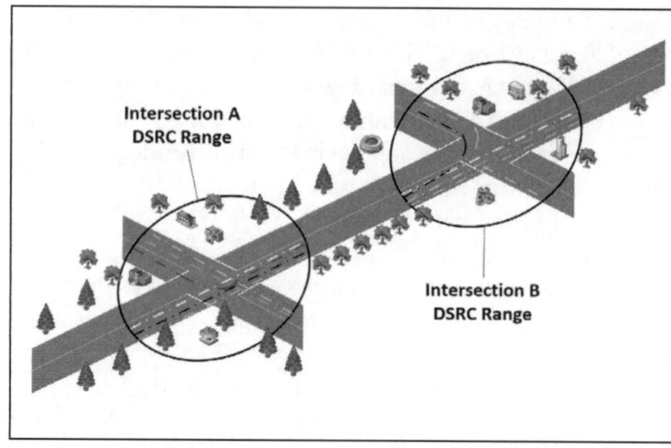

Figure 1. Vehicle trajectories.

(i.e., 5 trajectory points) is used when the trained model is applied to the test dataset.

Logistic Regression

Logistic regression (LR) models the boundary between two trajectory attributes using linear functions with the logit transformation

$$\log \frac{\Pr(Y = i|\mathbf{X} = \boldsymbol{x})}{\Pr(Y = K|\mathbf{X} = \boldsymbol{x})} = \beta_{i0} + \boldsymbol{\beta}_i^T \boldsymbol{x} \tag{1}$$

where $i = 1, \cdots, K - 1$. K represents the order of ID in the sample. Denoting the parameter vector $\boldsymbol{\theta} = \left\{\beta_{10}, \boldsymbol{\beta}_1, \cdots, \beta_{(K-1)0}, \boldsymbol{\beta}_{K-1}\right\}$, then, given a trajectory attribute, the probability of it being ID i is:

$$p_i(\boldsymbol{x}) = \Pr(Y = i|\boldsymbol{x}) = \begin{cases} \dfrac{\exp\left(\beta_{i0} + \boldsymbol{\beta}_i^T \boldsymbol{x}\right)}{1 + \sum_{j=1}^{K-1} \exp\left(\beta_{j0} + \boldsymbol{\beta}_j^T \boldsymbol{x}\right)} & i = 1, \cdots K - 1 \\ \dfrac{1}{1 + \sum_{j=1}^{K-1} \exp\left(\beta_{j0} + \boldsymbol{\beta}_j^T \boldsymbol{x}\right)} & i = K \end{cases} \tag{2}$$

The parameter estimation is based on a Maximum Likelihood Estimate (MLE) and the likelihood is $l(\boldsymbol{\theta}) = \sum_{v=1}^{n} \log p_{y_v}(\boldsymbol{\theta}; \boldsymbol{x}_v)$, in which n is the total number of trajectory attributes in the training set. Because the likelihood is of a complex form, it can be solved by the traditional numerical method, the Newton–Raphson algorithm (*34*). Given an unknown vehicle's trajectory attribute $\boldsymbol{x}$, the assigned ID is

$$ID = \arg\max_i p_i(\boldsymbol{x}) \tag{3}$$

Linear Discriminant Analysis

The LDA is another popular linear classification model. The assumption is that each trajectory's composing point density is multivariate Normal $N(\boldsymbol{\mu}_i, \boldsymbol{\Sigma})$ with the same covariance matrix $\boldsymbol{\Sigma}$. Let π_i be the prior probability of trajectory i $(i = 1, \cdots, K)$ and $g_i(\boldsymbol{x})$ be the ith trajectory-conditional density. Then the posterior ID probability is $\Pr(Y = i|\boldsymbol{x}) = g_i(\boldsymbol{x})\pi_i / \sum_{j=1}^{K} g_j(\boldsymbol{x})\pi_j$. The discriminant function (Equation [4]) for each trajectory ID is obtained using $\log\{\Pr(Y = i|\boldsymbol{x})/\Pr(Y = j|\boldsymbol{x})\}$ and some simple algebra.

$$f_i(\boldsymbol{x}) = \boldsymbol{x}^T\boldsymbol{\Sigma}^{-1}\boldsymbol{\mu}_i - \frac{1}{2}\boldsymbol{\mu}_i^T\boldsymbol{\Sigma}^{-1}\boldsymbol{\mu}_i + \log\pi_i \tag{4}$$

The parameter estimation is defined by

$$\hat{\pi}_i = \frac{n_i}{n} \tag{5}$$

$$\hat{\boldsymbol{\mu}}_i = \sum_{i=1}^{K} \frac{\boldsymbol{x}_i}{n_i} \tag{6}$$

$$\hat{\boldsymbol{\Sigma}} = \sum_{i=1}^{K} \frac{n_i - 1}{\sum_{j=1}^{K}(n_j - 1)} S_i \tag{7}$$

$$S_i = \frac{1}{n_i - 1}\sum\nolimits_{Y_v = i} (\boldsymbol{x}_v - \hat{\boldsymbol{\mu}}_i)(\boldsymbol{x}_v - \hat{\boldsymbol{\mu}}_i)^T \tag{8}$$

where S_i is i^{th} trajectory attribute's covariance matrix. Given an unknown vehicle's trajectory attribute $\boldsymbol{x}$, the assigned ID is

$$ID = \arg\max_i f_i(\boldsymbol{x}) \tag{9}$$

Quadratic Discriminant Analysis

The QDA model assumptions are similar to the LDA's and the only difference is that the trajectories' multivariate Normal $N(\mu_i, \Sigma_i)$ densities can have their own covariance matrix Σ_i. The corresponding discriminant function is

$$f_i(\boldsymbol{x}) = -\frac{1}{2}\log|\boldsymbol{\Sigma}_i| - \frac{1}{2}(\boldsymbol{x} - \boldsymbol{\mu}_i)^T\boldsymbol{\Sigma}_i^{-1}(\boldsymbol{x} - \boldsymbol{\mu}_i) + \log\pi_i \tag{10}$$

The parameter estimation procedure (Equations [5–7]) is applied to QDA, but the covariance matrix estimation is different:

$$\hat{\Sigma}_i = \frac{1}{n_i - 1}\sum\nolimits_{Y_v = i} (\boldsymbol{x}_v - \hat{\boldsymbol{\mu}}_i)(\boldsymbol{x}_v - \hat{\boldsymbol{\mu}}_i)^T \tag{11}$$

Linear Support Vector Machine

In the linear SVM literature, the boundary separating two trajectories is also in the linear form. The decision function is of the form $f(\boldsymbol{x}) = \beta_0 + \boldsymbol{\beta}^T\boldsymbol{x}$ and the parameter estimation is based on the idea of maximizing the margin between the two trajectories (*35*). Solving such a problem is an exercise in the convex optimization. Given that the two trajectories' IDs are coded as -1 and $+1$, the popular setup is

$$\min_{\beta_0, \boldsymbol{\beta}} \frac{1}{2}\|\beta\|^2 + C\sum_{i=1}^{N}\xi_i \tag{12}$$

subject to

$$\xi_i \geq 0, y_i(\boldsymbol{x}_i^T\boldsymbol{\beta} + \beta_0) \geq 1 - \xi_i, \forall i \tag{13}$$

where ξ_i is the non-negative slack variable allowing trajectory attributes to be on wrong side of the decision boundary, N is the total number of two training trajectories' attributes, and C is a cost parameter that controls the overlap and is decided by a 10-fold cross validation (*36*). A quadratic programming solution of $\boldsymbol{\beta}$ and β_0 using Lagrange multipliers was introduced by (*37*). The solution for $\boldsymbol{\beta}$ has the form $\hat{\boldsymbol{\beta}} = \sum_{i=1}^{N}\hat{\alpha}_i y_i \boldsymbol{x}_i$ in which $\hat{\alpha}_i$ is a positive Lagrange multiplier and the resulting decision function is expressed by

$$\hat{f}(\boldsymbol{x}) = \hat{\beta}_0 + \sum_{i=1}^{N}\hat{\alpha}_i y_i\langle\boldsymbol{x}, \boldsymbol{x}_i\rangle \tag{14}$$

where $\langle\cdot,\cdot\rangle$ is the inner product operator. Given an unknown vehicle's trajectory attribute x, the assigned ID is indicated by

$$ID = sign\left[\hat{f}(x)\right] \tag{15}$$

For $m \geq 2$ vehicle ID matching problems, a "one against one" approach is adopted and $m(m-1)/2$ binary decision functions are estimated. The final decision is made based on the most identified ID by all the binary classifiers.

Nonlinear Support Vector Machine

The nonlinear SVM allows more general decision surfaces rather than the linear boundaries of the linear SVM. This flexibility comes from the enlarged feature space consisting of the expanded basis function $h(\boldsymbol{x}) = (h_1(\boldsymbol{x}), h_2(\boldsymbol{x}), \cdots, h_M(\boldsymbol{x}))$. The corresponding estimated decision function is indicated by

$$\hat{f}(\boldsymbol{x}) = \hat{\beta}_0 + \sum_{i=1}^{N}\hat{\alpha}_i y_i\langle h(\boldsymbol{x}), h(\boldsymbol{x}_i)\rangle \tag{16}$$

where the inner product of expanded basis $\langle h(\boldsymbol{x}), h(\boldsymbol{x}_i)\rangle$is further defined by the kernel function $K(\boldsymbol{x}, \boldsymbol{x}_i)$. Two popular choices for the K function are implemented in this research, the polynomial kernel function by Equation (17) and the Gaussian kernel function by Equation (18).

The parameters d and σ are the tuning parameters determined using a 10-fold cross validation. Finally, the ID matching rule is the same as Equation (14).

$$K(\boldsymbol{x}, \boldsymbol{x}_i) = (1 + \langle \boldsymbol{x}, \boldsymbol{x}_i \rangle)^d \quad (17)$$

$$K(\boldsymbol{x}, \boldsymbol{x}_i) = \exp\left(-\|\boldsymbol{x} - \boldsymbol{x}_i\|^2/\sigma\right) \quad (18)$$

K Nearest Neighbor

The K nearest neigbor (KNN) classifier needs no model to fit. The idea of this matching mechanism is to assign the ID to vehicle trajectory attributes based on the closest K training examples in the spatial and temporal space. The attribute $\boldsymbol{x}$ is assigned to the ID most common among its K nearest neighbors. The metric to measure distances between attributes is the Euclidean distance. The tuning parameter K is decided by a 10-fold cross validation.

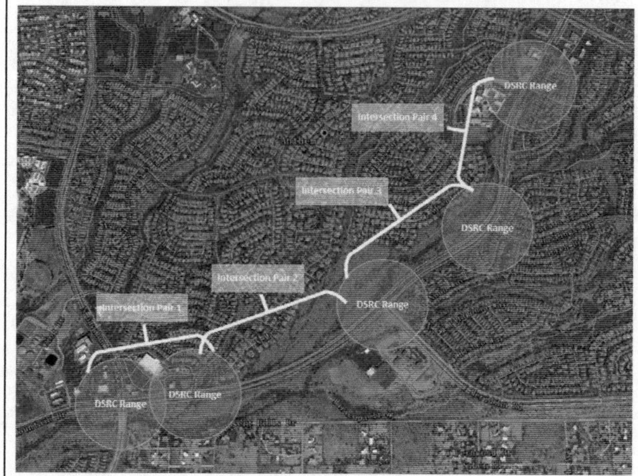

Figure 2. CV test corridor.

Result

Experiment Design

To investigate the matching accuracy, a 1-h simulation dataset was collected. The first reason for using simulation data is that there are only a few On Board units (OBUs; fewer than 10) in deployment. The small quantity of OBUs limits the experimental capability to investigate basic scenario impacts on matching accuracy. The second is that using simulation data makes it possible to collect BSM data across all kinds of condition combinations, some of which are rarely seen in the real world (e.g. 100% CV market penetration rate).

Besides the method factor, other factors include the traffic volume, CV market penetration rate, number of vehicles to be matched, and the location. Among those four factors, the traffic volume, CV market penetration, and number of vehicles to be matched are treated as random factors, indicating that if the experiment is conducted again, the treatment levels may be different. Three traffic volume levels are selected randomly: 200 vehicles per hour (vph) on the main street and 100 vph on side streets for the low volume, 400 vph on the main street and 200 vph on side streets for the medium volume, and 800 vph on the main street and 400 vph on side streets for the high volume. Different CV market penetration rates are: 20%, 60%, and 100%. Six numbers of vehicles are matched: 2, 5, 10, 30, 60, and 100. The method and the location are regarded as fixed factors. Four pairs of intersections are selected from the MMITTS CV test corridor at Anthem, Arizona (Figure 2).

Because of the traffic volume limitations, the number of vehicles to be matched and the traffic volume levels cannot be fully cross-linked in this experiment. For the low traffic volume case, only 2, 5, 10, and 30 vehicles are used for ID matching. For the medium traffic volume, the 60-vehicle level is also used. For the high traffic volume case, all the vehicle number levels are used. For some vehicle number and traffic volume combinations, there are no result data collected. This characteristic makes this experiment an unbalanced one. Vehicles' trajectory data are collected through the SIL technique in which the RSU applications collect BSMs sent by CVs in the VISSIM simulation environment (*31*). For each combination of the four factors, five replications of matching experiments are conducted by different random seeds in the R software environment.

The total number of matching records is 6300 and the average mismatching rates are summarized by factor in Table 1. A first look at the table reveals that the LR has the worst performance among all test methods. The LDA, QDA, and linear SVM have the lowest mismatching rates. More vehicles lead to higher mismatching rates. Differences of mismatching rates exist among subcategories of location, market penetration rate, and traffic volume. Because of the complexity and unbalanced characteristics of this experiment, a formal analysis is proposed below.

Because the mismatched number is a non-negative integer, for the purpose of investigating impacts of associated fixed and random factors, a Poisson mixed regression model is appropriate. The traditional standard Poisson regression usually suffers from the heterogeneity problem in count data. Overdispersion is the major heterogeneity form in Poisson distribution (*38*). In other words, only the fixed and random factors cannot guarantee to explain all the variations in the data. This extra

Table 1. Average Mismatching Rates

Method	Average	Number of vehicles to be matched	Average	Location	Average	Market penetration rate	Average	Traffic volume	Average
LR	0.82	2 Veh	0.10	Intersection Pair 1	0.17	20%	0.30	Low	0.23
LDA	0.14	5 Veh	0.17	Intersection Pair 2	0.44	60%	0.31	Medium	0.30
QDA	0.14	10 Veh	0.26	Intersection Pair 3	0.27	100%	0.31	High	0.36
Linear SVM	0.14	30 Veh	0.41	Intersection Pair 4	0.33				
Gaussian Kernel SVM	0.45	60 Veh	0.53						
Polynomial Kernel SVM	0.22	100 Veh	0.63						
KNN	0.21								

Note: Shaded cells have higher mismatching rates.

variation occurring in modeling count data is accounted by an additional random residual term. Equations (19) and (20) are proposed for modeling relations between the count response and the latent variable, which is further predicted by a linear model.

$$z_i \sim Poi(\lambda_i = \exp(l_i)) \tag{19}$$

$$l_i = \boldsymbol{x}_i^T\boldsymbol{\beta} + \boldsymbol{u}_i^T\boldsymbol{\eta} + e_i \tag{20}$$

where z_i represents the ith record of mismatched number, following the Poisson distribution with the rate λ_i. In the generalized linear model family, the rate λ_i is related to the latent variable l_i given that a canonical log link function is employed (*39*). Here variables $\boldsymbol{x}$ and $\boldsymbol{\beta}$ are reused. $\boldsymbol{x}_i$ is a 10×1 vector $\boldsymbol{x}_i = (1, x_{i1}, x_{i2}, \ldots, x_{i9})^T$ representing the intercept and fixed factor indicator variables and $\boldsymbol{\beta}$ is a 10×1 vector $\boldsymbol{\beta} = (\beta_0, \beta_1, \beta_2, \ldots, \beta_9)^T$ representing the corresponding effects. $\boldsymbol{u}_i$ is a 3×1 vector $\boldsymbol{u}_i = (u_{i1}, u_{i2}, u_{i3})^T$ representing three random factor indicator variables and $\boldsymbol{\eta}$ is a 3×1 vector $\boldsymbol{\eta} = (\eta_1, \eta_2, \eta_3)^T$ representing the corresponding effects. e_i is the residual term.

The MCMC sampling technique within the Bayesian hierarchical model framework is used for exploring both the fixed and random factors' effects. The work of (*40*) is followed and a multivariate normal distribution assumption (Equation [21]) is made for the effect and residual parameters ($\boldsymbol{\beta}$, $\boldsymbol{\eta}$ and $\boldsymbol{e} = (e_1, \cdots, e_{6300})^T$).

$$\begin{bmatrix} \boldsymbol{\beta} \\ \boldsymbol{\eta} \\ e \end{bmatrix} \sim N\left(\begin{bmatrix} \boldsymbol{\mu} \\ 0 \\ 0 \end{bmatrix} \begin{bmatrix} \boldsymbol{B} & 0 & 0 \\ 0 & \boldsymbol{G} & 0 \\ 0 & 0 & \boldsymbol{R} \end{bmatrix} \right) \tag{21}$$

where $\boldsymbol{\mu}$ is the prior mean for fixed effects with prior covariance matrix $\boldsymbol{B}$. For random factors and residual terms, their prior means are set to 0 and covariances are $\boldsymbol{G}$ and $\boldsymbol{R}$. Parameters can be Gibbs sampled and details can be found in (*40, 41*). Before proceeding to the sampling process, parameter prior distributions should be set up. For fixed factor, a normal prior is set with $\boldsymbol{\mu} = (0, \cdots, 0)^T$ and $\boldsymbol{B} = diag\{10^8, \cdots, 10^8\}$. For random and residual variance components ($\boldsymbol{G} = diag\{G_1, G_2, G_3\}$ and $\boldsymbol{R} = diag\{r, \cdots, r\}$), inverse gamma distributions are assigned with shape and scale parameters set to 0.001. The MCMC sample size is 6×10^6 and the first 6×10^5 samples are discarded as the burn-in process. Each parameter sample is collected every 500 iterations for the purpose of being identically and independently distributed.

Model Validation

The purpose of the Bayesian model validation is to check whether the parameter samples drawn by the MCMC can accurately represent their true posterior distributions.

Visualized Convergence Check. Parameter samples of posterior distributions are plotted in Figure 3, *a–n*. The constant variation and the unchanging mean are good signs of convergence of the sampling process. All the fixed effect and the residual parameters perform well, but the random effect parameter chains (Figure 3, *k, l,* and *m*) for the variances are spiky because of their posterior distributions having long right tails. It is helpful to plot them on a log scale to reduce the influence of a few outliers. In Figure 3, *o–q*, the parameters are behaving better. The visual check implies good convergence of posterior distributions. Additionally, the autocorrelation check (*42*) is used and shows evidence of samples being independently distributed.

Gelman–Rubin Test. The stricter validation approach relies on test statistics and sampling properties. The Gelman–Rubin test (*43*) is used. Several independent MCMC experiments are checked for whether they converge to the same posterior distribution. By comparing estimated between-chain and within-chain variances, the sampling chain's convergence is examined. Assuming there are M chains and each has a length of L, let $\hat{\gamma}_i$ be the sampled parameter value mean and $\hat{\sigma}_i^2$ be the sample variance of ith chain. Let $\hat{\gamma} = (1/M)\sum_{i=1}^{M}\hat{\gamma}_i$ be the overall sample posterior mean. Then the between-chain variance is defined by $V_B = [L/(M-1)]\sum_{i=1}^{M}(\hat{\gamma}_i - \hat{\gamma})^2$ and the within-chain variance is defined by $V_W = (1/M)\sum_{i=1}^{M}\hat{\sigma}_i^2$. An unbiased estimator of the marginal posterior variance of γ is defined by $\hat{V} = [(L-1)/L]V_W + [(M+1)/ML]V_B$. In (44), the potential scale reduction factor (PSRF) test statistic is defined by

$$R_c = \sqrt{\frac{\hat{d}+3}{\hat{d}+1}\frac{\hat{V}}{V_W}} \tag{22}$$

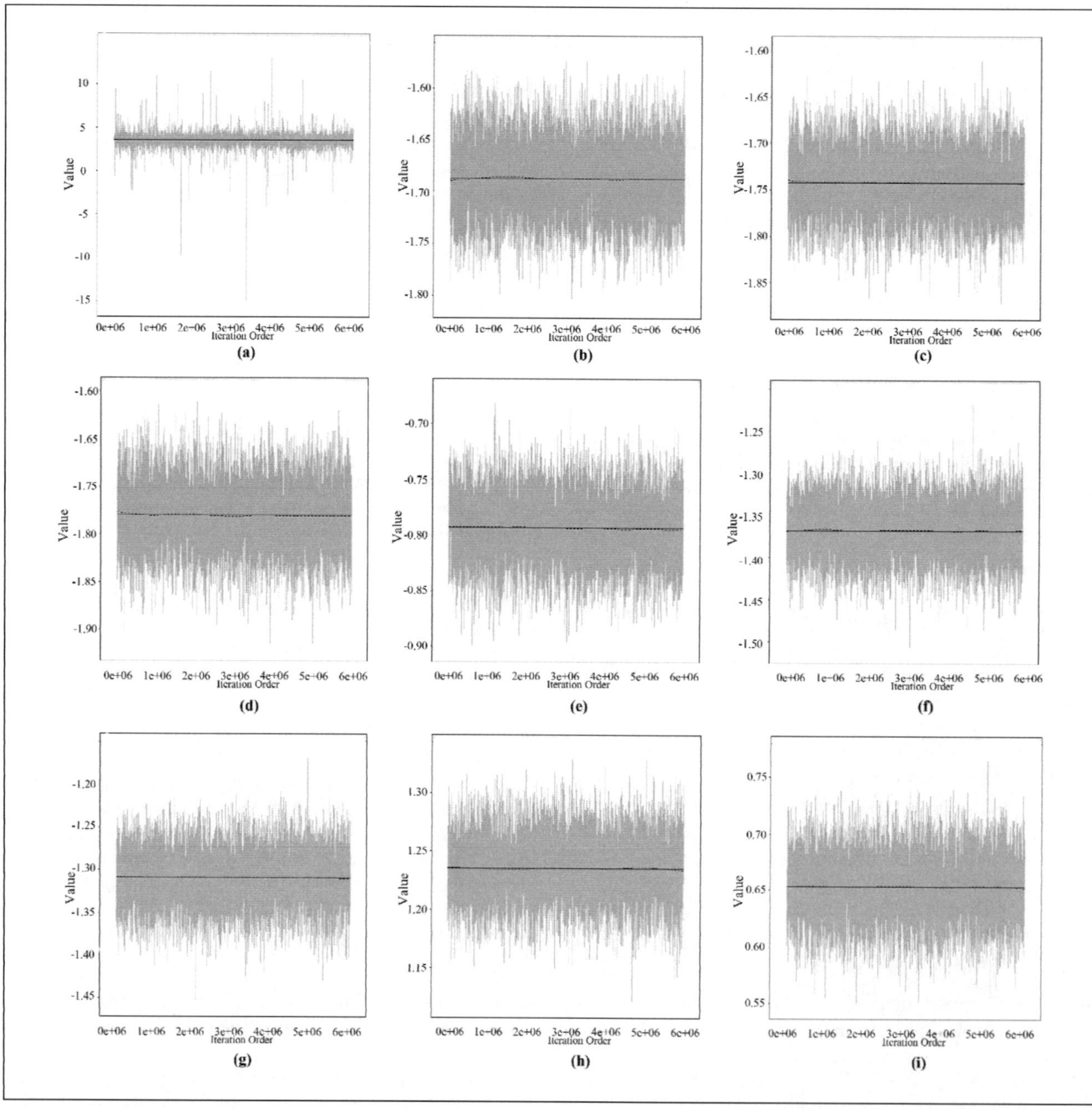

Figure 3. (Continued)

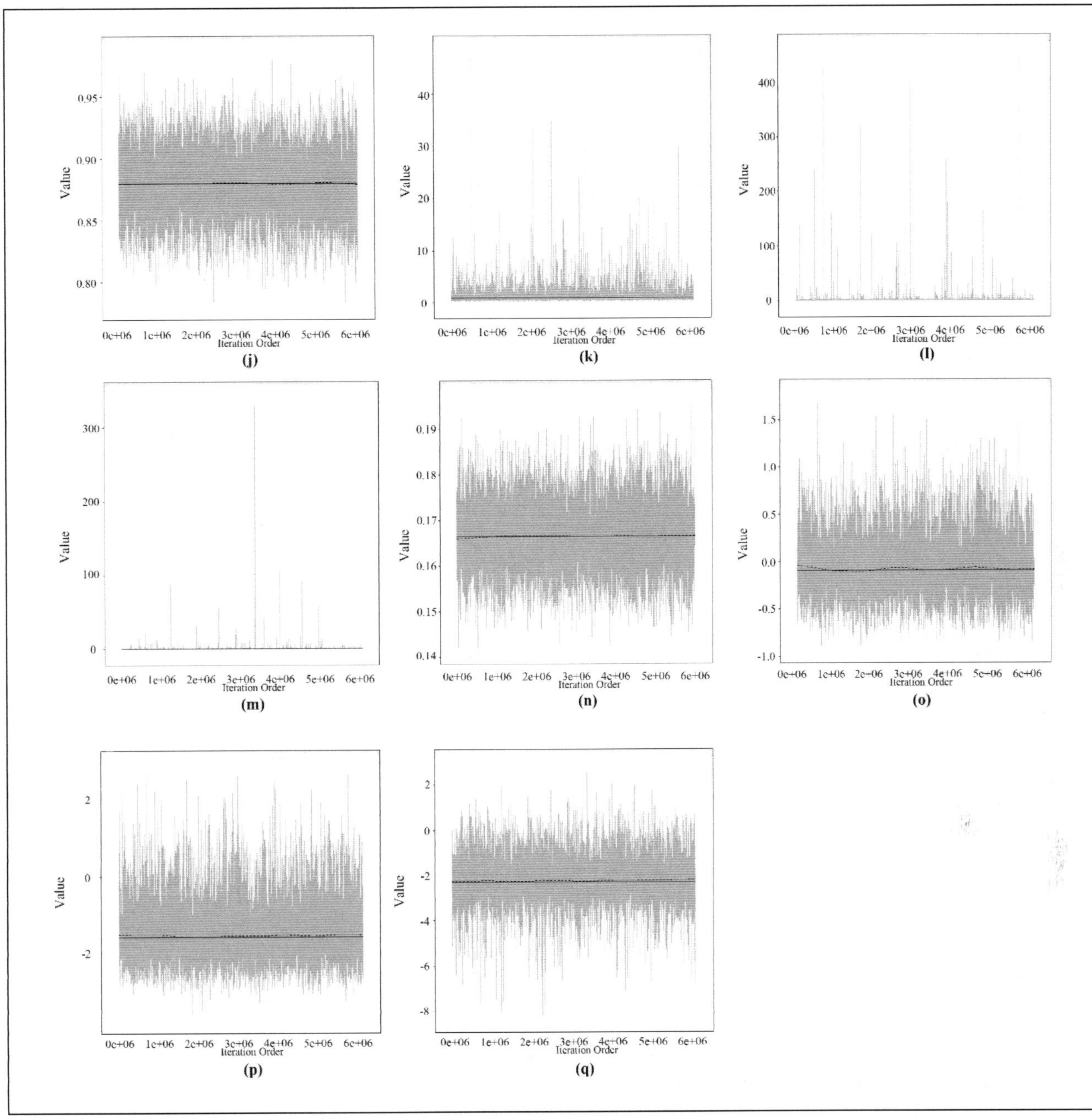

Figure 3. Time series of parameter traces, showing (*a*) intercept, (*b*) LDA, (*c*) QDA, (*d*) linear SVM, (*e*) Gaussian kernel SVM, (*f*) polynomial kernel SVM, (*g*) KNN, (*h*) Intersection Pair 2, (*i*) Intersection Pair 3, (*j*) Intersection Pair 4, (*k*) number of vehicles to be matched, (*l*) traffic volume, (*m*) market penetration rate, (*n*) residual, (*o*) number of vehicles to be matched (rescaled), (*p*) traffic volume (rescaled), and (*q*) market penetration rate (rescaled).

where $\hat{d}$ is an estimate of degree of freedom of a t distribution. If the M chains have converged to the target posterior distribution, then the value of R_c should be close to 1. Practically, if $R_c < 1.2$ for all parameters, it is safe to say the convergence has been reached. An upper limit is defined for monitoring the test statistic (*43*) as

$$R_u(\alpha) = \sqrt{\frac{\hat{d}+3}{\hat{d}+1}\left(\frac{L-1}{L}V_W + \frac{M+1}{M}q_{1-\alpha/2}\right)} \tag{23}$$

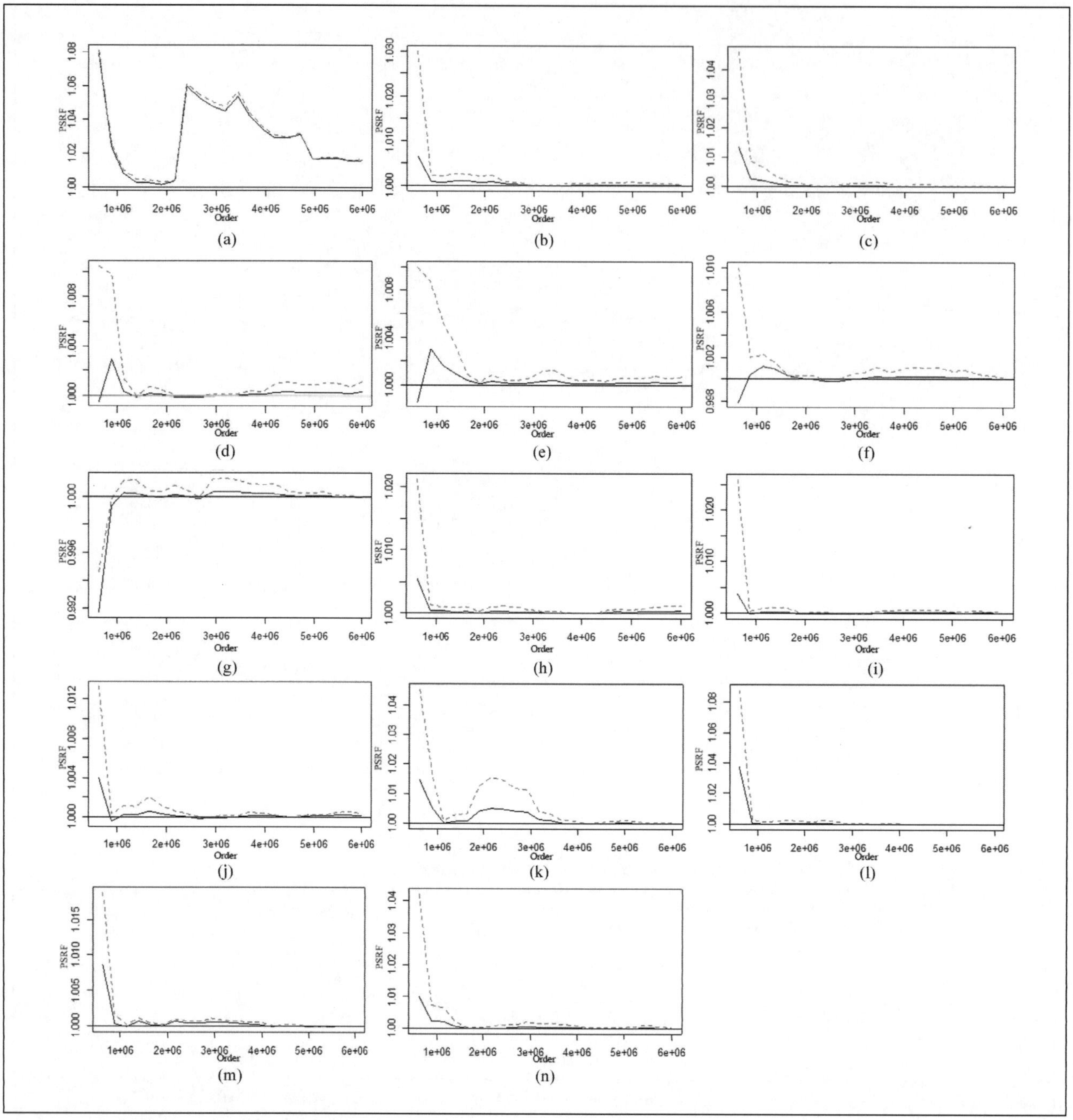

Figure 4. PSRF plot showing (*a*) intercept, (*b*) LDA, (*c*) QDA, (*d*) linear SVM, (e) Gaussian kernel SVM, (*f*) polynomial kernel SVM, (*g*) KNN, (*h*) Intersection Pair 2, (*i*) Intersection Pair 3, (*j*) Intersection Pair 4, (*k*) number of vehicles to be matched, (*l*) traffic volume, (*m*) market penetration rate, and (*n*) residual.

where $q_{1-\alpha/2}$ is the $(1-\alpha/2)$ quartile of an F distribution. By calculating the PSRF as the function of sampled parameter values, one can tell when the test statistic is really converged and not just randomly near 1. Figure 4 shows a plot ($M = 4, L = 6 \times 10^6$) of PSRFs (black solid lines) with the upper limits (red dashed lines) at a 95% confidence level. Note that the Gelman–Rubin convergence test assumes normality for the parameter's marginal posterior distribution. But for the Poisson regression model, there is no available conjugate normal posterior distribution. Therefore, those test statistics are calculated based on data normalized by the logarithm

Table 2. Posterior Distribution Summary

Factor type	Factor	Factor level	Sample median	Lower 95% CI	Upper 95% CI	Effective sample size
Fixed	Intercept	Intersection Pair 1 + LR	3.55	2.33	4.67	10513
	Method	LDA	−1.69	−1.75	−1.62	10800
		QDA	−1.74	−1.80	−1.68	10800
		Linear SVM	−1.78	−1.85	−1.71	9507
		Gaussian kernel SVM	−0.79	−0.85	−0.74	10800
		Polynomial kernel SVM	−1.37	−1.43	−1.31	9014
		KNN	−1.31	−1.37	−1.25	10334
	Location	Intersection Pair 2	1.24	1.18	1.28	10800
		Intersection Pair 3	0.65	0.60	0.71	10800
		Intersection Pair 4	0.88	0.83	0.93	10800
Random	Number of vehicles to be matched	NA	0.81	0.15	3.34	6573
	Traffic volume	NA	0.03	0.00	1.21	10800
	Market penetration rate	NA	0.01	0.00	0.34	10800
	Residual	NA	0.17	0.15	0.18	10800

transformation. All parameters' test statistics are approaching 1 quickly, indicating the convergence is achieved. Therefore, the Poisson mixed model parameter samples drawn by the MCMC are validated and are treated as posterior distributions for the further effect analysis.

Factor Effect Analysis

After obtaining effect parameters' sampled distributions, their contribution to the mismatching vehicle rate (per 100 vehicles) should be examined. Medians are used as the estimators. The highest posterior density interval is reported as the Bayesian equivalent confidence interval (CI). Table 2 shows the parameter medians, upper and lower limits with 95% CI and the effective sample sizes. Effective sample sizes are different because of the MCMC sampling rejection rule (*34*); if no sample is rejected then the sample size is $(6 \times 10^6 - 6 \times 10^5)/500 = 10800$.

The factor effects on mismatching rates differ among locations and methods. All fixed factors' CIs do not include 0, indicating their impacts are significant. For the default level combination (the Intercept), location Intersection Pair 1 and LR, the mean of the expected mismatching rate is 35 per 100 vehicles ($\exp(3.55) \approx 35$). Given the matching is conducted in the same intersection pair, all the other matching methods have better performance than LR, because their coefficients' sample medians have negative impacts on the mismatching rate. Specifically, the Linear SVM ($\exp(3.55 - 1.78) \approx 5.87$), the QDA ($\exp(3.55 - 1.74) \approx 6.11$) and the LDA ($\exp(3.55 - 1.69) \approx 6.42$) have the best top three accuracy performances. All the other locations have worse performance than Intersection Pair 1, as their medians all have positive impacts on the mismatching rate. The performance ranking is consistent with their within distances. For example, Intersection Pair 1 has the shortest within distance and the smallest mismatching rate if the method is kept unchanged, whereas the Intersection Pair 2 has the largest within distance and the largest mismatching rate (sample median value 1.24).

For the three random factors, their variances on the mismatching rate are the experiment designer's interest (*22*), and therefore their specific levels' impacts are not included in the model. The variance components of the traffic volume factor and the market penetration rate factor have the least impact on the mismatching rate, as their sample medians are 0.03 and 0.01. This suggests that if the ID matching is conducted in a different traffic volume situation or in a different CV market penetration rate scenario, given other factors are kept unchanged, there will be little difference in the result's accuracy. To the contrary, the variance component of the number of vehicles to be matched factor has a large influence on the mismatching rate with a median value 0.81. This suggests that conducting the CV ID matching with 10 vehicles will make a significant difference in the result's accuracy compared with doing it with 60 vehicles or 100 vehicles. Meanwhile, the residual's variance component is not negligible (sample median value 0.17), indicating that the data is not strictly Poisson distributed but with slight overdispersion.

Conclusion

The advantages of the proposed methods include three aspects. First, this paper demonstrates how to use the structured CV data rather than the unstructured image data for vehicle reidentification. Seven popular machine learning classification methods are implemented using the vehicle's trajectory attributes. Second, potential factors affecting matching accuracy are systematically investigated. A Poisson mixed regression model is proposed to analyze the effects of potential factors on matching accuracy. The analysis is conducted based on the Bayesian inference framework. The experiment result shows that the linear SVM, the QDA and the LDA are the three best techniques out of all the candidate machine learning methods discussed in this paper for CV ID matching. Additionally, the location factor and the number of vehicles to be matched factor have significant impacts on matching accuracy, whereas the traffic volume factor and the CV market penetration factor are not considered significant. Third, this research work can help with future CV based research and also serve as evidence for public concern. Note that the smallest average mismatching rate is 14%, achieved by the three mentioned machine learning techniques (Table 1). This possibility of using trajectory data to reidentify connected vehicle IDs with high accuracy renders researchers capable of assessing performance of CV based applications in the situation in which vehicle IDs are changing frequently. However, securer protection mechanisms for trajectory data should be investigated and the access to trajectory data should be carefully controlled.

One limitation when using simulation data is assuming that there are no new vehicles appearing in the test data set. For instance, when a new vehicle enters the traffic stream from a parking lot between intersections, the mismatching rate will likely increase. One mitigation measure to reduce the influence of newly appeared vehicles is to use static vehicle properties to do filtering. For example, if only 10 sedan vehicles appear in the first RSU range, then any other type of vehicle appearing at the next RSU must be a new vehicle which does not need ID matching. But such a mechanism cannot ensure new vehicles are filtered out every time given that they may have the same static properties as existing vehicles. The ultimate solution must be considered in the classification method itself. Zero shot learning (ZSL) techniques (*45*) are promising in this regard. Future work can include such considerations and other matching techniques.

Author Contributions

The authors confirm contribution to the paper as follows: study conception and design: Zuoyu Miao, K. Larry Head, Byungho Beak; data collection: Zuoyu Miao, K. Larry Head, Byungho Beak; analysis and interpretation of results: Zuoyu Miao, K. Larry Head, Byungho Beak; draft manuscript preparation: Zuoyu Miao, K. Larry Head, Byungho Beak. All authors reviewed the results and approved the final version of the manuscript.

References

1. FHWA. *FHWA Announces Vehicle-to-Infrastructure Guidance*. Federal Highway Administration, Washington, D.C. https://www.its.dot.gov/v2i/. Accessed January 19, 2017.
2. Juliana Gruenwald Henderson, K. J. *Connected Cars: Privacy, Security Issues Related to Connected, Automated Vehicles*. Federal Trade Commission, Washington, D.C. https://www.ftc.gov/news-events/events-calendar/2017/06/connected-cars-privacy-security-issues-related-connected.
3. Zeng, N., and J. D. Crisman. Vehicle Matching Using Color. *Proc., IEEE Conference on Intelligent Transportation System, ITSC'97*, Boston, MA, IEEE, New York, 1997, pp. 206–211.
4. Kogut, G. T., and M. Trivedi. Maintaining the Identity of Multiple Vehicles as They Travel through a Video Network. *Proc., 2001 IEEE Conference on Intelligent Transportation Systems*, Oakland, Calif., IEEE, New York, 2001. pp. 756–761.
5. MacCarley, C. A. *Video-Based Vehicle Signature Analysis and Tracking System Phase 2: Algorithm Development and Preliminary Testing*. MOU 350 Final Report to California PATH, California Polytechnique State University, San Luis Obispo, 2001.
6. Wang, S., L. Cui, D. Liu, R. Huck, P. Verma, J. J. Sluss, and S. Cheng. Vehicle Identification via Sparse Representation. *IEEE Transactions on Intelligent Transportation Systems*, Vol. 13, No. 2, 2012, pp. 955–962.
7. Li, P., and R. R. Souleyrette. A Generic Approach to Estimate Freeway Traffic Time Using Vehicle ID-Matching Technologies. *Computer-Aided Civil and Infrastructure Engineering*, Vol. 31, No. 5, 2016, pp. 351–365.
8. Haghani, A., M. Hamedi, K. Sadabadi, S. Young, and P. Tarnoff. Data Collection of Freeway Travel Time Ground Truth with Bluetooth Sensors. *Transportation Research Record: Journal of the Transportation Research Board*, 2010. 2160: 60–68.
9. Barceló, J., L. Montero, L. Marqués, and C. Carmona. Travel Time Forecasting and Dynamic Origin-Destination Estimation for Freeways Based on Bluetooth Traffic Monitoring. *Transportation Research Record: Journal of the Transportation Research Board*, 2010. 2175: 19–27.
10. Brennan, T. M. Jr., J. M. Ernst, C. M. Day, D. M. Bullock, J. V. Krogmeier, and M. Martchouk. Influence of Vertical Sensor Placement on Data Collection Efficiency from Bluetooth MAC Address Collection Devices. *Journal of Transportation Engineering*, Vol. 136, No. 12, 2010, pp. 1104–1109.
11. Hainen, A., J. Wasson, S. Hubbard, S. Remias, G. Farnsworth, and D. Bullock. Estimating Route Choice and Travel Time Reliability with Field Observations of Bluetooth Probe Vehicles. *Transportation Research Record: Journal of the Transportation Research Board*, 2011. 2256: 43–50.

12. Quayle, S., P. Koonce, D. DePencier, and D. Bullock. Arterial Performance Measures with Media Access Control Readers: Portland, Oregon, Pilot Study. *Transportation Research Record: Journal of the Transportation Research Board*, 2010. 2192: 185–193.
13. Richardson, J., B. Smith, M. Fontaine, and S. Turner. Network Stratification Method by Travel Time Variation. *Transportation Research Record: Journal of the Transportation Research Board*, 2011. 2256: 1–9.
14. Hofleitner, A., R. Herring, and A. Bayen. Arterial Travel Time Forecast with Streaming Data: A Hybrid Approach of Flow Modeling and Machine Learning. *Transportation Research Part B: Methodological*, Vol. 46, No. 9, 2012, pp. 1097–1122.
15. Charbonnier, S., A.-C. Pitton, and A. Vassilev. Vehicle Re-Identification with a Single Magnetic Sensor. *Proc., Instrumentation and Measurement Technology Conference (I2MTC), 2012 IEEE International*, Graz, Austria, IEEE, New York, 2012, pp. 380–385.
16. Sanchez, R. O., C. Flores, R. Horowitz, R. Rajagopal, and P. Varaiya. Arterial Travel Time Estimation Based on Vehicle Re-Identification Using Magnetic Sensors: Performance Analysis. *Proc., 2011 14th International IEEE Conference on Intelligent Transportation Systems (ITSC)*, Washington, D.C., IEEE, New York, 2011, pp. 997–1002.
17. Kwong, K., R. Kavaler, R. Rajagopal, and P. Varaiya. Arterial Travel Time Estimation Based on Vehicle Re-Identification Using Wireless Magnetic Sensors. *Transportation Research Part C: Emerging Technologies*, Vol. 17, No. 6, 2009, pp. 586–606.
18. Kavaler, R., K. Kwong, A. Raman, P. Varaiya, and D. Xing. Arterial Performance Measurement System with Wireless Magnetic Sensors. *First International Conference on Transportation Information and Safety (ICTIS)*, Wuhan, China,American Society of Civil Engineers, Reston, VA, 2011, pp. 20191–4400
19. Sivaraman, S., and M. M. Trivedi. Looking at Vehicles on the Road: A Survey of Vision-Based Vehicle Detection, Tracking, and Behavior Analysis. *IEEE Transactions on Intelligent Transportation Systems*, Vol. 14, No. 4, 2013, pp. 1773–1795.
20. Searle, S. R., and M. H. Gruber. *Linear Models*. John Wiley & Sons, New York, 2016.
21. Speed, F. M., R. R. Hocking, and O. Hackney. Methods of Analysis of Linear Models with Unbalanced Data. *Journal of the American Statistical Association*, Vol. 73, No. 361, 1978, pp. 105–112.
22. Montgomery, D. C. *Design and Analysis of Experiments*. John Wiley & Sons, Inc., Hoboken, NJ, 2017.
23. Box, G. E., and D. R. Cox. An Analysis of Transformations. *Journal of the Royal Statistical Society. Series B (Methodological)*, Vol. 26, No. 2, 1964, pp. 211–252.
24. Corbeil, R. R., and S. R. Searle. Restricted Maximum Likelihood (REML) Estimation of Variance Components in the Mixed Model. *Technometrics*, Vol. 18, No. 1, 1976, pp. 31–38.
25. Schall, R. Estimation in Generalized Linear Models with Random Effects. *Biometrika*, Vol. 78, No. 4, 1991, pp. 719–727.
26. Wolfinger, R., and M. O'Connell. Generalized Linear Mixed Models a Pseudo-Likelihood Approach. *Journal of statistical Computation and Simulation*, Vol. 48, No. 3–4, 1993, pp. 233–243.
27. Breslow, N. E., and D. G. Clayton. Approximate Inference in Generalized Linear Mixed Models. *Journal of the American Statistical Association*, Vol. 88, No. 421, 1993, pp. 9–25.
28. Pinheiro, J. C., and E. C. Chao. Efficient Laplacian and Adaptive Gaussian Quadrature Algorithms for Multilevel Generalized Linear Mixed Models. *Journal of Computational and Graphical Statistics*, Vol. 15, No. 1, 2006, pp. 58–81.
29. Bolker, B. M., M. E. Brooks, C. J. Clark, S. W. Geange, J. R. Poulsen, M. H. H. Stevens, and J.-S. S. White. Generalized Linear Mixed Models: A Practical Guide for Ecology and Evolution. *Trends in Ecology & Evolution*, Vol. 24, No. 3, 2009, pp. 127–135.
30. Gilks, W. R., S. Richardson, and D. J. Spiegelhalter. Introducing Markov Chain Monte Carlo. *Markov Chain Monte Carlo in Practice*, Vol. 1, 1996, p. 19.
31. University of Arizona, University of California PATH Program, Savari Networks Inc., and Econolite. *Multi-Modal Intelligent Traffic Signal System—Phase II: System Development, Deployment and Field Test*. University of Virginia Center for Transportation Studies. http://www.cts.virginia.edu/cvpfs_research/.
32. Kenney, J. B. Dedicated Short-Range Communications (DSRC) Standards in the United States. *Proceedings of the IEEE*, Vol. 99, No. 7, 2011, pp. 1162–1182.
33. Barth, M., and J. A. Farrell. *The Global Positioning System & Inertial Navigation*. McGraw-Hill, New York, NY, 1999, pp. 21–56.
34. Rizzo, M. L. *Statistical Computing with R*. CRC Press, Boca Raton, Fla., 2007.
35. Vapnik, V. *The Nature of Statistical Learning Theory*. Springer Science & Business Media, New York, 2013.
36. Geisser, S. The Predictive Sample Reuse Method with Applications. *Journal of the American Statistical Association*, Vol. 70, No. 350, 1975, pp. 320–328.
37. Friedman, J., T. Hastie, and R. Tibshirani. *The Elements of Statistical Learning*. Springer Series in Statistics. Springer, Berlin, 2001.
38. Cameron, A. C., and P. K. Trivedi. *Regression Analysis of Count Data*. Cambridge University Press, 2013.
39. Nelder, J. A., and R. J. Baker. *Generalized Linear Models*. Wiley Online Library, New York, 1972.
40. Hadfield, J. D. MCMC Methods for Multi-Response Generalized Linear Mixed Models: The MCMCglmm R Package. *Journal of Statistical Software*, Vol. 33, No. 2, 2010, pp. 1–22.
41. García-Cortés, L. A., and D. Sorensen. Alternative Implementations of Monte Carlo EM Algorithms for Likelihood Inferences. *Genetics Selection Evolution*, Vol. 33, No. 4, 2001, p. 443.
42. Cryer, J. D., and K.-S. Chan. Time Series Analysis with Applications in R. In *Springer Texts in Statistics*, Springer, New York, 2008. pp. 1 online resource (xiii, 491 p.).

43. Gelman, A., and D. B. Rubin. Inference from Iterative Simulation Using Multiple Sequences. *Statistical Science*, Vol. 7, No. 4, 1992, pp. 457–472.
44. Brooks, S. P., and A. Gelman. General Methods for Monitoring Convergence of Iterative Simulations. *Journal of Computational and Graphical Statistics*, Vol. 7, No. 4, 1998, pp. 434–455.
45. Xian, Y., B. Schiele, Z. Akata. Zero-shot Learning - The Good, the Bad and the Ugly. In *30th IEEE Conference on Computer Vision and Pattern Recognition* (pp. 3077–3086). Los Alamitos, CA: IEEE Computer Society, 2017. doi:10.1109/CVPR.2017.328.

The Standing Committee on Artificial Intelligence and Advanced Computing Applications (ABJ70) peer-reviewed this paper (18-03664).

Article

TRR
JOURNAL OF THE TRANSPORTATION RESEARCH BOARD

Transportation Research Record
2018, Vol. 2672(45) 173–184

DOI: 10.1177/0361198118794735
journals.sagepub.com/home/trr
SAGE

Network-Wide Vehicle Trajectory Prediction in Urban Traffic Networks using Deep Learning

Seongjin Choi[1], Hwasoo Yeo[1], and Jiwon Kim[2]

Abstract

This paper proposes a deep learning approach to learning and predicting network-wide vehicle movement patterns in urban networks. Inspired by recent success in predicting sequence data using recurrent neural networks (RNN), specifically in language modeling that predicts the next words in a sentence given previous words, this research aims to apply RNN to predict the next locations in a vehicle's trajectory, given previous locations, by viewing a vehicle trajectory as a sentence and a set of locations in a network as vocabulary in human language. To extract a finite set of "locations," this study partitions the network into "cells," which represent subregions, and expresses each vehicle trajectory as a sequence of cells. Using large amounts of Bluetooth vehicle trajectory data collected in Brisbane, Australia, this study trains an RNN model to predict cell sequences. It tests the model's performance by computing the probability of correctly predicting the next *k* consecutive cells. Compared with a base-case model that relies on a simple transition matrix, the proposed RNN model shows substantially better prediction results. Network-level aggregate measures such as total cell visit count and intercell flow are also tested, and the RNN model is observed to be capable of replicating real-world traffic patterns.

Large-scale mobility data that record detailed movements or trajectories of people and vehicles have become increasingly available in recent years. A "trajectory" in this study refers to a sequence of locations and the passage times describing the path that a vehicle follows along its journey. By applying data analytics to large-scale trajectory datasets, it is possible to discover complex patterns in human mobility and obtain deeper insights into travel behaviors and traffic dynamics, allowing road operators and transit agencies to identify opportunities to improve their systems. It is also possible to build predictive models for predicting movement patterns of travelers. This study addresses the problem of predicting individual trajectories in an urban network using a data-driven approach based on deep learning.

Network-wide trajectory prediction aims to predict movements of individual vehicles across the network by predicting where each vehicle may be going next at a given time based on where it is now and how it got there. Viewing each trajectory as a sequence of locations, in which locations can be defined at various spatial resolutions such as link level (e.g., links and intersections) and region level (e.g., areas and geographic subdivisions), the trajectory prediction can be considered as a sequence prediction problem, in which the aim is to predict the next location in a sequence given the previous locations visited.

Among various methods for sequence prediction, this study employs a deep learning method using recurrent neural networks (RNN). RNN (*1–3*) is a deep neural network system designed to make use of sequential information. Unlike other traditional deep neural network models, which assume independencies among all inputs (and outputs), RNN can capture temporal dependencies in sequential data and thus is suitable for performing tasks that require memories of previous events. As such, RNN has shown great performance in learning patterns in sequential data, particularly in areas of language modeling such as autotexting, text recommendation, speech recognition, and autotranslation systems.

Trajectory prediction has similarities with the problem of predicting words and sentences in language modeling. In sentence prediction, many sentence sets are collected

[1]Korea Advanced Institute of Science and Technology, Daejeon, South Korea
[2]The University of Queensland, Brisbane, Australia

Corresponding Author:
Address correspondence to Jiwon Kim: jiwon.kim@uq.edu.au

and words are extracted. A language model is trained to learn word sequence patterns from sentence data and to predict a certain word that would come next given a sequence of words. In the context of movement data, a set of locations in a network can be viewed as a set of words in a dictionary, and a trajectory (location sequence) as a sentence (word sequence), thereby linking the problem of predicting the next location in a trajectory to that of predicting the next word in a sentence. Motivated by this idea, this study applies an RNN method that has been successfully applied in sequence prediction to solve the problem of trajectory prediction by adapting it to urban movement data.

Literature Review

Predicting future trips based on previously visited locations has been widely studied in terms of predicting the location which a road user will visit next (*4–6*), the location where the user ends the trip (*7–12*), and the location which the user will visit in the next time interval (*13–17*). The first and the second aspects understand individual trips as a sequence of locations, like the one explained previously. The third predicts the location that the user will visit in the next time interval, which is usually set as an hour. This may be widely applicable because it is adapted to the temporal characteristic of mobility; however, it requires frequent updates of the user's actual location. Moreover, one of the major problems is that most of the trips end in less than 30 min to 1 h in urban areas as the majority of trips are for commuting or visiting a commercial area. Therefore, it is hard to distinguish whether the users are still traveling or have stopped. In fact, these methods are built to solve macroscopic trips rather than city-scaled problems. For example, in (*17*), the authors presented an N-gram model to predict the trip time, entry, and exit station. They used the Oyster entry and exit records data collected from London Underground, Overground, and National Rail.

There are several previous studies that used machine-learning models to predict the future location or the destination of a trip. One study (*5*) used mobility Markov chain to predict the next location of an individual. The research was based on observations of an individual's mobility so that the model had to be specified for each individual. This research, however, aims to build a trajectory prediction model for more general purposes and for both microscopic and macroscopic perspectives, so a more generalized model with aggregated data is used. Additionally, in (*6*), a hidden Markov model is used to predict pedestrian movement by using a GeoLife dataset. Hidden Markov model computes the latent state for each sequence, which maximizes the likelihood of the existence of an input sequence. Usually, the number of latent states or the number of clusters is given and hidden Markov model calibrates the transition matrix among the latent states and emission probability to decode latent states to observable sequences.

Some previous studies also tried to use artificial neural networks in trajectory prediction. Recent work by de Brébisson et al. includes a study on the prediction of taxi destination by using multilayer perceptron (MLP) (*18*). They represented the destination as a linearly weighted combination of predefined destination clusters. The result showed that the overall distance error is negligible; however, it is pointed out that it is difficult to predict unpopular destinations.

With the recent development of deep neural networks, including RNN models, and computation powers, there has been some research in the transportation field into predicting both the microscopic vehicle location for autonomous vehicles (*19*) and mobility sequences (*20, 21*). The research in (*19*) used RNN with long short-term memory (LSTM) to predict the vehicle movement in front of a subject vehicle. Endo et al. used RNN to predict destination (*20*). In this research, trajectory sequence is represented as a sequence of locations in a discretized grid space which is an arbitrary network partitioning. Therefore, in this study, network partitioning method based on the vehicle trajectory data was used.

Methodology

Representing Vehicle Trajectories as Sequence Data

Let $\mathrm{Tr} = [(x_1, y_1), (x_2, y_2), \cdots, (x_l, y_l)]$ represent a vehicle trajectory consisting of l data points, where data point (x_i, y_i) represents the longitude and latitude coordinates of the vehicle's i^{th} position. Typical trajectory datasets include the timestamp information for each data point, but this study will only consider the spatial path of each trajectory, as the goal is to predict the next location given the previous path, regardless of the time-of-day and travel time along the journey. Incorporating such temporal information in movement prediction will be explored in future research. Given massive amounts of vehicle trajectories, there will be an infinite number of possible data points used to describe all those trajectories because the longitude and latitude coordinates are continuous in space. To apply the concept of sentence prediction in language modeling to trajectory prediction, however, it is necessary to define a finite set of locations with which all the trajectories can be expressed, as one would define vocabulary for sentence prediction. As such, the study partitions an urban network into smaller regions or cells and expresses each trajectory in terms of a sequence of cells that it has passed. In partitioning the network into cells, the method presented in the authors' earlier work (*22, 23*) is used. In this method, data points

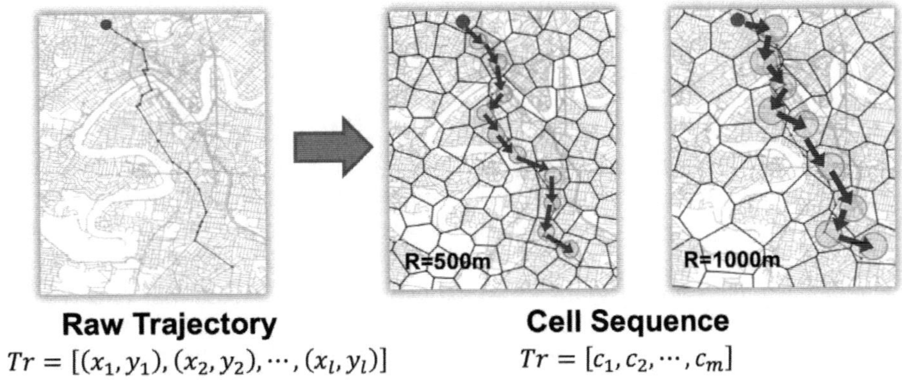

Figure 1. An illustration of the process of converting a raw trajectory to cell sequences.

in all the trajectories are combined and clustered in space based on a desired radius, denoted by R, so that for each spatial cluster the distance between the centroid of the point cluster and its farthest member point is approximately R. The centroid of each point cluster is estimated by finding the mean of the data points within the cluster. Once the centroids of all point clusters are obtained, a Voronoi tessellation method is used to construct cell boundaries (Voronoi polygons) using the centroid points as seeds.

Given N cells in the network, trajectory Tr can be mapped onto the underlying cells and expressed as the cell sequence $[c_1, c_2, \cdots, c_m]$, where c_j is the index of the jth visited cell $(1 \leq c_j \leq N)$ within trajectory Tr. Figure 1 illustrates the process of converting a raw trajectory to cell sequences. The length of the cell sequence, m, can be smaller than the original trajectory length, l, if two or more consecutive trajectory data points belong to the same cell (i.e., $m \leq l$). Moreover, a smaller number of cells would be needed to represent the same trajectory as the cell radius, R, increases. In addition to the cell sequence, two special tokens, Start_code and End_code, are added to the front and the back of the cell sequence to indicate the start and the end of the trip, respectively, where value 0 is used for Start_code and $N + 1$ is used for End_code. As a result, the original trajectory Tr is converted to the following cell sequence form:

$$\begin{aligned}\text{Tr} = [(x_1, y_1), (x_2, y_2), \cdots, (x_l, y_l)] \rightarrow \\ [\text{Start\_code}, c_1, c_2, \cdots, c_m, \text{End\_code}]\end{aligned} \quad (1)$$

Predicting Vehicle Trajectory using Recurrent Neural Network

Figure 2 shows the overall structure of an RNN model using a gated recurrent unit (GRU). The model consists of a sequence of connected GRU layers, where each layer i represents a GRU function unit that receives input (X_i) and a hidden state or "memory" from the previous GRU layer (s_{i-1}) and produces output ($\hat{Y}_i$) which is the prediction of the correct label (Y_i). At the beginning and the end of a cell sequence, Start_code and End_code are added to indicate the start and the end of the trip, respectively. As a result, the number of GRU layers corresponds to the length of the cell sequence of interest plus one, where each layer performs the operation of predicting one cell based on its previous cells. For instance, considering a trajectory that visits m cells, that is, $\text{Tr} = [\text{Start\_code}, c_1, c_2, \cdots, c_m, \text{End\_code}]$, the RNN model is set to have $m + 1$ GRU layers, where input sequence (X) and the associated label sequence (Y) for training the model are coded as:

$$\begin{aligned} X &= [X_0, X_1, X_2, \cdots, X_m] \equiv [\text{Start\_code}, c_1, c_2, \cdots, c_m] \\ Y &= [Y_0, Y_1, Y_2, \cdots, Y_m] \equiv [c_1, c_2, \cdots, c_m, \text{End\_code}] \end{aligned} \quad (2)$$

Inside each GRU, there are two functions called "reset gate" (r) and "update gate" (z); the reset gate decides how to combine new input (X_i) with memory (h) from previous computations and the update gate decides how much memory to keep from previous computations (*2, 3*). The mathematical expressions for the operation taking place within the GRU for the ith layer are:

$$\begin{aligned} z &= \sigma(X_i U^z + s_{i-1} W^z) \\ r &= \sigma(X_i U^r + s_{i-1} W^r) \\ h &= \tanh\left(X_i U^h + (s_{i-1} \circ r) W^h\right) \\ s_i &= (1 - z) \circ h + z \circ s_{i-1} \\ \hat{Y}_i &= \text{softmax}(V \circ s_i) \end{aligned} \quad (3)$$

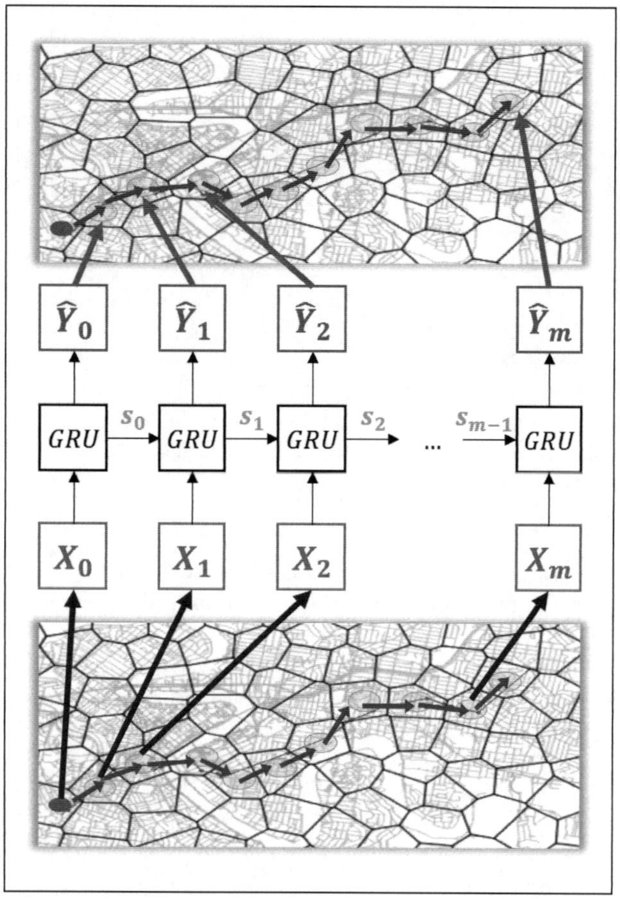

Figure 2. Overview of RNN model for trajectory prediction.

where

∘ represents element-wise multiplication between two vectors,

σ represents a sigmoid function to limit function output between 0 and 1,

X_i represents a one-hot vector of size $N + 1$ (N cells + Start_code) indicating the cell visited at the ith position of the cell sentence (or indicating the start of the trip in case of Start_code),

$\hat{Y}_i$ represents a real-valued vector of size $N + 1$ (N cells + End_code), where each value in $\hat{Y}_i$ represents the probability of visiting each cell (or the probability of terminating the trip in case of End_code) at the $(i + 1)^{th}$ position of the sequence,

s_{i-1} denotes the memory from previous computations or the hidden state computed in the GRU,

U and W are matrices of parameters used inside GRU units, where the superscripts on U and W indicate the functions in which these parameters are used, and

V is a vector of parameters used to translate the internal hidden state to a vector of probability of each cell being visited in the next step ($\hat{Y}_i$).

Equation 3 basically shows a series of computations to obtain Y from X, entailing computing the probability of each cell being visited in the next step ($\hat{Y}_i$) based on the hidden state at the current step (s_i), where this hidden state (s_i) is in turn computed based on the current cell (X_i) and the hidden state transferred from the previous step or the memory of the previous visited cells (s_{i-1}).

The next step is to define the loss function that calculates the loss or error in a model prediction; the aim is to minimize this during training. The cross-entropy loss function is used in this study as shown in Equation 4. By treating the full trajectory (and therefore its cell sequence Y) as one training case, the cross-entropy loss (L) is calculated by summing the errors at each step (cell) in the sequence over the entire trajectory using:

$$L(Y, \hat{Y}) = \sum_{i \in \{1, \cdots, m+1\}} -\frac{1}{N} \sum_{n \in N} \left(Y_{i,n} \log \hat{Y}_{i,n} + (1 - Y_{i,n}) \log(1 - \hat{Y}_{i,n}) \right) \quad (4)$$

where

m is the cell sequence length,

N is the number of cells,

$Y_{i,n}$ is the binary label set to 1 if cell n is visited at the $(i + 1)^{th}$ step in the sequence or 0 otherwise, and

$\hat{Y}_{i,n}$ denotes the probability of cell n being visited at the $(i + 1)^{th}$ step in the sequence.

To train the RNN model, a large quantity of cell sequence data is prepared in the format shown in Equation 2 and fed into the RNN. Because the lengths of input sequences are different, the number of GRU layers in the RNN model needs to be dynamically changed. The RNN model scans one input sequence at a time and adjusts the number of GRU layers so that the number of layers becomes the length of the given input sequence (m) plus one. As a result, every time a new input sequence with length m is fed into the model, the model constructs RNN with $m + 1$ GRU layers.

The model then calculates cross-entropy loss (L) between the correct label (Y) and predicted label probability based on the current parameter ($\hat{Y}$). The model uses the stochastic gradient descent (SGD) method (see, e.g., [24] and [25]) to update the model parameters in the direction of decreasing the loss. This process is repeated until the parameters converge. Once the model is trained and the parameters are fixed, the model can be used to predict the next cell for any given sequence of previous cells.

Model Performance Evaluation

Data

To evaluate the performance of the RNN-based trajectory prediction model, a case study was designed using vehicle trajectories collected from the Bluetooth sensors in Brisbane, Australia, provided by Queensland Department of Transport and Main Roads (TMR) and Brisbane City Council (BCC). The Bluetooth sensors installed in state-controlled roads and city intersections detect Bluetooth devices (e.g., in-vehicle navigation systems and mobile devices) passing the sensors and record their passage times. By tracking the identifier of each Bluetooth device, the trajectories of individual vehicles (or Bluetooth devices) can be constructed, where each trajectory represents a time-ordered sequence of Bluetooth sensor locations that a given vehicle passed. Vehicle trajectories containing a resting period (time without moving) of over 1 h are considered to have multiple trips and they are separated into multiple trajectories. This case study used the data from one day on 1 March, 2016, for training and the data from one day on 8 March, 2016, for testing. Each dataset contains approximately 350,000 trajectories per day.

Using the method described in Figure 1, the Brisbane network is partitioned into cells using three different cell sizes (300 m, 500 m, and 1000 m) expressed in terms of the desired cell radius (R). The network is represented as 5712 cells, 2204 cells, and 319 cells under $R = 300$ m, $R = 500$ m, and $R = 1000$ m, respectively. For each cell network, trajectories are mapped onto the underlying cells and represented as the associated cell sequences.

Base-Case Model

To better understand the performance of the proposed RNN-based model, a simple statistical model was designed as a base-case model for a comparison. In predicting the next cell in a cell sequence, the base-case model, which is here referred to as the transition matrix method (TRN), relies on the transition matrix that describes the probability of going to a particular cell from the current cell. The transition probabilities are estimated based on the historical data, where the transition probability from cell i to cell j, denoted by $p_{i \to j}$, is determined by computing the fraction of the outgoing flows from cell i to cell j given the total outgoing flows from cell i as:

$$p_{i \to j} = \frac{f_{i \to j}}{\sum_{n=1}^{N} f_{i \to n}} \quad (5)$$

where $f_{i \to j}$ is the intercell flow from cell i to cell j (i.e., vehicle flows passing cell i and cell j consecutively) and N is the number of cells in the network. In predicting where a vehicle will go next given a sequence of cells the vehicle has passed so far, TRN determines the next cell only based on the last cell of the input sequence and does not use any information about the vehicle's travel history. The key difference between RNN and TRN is thus that the former uses the memory of the previous cells visited and incorporates sequential characteristics in predicting the next cell, whereas the latter is memoryless and the next cell only depends on its immediate predecessor.

Results

Cross-Entropy Loss

Figure 3 shows the cross-entropy loss calculated during model training, plotted for different cell sizes. At each iteration, the parameters of the RNN model are updated based on the training dataset and the average loss ($\bar{L}$) is calculated for both training and testing datasets under the given parameters as shown in Figure 3, a and b, respectively. The RNN parameters converge after five

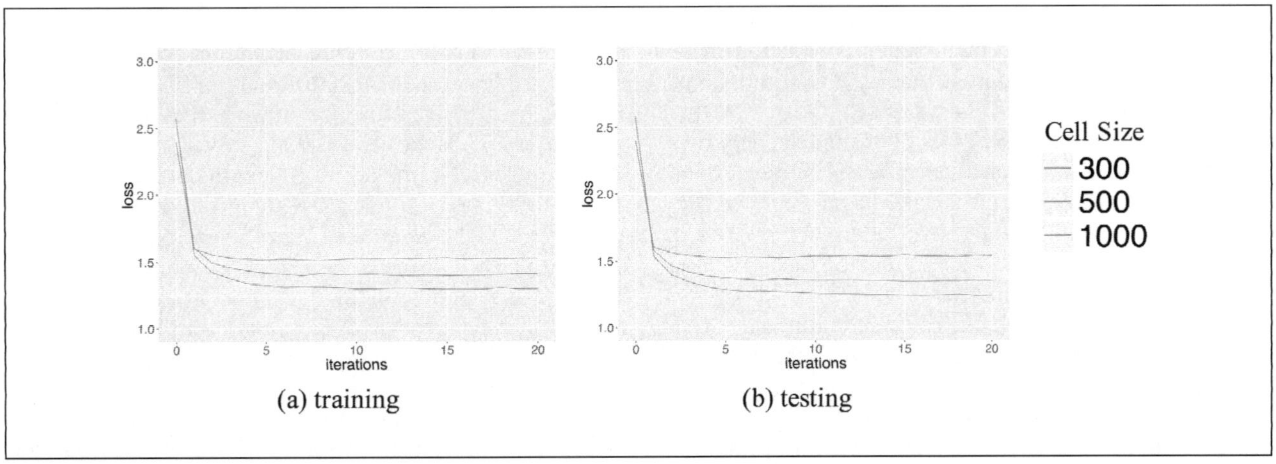

Figure 3. Cross-entropy loss at each training iteration: (a) average loss calculated from training dataset and (b) average loss calculated from testing dataset.

iterations in all cases and the loss decreases for both training and testing datasets, indicating that the model is trained properly and not overfitting to the training data. To further ensure the robustness of the model regardless of the choice of day-of-week from which data are collected and sample size of training and testing dataset, additional analyses were performed. The model was trained with a subset of the original training data containing 10,000 randomly sampled data from 1 March, 2016, and was tested with 10,000 randomly selected data from the whole of March 2016. For all test cases, the results showed similar results to those shown in Figure 3. Comparing plots across different cell sizes, the case with cell radius of 300 m produces the smallest loss. This may suggest that, the smaller the cell size, the better the RNN model is at recognizing sequential patterns and, therefore, the easier it is for the model to predict the next cell.

Validation

Cell Visit Count and Intercell Flow. One of the important applications of trajectory prediction is to anticipate how many vehicles will use a particular region in the network by predicting an individual vehicle's moving paths. For instance, when a major disruption (e.g., construction, special event) is anticipated for a certain region, road operators will want to identify who will travel to or pass through that region during the day to identify the potential impact of the disruption and provide the relevant travelers with more personalized and targeted information. To assess model performance in this regard, this section aims to evaluate the sequence prediction accuracy in term of cell-level (area-level) aggregated measures, namely cell visit count and intercell flow. Cell visit count measures the number of unique travelers who visit a particular cell during a given day and intercell flow (or cross-boundary flow) measures the total daily volume of vehicle traffic from one cell to another between neighboring cells.

The first step is to generate 700,000 synthetic trajectories (in terms of cell sequences) using RNN and TRN models for three different cell sizes (R=300 m, 500 m, and 1000 m). For both RNN and TRN models, only origin cells randomly sampled from the historical data (training set) are given, and the models are allowed to determine the remaining sequences. Based on the 700,000 cell sequences generated by each model for each cell size, the cell visit counts and intercell flows are computed for all cells. As a ground truth, the cell visit count and intercell flow measures from real data are also obtained using the training set. To compare the measures across these three cases (RNN, TRN, and real data), the measures are normalized with the number of trajectories. To exclude abnormal trajectories, a maximum trajectory length of 50 km, roughly equivalent to the longest straight-line distance covering the Brisbane network, is imposed. To reflect this condition, the maximum number of cells in a sequence is set to 80 for R = 300m, 50 for R = 500m, and 25 for R = 1000m in generating cell sequences. The historical data also contain unrealistically long cell sequences. These are presumed to be from taxi vehicles and ride-sharing vehicles that usually keep moving for a long period. Therefore, the limit is also applied to the historical data to discard long sequences.

A real-world trajectory often visits a certain cell more than once during its journey. It is important for a trajectory prediction model to mimic this behavior as realistically as possible because the number of cells in a sequence (sequence length) and the number of unique cells have different implications: the former gives information on traffic volume in a given cell whereas the latter gives information on the number of actual travelers who visit the cell. To measure the extent to which each cell sequence contains duplicate cells, cell re-visit ratio (D) is defined as:

$$D = \frac{m - m^u}{m} \times 100\% \tag{6}$$

where m is the length of a cell sequence expressed in terms of the number of cells and m^u is the number of unique cells excluding duplicate cells. The D measure is a relative measure and the D measure of RNN and TRN are compared with the D measure obtained from the real data. On average, for R = 300 m, 500 m, and 1000 m, real data show D = 18.6%, 20.2%, and 17.3% with m = 26.83, 17.05, and 9.29 and m^u = 21.85, 13.60, and 7.68, respectively. For trajectories generated by RNN, on average, D = 25.7%, 24.7%, and 19.9% for R = 300 m, 500 m, and 1000 m, respectively (with m = 18.38, 14.26, and 8.33 and m^u = 13.66, 10.74, and 6.67). Trajectories generated by TRN produce the average values of D = 62.8%, 56.7%, and 45.6% for R = 300 m, 500 m, and 1000 m, respectively (with m = 21.61, 15.80, and 8.89 and m^u = 8.03, 6.84, and 4.84). In terms of average sequence length (m), model-generated trajectories are on average shorter than real-world trajectories, showing that both models tend to end trips earlier than real-world trips. Between RNN and TRN, the average lengths from TRN are closer to the real-world case than those from RNN. However, TRN-generated trajectories contain many duplicate cells as can be seen from very high cell revisit ratios (D), resulting in unrealistic movement tendency that a trajectory repeatedly goes back and forth between a few cells instead of progressing toward its destination. On the other hand, RNN produces cell revisit ratios that are similar to the real data, suggesting that the ability to incorporate "memory" of the previously visited cells in RNN can prevent such an unrealistic "memoryless" behavior from happening.

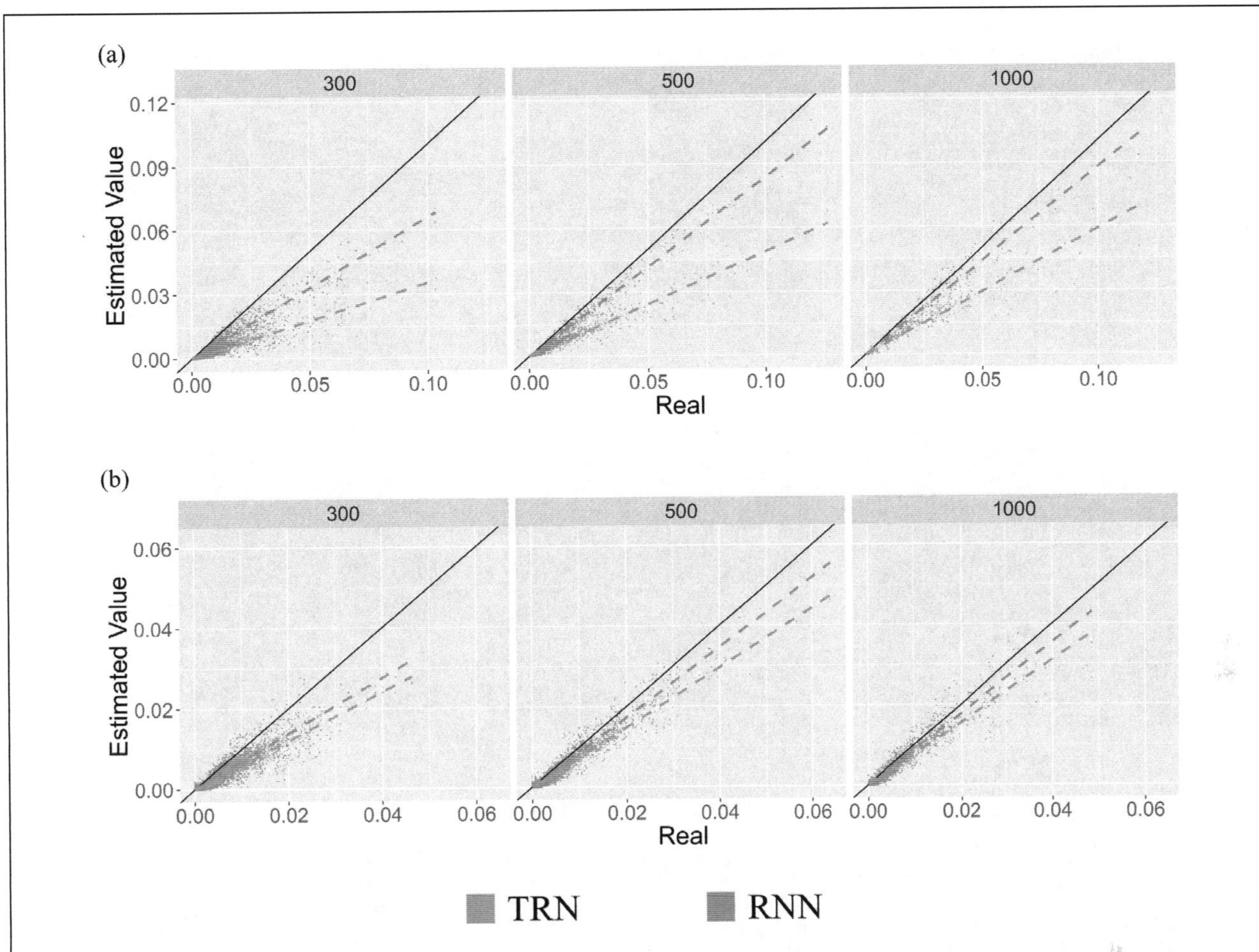

Figure 4. Cell-level measures computed using trajectories generated by RNN and TRN models: (*a*) normalized unique cell visit count per trajectory and (*b*) normalized inter-cell flow per trajectory.

Figure 4 shows the estimated cell visit count and inter-cell flow for TRN (blue dots) and RNN (red dots) with respect to the corresponding real-world measures from the historical data, where all measures are normalized with the number of trajectories. Each data point on the plots in Figure 4*a* represents the cell visit count for each cell and each data point in Figure 4*b* represents the inter-cell flow for each neighboring cell pair. Linear regression lines (dashed lines) are also shown, where the closer the regression line is to the 45° line ($y = x$), the closer the model prediction is to the reality. In all cases, the regression lines from RNN are closer to the 45° line than those from TRN, suggesting that RNN predicts cell-level aggregated measures more accurately than TRN.

For cell visit count, color maps are created to visualize the spatial distribution of cell measures across the Brisbane network. Figure 5 presents nine color maps created using the data points in Figure 4*a*; the results can be compared across different trajectory datasets (by column) and different cell sizes (by row). From the maps, it is clear that RNN (center) reflects the flow magnitudes and spatial patterns in the real data (left) much closely than TRN (right) does.

Sequence Prediction. This section focuses on a more direct performance measure of cell sequence prediction, namely correct prediction probability (CPP_k), which represents the probability of correctly predicting the next k cell(s) given previously visited cells. The procedure for obtaining CPP_k measures is shown in Figure 6.

For a prediction scenario, consider three scenarios of predicting the next single cell ($k = 1$), the next two cells ($k = 2$), and the next three cells ($k = 3$), respectively. With three k values, two models (RNN and TRN) and three cell sequence databases (one for each cell size R), a total of 18 cases are generated, each of which produces

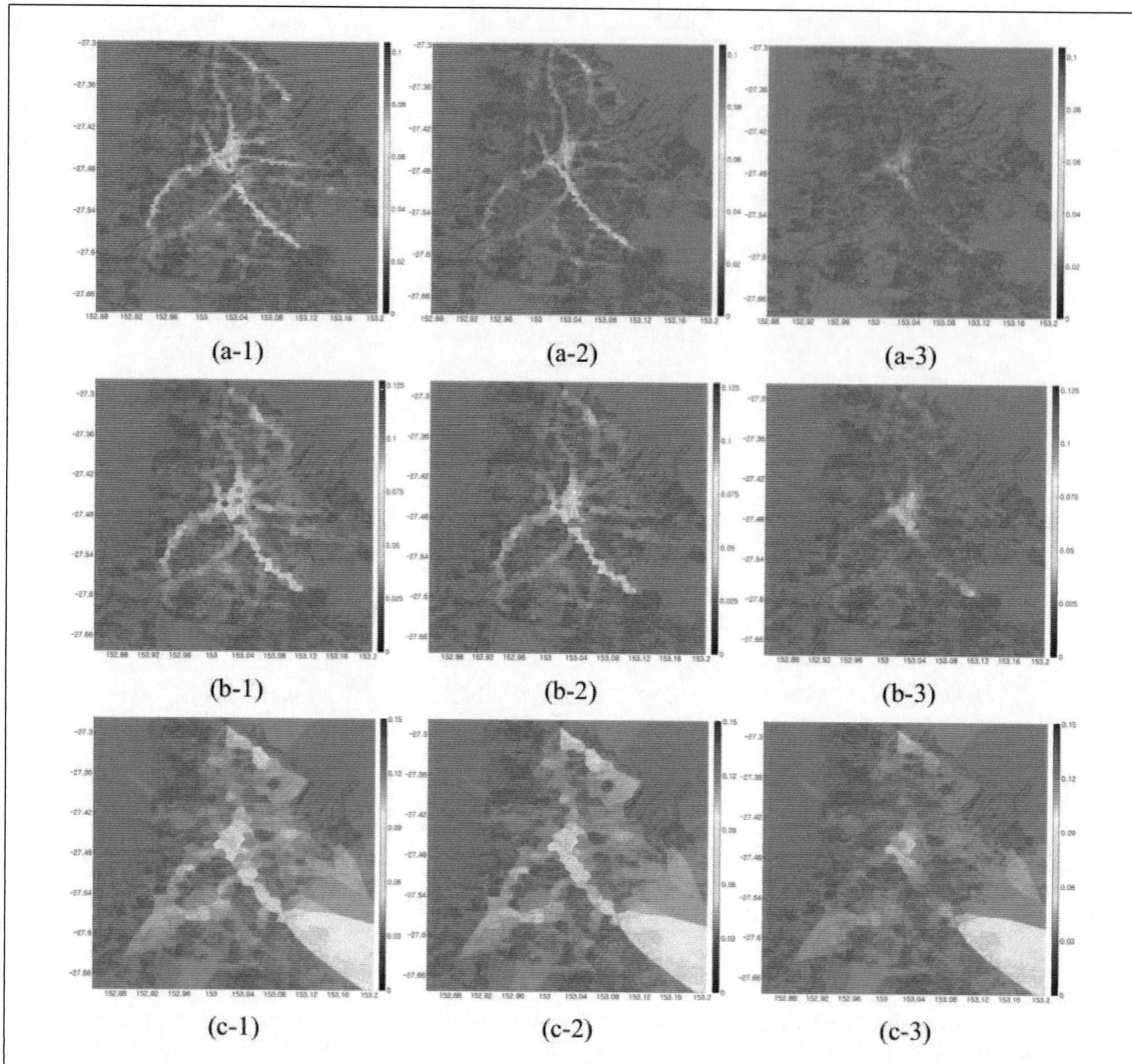

Figure 5. Color maps of normalized cell visit count for different cell sizes (rows) and different trajectory datasets (columns): (top) cell radius = 300 m; (middle) cell radius = 500 m; (bottom) cell radius = 1000 m; (left) real data; (center) RNN model; (right) TRN model.

the distribution of CPP_k over all combinations of actual sequence length (m) and given sequence length (i).

Figure 7, *a–c*, presents the CPP_k distributions for $k = 1, 2, \cdots, 3$, respectively, in terms of the complementary cumulative distribution function (CCDF). The CCDF of CPP_k, denoted by $\bar{F}_{CPP_k}(x)$, is defined as the probability that CPP_k is greater than x, that is, $\bar{F}_{CPP_k}(x) = P(CPP_k > x) = 1 - F_{CPP_k}(x)$, where $F_{CPP_k}(x)$ is the cumulative distribution function (CDF). For instance, in Figure 7*a*, the value of $\bar{F}_{CPP_1}(x = 0.75)$ for 1000 m RNN (blue curve) is approximately 0.375, meaning that 37.5% of the CPP_1 values produced via Figure 6 are greater than 0.75, that is, RNN predicted the next cell correctly 35% of the time with the probability higher than 0.75. Given this interpretation, it can be seen that, the closer the $\bar{F}_{CPP_k}(x)$ curve is toward the top right corner (1,1) and the farther the curve is away from the bottom left corner (0,0), the better the prediction performance is. The area under curve (AUC) of $\bar{F}_{CPP_k}(x)$ thus provides a good summary metric quantifying the

Algorithm 1 Measureing Correct Prediction Probability CPP_k

Input: Number of cells to predict (k), Number of predictions to make (N), Number of sample data (M) Prediction model (M), and real-world cell sequence data (DB) obtained from the training dataset

for all $m \in 3, ..., 30$ **do**
  $DB_m \leftarrow$ SUBSET(DB , $length(DB[i]) == m$)
    ▷ DB_m is subset of DB with sequence length of m
  $Tr \leftarrow$ RANDOMSAMPLE(DB_m, M)   ▷ Tr is random-sampled result of size M
  for all $i \in 1, ..., m-1$ **do**
    $CPP_1, ...CPP_k \leftarrow 0$ ▷ CPP_x is the x-th correct prediction probability
    for all $j \in 1, ...N$ **do**
      $[\hat{c}_{i+1}, ..., \hat{c}_{i+k}] \leftarrow$ PREDICT(M,$[c_1, ..., c_i]$,k)
        ▷ $\hat{c}_{i+x}$ is $i+x$-th prediction from model M with sequence $[c_1, ..., c_i]$
      for all $w \in 1, ...k$ **do**
        if $\hat{c}_{i+w} = c_{i+w}$ **then** CPP_w += 1/N
        end if
      end for
    end for
  end for
end for

Output: The set of CPP_k values = $[CPP_1, CPP_2, ...CPP_k]$

Figure 6. Algorithm for measuring correct prediction probability CPP_k.

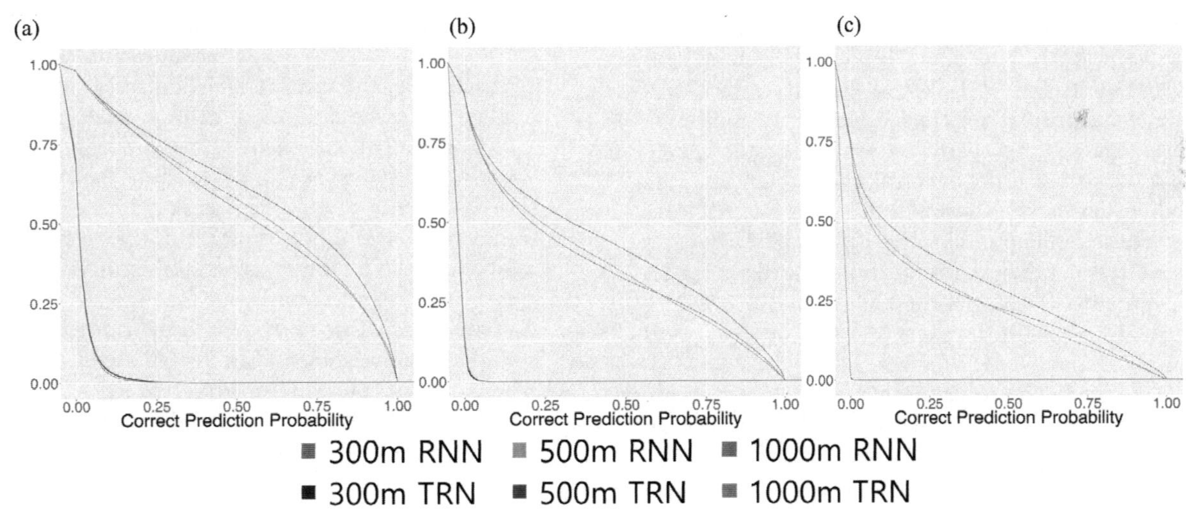

Figure 7. CCDF of correct prediction probability (CPP) for three prediction scenarios: (*a*) predicting next one cell ($k = 1$); (*b*) predicting next two cells ($k = 2$); and (*c*) predicting next three cells ($k = 3$).

prediction performance described by $\bar{F}_{\mathrm{CPP}_k}(x)$ with a single number. The value of AUC of $\bar{F}_{\mathrm{CPP}_k}(x)$ varies from 0 to 1, where 1 occurs when the curve passes (1,1) and 0 occurs when the curve passes (0,0).

As shown in Figure 7, in all cases, RNN performs much better than TRN as the $\bar{F}_{\mathrm{CPP}_k}(x)$ curves for RNN are always above those for TRN, where the curves of RNN predicting one or two consecutive cell sequences

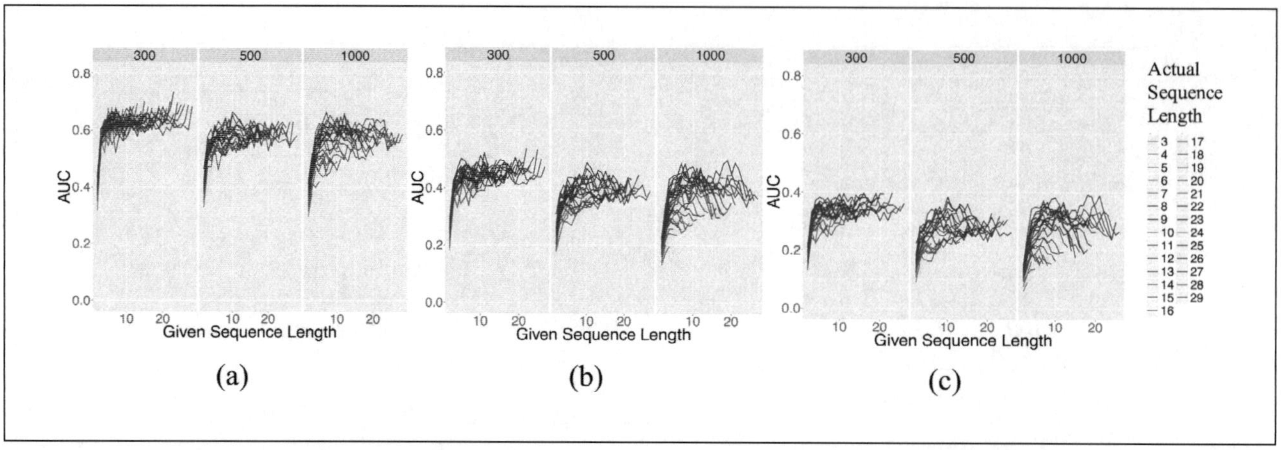

Figure 8. AUC (as a cell prediction performance measure) with respect to given sequence length and actual sequence length in (*a*) predicting next one cell ($k = 1$), (*b*) predicting next two cells ($k = 2$), and (*c*) predicting next three cells ($k = 3$).

are convex toward (1,1) and those of TRN are concave toward (0,0). In the case of predicting three consecutive cell sequences, although both curves for RNN and TRN are concave towards (0,0), RNN is still able to predict cell sequence when correct prediction probability (CPP) of TRN at this point reaches zero at all points. For the case of predicting the next single cell in Figure 7*a*, the average AUC of $\bar{F}_{\mathrm{CPP}_k}(x)$ across the three cell sizes is 0.0198 for TRN and 0.5096 for RNN. When it comes to predicting multiple consecutive cells ($k>1$), the performance difference between RNN and TRN becomes substantial as shown in Figure 7, *b–c*. The $\bar{F}_{\mathrm{CPP}_k}(x)$ curves for TRN exhibit a sharp drop at very low values of x, resulting in the AUC values being nearly zero. On the other hand, the $\bar{F}_{\mathrm{CPP}_k}(x)$ curves for RNN still show relatively high probabilities of having high CPP_k. For instance, for $k = 3$, there are still approximately 10% to 15% of cases for which RNN predicts the next three cells correctly with the probability higher than 0.75 (see $\bar{F}_{\mathrm{CPP}_3}[x = 0.75]$ in Figure 7*c*). Overall, the average AUC of $\bar{F}_{\mathrm{CPP}_k}(x)$ is 0.2849 for $k = 2$ and 0.1585 for $k = 3$.

Next, the prediction results from the RNN models are more closely examined, focusing on the impact of original sequence length (m) and given sequence length (i) on the prediction performance expressed by AUC. Within each combination of k and cell size R, the associated CPP_k observations are split into different m and i groups. A separate CCDF curve is then constructed for each group to compute AUC. Figure 8 shows six groups of AUC plots; each plot contains a set of AUC curves and each AUC curve represents a set of AUC values obtained for predicting cell sequences of certain length m. The x-axis represents the given sequence length i in predicting the next k cells. Some observations from Figure 8 are summarized as follows:

1. From each AUC curve (given a fixed m), AUC increases sharply as i increases up to a certain point (e.g., $i = 5$ or 6), but remains stable afterward. This means that until a certain point, giving a longer initial sequence helps in predicting the next cells (e.g., predicting the third cell given two previous cells is easier than predicting the second cell given one previous cell). Beyond that point, however, the length of initial sequence does not significantly affect prediction performance.
2. From each AUC plot (across different m), AUC increases as m increases. This indicates that, even when the same number of cells is given, the prediction performance depends on the length of the actual cell sequence (e.g., predicting the third cell in a sequence of 10 cells is easier than predicting the third cell within a sequence of five cells).
3. Across AUC plots (across different k), AUC decreases as k increases, which is expected because predicting further steps toward the future is more difficult than predicting the immediate next step.
4. Across AUC plots (across different R), AUC does not significantly change with different cell sizes. There is, however, a tendency of decreasing AUC and increasing variance in AUC values when spatial resolution is low (e.g., $R = 1000\,\mathrm{m}$).

Conclusion

The overall goal of this research is to leverage massive amounts of urban movement data, which have become increasingly available nowadays, to better understand city mobility dynamics and enhance the design and operations of transportation systems. Of particular interest is the ability to predict individual vehicles' movements—at least in terms of a sequence of aggregated spatial locations—and therefore anticipate the flow of vehicles at a given location and time more accurately. This study showed a promising direction toward achieving this ability by applying deep learning with RNN on vehicle trajectory data. As a way to represent complex vehicle trajectories as simpler location sequences, this study proposes a method to partition the network into cells so that entire vehicle movements can be expressed in terms of combinations of a finite set of cells. Mapping trajectories onto cells not only reduces computational complexity but also allows working with multisource and multiresolution trajectories. The study tests different cell sizes and provides discussions on the impacts of cell size on trajectory prediction performance. Using large amounts of Bluetooth vehicle trajectory data collected in Brisbane, Australia, this study trains an RNN model to predict cell sequences. Model performance is tested by computing the probability of correctly predicting the next k consecutive cells. Compared with a base-case model that relies on a simple transition matrix, the proposed RNN model shows substantially better prediction results. Network-level aggregate measures such as total cell visit count and intercell flow are also tested and it is observed that the RNN model can replicate real-world traffic patterns.

In summary, the contribution of this study is the development of a novel network-wide trajectory prediction framework that entails (i) transforming raw trajectories into location sequence data using the proposed cell construction method; (ii) applying the RNN model to learn and predict trajectory sequence patterns by recognizing the similarity between trajectory sequence prediction and language modeling, in which RNN showed a great success; and (iii) proposing different performance measures, at both individual sequence level (e.g., CPP of predicting next cells in a sequence) and aggregated region level (i.e., cell visit count, intercell flow, network-wide usage pattern), to evaluate and demonstrate the application of the proposed model from different angles.

Acknowledgments

This work was supported by the Institute for Information & Communications Technology Promotion (IITP) grant funded by the Korea government (MSIT; No. 2017-0-00121, Development of a Traffic Predictive Simulation SW for Improving the Urban Traffic Congestion). This research was also supported by Queensland Department of Transport and Main Roads (TMR) under the Transport Academic Partnership (TAP) agreement with the University of Queensland. The authors would like to thank TMR and Brisbane City Council (BCC) for access to the data on which the paper is based.

Author Contributions

The authors confirm contribution to the paper as follows: study conception and design: Choi, Yeo, Kim; data collection: Kim; analysis and interpretation of results: Choi, Yeo, Kim; draft manuscript preparation: Choi, Kim. All authors reviewed the results and approved the final version of the manuscript.

References

1. Hochreiter, S., and J. Schmidhuber. Long Short-Term Memory. *Neural Computation*, Vol. 9, No. 8, 1997, pp. 1735–1780.
2. Cho, K., B. Van Merriënboer, C. Gulcehre, D. Bahdanau, F. Bougares, H. Schwenk, and Y. Bengio. Learning Phrase Representations Using RNN Encoder-Decoder for Statistical Machine Translation. *arXiv preprint arXiv:1406.1078*, 2014,
3. Chung, J., C. Gulcehre, K. Cho, and Y. Bengio, Empirical Evaluation of Gated Recurrent Neural Networks on Sequence Modeling. *arXiv preprint arXiv:1412.3555*, 2014.
4. Noulas, A., S. Scellato, N. Lathia, and C. Mascolo. Mining User Mobility Features for Next Place Prediction in Location-Based Services. *Proc., 2012 IEEE 12th International Conference on Data Mining*, Brussels, Belgium, IEEE, 2012.
5. Gambs, S., M.-O. Killijian, and M. N. del Prado Cortez. Next Place Prediction Using Mobility Markov Chains. *Proc., 1st Workshop on Measurement, Privacy, and Mobility*, Bern, Switzerland, ACM, New York, NY, 2012.
6. Mathew, W., R. Raposo, and B. Martins. Predicting Future Locations with Hidden Markov Models. *Proc., 2012 ACM Conference on Ubiquitous Computing*, Pittsburgh, PA, ACM, New York, NY, 2012.
7. Krumm, J., and E. Horvitz. Predestination: Inferring Destinations from Partial Trajectories. *Proc., International Conference on Ubiquitous Computing*, Springer, Berlin, Heidelberg, 2006, pp. 243–260.
8. Krumm, J., and E. Horvitz. Predestination: Where Do You Want to Go Today? *Computer*, Vol. 40, No. 4, 2007, pp. 105–107.
9. Horvitz, E., and J. Krumm. Some Help on the Way: Opportunistic Routing under Uncertainty. *Proc., 2012 ACM conference on Ubiquitous Computing, ACM*, 2012, pp. 371–380.
10. Xue, A. Y., J. Qi, X. Xie, R. Zhang, J. Huang, and Y. Li Solving the Data Sparsity Problem in Destination Prediction. *The VLDB Journal—The International Journal on Very Large Data Bases*, Vol. 24, No. 2, 2015, pp. 219–243.
11. Ziebart, B. D., A. L. Maas, A. K. Dey, and J. A. Bagnell. Navigate Like a Cabbie: Probabilistic Reasoning from Observed Context-Aware Behavior. *Proc., 10th International Conference on Ubiquitous Computing*, ACM, 2008, pp. 322–331.
12. Marmasse, N., and C. Schmandt. A User-Centered Location Model. *Personal and Ubiquitous Computing*, Vol. 6, No. 5–6, 2002, pp. 318–321.
13. Hawelka, B., I. Sitko, P. Kazakopoulos, and E. Beinat. Collective Prediction of Individual Mobility Traces with

Exponential Weights. *arXiv preprint arXiv:1510.06582*, 2015.
14. Alhasoun, F., M. Alhazzani, F. Aleissa, R. Alnasser, and M. C. Gonzalez. City Scale Next Place Prediction from Sparse Data through Similar Strangers. *Proc., Knowledge Discovery and Data Mining Workshop*, Halifax, Nova Scotia, Canada, 2017.
15. Lu, X., E. Wetter, N. Bharti, A. J. Tatem, and L. Bengtsson. Approaching the Limit of Predictability in Human Mobility. *Scientific Reports*, Vol. 3, 2013, p. 2923.
16. Calabrese, F., G. D. Lorenzo, and C. Ratti. Human Mobility Prediction Based on Individual and Collective Geographical Preferences. *Proc., 13th International IEEE Conference on Intelligent Transportation Systems (ITSC)*, Funchal, Portugal, IEEE, New York, NY, 2010.
17. Zhao, Z., H. N. Koutsopoulos, and J. Zaho. Mobility as a Language: Predicting Individual Mobility in Public Transportation Using N-Gram Models. Presented at 96th Annual Meeting of the Transportation Research Board, Washington, D.C., 2017.
18. De Brébisson, A., É. Simon, A. Auvolat, P. Vincent, and Y. Bengio. Artificial Neural Networks Applied to Taxi Destination Prediction. *arXiv preprint arXiv:1508.00021*, 2015.
19. Kim, B., C. M. Kang, S. H. Lee, H. Chae, J. Kim, C. C. Chung, and J. W. Choi. Probabilistic Vehicle Trajectory Prediction over Occupancy Grid Map via Recurrent Neural Network. *arXiv preprint arXiv:1704.07049*, 2017.
20. Endo, Y., K. Nishida, H. Toda, and H. Sawada. Predicting Destinations from Partial Trajec- tories Using Recurrent Neural Network. In *Pacific-Asia Conference on Knowledge Discovery and Data Mining*, Springer, 2017, pp. 160–172.
21. Liu, Q., S. Wu, L. Wang, and T. Tan. Predicting the Next Location: A Recurrent Model with Spatial and Temporal Contexts. *Proc., 30th AAAI Conference on Artificial Intelligence*, 2016, pp. 194–200.
22. Kim, J., K. Zheng, S. Ahn, M. Papamanolis, and P. Chao. Graph-Based Analysis of City-Wide Traffic Dynamics Using Time-Evolving Graphs of Trajectory Data. *Proc., 38th Australasian Transport Research Forum (ATRF)*, Melbourne, Victoria, Australia, 2016.
23. Kim, J., K. Zheng, J. Corcoran, S. Ahn, and M. Papamanolis. Trajectory Flow Map: Graph-Based Approach to Analysing Temporal Evolution of Aggregated Traffic Flows in Large-Scale Urban Networks. Presented at 96th Annual Meeting of the Transportation Research Board, Washington, D.C., 2017.
24. Bengio, Y., N. Boulanger-Lewandowski, and R. Pascanu. Advances in Optimizing Recurrent Networks. *Proc., 2013 IEEE International Conference on Acoustics, Speech and Signal Processing*, Vancouver, BC, IEEE, New York, NY, 2013, pp. 8624–8628.
25. Pascanu, R., T. Mikolov, and Y. Bengio. On the Difficulty of Training Recurrent Neural Networks. *Proc., International Conference on Machine Learning*, Atlanta, GA, 2013, pp. 1310–1318.

The Standing Committee on Artificial Intelligence and Advanced Computing Applications (ABJ70) peer-reviewed this paper (18-03692).

The authors of course remain solely responsible for the content of this paper.

TRR
JOURNAL OF THE TRANSPORTATION RESEARCH BOARD

Article

Transportation Research Record
2018, Vol. 2672(45) 185–195
© National Academy of Sciences:
Transportation Research Board 2018
Article reuse guidelines:
sagepub.com/journals-permissions
DOI: 10.1177/0361198118795010
journals.sagepub.com/home/trr
SAGE

Machine Learning Approach to Short-Term Traffic Congestion Prediction in a Connected Environment

Amr Elfar[1], Alireza Talebpour[2], and Hani S. Mahmassani[1]

Abstract

Traffic congestion is a complex phenomenon triggered by a combination of multiple interacting factors. One of the main factors is the disturbances caused by individual vehicles, which cannot be identified in aggregate traffic data. Advances in vehicle wireless communications present new opportunities to measure traffic perturbations at the individual vehicle level. The key question is whether it is possible to find the relationship between these perturbations and shockwave formation and utilize this knowledge to improve the identification and prediction of congestion formation. Accordingly, this paper explores the use of three machine learning techniques, logistic regression, random forests, and neural networks, for short-term traffic congestion prediction using vehicle trajectories available through connected vehicles technology. Vehicle trajectories provided by the Next Generation SIMulation (NGSIM) program were utilized in this study. Two types of predictive models were developed in this study: (1) offline models which are calibrated based on historical data and are updated (re-trained) whenever significant changes occur in the system, such as changes/updates to the infrastructure, and (2) online models which are calibrated using historical data and updated regularly using real-time information on prevailing traffic conditions obtained through V2V/V2I communications. Results show that the accuracy of the models built in this study to predict the congested traffic state can reach 97%. The models presented can be used in various potential applications including improving road safety by warning drivers of upcoming traffic slowdowns and improving mobility through integration with traffic control systems.

Traffic congestion is a complex phenomenon triggered by a combination of multiple interacting factors (*1, 2*). Studies of real-world traffic congestion have identified three main factors as the main causes of traffic breakdowns (*1, 3*):

1. High traffic loads, where the traffic demand exceeds the average sustainable throughput on a road section
2. Local capacity reductions (or "bottlenecks"), which may be permanent, such as on-ramps and off-ramps, or temporary such as traffic accidents or slow-moving vehicles
3. Disturbances caused by individual drivers, which refers to temporary perturbations in the traffic flow, such as those caused by lane-change maneuvers, abrupt braking, or long-lasting overtaking maneuvers of trucks

The first two factors, acting alone or jointly, generally lead to high density levels, at which the traffic flow becomes unstable, and less able to absorb perturbations due to the third factor, resulting in greater likelihood of flow breakdown (*4, 5*). High traffic loads and bottlenecks can be readily measured and analyzed using aggregated traffic data from traditional sensors (e.g., loop detectors and radars), or directly from the infrastructure configuration (for the second factor) such as on-ramps and lane drops. Disturbances in traffic flow, however, are difficult to observe using aggregated data (*3*) as they occur at the level of individual vehicles, and therefore are more difficult to control using techniques that rely on traditional sensor measurements.

[1]Civil and Environmental Engineering, Northwestern University Transportation Center, Evanston, IL
[2]Zachry Department of Civil Engineering, Texas A&M University, College Station, TX

Corresponding Author:
Address correspondence to Hani S. Mahmassani:
masmah@northwestern.edu

Advances in vehicle wireless communications present new opportunities to measure traffic perturbations at the individual vehicle level and to improve the identification and prediction of congestion patterns (*6–11*). This new technology enables vehicles and infrastructure to exchange vehicle trajectory information (e.g., location, speed, and acceleration) through vehicle-to-vehicle (V2V) and vehicle-to-infrastructure (V2I) communications. The detailed trajectories can then be used to track finer patterns in vehicle movements (including the perturbations in traffic flow) and understand traffic flow dynamics in order to improve traffic prediction and control.

The key question is whether it is possible to utilize this detailed information to improve the identification and prediction of congestion formation. Machine learning is one way to take advantage of a large amount of information that can be generated by connected vehicles. These techniques have been studied in the literature for a wide range of transport applications, such as traffic operations, planning, and safety (*12*). One of the potential applications of machine learning that has not been extensively explored in the literature is predicting traffic congestion using vehicle trajectories. Some studies have investigated congestion prediction with machine learning using other emerging data sources such as GPS records from mobile devices. For example, Thianniwet et al. (*13*) used a decision tree algorithm to classify road congestion by analyzing the movement patterns of vehicles collected through phone GPS. In another study, by Pattara-Atikom and Peachavanish (*14*), the authors used neural networks to estimate traffic congestion from cell dwell time (the duration for which a cell is registered to a base station before moving to another base). Other studies conducted a time-series analysis to forecast traffic flow using neural networks (*15, 16*), or hybrids of multiple techniques (*15, 17–20*). For instance, in a study by Vlahogianni et al. (*20*), traffic patterns were recognized using clustering and then flow was forecast using neural networks.

The main contribution of this paper is to explore the capability of machine learning algorithms to predict short-term traffic congestion using vehicle trajectories available through connected vehicles. This is motivated by the various potential applications of these techniques for developing predictive traffic control algorithms in a connected environment.

The study utilizes logistic regression, random forests, and neural networks to develop two types of predictive models: (a) offline models which are calibrated based on historical data and are updated (re-trained) whenever significant changes occur in the system, such as changes/updates to the infrastructure, and (b) online models which are calibrated using historical data and updated regularly using real-time information on prevailing traffic conditions obtained through V2V/V2I communications.

The paper is organized as follows: the next section provides a brief description of the vehicle trajectory data set used in this study. Following that, the Methodology section describes how traffic properties were estimated from vehicle trajectories. Moreover, this section presents the methodology for traffic state identification using K-means clustering and the two types of models developed in this study. After that, the Results section discusses the prediction accuracy of these models and their potential applications. Finally, the Conclusion section summarizes the main findings of this study.

Data

The vehicle trajectory dataset used to build and test the predictive models was obtained from the Next Generation SIMulation (NGSIM) data for the US-101 Freeway in Los Angeles, California (*21*). The study segment is approximately 2100-ft long and contains five main lanes, one on-ramp, and one off-ramp. The trajectory dataset includes information on the vehicles' speed, acceleration, location, and headways at a 0.1-second resolution.

Methodology

Estimating Traffic Properties

This study assumes a connected environment where vehicles share their detailed trajectories (i.e., location, speed, acceleration) through V2V and V2I communications. In order to estimate the changes in traffic properties along the study segment of US-101, the 2100-ft segment was divided into ten 200-ft sections. Traffic flow, density, and mean speed were estimated for each section over 10-second time steps using Edie's generalized definitions (*22, 23*):

$$q(A) = \frac{d(A)}{|A|} \tag{1}$$

$$k(A) = \frac{t(A)}{|A|} \tag{2}$$

$$v(A) = \frac{d(A)}{t(A)} \tag{3}$$

where $q(A)$, $k(A)$, and $v(A)$ denote flow, density, and mean speed for observed vehicles in section A, respectively. $d(A)$ and $t(A)$ represent the total distance traveled and total time spent by all vehicles in section A, respectively. Finally, $|A|$ is the area covered by section A (200-ft section multiplied by a 10-sec interval).

The speed standard deviation (SSD) of individual vehicles is the core of the predictive models in this study

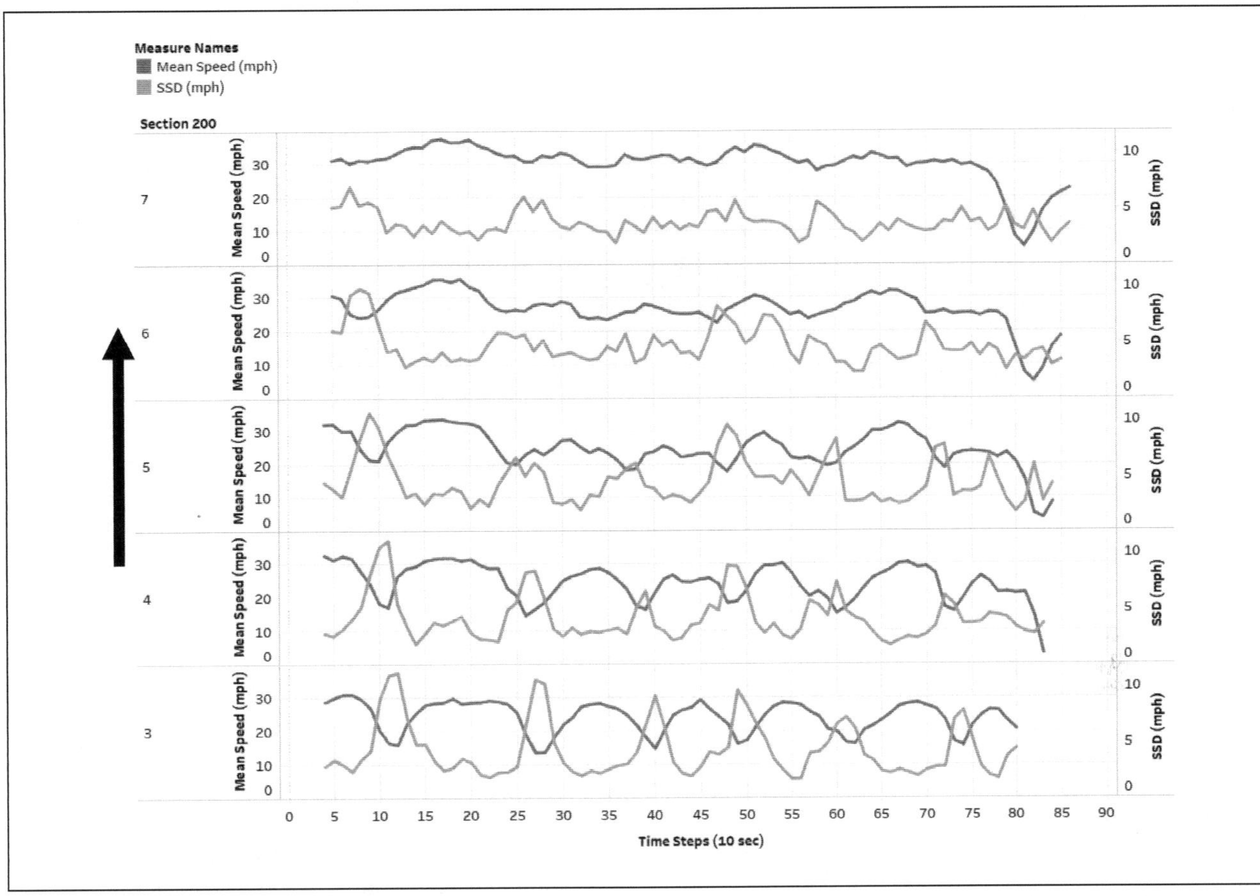

Figure 1. SSD waves as an indicator of traffic shockwaves for the NGSIM US101 segment during the period from 7:50 to 8:05 a.m., direction of travel is from section 3 to 7.

and was selected as a measure of perturbations in traffic flow at the individual vehicle level. SSD was estimated for each section and time step as follows: (a) the average speed over the 10-second time steps was calculated for each vehicle passing a specific section, and (b) the standard deviation was estimated for the average speeds of all individual vehicles in that section. The next section discusses how SSD can be used to predict congestion.

Speed Standard Deviation as an Indicator for Congestion

Disturbances caused by individual drivers constitute one of the main factors that cause traffic breakdown. The SSD of individual vehicles is one way to measure traffic perturbation (*24*) using detailed vehicle trajectories and is one of the main explanatory variables of the predictive models built in this study. SSD was particularly selected in this study because: (a) it has been suggested in the literature using macroscopic (*25, 26*) and microscopic (*27*) traffic models that the increase in SSD among individual vehicles is a prelude to traffic breakdown, and (b) SSD can be estimated at fine temporal and spatial scales (e.g., every 200-ft at 10-second intervals) using vehicle trajectories of fully or partially connected traffic stream.

Figure 1 shows the changes in the SSD (orange line, right y-axis) over 10-second time steps in addition to the changes in mean speed (blue line, left y-axis) for sections 3 through 7 of US-101 during the morning period from 7:50 to 8:05 a.m. This figure shows a pattern where a significant reduction in speed is associated with a peak in the SSD for all the shockwaves that are shown in sections 3 through 5. Furthermore, this figure shows that the disruption in SSD starts before shockwave formation and propagation. This is illustrated in section 7 of Figure 1, where SSD starts to increase before shockwaves form in section 6 and propagate to sections 5 through 3.

Identifying Traffic States using K-Means Clustering of the Fundamental Diagram

Identifying the traffic state from vehicle trajectories is the first step in building the predictive models in this study. The traffic state will be used later as the

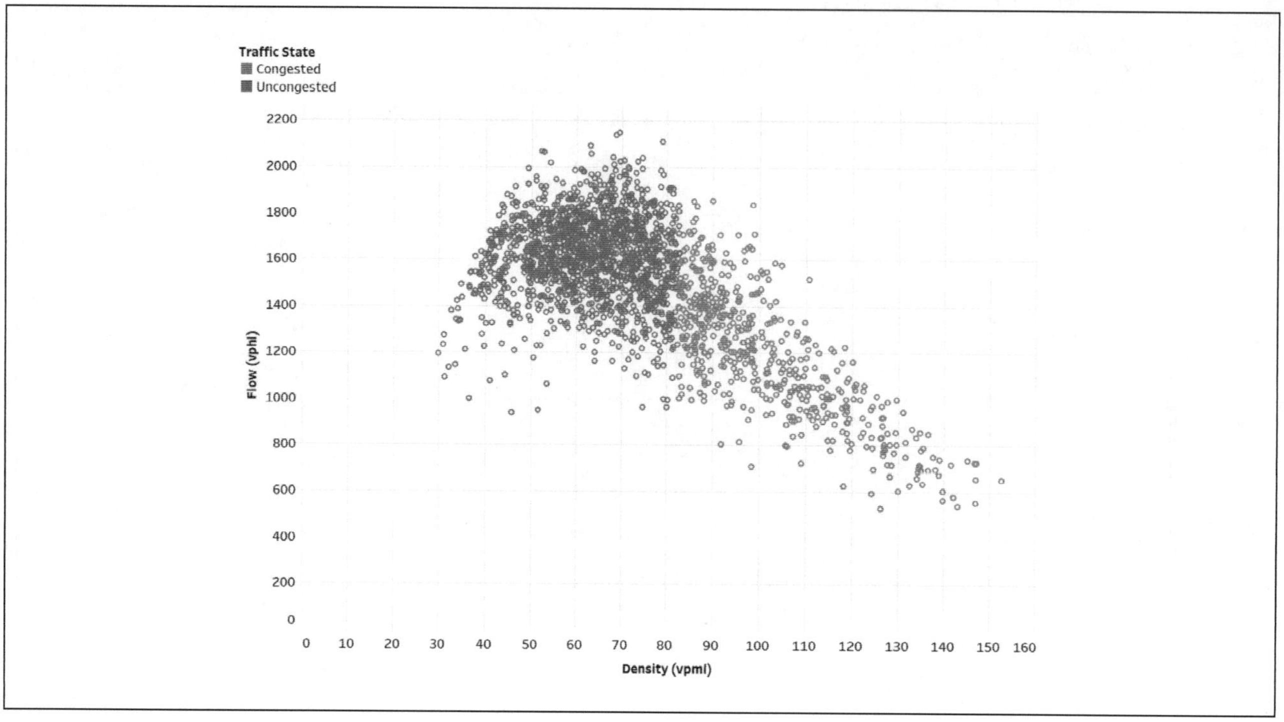

Figure 2. Traffic states for the study segment of US-101.

dependent variable, which will be classified in the models. For simplicity, two traffic states were identified: uncongested (blue dots) and congested (red dots), as shown in Figure 2. A K-means clustering algorithm was used to identify the two states by specifying two clusters. The K-means clustering algorithm segments data points into clusters (groups) where the total distances calculated from each point to its respective cluster center is minimized. As density and flow values are significantly different, the values were scaled to be between 0 and 1 before running the algorithm. Then, the density values were multiplied by a factor of five (arbitrarily chosen) to give density values a higher weight and force the algorithm to split the clusters based on density and define the critical density at which traffic breaks down. The critical density estimated using the K-means algorithm is around 80 vpmpl, which was also shown by a previous study using the NGSIM US-101 dataset (*28*).

Temporally and Spatially Lagged Variables to Build Predictive Models

Mean speed and the SSD are the independent variables (explanatory variables) used in the models to predict the traffic state because: (a) both can be calculated accurately for a sample of connected vehicles, and (b) congestion is associated with high traffic perturbations (high SSD, see Figure 1) and low mean speed. While flow and density are also good indicators of the traffic state, neither property can be estimated accurately from vehicle trajectories of partially connected traffic as only connected vehicles will be counted.

The models developed in this study are temporally lagged models, which are a type of time-series models calibrated/trained to predict current values of the dependent variable, the traffic state, in this case, using lagged (past-values) of the explanatory variables. Thus, the models predict future values of the traffic state, our goal, when current values of the mean speed and SSD (the explanatory variables) are observed. The values of mean speed and SSD were lagged 10 and 20 seconds to calibrate models with 10-second and 20-second prediction horizons which are intended for short-term prediction applications such as warning drivers of upcoming traffic slowdown, ramp metering, or speed harmonization.

In addition to temporally lagging SSD, the variable was spatially lagged one section downstream as the disruption in SSD waves starts downstream of the section where the shockwave (reduction in speed) begins to form. This is illustrated in Figure 1 where SSD waves in section 7 are signaling a shockwave formation that starts in section 6 and propagates upstream to sections 5 through 3. Table 1 provides a summary of the variables used in this study, their description, the market penetrations at which they were estimated, and the type of models they were used in. Market penetration refers to the

Table 1. Summary of the Variables Used in the Predictive Models

Variable	Description	Market penetrations	Models
Dependent variable			
Traffic state	Binary: the state of traffic whether congested or uncongested as identified using K-means clustering	100%	All
Explanatory variables			
Lagged mean speed in current section (10 s)	Continuous: the average speed of individual vehicles in current section, lagged 10 seconds	30%, 50%, 100%	10 seconds predictive models
Lagged mean speed in downstream section (10 s)	Continuous: the average speed of individual vehicles in the next downstream section, lagged 10 seconds	30%, 50%, 100%	10 seconds predictive models
Lagged speed standard deviation in downstream section (10 s)	Continuous: the SSD of individual vehicles in the next downstream section, lagged 10 seconds	30%, 50%, 100%	10 seconds predictive models
Lagged mean speed in current section (20 s)	Continuous: the average speed of individual vehicles in current section, lagged 20 seconds	30%, 50%, 100%	20 seconds predictive models
Lagged mean speed in downstream section (20 s)	Continuous: the average speed of individual vehicles in the next downstream section, lagged 20 seconds	30%, 50%, 100%	20 seconds predictive models
Lagged speed standard deviation in downstream section (20 s)	Continuous: the SSD of individual vehicles in the next downstream section, lagged 20 seconds	30%, 50%, 100%	10 seconds predictive models

percentage of connected vehicles that are able to transmit their trajectories.

Offline Predictive Models

Offline predictive models are the first type of models calibrated and tested in this study. These models are built using historical data and updated whenever new data is available or as needed. The models are built using a training dataset that contains observed traffic states and a number of explanatory variables, such as mean speed and SSD in this case. Then, the models are used to predict the traffic state with the explanatory variables only.

The relevancy of data is key for the accuracy of offline models. For example, in order to predict the traffic state for a freeway on a summer day, the data used to build this model should cover a sample of summer days for the same freeway or a very similar one. If the data set used to build the model contains winter dates only, the accuracy will drop significantly as traffic dynamics are highly affected by weather conditions. Ideally, in practice, historical data should cover a wide range of traffic states over different days of the week (weekdays, weekends) at different times (peak and off-peak hours) and weather conditions.

As the NGSIM US-101 data set is only available for a short time period (7:50 to 8:35 a.m.) and for a single day, the whole data set was used to train the offline models and the accuracy was tested using K-fold cross-validation, which is suitable for small data sets when a separate test set is not available (*29*). The method will be discussed further in the section.

Three types of machine learning techniques were tested in this paper: binary logistic regression, random forests, and artificial neural networks. The binary logistic regression model is a simple model that can help explain the relationships among the variables. Random forests and artificial neural networks, on the other hand, are among the best machine learning classifiers (*30*) and are widely used for different transportation-related applications (*12*).

Online (Real-Time) Predictive Models

Online models are the second type of models tested in this study. These models are built using historical data and updated (re-trained) regularly using real-time information on prevailing traffic conditions. The rate at which models are updated depends on how frequently traffic conditions are changing. During rush-hours for example, the models can be updated at a higher rate than during off-peak hours.

In this paper, the NGSIM dataset was used to simulate online models that are updated every five minutes. A high re-training rate was selected because the available NGSIM

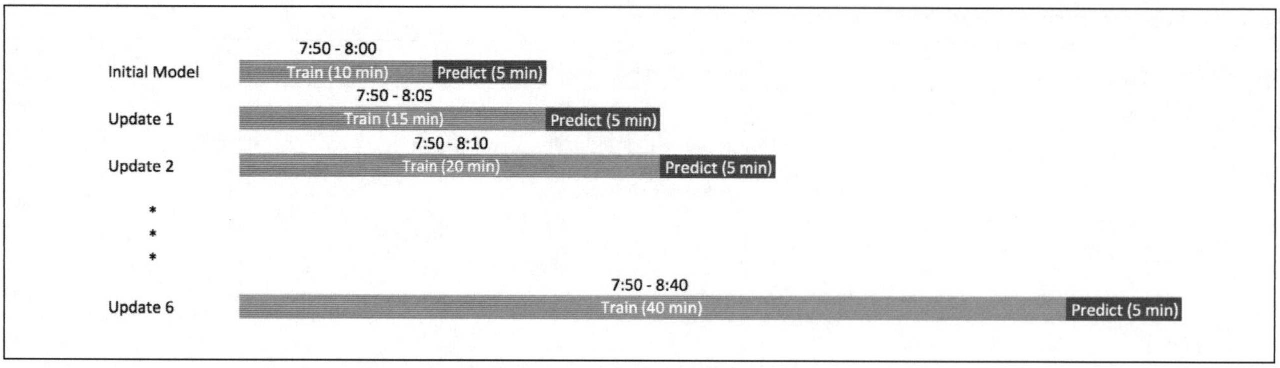

Figure 3. Model updating process for online predictive models using NGSIM data.

data covers a very short time period during the morning rush hour (45 minutes, 7:50 to 8:35 a.m.). We assumed that the historical data was similar to the first 10-min (7:50 to 8:00 a.m.) of the total time period (7:50 to 8:35 a.m.). The model was initially built using data for the first 10-minute period to predict the following 5-min interval (8:00 to 8:05 a.m.). Then, assuming that five minutes had passed, the models were re-trained using the historical data (first 10 minutes) in addition to the new information received for the past five minutes. The model updating process was repeated until the whole time period was predicted (excluding the first 10 minutes). The updating process is illustrated in Figure 3. As in the case of offline models, three types of machine learning techniques were used to build online models: binary logistic regression, random forests, and artificial neural networks.

Note that training the model was assumed to be instantaneous. In practice, however, training can take from seconds to a few minutes depending on the size of the dataset and the machine learning technique used. This needs to be taken into consideration when the updating process is designed. Another point to note is that the cumulative dataset up until the update point was used to retrain the models, which was not an issue since the available set is relatively small. A more efficient way is critical in practice and implementation to build these models for large datasets. One solution is to train them using a representative sample of the data.

Model Accuracy and K-Fold Cross-Validation

The accuracy of the models tested in this paper is evaluated by three metrics:

1. Overall accuracy: the percentage of traffic states correctly predicted
2. Congested state prediction accuracy: the percentage of the congested states correctly predicted
3. Uncongested state prediction accuracy: the percentage of the uncongested states correctly predicted

In order to have a meaningful estimate of a predictive model's accuracy, the model should be tested on a separate dataset (new observations) that was not used to train it (*29*). In the case of offline models, however, a separate data set was not available for testing. Therefore, a K-fold cross-validation technique was used to estimate the accuracy of the models.

K-fold cross-validation is a technique that is used to estimate the prediction accuracy (test error term) of a statistical learning method, such as the machine learning models used in this paper, by holding out a subset of the training data set from the fitting process, and then applying the models to predict those held out observations.

Using K-fold cross-validation to estimate the accuracy of the offline models, the data set was divided into five groups (folds) of equal size. The first group was treated as a validation group, and the remaining four groups were used to build the models. The models were then tested using observations of the first group, and the prediction accuracy (three metrics) was estimated. The process was repeated five times, a typical k-value in practice for this method (*29*), wherein each time a different group of observations was treated as a validation group. The K-fold cross-validation accuracy was then calculated by taking the average of the five different values estimated for each group. In the case of online models, the accuracy was estimated directly using the predicted observations since those models predict new observations that were not used to fit the models (see Figure 3).

Predictive Models with Partial Connectivity

In the case of partially connected traffic, mean speed and SSD (the explanatory variables) can be calculated using vehicle trajectories assuming that a minimum market

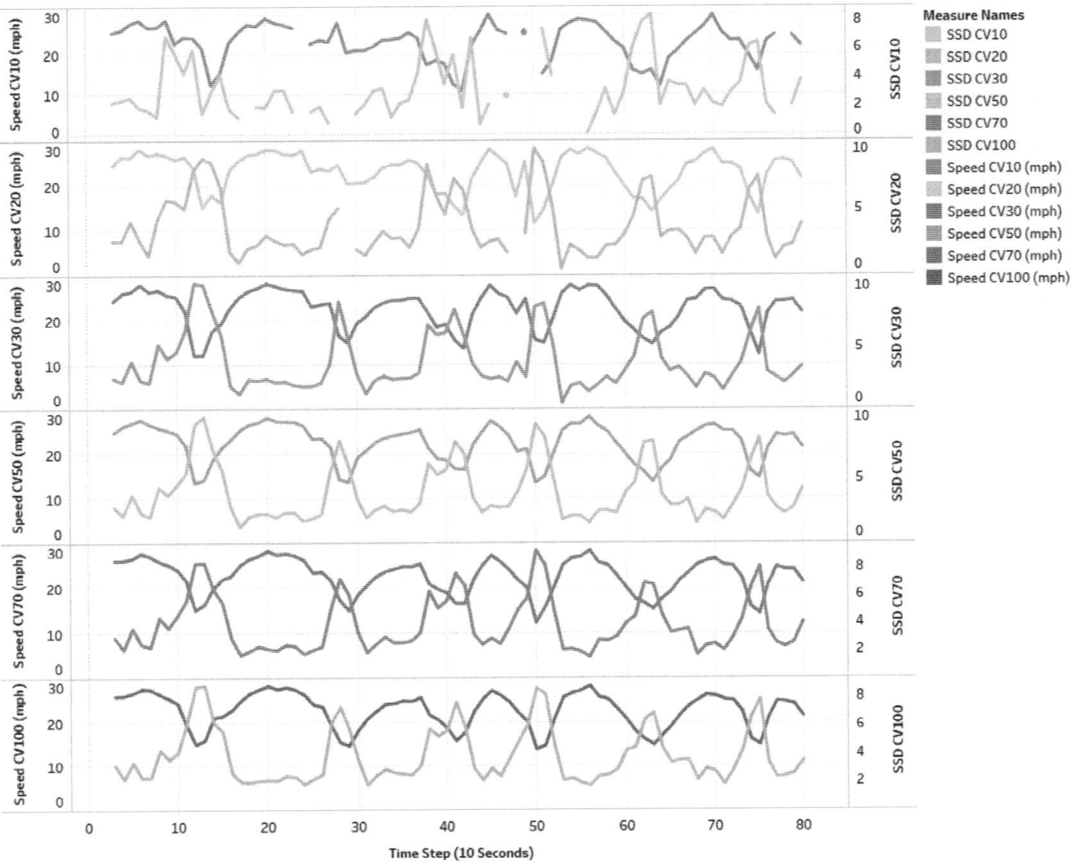

Figure 4. Mean speed and SSD estimated at different market penetrations for section 2 during the period from 7:50 to 8:05 a.m.

penetration is required. This is illustrated in Figure 4, where the mean speed and standard deviation are estimated at different market penetration rates (i.e., 30%, 50%, and 100%). The graph shows that below 30% market penetration, mean speed and SSD could not be estimated for some time steps due to incomplete trajectories (there exist sections without any detected vehicles) or they are inaccurate. For higher market penetrations, mainly above 30%, the figure shows that mean speed and SSD estimates are very close to the true value (estimated at 100% connectivity) and therefore can be used to predict the traffic state.

Note that estimating flow and density directly for partially connected traffic will not be accurate, as only connected vehicles in the stream will be counted. Therefore, in order to use partial connectivity models, the trajectory data obtained from connected vehicles should be supported by extra information on density and flow to identify the observed traffic states (the dependent variable) and build the predictive models. This can be done through developing an inverse function to estimate density from mean speed, adjust the density/flow values directly estimated from trajectories, or collecting extra data on density using other methods such as cameras. This might be easier to implement in the case of offline models as data collection is less frequent than online models.

Specifications of the Machine Learning Techniques

The three machine learning techniques were tested using the statistical software R (*31*). The binary logistic regression was calibrated using the generalized linear model function in R. The cut-off probability beyond which the state was classified as congested was set to 50%. For the random forest models, the randomForest package (*32*) was used for calibration. Five hundred trees were selected to build the models as there had not been any improvement in the models' accuracy when a higher number of trees were tested. Finally, the nnet package (*33*) was used

Table 2. Comparison of Offline Models with Full Connectivity

Model	Overall accuracy	Congested state prediction accuracy	Uncongested state prediction accuracy
Logistic 10 s	93%	96%	85%
Logistic 20 s	91%	95%	79%
Random forest 10 s	93%	95%	85%
Random forest 20 s	90%	94%	77%
Neural network 10 s	89%	97%	68%
Neural network 20 s	90%	95%	78%

to calibrate neural network models using one hidden layer.

Results and Analysis

Offline Predictive Models with Full Connectivity

As discussed earlier, three types of machine learning techniques (binary logistic regression, random forest, and artificial neural networks) were used to build offline models. In this case, a fully connected environment was assumed in which all vehicles can exchange their trajectories using V2V/V2I communication technology. In addition, two prediction horizons were tested for these models: 10 and 20 seconds, which are intended for short-term prediction applications such as warning drivers of upcoming traffic slowdown, ramp metering, or speed harmonization.

Table 2 compares the different offline models built in this study with the assumption of full connectivity. The comparison is based on the three metrics defined in the Methodology section (overall accuracy, congested state prediction accuracy, and uncongested state prediction accuracy).

In terms of overall prediction accuracy, the table shows that the three techniques have high accuracy with values that range from 89% to 93% and that models with the shorter prediction horizon, 10 seconds, have higher accuracy than those with 20-second prediction horizon. Regarding congestion prediction, the three techniques performed well with accuracy that ranges from 94% to 97%. Additionally, the 10-second prediction horizon models performed marginally better (1% to 2% improvement) than the 20-second models. Finally, the prediction of the uncongested state was less accurate than the prediction of the congested state. This could be due to the noise in SSD waves which may imply congestion. The uncongested state prediction accuracy of the models was 68% to 85%. Out of the three machine learning techniques used, logistic regression and random forests performed slightly better than neural networks. In terms of training speed, logistic regression was the fastest. However, the differences were marginal as the dataset was small.

Offline Predictive Models with Partial Connectivity

Table 3 compares the three types of offline predictive models at low, medium, and full market penetrations (30%, 50%, and 100% respectively). The table shows that the predictive models perform well for partially connected traffic streams with an overall accuracy that ranges from 88% to 92% for both low and medium connectivity cases. This shows that full connectivity is not required to predict the traffic state using the above-mentioned techniques even though prediction accuracy improves when more vehicles in the traffic stream are connected. However, a minimum market penetration above 30% is required to estimate the values of mean speed and SSD as shown previously in Figure 4.

As in the case of full connectivity, the models built assuming partial connectivity predicted the congested state better than the uncongested state due to the noise in SSD. The accuracy range for predicting the congested state was 92% to 96% while the accuracy range for predicting the uncongested state was 68% to 83%. In addition, 10-second prediction models, in general, performed better than 20-second prediction models, as in the case of full connectivity models.

Online Predictive Models

Table 4 compares the online predictive modes built using logistic regression, random forest, and neural networks techniques. The models used the first 10 minutes of the NGSIM data available as historical data and were updated every five minutes in real-time by adding the extra five minutes as new observations to the training dataset. The model updating process is illustrated in Figure 3. In addition, the online models were built under the assumption of full connectivity. As mentioned in the Methodology section, flow and density of traffic, which cannot be estimated at partial market penetrations, is required to identify the traffic states (the dependent variable) in new observations before adding them to the training dataset and update/retrain the models.

Table 4 shows that the overall accuracy of the predictive models is in the range of 86% to 91% which is slightly lower than the prediction accuracy of the offline

Table 3. Comparison of Offline Models with Partial Connectivity

Model	Connectivity	Overall accuracy	Congested state prediction accuracy	Uncongested state prediction accuracy
Logistic 10s	30%	91%	96%	79%
Logistic 10s	50%	92%	95%	82%
Logistic 10s	100%	93%	96%	85%
Logistic 20s	30%	88%	94%	73%
Logistic 20s	50%	89%	94%	76%
Logistic 20s	100%	91%	95%	79%
Neural network 10s	30%	91%	95%	80%
Neural network 10s	50%	92%	95%	83%
Neural network 10s	100%	89%	97%	68%
Neural network 20s	30%	88%	94%	71%
Neural network 20s	50%	89%	94%	75%
Neural network 20s	100%	90%	95%	78%
Random forest 10s	30%	91%	95%	80%
Random forest 10s	50%	92%	95%	82%
Random forest 10s	100%	93%	95%	85%
Random forest 20s	30%	86%	92%	70%
Random forest 20s	50%	88%	93%	73%
Random forest 20s	100%	90%	94%	77%

Table 4. Comparison of Online Predictive Model

Model	Overall accuracy	Congested state prediction accuracy	Uncongested state prediction accuracy
Logistic 10 s online	91%	93%	87%
Logistic 20 s online	89%	94%	80%
Neural network 10 s online	90%	93%	86%
Neural network 20 s online	88%	94%	76%
Random forest 10 s online	91%	94%	85%
Random forest 20 s online	86%	93%	70%

models tested before. This also applies to the prediction accuracy of the congested state where the range is 93% to 94% compared with 94% to 97% in the case of offline models.

The results do not necessarily mean that online models perform worse than offline models. In fact, the results are influenced by the limited data set in this analysis. In the case of online models, the models were initially trained using only the first 10 minutes of the 45-minute period available in the data set and were used to predict the following five minutes. The models were then updated/retrained every five minutes while in each update new observations were added to the training set. In other words, online models were built to improve over time to simulate a real-world application. Offline models, on the other hand, were built assuming that the whole 45 minutes of data was available as historical data and were all used to train the models once rather than build an initial model with a small dataset and improve it over time. Ideally, in practice, online models would use historical data as a basis for training and real-time information to improve the models.

Conclusion

Connected vehicles technology provides new opportunities to improve traffic congestion identification and prediction by generating detailed information on individual vehicle movements through V2V/V2I communications. Machine learning is one way to take advantage of the huge amount of information that can be generated by connected vehicles. Therefore, this paper presented three machine learning techniques that can be used for short-term traffic congestion prediction using vehicle trajectories that would be available for connected vehicles. Two types of predictive models were developed: (a) offline models that are built using historical data only, and (b) online models that are updated in real-time.

The analysis shows that the prediction accuracy of the congested state by these models is acceptable for the example data considered. Furthermore, the results show that congestion can be predicted accurately in the case of partially connected traffic streams, which is important in practice as traffic is unlikely to be fully connected, at least in the early deployment of this new technology.

The models have various safety and traffic performance applications. For instance, the models can be used to warn drivers of traffic slowdowns ahead to prevent potential accidents. In terms of traffic operations, the models can be integrated with traffic control algorithms to enhance their performance. For example, the models can be used to develop a predictive speed harmonization algorithm to predict the location of congestion and set the speed limit upstream of it (*34*), or to detect shockwave formation and provide queue warning information upstream (*35*). The prediction horizons of these models can be very short for some other applications, given that the prediction accuracy degrades for longer horizons using the current structure of the models. However, the analysis could be extended to predict longer horizons and multiple traffic states given vehicle trajectory data that covers a long duration and a wide range of states, though this matter would need to be empirically tested. Furthermore, as is the case with "black box" statistical methods that are not derived from fundamental theory, the ability to predict changes and states that depart from what had been observed during the training period degrades rapidly. Nonetheless, the high degree of predictability achieved for very short-term forecasts suggests that a productive approach would be to combine the advantages of machine learning techniques with those of approaches based on fundamental traffic concepts and theories.

Author Contributions

All authors, Amr Elfar, Alireza Talebpour, and Hani S. Mahmassani, contributed to all aspects of the study: conception and design, data collection, analysis and interpretation of results, and manuscript preparation. All authors reviewed the results and approved the final version of the manuscript.

References

1. Helbing, D., M. Treiber, A. Kesting, and M. Schönhof. Theoretical vs. Empirical Classification and Prediction of Congested Traffic States. *The European Physical Journal B-Condensed Matter and Complex Systems*, Vol. 69, No. 4, 2009, pp. 583–598.
2. Treiber, M., A. Hennecke, and D. Helbing. Congested Traffic States in Empirical Observations and Microscopic Simulations. *Physical Review E*, Vol. 62, No. 2, 2000, p. 1805.
3. Treiber, M., and A. Kesting. *Traffic Flow Dynamics: Data, Models and Simulation*. Springer-Verlag, Berlin and Heidelberg, Germany, 2013.
4. Dong, J., and H. Mahmassani. Flow Breakdown and Travel Time Reliability. *Transportation Research Record: Journal of the Transportation Research Board*, 2009. 2124: 203–212.
5. Kim, J., H. Mahmassani, and J. Dong. Likelihood and Duration of Flow Breakdown: Modeling the Effect of Weather. *Transportation Research Record: Journal of the Transportation Research Board*, 2010. 2188: 19–28.
6. Mahmassani, H. S. 50th Anniversary Invited Article—Autonomous Vehicles and Connected Vehicle Systems: Flow and Operations Considerations. *Transportation Science*, Vol. 50, No. 4, 2016, pp. 1140–1162.
7. Talebpour, A. *Modeling Driver Behavior in a Connected Environment: Integration of Microscopic Traffic Simulation and Telecommunication Systems*. PhD thesis, Northwestern University, Evanston, IL, 2015.
8. Mudge, R., H. Mahmassani, R. Haas, A. Talebpour, and L. Carroll. *Work Zone Performance Measurement using Probe Data*. FHWA-HOP-13-043. Federal Highway Administration, Washington, D.C., 2013.
9. Talebpour, A., H. Mahmassani, F. Mete, and S. Hamdar. Near-Crash Identification in a Connected Vehicle Environment. *Transportation Research Record: Journal of the Transportation Research Board*, 2014. 2424: 20–28.
10. Talebpour, A., and H. S. Mahmassani. Influence of Connected and Autonomous Vehicles on Traffic Flow Stability and Throughput. *Transportation Research Part C: Emerging Technologies*, Vol. 71, 2016, pp. 143–163.
11. Talebpour, A., H. S. Mahmassani, and F. E. Bustamante. Modeling Driver Behavior in a Connected Environment: Integrated Microscopic Simulation of Traffic and Mobile Wireless Telecommunication Systems. *Transportation Research Record: Journal of the Transportation Research Board*, 2016. 2560: 75–86.
12. Karlaftis, M. G., and E. I. Vlahogianni. Statistical Methods versus Neural Networks in Transportation Research: Differences, Similarities and Some Insights. *Transportation Research Part C: Emerging Technologies*, Vol. 19, No. 3, 2011, pp. 387–399.
13. Thianniwet, T., S. Phosaard, and W. Pattara-Atikom. Classification of Road Traffic Congestion Levels from GPS Data using a Decision Tree Algorithm and Sliding Windows. *Proc., World Congress on Engineering*, International Association of Engineers, London, UK, 2009, pp. 1–3.
14. Pattara-Atikom, W., and R. Peachavanish. Estimating Road Traffic Congestion from Cell Dwell Time using Neural Network. *Proc., 2007 7th International Conference on ITS Telecommunications*, Sophia Antipolis, France, IEEE, New York, NY, 2007, pp. 1–6.
15. Vlahogianni, E. I., M. G. Karlaftis, and J. C. Golias. Optimized and Meta-Optimized Neural Networks for Short-Term Traffic Flow Prediction: A Genetic Approach. *Transportation Research Part C: Emerging Technologies*, Vol. 13, No. 3, 2005, pp. 211–234.
16. Lingras, P., S. C. Sharma, P. Osborne, and I. Kalyar. Traffic Volume Time-Series Analysis According to the Type of Road Use. *Computer-Aided Civil and Infrastructure Engineering*, Vol. 15, No. 5, 2000, pp. 365–373.
17. Jiang, X., and H. Adeli. Dynamic Wavelet Neural Network Model for Traffic Flow Forecasting. *Journal of Transportation Engineering*, Vol. 131, No. 10, 2005, pp. 771–779.
18. Yin, H., S. C. Wong, J. Xu, and C. Wong. Urban Traffic Flow Prediction using a Fuzzy-Neural Approach. *Transportation Research Part C: Emerging Technologies*, Vol. 10, No. 2, 2002, pp. 85–98.

19. Chen, H., S. Grant-Muller, L. Mussone, and F. Montgomery. A Study of Hybrid Neural Network Approaches and the Effects of Missing Data on Traffic Forecasting. *Neural Computing & Applications*, Vol. 10, No. 3, 2001, pp. 277–286.
20. Vlahogianni, E. I., M. G. Karlaftis, and J. C. Golias. Temporal Evolution of Short-Term Urban Traffic Flow: A Nonlinear Dynamics Approach. *Computer-Aided Civil and Infrastructure Engineering*, Vol. 23, No. 7, 2008, pp. 536–548.
21. Alexiadis, V., J. Colyar, J. Halkias, R. Hranac, and G. McHale. The Next Generation Simulation Program. *Institute of Transportation Engineers. ITE Journal*, Vol. 74, No. 8, 2004, p. 22.
22. Edie, L. C. Car-Following and Steady-State Theory for Noncongested Traffic. *Operations Research*, Vol. 9, No. 1, 1961, pp. 66–76.
23. Edie, L. C. Discussion of Traffic Stream Measurements and Definitions. *Proc., 2nd International Symposium on the Theory of Traffic Flow*, London, UK, 1963, pp. 139–154.
24. Treiber, M., A. Kesting, and D. Helbing. Understanding Widely Scattered Traffic Flows, the Capacity Drop, and Platoons as Effects of Variance-Driven Time Gaps. *Physical Review E*, Vol. 74, No. 1, 2006, p. 016123.
25. Kühne, R. Macroscopic Freeway Model for Dense Traffic-Stop-Start Waves and Incident Detection. *Proc., 9th International Symposium on Transportation and Traffic Theory*, Utrecht, the Netherlands, VNU Science Press, Delft, the Netherlands, 1984, pp. 21–42.
26. Kühne, R. Freeway Speed Distribution and Acceleration Noise—Calculations from a Stochastic Continuum Theory and Comparison with Measurements. *Proc., 10th International Symposium on Transportation and Traffic Theory*, Cambridge, MA, Elsevier, 1987, pp. 119–137.
27. Krauß, S., P. Wagner, and C. Gawron. Metastable States in a Microscopic Model of Traffic Flow. *Physical Review E*, Vol. 55, No. 5, 1997, p. 5597.
28. Lu, X.-Y., P. Varaiya, and R. Horowitz. Fundamental Diagram Modelling and Analysis Based NGSIM Data. *International Federation of Automatic Control (IFAC) Proceedings Volumes*, Vol. 42, No. 15, 2009, pp. 367–374.
29. James, G., D. Witten, T. Hastie, and R. Tibshirani. *An Introduction to Statistical Learning: With Applications in R*. Springer, New York, NY, 2013.
30. Caruana, R., and A. Niculescu-Mizil. An Empirical Comparison of Supervised Learning Algorithms. *Proc., 23rd International Conference on Machine Learning*, ACM, New York, NY, 2006, pp. 161–168.
31. Team, R. C. *R: A Language and Environment for Statistical Computing*. R Foundation for Statistical Computing, Vienna, Austria, 2014.
32. Liaw, A., and M. Wiener. Classification and Regression by randomForest. *R News*, Vol. 2, No. 3, 2002, pp. 18–22.
33. Venables, W. N., and B. D. Ripley. *Modern Applied Statistics with S-PLUS*. Springer Science & Business Media, New York, NY, 2013.
34. Talebpour, A., H. Mahmassani, and S. Hamdar. Speed Harmonization: Evaluation of Effectiveness under Congested Conditions. *Transportation Research Record: Journal of the Transportation Research Board*, 2013. 2391: 69–79.
35. Elfar, A., A. Talebpour, and H. S. Mahmassani. Analyzing Queue Warning in Connected Environment: A Micro Simulation Approach. Presented at 95th Annual Meeting of the Transportation Research Board, Washington, D.C., 2016.

The Standing Committee on Artificial Intelligence and Advanced Computing Applications (ABJ70) peer-reviewed this paper (18-03978).

TRR
JOURNAL OF THE TRANSPORTATION RESEARCH BOARD

Article

Transportation Research Record
2018, Vol. 2672(45) 196–209

DOI: 10.1177/0361198118796938
journals.sagepub.com/home/trr
SAGE

Traffic Incident Duration Estimation Based on a Dual-Learning Bayesian Network Model

Haozhe Cong[1], Cong Chen[2], Pei-Sung Lin[2], Guohui Zhang[3], John Milton[4], and Ye Zhi[5]

Abstract

Highway traffic incidents induce a significant loss of life, economy, and productivity through injuries and fatalities, extended travel time and delay, and excessive energy consumption and air pollution. Traffic emergency management during incident conditions is the core element of active traffic management, and it is of practical significance to accurately understand the duration time distribution for typical traffic incident types and the factors that influence incident duration. This study proposes a dual-learning Bayesian network (BN) model to estimate traffic incident duration and to examine the influence of heterogeneous factors on the length of duration based on expert knowledge of traffic incident management and highway incident data collected in Zhejiang Province, China. Fifteen variables related to three aspects of traffic incidents, including incident information, incident consequences, and rescue resources, were included in the analysis. The trained BN model achieves favorable performance in several areas, including classification accuracy, the receiver operating characteristic (ROC) curve, and the area under curve (AUC) value. A classification matrix, and significant variables and their heterogeneous influences are identified accordingly. The research findings from this study provide beneficial reference to the understanding of decision-making in traffic incident response and process, active traffic incident management, and intelligent transportation systems.

Emergency incidents on transportation networks, including traffic crashes, natural disasters, infrastructure failures, and so forth, have critical influence on traffic safety and operation efficiency, and impose significant threats to human well-being and social welfare. Transportation emergency response processes involve a significant amount of social resources and pose tremendous challenges to transportation planners and engineers (*1*). Therefore, an efficient and cost-effective transportation emergency response mechanism and management is of practical importance to minimize social resource waste and ensure robust transportation networks. A significant number of studies have been conducted to investigate traffic emergency response mechanisms, and considerable national guidelines and peer studies have been developed to standardize and normalize transportation emergency response processes (*2–7*). For example, the *Highway Safety Design and Operations Guide*, published by the American Association of State Highway and Transportation Officials (AASHTO) (*2*), summarizes the regulations and technologies of highway network planning, design, and operation, including traffic emergency management. Significant peer studies also have been conducted regarding accurate traffic incident prediction and emergency response optimization (*1, 7–11*). For instance, Bahar et al. (*7*) proposed an integrated safety management process to improve the level of traffic safety management in disaster situations on a highway network. Smith (*8*) applied a decision tree method to model incident duration data and concluded that the decision tree model had an accuracy of 50% based on accident clearance times. Recently, Chakraborty et al. (*9*) developed an algorithm based on dynamic signal timing, traffic condition, and emergency vehicle security levels to

[1]Department of Traffic Safety Education, Road Traffic Safety Research Center of the Ministry of Public Security, Beijing, China
[2]Center for Urban Transportation Research, University of South Florida, Tampa, FL
[3]Department of Civil and Environmental Engineering, University of Hawaii at Manoa, Honolulu, HI
[4]Washington State Department of Transportation, Olympia, WA
[5]Department of Traffic Management Information, Road Traffic Safety Research Center of the Ministry of Public Security, Beijing, China

Corresponding Author:
Address correspondence to Cong Chen: congchen1@cutr.usf.edu

optimize emergency vehicle dispatch and minimize vehicle waiting time in traffic emergency response.

As is widely known, statistical and regression analysis, mostly using discrete choice models, is the major approach to traffic incident prediction and safety-related analysis. However, these models were developed based on certain assumptions regarding model structure and parameter distribution (i.e., linear relationships, ordered response values, etc.), which may not always hold for a typical dataset of research interest. Systematical result bias may, therefore, be introduced. A Bayesian network (BN) model was developed based on graphical topology and probability inference to explore the interactional relationship among variables represented with conditional probabilities without posing any pre-defined assumptions (*12*, *13*). BN has been widely used in transportation research to address a large variety of issues, including traffic safety, traffic operation efficiency, infrastructure management, and so forth (*11–23*). For example, Ozbay and Noyan (*11*) applied BNs to predict incident clearance durations and concluded that the methodology enabled decision-makers to make real-time decisions by giving them the flexibility of employing the probabilistic inference capabilities of BNs. Chen et al. (*13*) developed a nested-logit BN model to examine road user preference distribution on toll road utilization. Liang and Lee (*16*) developed a hybrid approach integrating dynamic BN and supervised clustering for driver cognitive distraction and effectively revealed a hidden relationship between driver performance and cognitive status. Castillo et al. (*18*) predicted traffic flow status using BN based on origin-destination (O-D) pairs and link flow information. Wu and Yang (*22*) modeled commuter travel mode choice via a BN model and concluded that commute distance has significant influence on travel mode choice. To the best of our knowledge, the study by Ozbay and Noyan (*11*) was the only study to enrich the application of BN in traffic incident clearance duration and provide a solid foundation for this research topic, where it focused on the applicability of BN in traffic incident management, but the heterogeneous influence of factors on the total duration length was not fully investigated. Therefore, we conducted this study to fill this gap.

The studies mentioned above provide comprehensive and in-depth understandings of traffic incident response management and BN applications in transportation research. This study aimed to develop a BN approach to estimate traffic incident duration time and explore the heterogeneous influence of contributing factors on the length of incident duration. A dual-learning process based on the studied dataset and expert knowledge and experience in traffic incident management was applied in the model training procedure. The heterogeneous influence of contributing factors on traffic incident duration were also assessed. A BN model is not only able to explore the relationship between traffic incident response duration time and contributing variables, but can also reveal the interactions among the contributing variables, which will provide additional benefits in effective traffic incident response and management. The remainder of this paper is organized as follows: the descriptions of data source, nature, and process are provided in the next section; the methodology section introduces the methods applied in this study. This is followed by analysis of results and discussion. An overview of the entire research effort is found in the last section.

Data Description and Preprocessing

Traffic incident duration is usually defined as the time between incident occurrence and roadway clearance (*24*). The duration of an incident comprises three components—reporting time (time duration between incident occurrence and incident notification), response time (time duration between incident notification and response unit arrival), and clearance time (time duration between response unit arrival and incident clearance). See Figure 1.

In this research, the studied dataset was obtained from the Freeway Traffic Monitoring Center of Zhejiang Province Public Security Bureau and includes two years of traffic incident data collected from 2008 to 2010 on a 78 km (48.5 mile) eight-lane divided highway section with a speed limit of 120 km/h (75 mph). After incomplete records were removed, a total of 1,174 incident records were included in this study. For each incident, the notification time was defined as the instance when the traffic monitoring center received the traffic incident call from the scene or the incident was confirmed by the police onsite after it was detected by a video-based automatic incident detection system. It is understandable that it is often impossible to obtain the accurate time of incident occurrence. Therefore, based on practical considerations, for each incident, the beginning time of its duration was the instance of incident notification and the end time was when the clearance was completed. With the above definitions, the dependent variable in this paper, traffic incident duration (in minutes) includes two components—response time and clearance time—and is classified into nine categories based on 30-minute equal intervals to improve modeling efficiency (*11*), ranging from 30 min or less ((0,30] min) to longer than 4 hours ((240, $+\infty$) min). The recorded information contained the following variables: weather, alarm time, incident style, number of lanes occupied, number of vehicles involved, number of trucks involved, number of cars involved, number of injuries, number of fatalities, number of persons stuck in car, rescue vehicle type needed, and incident duration,

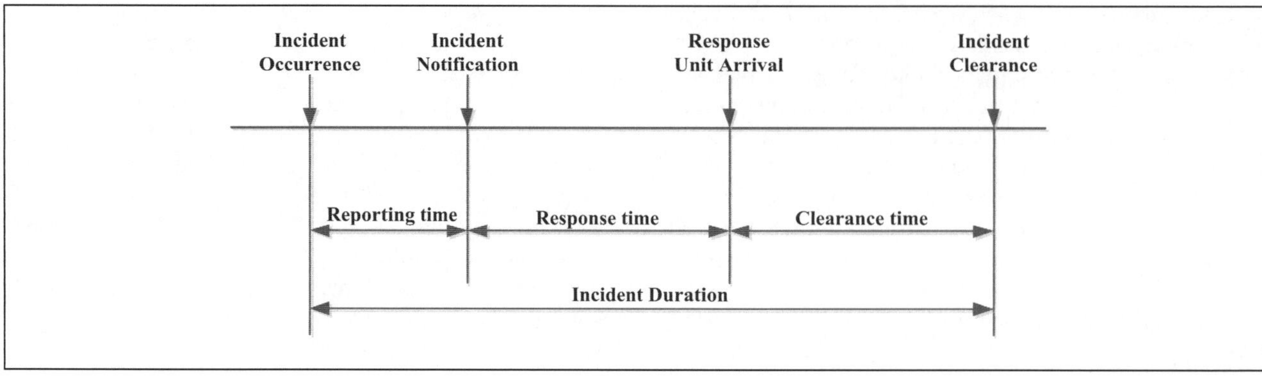

Figure 1. Traffic incident duration definition.

which could be categorized into three types—incident information, incident consequences, or rescue resources. Detailed information on variable definitions and value distributions in the studied dataset is illustrated in Table 1.

Methodology

Bayes' Theorem

The core concept of Bayesian analysis is to represent uncertainties with statistical probability, and Bayes' theorem is the theoretical foundation of all Bayesian analysis. Bayes' theorem is a formula to describe how to calculate the posterior probability of a hypothesis based on given evidence and the prior probability of the hypothesis. Given θ representing the hypothesis and a vector x representing the observed evidence, the Bayes' theorem could be described as follows:

$$p(\theta|x) = \frac{\pi(\theta) \times p(x|\theta)}{p(x)} = \frac{\pi(\theta) \times p(x|\theta)}{\int \pi(\theta) \times p(x|\theta) d\theta} \quad (1)$$

where $p(\theta|x)$ is the posterior probability of θ given the evidence x; $\pi(\theta)$ is the prior probability for the hypothesis θ; $p(x)$ is the marginal probability for x, meaning the probability of the evidence x being true; $p(x|\theta)$ is the conditional probability of x given the hypothesis θ, and $\frac{p(x|\theta)}{p(x)}$ is the support on the hypothesis θ from evidence x.

Bayesian Network

A Bayesian network (BN) is a special Bayesian method that integrates graphical topology and Bayesian probability inference to explore the relationships among variables. A BN model has been described in several previous studies (*12*, *13*, *20*) and is summarized as follows. Given a set of variables $V = \{X_1, X_2, \ldots, X_n\}$, $n > 1$, a BN model is represented by a directed acyclic structure B_s consisting of nodes representing the above set of variables and directed arcs among these variables representing their statistical dependence. In correspondence with the BN structure B_s is a complete set of probability tables $B_p = \{P(X_i|pr(X_i)), X_i \in X\}$ that contains all the probabilistic dependence information among these variables, where $pr(X_i)$ is the set of parent variables of X_i in B_s and $i = 1,2,\ldots,n$. The joint probability distribution $P(X_1, X_2, \ldots, X_n)$ is expressed as follows:

$$P(X) = P(X_1, X_2, \ldots, X_n) = \prod\nolimits_{X_i \in V} p(X_i|pr(X_i)) \quad (2)$$

To use BN as a classification model for traffic incident duration Y, given a set of explanatory $X = \{X_1, X_2, \ldots, X_k\}$, $k = n - 1$, the predicted variable Y, and an evidence data set D (X, Y) where $V = (X, Y)$ is the complete set of variables, it is necessary to calculate the maximum conditional probability $P(Y|X)$, where

$$P(Y|X) = \frac{P(X, Y)}{P(X)} = \frac{P(V)}{P(X)} \propto P(V) = \prod\nolimits_{X_i \in V} p(X_i|pr(X_i)) \quad (3)$$

There are generally two ways to develop a BN model: 1) model development based on expert knowledge and experience in the field of endeavor (traffic incident and management in this research) (*25*, *26*); and 2) model development based on the available dataset (*13*, *27*). However, there exist disadvantages for each of these two methods. Model development based on expert knowledge and experience will inevitably introduce expert subjectivity and may not be able to fully reveal the hidden relationships among variables given the understanding bias potentially introduced. On the other hand, although model training solely based on the available dataset could reveal variable dependence hidden in the dataset, redundant information or non-logical cause-effect relationships are difficult to avoid due to the intrinsic limitations in the dataset such as limited data size or biased data sampling. Therefore, in this research, these two

Table 1. Variable Definitions and Descriptions

Variable	Description	Codes/values	No injury	Percentage (%)	Injury	Percentage (%)	Fatality	Percentage (%)	Total
Weather	Weather	1: Sunny	543	67.54	245	30.47	16	1.99	804
		2: Cloudy	100	68.49	41	28.08	5	3.42	146
		3: Rainy	145	69.38	59	28.23	5	2.39	209
		4: Snowy	11	73.33	4	26.67	0	0	15
Alarm time	Reporting time span	1: Daytime peak[a]	278	68.64	123	30.37	4	0.99	405
		2: Daytime nonpeak	303	68.4	130	29.35	10	2.26	443
		3: Night	218	66.87	96	29.45	12	3.68	326
Incident style	Incident category	1: Rear-end	298	52.84	247	43.62	20	3.55	564
		2: Overturn	133	66.17	66	32.84	2	1	201
		3: Hit mid guardrail	92	80	20	17.39	3	2.61	115
		4: Break down	105	100	0	0	0	0	105
		5: Scratch	29	93.55	2	6.45	0	0	31
		6: Tire flat	51	98.08	1	1.92	0	0	52
		7: Hit side guardrail	25	75.76	8	24.24	0	0	33
		8: Spilled	69	94.52	3	4.11	1	1.37	73
Lanes	Number of lanes occupied	1: One lane	519	73.83	175	24.89	9	1.28	703
		2: Two lanes	244	61.15	142	35.59	13	3.26	399
		3: Three lanes	29	45.31	31	48.44	4	6.25	64
		4: Four lanes	7	100	0	0	0	0	7
		5: Road closed	1	100	0	0	0	0	1
Stuck in car	Number of people stuck in car	0: None	799	72.05	295	26.6	15	1.35	1109
		1: One person	6	13.33	33	73.33	6	13.33	45
		2: Two persons	0	0	11	73.33	4	26.67	15
		3: >2 persons	0	0	3	75	1	25	4
Vehicles	Number of vehicles involved	0: None	62	98.41	1	1.59	0	0	63
		1: One vehicle	404	81.12	90	18.07	4	0.8	498
		2: Two vehicles	261	54.26	202	42	18	3.74	481
		3: Three vehicles	42	53.85	32	41.03	4	5.13	78
		4: Four vehicles	21	72.41	8	27.59	0	0	29
		5: >4 vehicles	11	44	14	56	0	0	25
Trucks	Number of trucks involved	0: None	522	70.83	212	28.77	3	0.41	737
		1: One truck	262	74.22	83	23.51	8	2.27	353
		2: Two trucks	17	22.97	44	59.46	13	17.57	74
		3: Three trucks	2	25	4	50	2	25	8
		4: >3 trucks	1	50	1	50	0	0	2
Cars	Number of passenger cars involved	0: None	288	74.81	81	21.04	16	4.16	385
		1: One car	237	67.14	108	30.59	8	2.27	353
		2: Two cars	204	61.82	124	37.58	2	0.61	330
		3: Three cars	40	68.97	18	31.03	0	0	58
		4: Four cars	18	75	6	25	0	0	24
		5: >4 cars	11	45.83	13	54.17	0	0	24
Injury	Number of injuries	0: None	814	99.51	0	0	4	0.49	818
		1: One person	0	0	196	95.61	9	4.39	205
		2: Two persons	0	0	85	91.4	8	8.6	93
		3: Three persons	0	0	34	89.47	4	10.53	38
		4: Four persons	0	0	11	91.67	1	8.33	12
		5: >4 persons	0	0	8	100	0	0	8
Death	Number of deaths	0: None	814	70.91	334	29.09	0	0	1148
		1: One person	0	0	0	0	23	100	23
		2: >1 person	0	0	0	0	3	100	3
Trailer	If need a trailer	0: Don't need	465	73.81	157	24.92	8	1.27	630
		1: Need	334	61.4	192	35.29	18	3.31	544
Ambulance	If need an ambulance	0: Don't need	797	96.84	26	3.16	0	0	823
		1: Need	15	4.27	310	88.32	26	7.41	351
Crane	If need a crane	0: Don't need	646	69.09	273	29.2	16	1.71	935
		1: Need	153	64.02	76	31.8	10	4.18	239
Transfer[b]	If need a transfer	0: Don't need	693	68.75	297	29.46	18	1.79	1008
		1: Need	106	63.86	52	31.33	8	4.82	166
Tire	If need a tire vehicle	0: Don't need	566	70.14	228	28.25	13	1.61	807
		1: Need	233	63.49	121	32.97	13	3.54	367

(continued)

Table 1. (continued)

Variable	Description	Codes/values	No injury	Percentage (%)	Injury	Percentage (%)	Fatality	Percentage (%)	Total
Duration class	Duration level	1: (0,30] min	241	93.05	18	6.95	0	0	259
		2: (30,60] min	312	65.82	161	33.97	1	0.21	474
		3: (60,90] min	119	60.1	78	39.39	1	0.51	198
		4: (90,120] min	52	52	40	40	9	9	100
		5: (120,150] min	36	54.55	24	36.36	6	9.09	66
		6: (150,180] min	15	53.57	9	32.14	4	14.29	28
		7: (180,210] min	8	61.54	3	23.08	2	15.38	13
		8: (210,240] min	10	76.92	3	23.08	0	0	13
		9: (240, $+\infty$) min	8	34.78	12	52.17	3	13.04	23

[a]Daytime peak = 8:00–10:59 and 15:00–18:59; Daytime nonpeak = 5:00–7:59, 11:00–14:59; and 19:00–21:59); Night = 22:00–4:59.
[b]A transfer vehicle is used jointly with a crane to tow away severely damaged vehicles that could not be towed by a trailer.

methods were integrated into the BN model development process, where model training through dataset was used as the major approach for BN model structure training, and expert knowledge and judgment were applied as an assistive approach to further refine the trained BN model to ensure the accuracy and effectiveness of the modeling process.

BN Structure Learning Algorithm and Parameter Training Method

In this research, the greedy thick thinning (GTT) structure learning algorithm was used to train the BN structure, which is basically a two-step process. GTT starts with an empty graph and repeatedly adds directed arcs among variables that maximally increase the marginal likelihood of the BN structure B_s given the dataset D, $P(B_s|D)$, until no more arc can be added to produce a positive increase; this arc-adding phase is the thickening phase. Next, the algorithm repeatedly examines and removes arcs until no arc deletion can result in a positive increase in $P(B_s|D)$; this process is the thinning phase (*28*). There are two parameters to define the GTT algorithm in this study. First, the K2 searching algorithm was used as the prior, which is a heuristic search procedure developed by Cooper and Herskovits (*29*) that starts with a node having no parents and then adds incrementally that parent whose addition most increases the probability of the resulting structure. The whole process stops until the probability of the result structure is no longer increased, and $P(B_s|D)$ is maximized. In addition, the maximum number of parents was defined, since an excessive number of parent nodes will cause model overfitting (*12*). In this study, the maximum number of parent nodes was limited to 3.

After the BN structure B_s was obtained through the above search process, eight experts from universities and national or provincial governmental agencies, who had solid backgrounds and experience in traffic incident management, participated in the model refinement process. The model structure adjustments were made based on the findings in previous research (*11*, *23*, *30*) and discussions among these experts based on their knowledge and experience, and agreement needed to be made among experts on these adjustments. When the BN structure is finalized, conditional probability tables can be inferred through the maximum likelihood estimation (MLE) method. The basic principle of MLE is to find the parameter that maximizes the likelihood function. The MLE method is a mature estimation method that has been well described in peer studies, and therefore the methodological details are not duplicated (*12*, *13*).

Research Result Analysis and Discussion

Model Performance

This research employed GeNIe software, developed and maintained by BayesFusion, LLC, to build the BN structure and estimate the parameters (*28*). The entire dataset was divided into a training dataset and a test dataset based on different splitting ratios (6:4, 7:3, and 8:2), classified by proportionate stratified sampling based on nine levels of incident duration, to examine model performance at different levels. The approach of stratified sampling set the training and testing datasets using approximately the same proportion of duration distribution. After trying different splitting ratios, it was found that the modeling estimation process based on the 7:3 splitting ratio yielded the highest percentage of correctly classified records in the testing dataset, and the difference in the percentages of correctly classified records was smallest between the training and testing datasets, indicating that the model training and testing results under this splitting ratio were the least biased. Finally, based

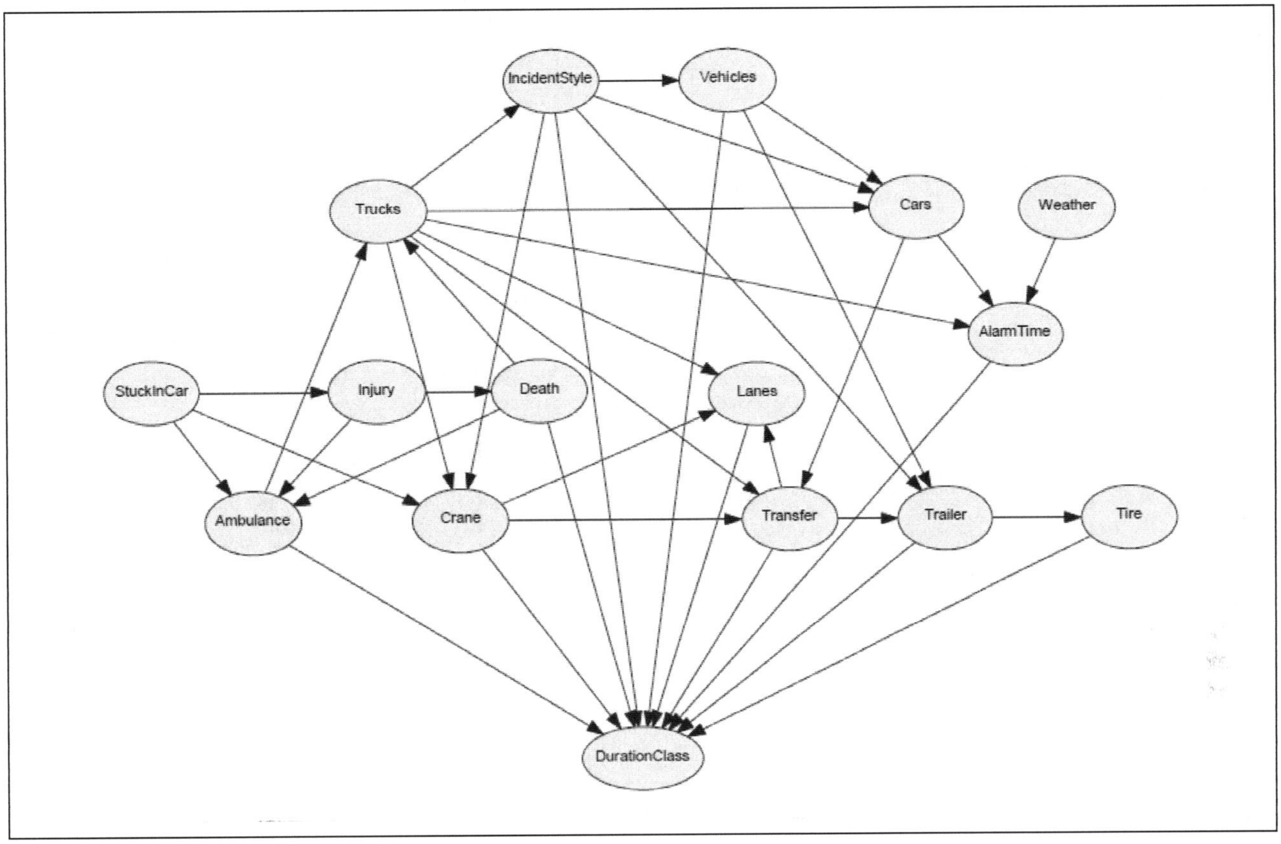

Figure 2. Trained BN structure.

on the 7:3 splitting ratio, 821 samples were used for BN structure learning and 353 samples were used for model validation and performance testing. The BN structure was trained in the dual-learning process based on a combination of prior knowledge and the GTT training algorithm, and the conditional probability table was optimized based on the MLE method following the trained BN structure. The trained BN structure is shown in Figure 2.

The overall BN estimation results for both the training and testing datasets are shown in Tables 2 and 3 detailing the classification accuracy, precision and recall of the test data classified by the nine levels of incident duration. In this study, for each incident duration level, the classification accuracy, also known as true positive (TP) rate, is defined as the proportion of the correctly classified records in all the observed records in that level. Precision and recall are two additional statistical measurements for model classification performance (*31*). Precision is defined as the fraction of correctly classified records in all the records that are classified as this level, including those incorrectly classified as this level, and recall has the same definition as classification accuracy. As shown in Table 2, the overall classification accuracies of this trained BN are 74.91% and 73.65% for the training and testing datasets, respectively, showing promising estimation capability of the trained model. The difference between the classification accuracies for the training and testing datasets is around 1%, indicating that the trained network is able to explain and model the testing data well. Specifically, as is shown in Table 3, the model performs best at two levels (180, 210] min and (210, 240] min), both of which yield an accuracy of 1.0000 (100.00%) and produce the lowest accuracy at (60, 90] min, and (90, 120] min and (150,180] min, but still with an accuracy equal to 0.6667 (66.67%). In terms of precision, the trained BN has the highest precision of 1.0000 for (150, 180] min and the lowest precision of 0.6383 for (0, 30] min. These results indicate that the trained BN structure not only demonstrates an overall promising performance, but also performs favorably or acceptably at each indication duration level. Table 4 shows the classification confusion matrix for the testing dataset regarding the nine incident duration levels and elaborates the model performance at each level with more details. As discussed previously, the performance at each incident duration level is promising based on the number of exact matching records, as highlighted by the numbers lying

Table 2. BN Estimation Results

	Training		Testing	
	Number	Percentage	Number	Percentage
Correctly classified instances	615	74.91%	260	73.65%
Incorrectly classified instances	206	25.09%	93	26.35%
Total number of instances	821		353	

Table 3. BN Estimation Results by Duration Level on Testing Dataset

Duration level (unit: min)	No. correct predictions	No. observed records	No. predicted records	Accuracy (Recall)	Precision
(0,30]	60	78	94	0.7692	0.6383
(30,60]	106	142	143	0.7465	0.7413
(60,90]	40	60	48	0.6667	0.8333
(90,120]	20	30	24	0.6667	0.8333
(120,150]	13	18	20	0.7222	0.6500
(150,180]	6	9	6	0.6667	1.0000
(180,210]	5	5	6	1.0000	0.8333
(210,240]	3	3	4	1.0000	0.7500
(240, +∞)	7	8	8	0.8750	0.8750

Table 4. BN Classification Confusion Matrix for Testing Dataset

		Predicted duration								
	Duration level (min)	(0,30]	(30,60]	(60,90]	(90,120]	(120,150]	(150,180]	(180,210]	(210,240]	(240, +∞)
Observed duration	(0,30]	**60**	16	1	1	0	0	0	0	0
	(30,60]	29	**106**	4	0	3	0	0	0	0
	(60,90]	4	15	**40**	0	0	0	0	1	0
	(90,120]	0	5	1	**20**	3	0	0	0	1
	(120,150]	1	0	2	2	**13**	0	0	0	0
	(150,180]	0	1	0	0	1	**6**	1	0	0
	(180,210]	0	0	0	0	0	0	**5**	0	0
	(210,240]	0	0	0	0	0	0	0	**3**	0
	(240, +∞)	0	0	0	1	0	0	0	0	**7**

on the diagonal line. Additionally, there are also 72 observations (highlighted in light green in Table 4) whose classification is within one level of variation, which could also be considered as reliable and acceptable under certain scenarios (i.e., incidents without any injuries or fatalities). With these records taken into consideration, the overall acceptable accuracy increases to 0.9405, showing reliable model performance.

In BN modeling analysis, the receiver operating characteristic (ROC) curve has been widely used to demonstrate the performance of the trained model in each response level in a graphical way, and the area under curve (AUC) value is the area enclosed by the ROC curve and the axis boundary and serves as a numerical measurement to present overall model performance (*13, 27*). The ROC curve was generally proposed to evaluate the BN model performance on binary outcome based on the TP rate and false positive (FP) rate extracted from the classifications and illustrates the trade-off between TP and FP rates. For a BN classification model with multiple response levels (i.e., nine incident duration levels in this research), an ROC curve could be produced for each response level where the TP and FP cases are defined accordingly, and the AUC value specific to each response level is calculated separately (*32, 33*). The AUC values range between 0.5 and 1, where AUC = 0.5 indicates a random guess without any educated inference (i.e., tossing a coin) and AUC = 1 indicates a perfect

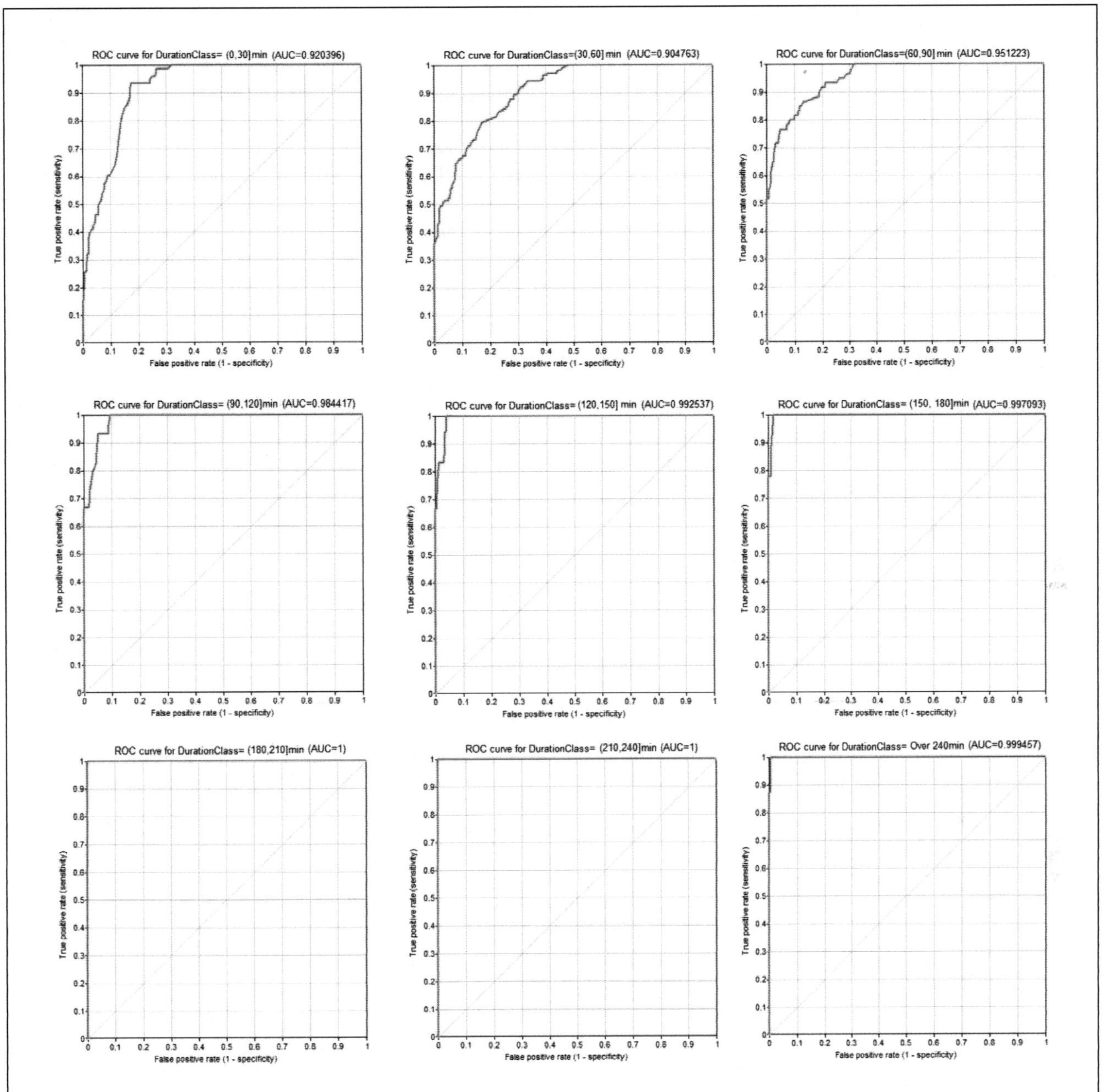

Figure 3. ROC curves and AUC values for nine incident duration levels.

classification. Figure 3 shows the ROC curve along with the corresponding AUC value for each of the nine incident duration levels. It shows that an ROC curve above the diagonal line was produced for each of these nine incident duration levels, indicating that the trained BN reveals the cause-effect relationships among these factors and effectively uses them in incident duration estimation. In addition, it shows that all AUC values for each of the nine incident duration levels is above 0.9, with that for (180, 210] min and (210, 240] min being equal to 1, showing that the trained model produced promising performance for each duration level, based on (*34*). For the entire testing dataset, the overall AUC can be calculated as a weighted average for each response outcome category (*35*), which has been used in previous studies (*12*, *27*). In this study, the overall AUC value was equal to 0.905, demonstrating that the trained model was effective in estimating traffic incident duration levels.

Many different models have been used for traffic incidents or crash analysis, and the multinomial logit (MNL) model is one of the most prevalent models. Given the large amount of response categories (incident duration levels), model performance comparisons were conducted between the proposed BN approach and the MNL model using the same training and testing datasets. It was found that the trained MNL model was able to correctly classify only 45.04% (159 records) of all the records in the testing dataset, which is significantly lower than the classification accuracy in Table 2. Therefore, the trained BN model outperforms the MNL model and is more effective in estimating traffic incident durations.

Variable Influence Analysis

In addition to the BN model structure, another indispensable component for a BN model is a set of probability tables containing interdependence information among the response variable and the explanatory variables. As mentioned, once the BN model structure is trained, the probability tables can be estimated accordingly using pre-set estimation methods (*13*). Based on the final model trained in Figure 2 and the probability tables inferred, the posterior probability for each value of each variable after evidence setting is shown in Table 5, and each line presents the posterior probability for each incident duration after the probability of corresponding value (i.e., Weather = Sunny) is set as 1.0.

Since all factors have been converted to indicator or multi-categorical variables, to calculate the influence of these variables in incident duration estimation, a variable influence analysis was conducted in this research based on the direct pseudo-elasticity of the probability approach that was used in peer studies (*13*, *20*), and the variable influence could be inferred based on Table 5. For indicator variables with a binary outcome, the factor influence with respect to each traffic incident duration is defined by the percentage change in probability when the indicator variable is changed from 0 to 1 as follows:

$$E_x^{P_i} = \frac{P_i[x = 1] - P_i[x = 0]}{P_i[x = 0]} \quad (4)$$

where $E_x^{P_i}$ represents the percentage change in probability regarding traffic incident duration level i from the indicator variable x, and $x = 0$ is set as the base condition. For a variable with multiple categorical values, the variable influence analysis was conducted for each of its values, where one of its values was set as the base condition, and the relative variable influence for each other value on all categories was calculated accordingly.

The variable influence analysis results from Equation 4 are shown in Table 6. It is noted that the statistics highlighted in bold in Table 6 indicate percentage change larger than 100% in either increase or decrease. Based on the trained BN structure in Figure 2 and the probability inference and variable influence results presented in Tables 5 and 6, significant variables for different incident duration levels could be identified.

As shown in Tables 5 and 6, weather was not found to significantly affect incident duration, since only minor percentage changes were identified. This is reasonable since the studied data were collected in regions where rain occurs regularly in summer and snowy weather is rarely expected in winter. Therefore, local road users are more cautious in adverse weather and government agencies are more prepared for prompt response if snow or rain is forecast. It was found in the results that, compared with night time, the probabilities for (120, 150] min, (150, 180] min, (180, 210] min, (210, 240] min, and (240, $+\infty$) min increase by 49.38%, 53.85%, 86.67%, 133.33%, and 50.00% during daytime peak hours, respectively. These results are understandable, in that there is often heavy traffic during daytime peak hours and it usually takes longer for an incident response team to arrive at the scene and clear the incident.

Incident types show various effects on different incident duration levels. For instance, it was found that hitting a median guardrail tends to increase overall incident duration time, as shown by the highlighted statistics in Table 6, although other incident types show mixed effects on different incident levels. The primary incident duration level can be identified by the posterior probability inference results in Table 5. It was found that most incident types have the largest posterior probability in either (30, 60] min or (60, 90] min, showing that most incidents can be cleared in 30–90 min. Similarly, the number of lanes occupied by the incident also shows various effects on incident duration levels. It was found that when more than one lane was occupied or the road was closed during the incident, the probabilities for (30, 60] min and (60, 90] min unanimously increased, whereas the probabilities for longer duration levels decreased for the majority of these cases. These results are consistent with common sense that when multiple lanes are occupied, severe congestion occurs. Therefore, the incident response team at the scene is motivated to clear the incident as soon as possible to resume normal traffic.

The quantitative variables, including number of people stuck in a vehicle, number of vehicles involved, number of trucks involved, and number of passenger cars involved, have varied influence on traffic incident durations. For example, it was found that one or more persons stuck in a vehicle tends to increase the probabilities of most incident duration levels of 1–4 hours, although most of these increases are not significant. In addition, it was found that an increased number of trucks involved significantly increases the probabilities of incident

Table 5. Posterior Probability Inference Results for Trained BN Model

Variable	Codes/values	(0,30]	(30,60]	(60,90]	(90,120]	(120,150]	(150,180]	(180,210]	(210,240]	(240, + ∞)
Weather	1: Sunny	0.014	0.199	0.425	0.180	0.086	0.056	0.019	0.009	0.012
	2: Cloudy	0.017	0.199	0.425	0.175	0.090	0.054	0.019	0.008	0.013
	3: Rainy	0.017	0.208	0.425	0.177	0.086	0.049	0.018	0.008	0.012
	4: Snowy	0.018	0.209	0.419	0.174	0.091	0.051	0.019	0.008	0.013
Alarm time	1: Daytime peak	0.021	0.163	0.361	0.196	0.121	0.080	0.028	0.014	0.018
	2: Daytime nonpeak	0.013	0.224	0.466	0.160	0.075	0.032	0.016	0.006	0.009
	3: Night	0.019	0.210	0.431	0.174	0.081	0.052	0.015	0.006	0.012
Incident style	1: Rear-end	0.012	0.367	0.308	0.161	0.056	0.067	0.018	0.005	0.006
	2: Overturn	0.015	0.157	0.474	0.195	0.085	0.048	0.016	0.005	0.006
	3: Hit mid guardrail	0.021	0.102	0.374	0.218	0.166	0.063	0.021	0.014	0.022
	4: Break down	0.016	0.234	0.463	0.144	0.048	0.052	0.023	0.007	0.014
	5: Scratch	0.011	0.394	0.362	0.116	0.054	0.027	0.020	0.004	0.013
	6: Tire flat	0.010	0.245	0.411	0.136	0.112	0.040	0.020	0.018	0.008
	7: Hit side guardrail	0.036	0.324	0.280	0.170	0.059	0.079	0.022	0.011	0.018
	8: Spilled	0.013	0.242	0.486	0.091	0.057	0.046	0.048	0.008	0.008
Lanes	1: One lane	0.032	0.021	0.254	0.412	0.115	0.082	0.037	0.029	0.017
	2: Two lanes	0.008	0.242	0.462	0.149	0.079	0.034	0.012	0.003	0.011
	3: Three lanes	0.030	0.147	0.383	0.218	0.099	0.072	0.027	0.011	0.012
	4: Four lanes	0.038	0.114	0.270	0.186	0.146	0.131	0.041	0.041	0.033
	5: Road closed	0.060	0.063	0.257	0.240	0.143	0.115	0.044	0.040	0.038
Stuck in car	0: None	0.016	0.210	0.423	0.172	0.089	0.052	0.018	0.008	0.013
	1: One person	0.036	0.070	0.441	0.234	0.106	0.058	0.031	0.012	0.012
	2: Two persons	0.043	0.052	0.433	0.232	0.119	0.062	0.035	0.012	0.012
	3: >2 persons	0.047	0.039	0.453	0.212	0.124	0.061	0.042	0.011	0.011
Vehicles	0: None	0.009	0.403	0.304	0.137	0.045	0.075	0.016	0.005	0.006
	1: One vehicle	0.019	0.227	0.378	0.162	0.101	0.058	0.022	0.011	0.022
	2: Two vehicles	0.018	0.149	0.462	0.208	0.088	0.050	0.012	0.006	0.006
	3: Three vehicles	0.012	0.171	0.535	0.122	0.093	0.029	0.027	0.005	0.006
	4: Four vehicles	0.011	0.152	0.528	0.145	0.063	0.032	0.059	0.005	0.005
	5: > 4 vehicles	0.012	0.353	0.438	0.112	0.031	0.032	0.012	0.005	0.005
Trucks	0: None	0.007	0.239	0.489	0.159	0.065	0.026	0.008	0.002	0.004
	1: One truck	0.029	0.158	0.313	0.199	0.126	0.093	0.033	0.019	0.030
	2: Two trucks	0.053	0.060	0.319	0.221	0.146	0.115	0.049	0.019	0.017
	3: Three trucks	0.095	0.036	0.264	0.157	0.196	0.133	0.072	0.020	0.027
	4: > 3 trucks	0.016	0.129	0.489	0.201	0.105	0.047	0.011	0.001	0.002
Cars	0: None	0.032	0.193	0.291	0.183	0.122	0.097	0.034	0.018	0.030
	1: One car	0.012	0.231	0.431	0.171	0.085	0.042	0.014	0.006	0.008
	2: Two cars	0.010	0.176	0.521	0.188	0.068	0.027	0.006	0.002	0.002
	3: Three cars	0.007	0.183	0.551	0.119	0.088	0.020	0.024	0.003	0.004
	4: Four cars	0.010	0.168	0.530	0.145	0.061	0.027	0.052	0.003	0.004
	5: > 4 cars	0.012	0.352	0.425	0.115	0.038	0.034	0.013	0.005	0.006
Injury	0: None	0.009	0.265	0.401	0.162	0.077	0.052	0.014	0.007	0.013
	1: One person	0.034	0.054	0.485	0.210	0.118	0.050	0.028	0.010	0.011
	2: Two persons	0.037	0.062	0.459	0.204	0.122	0.059	0.032	0.011	0.012
	3: Three persons	0.042	0.057	0.457	0.198	0.124	0.063	0.038	0.010	0.012
	4: Four persons	0.037	0.079	0.455	0.197	0.117	0.059	0.034	0.010	0.012
	5: > 4 persons	0.029	0.043	0.515	0.215	0.118	0.039	0.020	0.009	0.010
Death	0: None	0.014	0.206	0.431	0.178	0.087	0.048	0.015	0.008	0.012
	1: One person	0.156	0.017	0.089	0.053	0.210	0.242	0.192	0.019	0.022
	2: > 1 person	0.059	0.071	0.065	0.089	0.308	0.219	0.059	0.066	0.064
Trailer	0: Don't need	0.017	0.242	0.414	0.175	0.070	0.050	0.012	0.007	0.013
	1: Need	0.017	0.155	0.435	0.175	0.113	0.056	0.027	0.010	0.012
Ambulance	0: Don't need	0.008	0.271	0.400	0.161	0.075	0.052	0.013	0.007	0.013
	1: Need	0.039	0.041	0.478	0.209	0.124	0.055	0.032	0.011	0.011
Crane	0: Don't need	0.010	0.244	0.475	0.143	0.075	0.035	0.011	0.003	0.006
	1: Need	0.045	0.035	0.220	0.303	0.149	0.125	0.052	0.030	0.041
Transfer	0: Don't need	0.010	0.233	0.476	0.170	0.070	0.027	0.009	0.002	0.003
	1: Need	0.065	0.002	0.094	0.207	0.217	0.214	0.083	0.045	0.072
Tire	0: Don't need	0.018	0.223	0.419	0.163	0.091	0.050	0.016	0.005	0.013
	1: Need	0.016	0.154	0.434	0.202	0.086	0.059	0.025	0.014	0.011

Table 6. Variable Influence on Incident Duration (Unit: %)

Variable	Codes/values	(0,30]	(30,60]	(60,90]	(90,120]	(120,150]	(150,180]	(180,210]	(210,240]	(240, +∞)
Weather	1: Sunny[a]	0.00	0.00	0.00	0.00	0.00	0.00	0.00	0.00	0.00
	2: Cloudy	21.43	0.00	0.00	−2.78	4.65	−3.57	0.00	−11.11	8.33
	3: Rainy	21.43	4.52	0.00	−1.67	0.00	−12.50	−5.26	−11.11	0.00
	4: Snowy	28.57	5.03	−1.41	−3.33	5.81	−8.93	0.00	−11.11	8.33
Alarm time	1: Daytime peak	10.53	−22.38	−16.24	12.64	49.38	53.85	86.67	**133.33**	50.00
	2: Daytime nonpeak	−31.58	6.67	8.12	−8.05	−7.41	−38.46	6.67	0.00	−25.00
	3: Night[a]	0.00	0.00	0.00	0.00	0.00	0.00	0.00	0.00	0.00
Incident style	1: Rear-end	−7.69	51.65	−36.63	76.92	−1.75	45.65	−62.50	−37.50	−25.00
	2: Overturn	15.38	−35.12	−2.47	**114.29**	49.12	4.35	−66.67	−37.50	−25.00
	3: Hit mid guardrail	61.54	−57.85	−23.05	**139.56**	**191.23**	36.96	−56.25	75.00	**175.00**
	4: Break down	23.08	−3.31	−4.73	58.24	−15.79	13.04	−52.08	−12.50	75.00
	5: Scratch	−15.38	62.81	−25.51	27.47	−5.26	−41.30	−58.33	−50.00	62.50
	6: Tire flat	−23.08	1.24	−15.43	49.45	96.49	−13.04	−58.33	**125.00**	0.00
	7: Hit side guardrail	**176.92**	33.88	−42.39	86.81	3.51	71.74	−54.17	37.50	**125.00**
	8: Spilled[a]	0.00	0.00	0.00	0.00	0.00	0.00	0.00	0.00	0.00
Lanes	1: One lane[a]	0.00	0.00	0.00	0.00	0.00	0.00	0.00	0.00	0.00
	2: Two lanes	−75.00	**1052.38**	81.89	−63.83	−31.30	−58.54	−67.57	−89.66	−35.29
	3: Three lanes	−6.25	**600.00**	50.79	−47.09	−13.91	−12.20	−27.03	−62.07	−29.41
	4: Four lanes	18.75	**442.86**	6.30	−54.85	26.96	59.76	10.81	41.38	94.12
	5: Road closed	87.50	**200.00**	1.18	−41.75	24.35	40.24	18.92	37.93	**123.53**
Stuck in car	0: None[a]	0.00	0.00	0.00	0.00	0.00	0.00	0.00	0.00	0.00
	1: One person	**125.00**	−66.67	4.26	36.05	19.10	11.54	72.22	50.00	−7.69
	2: Two persons	**168.75**	−75.24	2.36	34.88	33.71	19.23	94.44	50.00	−7.69
	3: Three persons	**193.75**	−81.43	7.09	23.26	39.33	17.31	**133.33**	37.50	−15.38
Vehicles	0: None[a]	**0.00**	0.00	0.00	0.00	0.00	0.00	0.00	0.00	0.00
	1: One vehicle	**111.11**	−43.67	24.34	18.25	**124.44**	−22.67	37.50	**120.00**	**266.67**
	2: Two vehicles	**100.00**	−63.03	51.97	51.82	95.56	−33.33	−25.00	20.00	0.00
	3: Three vehicles	33.33	−57.57	75.99	−10.95	**106.67**	−61.33	68.75	0.00	0.00
	4: Four vehicles	22.22	−62.28	73.68	5.84	40.00	−57.33	**268.75**	0.00	−16.67
	5: >4 vehicles	33.33	−12.41	44.08	−18.25	−31.11	−57.33	−25.00	0.00	−16.67
Trucks	0: None[a]	0.00	0.00	0.00	0.00	0.00	0.00	0.00	0.00	0.00
	1: One truck	**314.29**	−33.89	−35.99	25.16	93.85	**257.69**	**312.50**	**850.00**	**650.00**
	2: Two trucks	**657.14**	−74.90	−34.76	38.99	**124.62**	**342.31**	**512.50**	**850.00**	**325.00**
	3: Three trucks	**1257.14**	−84.94	−46.01	−1.26	**201.54**	**411.54**	**800.00**	**900.00**	**575.00**
	4: >3 trucks	**128.57**	−46.03	0.00	26.42	61.54	80.77	37.50	−50.00	−50.00
Cars	0: None[a]	0.00	0.00	0.00	0.00	0.00	0.00	0.00	0.00	0.00
	1: One car	−62.50	19.69	48.11	−6.56	−30.33	−56.70	−58.82	−66.67	−73.33
	2: Two cars	−68.75	−8.81	79.04	2.73	−44.26	−72.16	−82.35	−88.89	−93.33
	3: Three cars	−78.13	−5.18	89.35	−34.97	−27.87	−79.38	−29.41	−83.33	−86.67
	4: Four cars	−68.75	−12.95	82.13	−20.77	−50.00	−72.16	52.94	−83.33	−86.67
	5: >4 cars	−62.50	82.38	46.05	−37.16	−68.85	−64.95	−61.76	−72.22	−80.00
Injury	0: None[a]	0.00	0.00	0.00	0.00	0.00	0.00	0.00	0.00	0.00
	1: One person	**277.78**	−79.62	20.95	29.63	53.25	−3.85	**100.00**	42.86	−15.38
	2: Two persons	**311.11**	−76.60	14.46	25.93	58.44	13.46	**128.57**	57.14	−7.69
	3: Three persons	**366.67**	−78.49	13.97	22.22	61.04	21.15	**171.43**	42.86	−7.69
	4: Four persons	**311.11**	−70.19	13.47	21.60	51.95	13.46	**142.86**	42.86	−7.69
	5: >4 persons	**222.22**	−83.77	28.43	32.72	53.25	−25.00	42.86	28.57	−23.08
Death	0: None[a]	0.00	0.00	0.00	0.00	0.00	0.00	0.00	0.00	0.00
	1: One person	**1014.29**	−91.75	−79.35	−70.22	**141.38**	**404.17**	**1180.00**	**137.50**	83.33
	2: >1 person	**321.43**	−65.53	−84.92	−50.00	**254.02**	**356.25**	**293.33**	**725.00**	**433.33**
Trailer	0: Don't need[a]	0.00	0.00	0.00	0.00	0.00	0.00	0.00	0.00	0.00
	1: Need	0.00	−35.95	5.07	0.00	61.43	12.00	**125.00**	42.86	−7.69
Ambulance	0: Don't need[a]	0.00	0.00	0.00	0.00	0.00	0.00	0.00	0.00	0.00
	1: Need	**387.50**	−84.87	19.50	29.81	65.33	5.77	**146.15**	57.14	−15.38
Crane	0: Don't need[a]	0.00	0.00	0.00	0.00	0.00	0.00	0.00	0.00	0.00
	1: Need	**350.00**	−85.66	−53.68	**111.89**	98.67	**257.14**	**372.73**	**900.00**	**583.33**
Transfer	0: Don't need[a]	0.00	0.00	0.00	0.00	0.00	0.00	0.00	0.00	0.00
	1: Need	**550.00**	−99.14	−80.25	21.76	**210.00**	**692.59**	**822.22**	**2150.00**	**2300.00**
Tire	0: Don't need[a]	0.00	0.00	0.00	0.00	0.00	0.00	0.00	0.00	0.00
	1: Need	−11.11	−30.94	3.58	23.93	−5.49	18.00	56.25	**180.00**	−15.38

Note: Bold font indicates an increase in overall incident duration time.
[a]Base condition.

duration to 2 hours and more ((120, 150] min, (150, 180] min, (180, 210] min, (210, 240] min, and (240, $+\infty$) min). These results are reasonable since an increased number of trucks involved in incidents generally indicates an increased number of injuries and fatalities as well as property damage. Therefore, it takes more time to process and clear the incident. It also was found that one or more trucks involved also significantly increases the probability of (0, 30] min, which could be because an incident with more trucks involved is generally considered more severe and with higher priority in response timeliness, and the incident response team reacts more promptly to clear the incident after sufficient investigation and processing. However, Table 5 shows that the probability for (0, 30] min is still relatively low, indicating the low likelihood of such incidents being cleared in less than 30 min.

The numbers of injuries and fatalities in an incident also have significant influence on different incident duration levels. It was found that, compared with incidents with no injuries, one or more injuries in the incident significantly increase the probability of the incident being cleared within 30 min. This is reasonable since the incident response team will have the injured properly treated as quickly as possible. However, Table 5 shows that these incidents are unlikely to be cleared within 30 min, as shown by the corresponding low probabilities. One or more injuries in the incident overall also increase the probability of incident duration levels within the 30 min to 4 hour range, especially for the duration within (180, 210] min. The possible reason for this finding, similar to those discussed before, is that it takes more time for the incident response team to address the injured and clear the incident scene. The results for the number of fatalities in Tables 5 and 6 can be interpreted similarly.

Four variables regarding the use of large equipment in incident clearance, including trailers, ambulances, cranes, and transfer vehicles, also have significant influence on certain incident duration levels. It was found that, overall, the use of large equipment tends to increase the probability of longer incident duration levels. For example, the use of a crane significantly increases the probabilities for (90, 120] min, (150, 180] min, (180, 210] min, (210, 240] min, and (240, $+\infty$) min. These results are expected, as it takes considerable time for a crane to arrive on site and be properly deployed, and more time is also needed in other related procedures until the incident is cleared. It should be noted that the use of a crane also significantly increases the probability of (0, 30] min, which could probably explain why the incident response team reacts more promptly for severe incidents that require large equipment. Results for other variables can be explained accordingly. Finally, it was found that a flat tire tends to extend incident duration overall, because it is of lower priority in incident management than those with casualties, and it takes time to send the nearest response team to provide necessary road assistance. Overall, the trained BN model provides reasonable results.

Conclusion

Highway traffic incidents not only cause different levels of traffic congestion but can also lead to traffic accidents and secondary accidents, inducing significant life, economy, and productivity loss in terms of injuries and fatalities, extended travel time and delay, and excessive energy consumption and air pollution. Traffic emergency management under incident conditions is the core element of active traffic management, and it is of practical significance to accurately understand the duration time distribution for typical traffic incident types and the factors that influence incident duration, which provides critical inference to requirements and scheduling of emergency resources, decision-making optimization, and intelligent transportation management. Based on expert knowledge of traffic incident management and traffic incident data collected on highways in Zhejiang Province, China, this study proposes a dual-learning Bayesian network model to estimate traffic incident duration and examine the influence of heterogeneous factors on the length of duration. Nine incident duration levels were defined based on 30-minute intervals and a variety of factors regarding incident and external environment characteristics. The trained BN model demonstrated promising performance based on several model performance measurements, including classification accuracy, precision and recall, ROC curve, and AUC value and the classification matrix.

Probability inference and variable influence analyses results reveal that these variables have heterogeneous influence on incident duration levels. It was found that incidents that occur during daytime peak hours tend to have longer duration due to peak-hour traffic and congestion. Incident types have varied influence on different incident duration levels, and it was found that most types of incidents can be cleared in 30–90 min. An increased number of occupied lanes is associated with longer traffic incident duration due to reduced road capacity and increased congestion. Increases in the number of people stuck in a car, number of trucks involved, and number of injuries and fatalities tend to extend total incident duration due to heavy casualties and property damage that may be associated with the incident. However, they also significantly increase the likelihood of incident duration of within 30 min, but the likelihood of such incidents being cleared in less than 30 min is still low. The use of large equipment for incident clearance, including trailers, ambulances, cranes, and transfer vehicles, has significant influence in extending incident duration, since this equipment is generally used in severe incidents and it takes additional time to properly set up the equipment, handle casualties and damaged property, and clear the scene. Finally, it was found that a flat tire tends to increase incident duration due to its lower priority in incident management than those with casualties.

Overall, these findings provide beneficial reference to the understanding of decision-making in traffic incident response, traffic incident processes, and intelligent transportation management. Further research is desirable to enhance the BN model performance and better understand the relationship among variables by including more comprehensive data in both spatial and temporal domains, and more advanced BN structure training methods could be included for performance evaluation and comparison.

Acknowledgments

This work was supported by the National Natural Science Foundation of China under the Grant Project "Modeling Intra-urban Human Traffic Trip-activity Chain Character Patterns: Evidence from the Multi-source Trajectory Data," Project Grant No.41601434, and the China Scholarship Council (CSC, 201600930049).

Author Contributions

The authors confirm contribution to the paper as follows: study conception and design: Haozhe Cong, Cong Chen, Guohui Zhang; data collection: Haozhe Cong, Zhi Ye; analysis and interpretation of results: Haozhe Cong, Cong Chen, Pei-Sung Lin, Guohui Zhang; draft manuscript preparation: Haozhe Cong, Cong Chen, Pei-Sung Lin, Guohui Zhang, John Milton, Zhi Ye. All authors reviewed the results and approved the final version of the manuscript.

References

1. Yuan, S., S. A. Chun, B. Spinelli, Y. Liu, H. Zhang, and N. R. Adam. Traffic Evacuation Simulation Based on Multi-Level Driving Decision Model. *Transportation Research Part C: Emerging Technologies*, Vol. 78, 2017, pp. 129–149.
2. American Association of State Highway and Transportation Officials. *Highway Safety Design and Operations Guide*, American Association of State Highway and Transportation Officials, Washington, D.C., 1997.
3. Federal Highway Administration. *Traffic Incident Management Handbook*. Federal Highway Administration, Washington, D.C., 2010.
4. Golob, T. F., W. W. Recker, and J. D. Leonard. An Analysis of the Severity and Incident Duration of Truck--Involved Freeway Accidents. *Accident Analysis & Prevention*, Vol. 19, No. 5, 1987, pp. 375–395.
5. Zografos, K. G., T. Nathanail, and P. Michalopoulos. Analytical Framework for Minimizing Freeway-Incident Response Time. *Journal of Transportation Engineering*, Vol. 119, No. 4, 1993, pp. 535–549.
6. Zografos, K. G., K. N. Androutsopoulos, and G. M. Vasilakis. A Real-Time Decision Support System for Roadway Network Incident Response Logistics. *Transportation Research Part C: Emerging Technologies*, Vol. 10, No. 1, 2002, pp. 1–18.
7. Bahar, G., M. Masliah, C. Mollett, and B. Persaud. *NCHRP Report 501: Integrated Safety Management Process*. Transportation Research Board of the National Academies, Washington, D.C., 2003.
8. Smith, K., and B. L. Smith. *Forecasting the Clearance Time of Freeway Accidents*. Center for Transportation Studies, University of Virginia, Charlottesville, VA, 2001.
9. Chakraborty, P. S., A. Tiwari, and P. R. Sinha. Adaptive and Optimized Emergency Vehicle Dispatching Algorithm for Intelligent Traffic Management System. *Procedia Computer Science*, Vol. 57, 2015, pp. 1384–1393.
10. Chen, A. Y., and T.-Y. Yu. Network Based Temporary Facility Location for the Emergency Medical Services Considering the Disaster Induced Demand and the Transportation Infrastructure in Disaster Response. *Transportation Research Part B: Methodological*, Vol. 91, 2016, pp. 408–423.
11. Ozbay, K., and N. Noyan. Estimation of Incident Clearance Times Using Bayesian Networks Approach. *Accident Analysis & Prevention*, Vol. 38, No. 3, 2006, pp. 542–555.
12. Chen, C., G. Zhang, R. Tarefder, J. Ma, H. Wei, and H. Guan. A Multinomial Logit Model-Bayesian Network Hybrid Approach for Driver Injury Severity Analyses in Rear-End Crashes. *Accident Analysis and Prevention*, Vol. 80, 2015, pp. 76–88.
13. Chen, C., G. Zhang, H. Wang, J. Yang, P. J. Jin, and C. M. Walton. Bayesian Network-Based Formulation and Analysis for Toll Road Utilization Supported by Traffic Information Provision. *Transportation Research Part C: Emerging Technologies*, Vol. 60, 2015, pp. 339–359.
14. Gregoriades, A., and K. C. Mouskos. Black Spots Identification through a Bayesian Networks Quantification of Accident Risk Index. *Transportation Research Part C: Emerging Technologies*, Vol. 28, 2013, pp. 28–43.
15. de Oña, J., G. López, R. Mujalli, and F. J. Calvo. Analysis of Traffic Accidents on Rural Highways Using Latent Class Clustering and Bayesian Networks. *Accident Analysis & Prevention*, Vol. 51, 2013, pp. 1–10.
16. Liang, Y., and J. D. Lee. A Hybrid Bayesian Network Approach to Detect Driver Cognitive Distraction. *Transportation Research Part C: Emerging Technologies*, Vol. 38, 2014, pp. 146–155.
17. Sun, J., and J. Sun. A Dynamic Bayesian Network Model for Real-Time Crash Prediction Using Traffic Speed Conditions Data. *Transportation Research Part C: Emerging Technologies*, Vol. 54, 2015, pp. 176–186.
18. Castillo, E., J. M. Menéndez, and S. Sánchez-Cambronero. Predicting Traffic Flow Using Bayesian Networks. *Transportation Research Part B: Methodological*, Vol. 42, No. 5, 2008, pp. 482–509.

19. Li, D., T. Miwa, and T. Morikawa. Modeling Time-of-Day Car Use Behavior: A Bayesian Network Approach. *Transportation Research Part D: Transport and Environment*, Vol. 47, 2016, pp. 54–66.
20. Chen, C., Y. Chen, J. Ma, G. Zhang, and C. M. Walton. Driver Behavior Formulation in Intersection Dilemma Zones with Phone Use Distraction via a Logit-Bayesian Network Hybrid Approach. *Journal of Intelligent Transportation Systems: Technology, Planning, and Operations*, Vol. 22, No. 4, 2018, pp. 311–324.
21. Chen, C., G. Zhang, H. Huang, J. Wang, and R. A. Tarefder. Examining Driver Injury Severity Outcomes in Rural Non-Interstate Roadway Crashes Using a Hierarchical Ordered Logit Model. *Accident Analysis & Prevention*, Vol. 96, 2016, pp. 79–87.
22. Wu, J., and M. Yang. Modeling Commuters? Travel Behavior by Bayesian Networks. *Procedia - Social and Behavioral Sciences*, Vol. 96, 2013, pp. 512–521.
23. Zhao, W., L. Xu, J. Bai, M. Ji, and T. Runge. Sensor-Based Risk Perception Ability Network Design for Drivers in Snow and Ice Environmental Freeway: A Deep Learning and Rough Sets Approach. *Soft Computing*, Vol. 22, No. 5, 2018, pp. 1457–1466.
24. Garib, A., A. E. Radwan, and H. Al-Deek. Estimating Magnitude and Duration of Incident Delays. *Journal of Transportation Engineering*, Vol. 123, No. 6, 1997, pp. 459–466.
25. Simoncic, M. A Bayesian Network Model of Two-car Accidents. *Journal of Transportation and Statistics*, Vol. 7, No. 2/3, 2004, pp. 13–25.
26. Gyftodimos, E., and P. A. Flach. Hierarchical Bayesian Networks: An Approach to Classification and Learning for Structured Data. In *Proc., Methods and Applications of Artificial Intelligence: Third Hellenic Conference on AI, SETN 2004* (Vouros, G. A., and T. Panayiotopoulos, eds.), Samos, Greece, May 5–8, 2004, Springer, Berlin, Heidelberg, Germany, pp. 291–300.
27. de Oña, J., R. O. Mujalli, and F. J. Calvo. Analysis of Traffic Accident Injury Severity on Spanish Rural Highways Using Bayesian Networks. *Accident Analysis & Prevention*, Vol. 43, No. 1, 2011, pp. 402–411.
28. Druzdzel, M. J., and T. Sowinski. *GeNIe Modeler User Manual*. Pittsburgh, PA, 2017.
29. Cooper, G., and E. Herskovits. A Bayesian Method for the Induction of Probabilistic Networks from Data. *Machine Learning*, Vol. 9, No. 4, 1992, pp. 309–347.
30. Chen, C., Q. Wu, G. Zhang, X. C. Liu, and P. D. Prevedouros. Extracting Arterial Access Density Impacts on Safety Performance Based on Clustering and Computational Analysis. *Journal of Transportation Engineering, Part A: Systems*, Vol. 144, No. 4, 2018, p. 4018008.
31. Chen, C., G. Zhang, Z. Qian, R. A. Tarefder, and Z. Tian. Investigating Driver Injury Severity Patterns in Rollover Crashes Using Support Vector Machine Models. *Accident Analysis & Prevention*, Vol. 90, 2016, pp. 128–139.
32. Fawcett, T. ROC Graphs: Notes and Practical Considerations for Researchers. *Machine Learning*, 2003, pp. 1–38.
33. Chen, C., G. Zhang, J. Yang, J. C. Milton, and A. "Dely" Alcántara. An Explanatory Analysis of Driver Injury Severity in Rear-End Crashes Using a Decision table/Naïve Bayes (DTNB) Hybrid Classifier. *Accident Analysis & Prevention*, Vol. 90, 2016, pp. 95–107.
34. Tape, T. G. Interpretation of Diagnostic Tests. *Annal of Interval Medicine*, Vol. 135, No. 1, 2001, p. 72.
35. Provost, F., and P. Domingos. *Well-Trained PETs: Improving Probability Estimation Trees*. Technical Report CeDER #IS-00-04. New York University, New York, NY, 2001.

The Standing Committee on Artificial Intelligence and Advanced Computing Applications (ABJ70) peer-reviewed this paper (18-04305).

Article

TRR
JOURNAL OF THE TRANSPORTATION RESEARCH BOARD

Transportation Research Record
2018, Vol. 2672(45) 210–221

DOI: 10.1177/0361198118796963
journals.sagepub.com/home/trr
SAGE

Classification of Distracted Driving Based on Visual Features and Behavior Data using a Random Forest Method

Ying Yao[1], Xiaohua Zhao[1], Hongji Du[1], Yunlong Zhang[2], and Jian Rong[1]

Abstract

This research is to explore the relationship between a driver's visual features and driving behaviors of distracted driving, and a random forest (RF) method is developed to classify driving behaviors and improve the accuracy of detecting distracted driving. Drivers were required to complete four distraction tasks while they followed simulated vehicles in the experiment. In data analysis, the features of distracted driving behaviors are first described, and the visual data are classified into three distraction levels based on the AttenD algorithm. Based on the collected data, this paper shows the relationship between visual features and driving behavior. Significant differences are discovered between different distraction tasks and distraction levels. Additionally, driving behavior data is used to build an RF model to classify distracted driving into three levels. Results demonstrate that this model is feasible to capture the classification of distraction and its accuracy for each distraction task is over 90%. Areas under receiver operating characteristic curve calculated through error-correcting output codes are mainly around 0.9, indicating good reliability. With this classification method, distraction levels could be classified with vehicle operation characteristics. The model established by this method could detect distractions in actual driving through the detection of driving behavior without the need of eye tracking systems.

With the ever-increasing use of electronic devices, distracted driving has become more common in recent years. Distracted driving is one of the primary reasons of traffic crashes. According to National Highway Traffic Safety Administration (NHTSA), 3196 fatal crashes (10% of all fatal crashes) that occurred on U.S. roadways involved distraction in 2015 (*1*). Additionally, in 2014 there were about 3.10 million traffic crashes (47.22%) caused by distracted driving in China (*2*). As a result, reducing distracted driving behaviors has become one of the most important tasks regarding transportation safety.

Presently, many countries are fully aware of the danger of distracted driving and have implemented relevant countermeasures. First, in terms of driving behavior, more than 50 countries, including China, have developed laws to ban the use of cell phones while driving. Second, from the aspect of environmental strategies, Stutts et al. (*3*) introduced two most effective measures – installing road shoulders or centerline rumble strips and improving access to stopping and resting areas. Third, with regard to the vehicular countermeasures, NHTSA (*4*) established visual–manual guidelines and issued detailed regulations of in-vehicle electronic devices. Meanwhile, Toyota and some other vehicle manufacturers have developed driver assistance systems (*5*). For example, if the glance-monitoring system recognizes that the driver was not looking ahead when the radar detects a potential crash, the driver will receive a warning.

There have been many studies regarding the effects of distracted driving in terms of driving behaviors and visual features. Horrey et al. (*6*) conducted a meta-analysis, which found that phone use was associated with definite costs to driving performance, but these costs were to measures of response time and not to the measures of lane-keeping or tracking performance. Wang et al. (*7*) conducted an experiment, which indicated that there was longer task completion time, higher probability of driving in a wrong route, larger lane departure standard deviation, and shorter vehicle overtaking distance with

[1]Beijing Engineering Research Center of Urban Transport Operation Guarantee, College of Metropolitan Transportation, Beijing University of Technology, Beijing, P.R. China
[2]Zachry Department of Civil Engineering, Texas A&M University, College Station, TX

Corresponding Author:
Address correspondence to Xiaohua Zhao: zhaoxiaohua@bjut.edu.cn

mobile phone use while driving. Victor et al. (*8*) conducted a study on driver's visual features of performing in-vehicle tasks while driving, and the results revealed that as the visual task became more difficult, drivers looked less at the road central area ahead and looked at the display more often, for longer periods, and for more varied durations.

Detection of distracted driving has always been one of the focuses in safe driving. The detection is generally performed through driver and/or vehicle operation data and visual data. Classification is the key issue in the process of distraction detection. Based on driver and/or vehicle operation data, Kutila et al. (*9*) used support vector machine (SVM) to detect driver distraction with data captured under real traffic conditions, resulting in an accuracy of 65–80%. Yang et al. (*10*) captured the sequence of driving characteristics to assign the driving behavior to either normal or distraction category, resulting in an accuracy above 70%, on the basis of Gaussian Mixture model (GMM). Based on visual data, Kircher developed an AttenD algorithm to detect distraction using eye-based indicators like time off the road and/or glance frequency (*11*). However, both ways to detect distracted driving have their own disadvantages. For driver and/or vehicle operation data, the detection accuracy of 60–80% is not so high compared to past studies. For visual detection, although it has higher accuracy, the equipment cost is too high, and the installation and calibration are difficult; additionally, the system is susceptible to the influence of complex light.

The objective of this paper is: 1) to explore the relationship between drivers' visual features and driving behaviors of distracted driving; 2) to develop a random forest (RF) method to classify driving behaviors and improve the accuracy of detecting distracted driving. The RF method is established based on driving behavior, and the classification criterion is developed with the analysis of accompanying visual features. By conducting an experiment in driving simulators, three distraction levels based on visual features are defined. Distraction levels are further integrated with drivers' operating data to predict and classify distraction levels of drivers, and the accuracy of the model in different distraction tasks is compared. Based on the methodology established in this paper, the model will detect distracted driving at a higher accuracy without the use of an eye tracking system, since all the model inputs are driver behavior data.

Methods

There are three main methods mentioned in this paper. The process of classification will be introduced later.

AttenD Algorithm

AttenD algorithm is a real-time visual distraction detection algorithm developed by Kircher et al. (*12*). The algorithm takes both glance duration and repeated glances into account. The general idea of the AttenD algorithm is that the driver has a buffer time up to two seconds, which gets depleted in real time when the driver looks away from the field relevant for driving (FRD). When the gaze direction is redirected towards the FRD again, the buffer starts filling up after a latency period of 0.1 s. When the driver glances at the mirrors or the speedometer, the buffer starts decreasing after a latency period of 1 s. Finally, when the buffer becomes 0 s, it is defined as distracted driving. An example of expression of the AttenD algorithm is shown in Figure 3*a*.

This method is used in this paper to identify distracted driving (see the Feature Description section).

Random Forest

The RF method is an effective tool in prediction and classification (*13*). RF classifier provides superior accuracy and overcomes the problem of over-fitting of decision tree. Also, RF can process data that has multiple features and its operation speed is faster than many other classification methods, making it capable of instant detection of distractions in real-time applications. Several other classification methods were investigated for classifying driving behavior. Logistic regression is generally used to solve linear classification problems which is not suitable for this problem. Linear and nonlinear classifications could be solved with SVM, but SVM was insensitive to the dimensions and the number of training sets, and its efficiency is relatively low when dealing with large data sets. Because of the large amount of high-dimensional driving behavior data, RF method was selected.

The RF method is a way of averaging multiple deep-decision trees, trained on different parts of the same training set, with the goal of reducing the variance. This comes at the expense of a small increase in the bias and some loss of interpretability, but generally greatly boosts the performance in the final model. The training algorithm for RF applies the general technique of bootstrap aggregating, or bagging, to tree learners. For each tree, training data is sampled with replacement, and the out-of-bag (OOB) sample is about 37%. Besides, for nodes of one tree, the feature is sampled without replacement according to proportion. In each decision tree, Gini coefficient is used to choose a predictor to make a binary split on that node. After training, predictions for unseen samples can be made by taking the majority vote in the case of decision trees.

Based on the classification results of the AttenD algorithm, we can categorize driving behaviors into different distraction levels (see the Classification section).

Error-Correcting Output Codes

Multiclass learning problem is defined as an unknown function $f(x)$ whose range is a discrete set containing classes with more than 2 values. Error-correcting output codes (ECOC) can convert multiclass learning problems to binary classification problems (*14*). This algorithm can provide reliable class probability estimates. Assuming there is a classification problem of K classes; each category includes a binary encoding whose length is L, and then a codebook of $K \times L$ is formed. It is noticeable that each row is distinct, and each class will have a unique code accordingly. Therefore, classification problems of K classes are converted to L binary classification problems.

The ECOC method is used in the evaluation of classification results of distracted driving (see the Evaluation Indicators and Methods section).

Experiment Design

Subjects

A total of 34 drivers participated in this experiment. All of them owned Chinese Class C driver's licenses with an age range of 19–55 years (Mean = 33.9, SD = 13.0) and had 1–30 years of driving experience (Mean = 10.0, SD = 8.9). Subjects included 14 students/young drivers, 9 office clerks, and 11 taxi drivers. Among them, there were 23 males and 11 females. This experiment was conducted in April 2016, and female subjects accounted for a small number of the total participants in this experiment to reflect the demographic characteristics of licensed drivers in China at that time (*15*). Subjects were informed of risks and provided consent.

Apparatus

Distracted driving is dangerous driving behavior; it is difficult to conduct field experiments and to control experimental parameters. Therefore, a driving simulator is chosen to complete the distracted driving experiment.

The driving simulator experiment is performed using the fixed-base driving simulator in the Key Lab of Traffic Engineering at the Beijing University of Technology. The hardware of this simulator consists of a renovated real car, computers, and audio/video equipment. The road scenario is projected onto three big screens, providing the driver with a 130-degree wide-angle field of view. The sampling period of simulator is 8 Hz. The data collected by this simulator includes throttle and brake pedal force, acceleration, velocity, lateral lane position, etc.

The performance of the driving simulator has been well calibrated and validated in field tests in a series of previous studies. About 200 subjective questionnaires were collected from different experiments using the same driving simulator. Evaluation items include the accelerator, brake, speed perception and so on. The ratings ranged from 0 (representing "not at all similar to the real world") to 10 (representing "extremely similar to the real world"); and the ratings of all items were higher than 7.5 (*16*). Therefore, this driving simulator environment has been demonstrated to be able to represent real-world conditions realistically.

The eye tracking system in this experiment is SMI ETG^{TM}, which is a type of spectacles. This device uses three points to calibrate, its resolution is 1280 × 960 p. The sampling frequency is 60 Hz, and the tracking range is 80° horizontally and 60° in the vertical direction. The real time data collected by this eye tracker includes eye position, pupil size, and indicators of fixation, saccade, and blink.

Scenarios

The experiment selects the Jingcheng Freeway in Beijing as the experiment roadway and built virtual scenarios by 3DMax software. The roadway in the simulated scenario is an eight-lane expressway with four lanes in each direction, and the width of each lane is 4 m.

In order to control the speed of different drivers, the experiment designs the vehicle shown in Figure 1*b* as event set to detect the visual distraction of drivers. The length of experiment section is about 3 km (Figure 1*a*), and the speed limit of this road is 80 km/h. A truck is set to drive at a speed of 80 km/h in the simulator, and drivers are instructed to follow the truck. In the first 1.5 km, drivers accelerate and catch up with the forward truck. Afterwards drivers would receive distraction instructions, and they perform the task based on the instructions while they keep driving and following the truck continuously until the road ended. Figure 1*c* illustrates the event setting of this experiment.

Distraction Tasks

According to the most common visual distracted behaviors in recent years, four tasks are designed in this experiment, including sending voice and word messages through the WeChat App, tuning the radio, and using the navigation system. Phone calling is not taken into consideration in this experiment because it mostly represents as a cognitive distraction.

Drivers are required to perform the tasks at the point of distraction instruction:

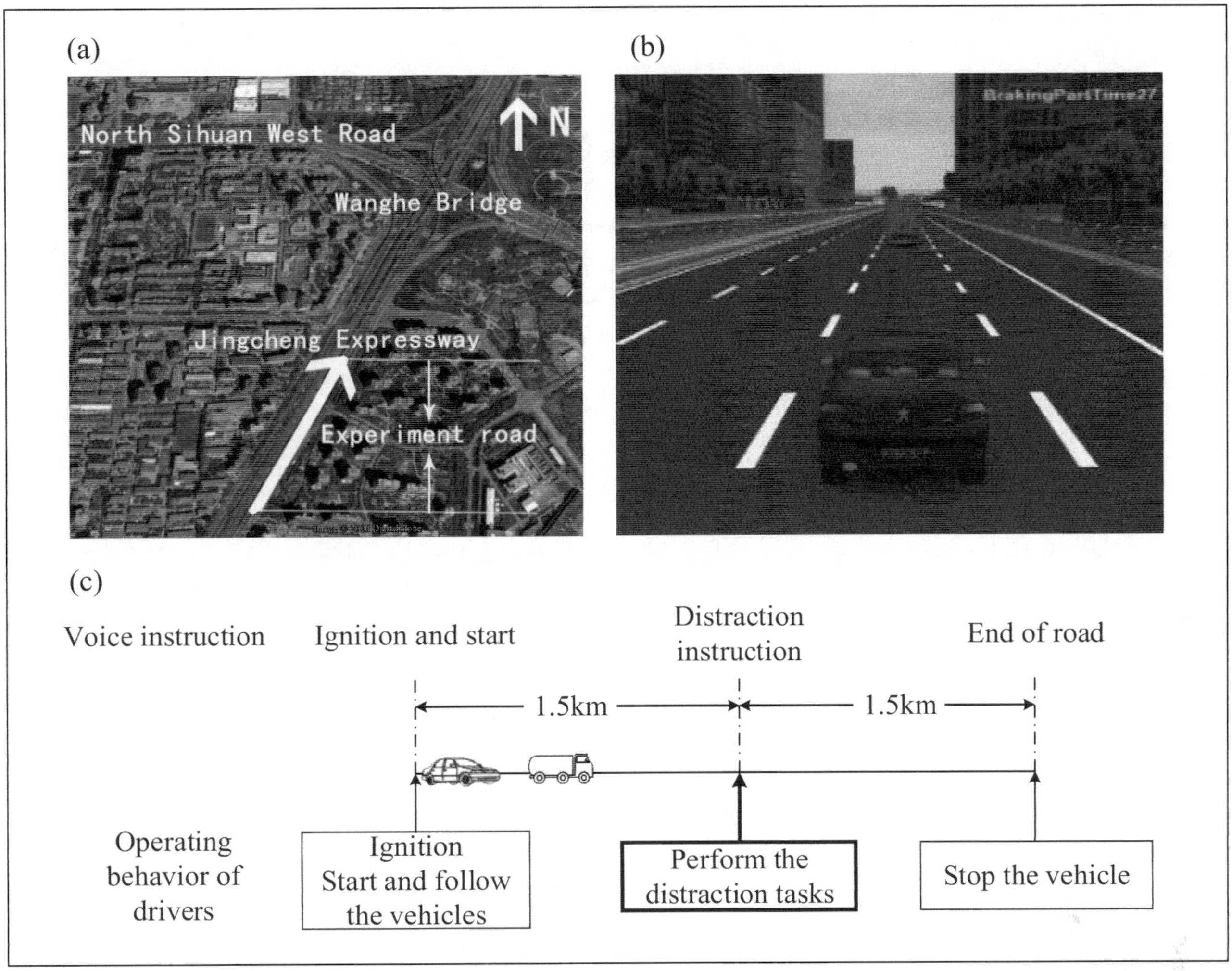

Figure 1. Design of experiment road: (*a*) road section; (*b*) virtual scenarios; (*c*) event setting.

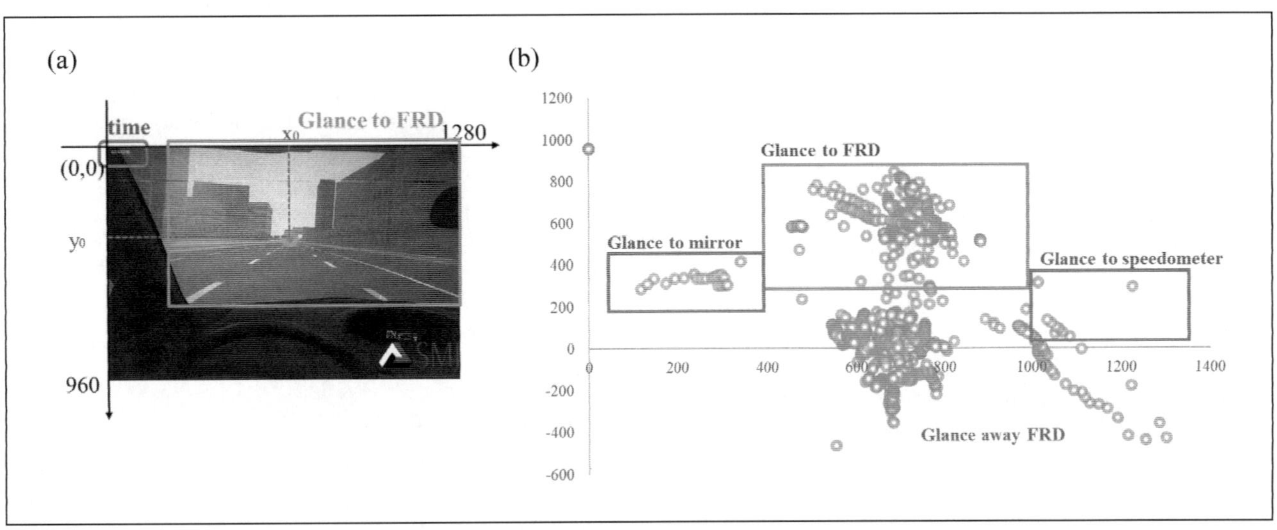

Figure 2. Partition method of visual data: (*a*) eye tracking video and corresponding coordinate; (*b*) critical areas.

Table 1. Distribution of Vehicle and Drivers' Operating Data

Distraction tasks	Speed (km/h)		Acceleration (m/s^2)	LP (m)		FD (m)	
	Mean	SD	SD	Mean	SD	Mean	SD
Normal driving	71.22	6.37	0.25	0.39	0.40	57.65	35.65
Voice message	64.22	8.94	0.45	0.46	0.46	86.06	59.10
Word message	67.84	9.42	0.42	0.44	0.42	98.42	75.96
Tuning the radio	63.69	8.13	0.49	0.42	0.42	71.41	46.56
Navigation	63.42	8.75	0.48	0.43	0.43	80.79	52.08

- For the tasks of voice and word message, operators send messages to driver's mobile phone at the point. Then drivers have to reply to this message following the instructions. The cell phone used in the experiment is a normal smartphone with an Android operating system.
- The task of tuning the radio requires drivers to tune the radio to a certain channel. The radio of this simulator is a push-button tuner. One push would lead to an increase of 0.05 MHz, and the initial frequency of the radio is 87.5 MHz.
- For the task of navigation, drivers would drive in a normal state while there is a pad playing the recorded videos in which the content is the real-world navigation of the experiment section. Drivers would receive some voice navigation instructions at the points of distraction instructions. This experiment is only concerned about driver's behavior when drivers receive voice navigation.

The prompts drivers received at the point of distraction instruction are as follows:

- Voice message: Please reply with this message in voice: I'm driving to Beijing University of Technology.
- Word message: Please reply by typing in today's date in four-digit.
- Tuning the radio: Please turn on the radio and tune to 87.9 MHz, then turn off the radio.
- Navigation: There is a High-Speed Camera ahead, speed limit is 80 km/h, and the distance to the camera is 500 m.

The scenarios are divided into five types, including four kinds of distraction scenarios and one control group with no distraction instructions. The experiment scenarios are counterbalanced for each driver.

Procedures

The basic process of the experiment is as follows:

- First, participants are provided five minutes to drive in a practice training scenario to become familiar with driving simulator.
- Basic information about the drivers is collected, including age, occupation, driving experience, annual mileage traveled, etc.
- Formal experiment begins. Each participant drives five scenarios with the same road, and each scenario is driven once. Four distraction tasks are assigned in four scenarios respectively. The order of the scenario for each participant is random.
- Participants fill out the subjective questionnaire which includes the simulator experience, dizziness level, etc.

Data Description

Data Preprocessing

Firstly, SMI tool as visual data analysis software is used to output videos including fixation point and corresponding data in text file. Secondly, according to the video content, the times of starting distraction instruction and completing the tasks are recorded. On the basis of recording time, corresponding distraction data are extracted from the text data and coordinate distribution is observed; data are accordingly divided into glance to FRD, glance away from FRD, and glance to speedometer/mirror (see Figure 2). Finally, some invalid data are removed, and the dropout rate is about 0.1%–0.7%.

In order to intercept and analyze visual data based on the AttenD algorithm, some data processing tools are explored. Researchers define the glance ranges of each driver, and finally implement the calculation of the AttenD algorithm.

Feature Description

Table 1 presents vehicle and drivers' operating behaviors under different distraction tasks. SD means the standard deviation of the data. When drivers are distracted, they would slow down and the following distance (FD) would

Table 2. Significance Testing of Driving Behaviors in Different Tasks

	Speed		Acceleration		LP		FD	
Sample1 - Sample2	Test statistic	Sig.	Test statistic	Sig.	Test statistic	Sig.	Test statistic	Sig.
Voice message - word message	−5656.5	.000	−1098.0	.000	−307.0	1.000	−896.7	.000
Voice message - tuning the radio	−1436.6	.000	−4637.7	.000	−1560.0	.000	−3331.3	.000
Voice message - navigation	−1460.6	.000	−1963.1	.000	−1188.7	.000	−688.64	.000
Word message - tuning the radio	−7093.1	.000	−5735.8	.000	−1867.1	.000	−4228.0	.000
Word message - navigation	−7117.1	.000	−3061.2	.000	−1495.8	.000	−1585.4	.000
Tuning the radio - navigation	−24.0	.000	2674.6	.000	371.2	1.000	2642.6	.000

increase, and the fluctuation of speed and acceleration would become larger. Thus, it is predicted that distracted driving would noticeably decrease the efficiency of the road. Distraction also causes problems such as driving skills declining and attention decreasing, manifest in the wider lane position (LP) and higher SD of LP.

Because of the unequal variances of the data, Kruskal–Wallis test is used to test if these indicators have significant differences between the overall distributions. Results reflect that there are significant differences in all indicators between different tasks. Table 2 describes the results of pairwise comparison based on different indicators of driving behaviors in various distraction tasks. Except for two cases, almost all indicators are significantly different between the pair groups, which means that different tasks could be separated to categorize the levels of distraction.

Visual characteristics of all drivers are categorized in real time based on the AttenD algorithm. There are three categories: distraction, semi-distraction, and non-distraction. "Distraction" (Dis) refers to the distraction defined by the AttenD algorithm, namely the attention buffer has decreased to 0 s. "Semi-distraction" (Semi) is represented as incomplete distraction, which means that the attention buffer is between 0 and 2 s. "Non-distraction" (Non) is regarded as normal driving, which means that the attention buffer is 2 s. Figure 3*a* depicts criterion of classification of three distraction categories, and Figure 3*b* shows a classification example of a driver sending voice messages while driving.

Based on this classification scheme, the vehicle and drivers' behaviors under different distraction levels can be described. Figure 4 illustrates the distributions of selected driving behaviors at three distraction levels.

Figure 4*a* illustrates driving speed at three distraction levels (test statistic = 4191.577, $p < 0.001$), which implies that drivers typically decelerate when they are visually distracted.

Drivers' accelerations at three distraction levels are presented in Figure 4*b* and Figure 4*c* (test statistic = 2077.868, $p < 0.001$). When drivers are distracted, they would be influenced to limit their acceleration tendency and conduct more brake pedal operations. Drivers' tension mainly manifests with the SD of the acceleration. Results reveal that drivers would feel more nervous when they are distracted since the SD of the acceleration becomes larger.

Drivers' LP at three distraction levels are shown in Figure 4*d* (test statistic = 884.062, $p < 0.001$). LP denotes the distance between the vehicle and the center line of the lane. Drivers tend to deviate from the center line when they are distracted. Lane position standard deviation (LPSD) is a primary indicator of driving impairment. Semi-distraction produces more severe impairments to drivers, which means that the start and the end of distraction impair the driving behavior more seriously.

Figure 4*e* presents drivers' FD at three distraction levels. Distracted driving leads to a longer FD, likely reducing the traffic efficiency or capacity of the road (test statistic = 112.050, $p < 0.001$).

Figure 4 shows that when distracted drivers reduce their speeds both longitudinal control and lateral control of the vehicles are impaired, whereas semi-distraction produces more severe lateral impairments to drivers. Meanwhile, there are significant differences between five distraction tasks and three distraction levels in driving behavior indicators including speed, acceleration, LP and FD ($p < 0.05$). For example, the impact on the distribution of negative acceleration (Figure 4*c*) may not be as pronounced as the impacts to the other indicators. These results imply that these driving behavior parameters could be used to separate different distraction levels.

Results and Discussion

Classification

RF is a fast and effective method for classification that corrects the decision trees' habit of over fitting to their training set. The goal of RF method in this paper is to create a trained model of driving behaviors which classifies distraction into different levels, including

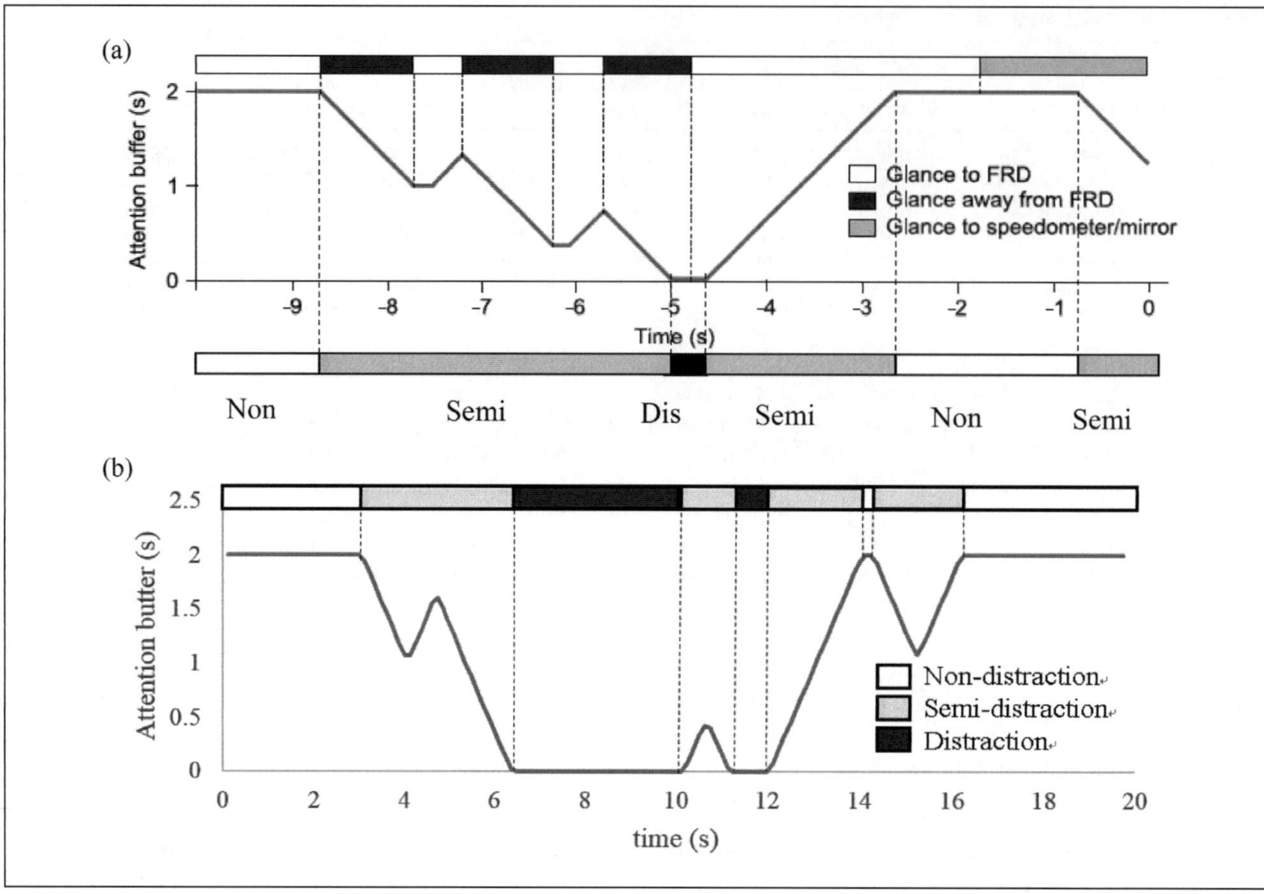

Figure 3. Description of distraction classification: (*a*) three distraction categories; (*b*) driver sending voice messages while driving.

distraction, semi-distraction and non-distraction. According to the classification results of visual features in this paper, the times corresponding to driving behaviors are identified. The integral data set is at a frequency of 8 times per second and consists of all the driver behavior data of each distraction task and its corresponding distraction level. Moreover, half of data set is selected randomly as training data, the other 25% as validation data (not including training data), and the rest as testing data to predict the distraction levels, and finally the verification is conducted. The process of classification is depicted in Figure 5.

The variation of OOB error along with the changes in the number of decision trees is introduced in Figure 6*a*. It is converged when there are 20 trees in the RF for each distraction task. Results demonstrate that the OOB errors in this classification of tuning the radio are larger, and distraction detection of voice message sending, word message sending, and navigation are more precise.

Three-fold cross-validation method is used to examine if the model has an overfitting issue (*17*). Figure 6*b* shows the average accuracy of the model as the number of trees increases from 1 to 100. Results show that with the increase of the number of trees, the average accuracy tends to be stable. It illustrates that this model with 50 trees did not over fit.

Evaluation Indicators and Methods

In terms of evaluating the accuracy of RF model, there are three commonly used indicators, including accuracy, kappa statistic, and the area under receiver operating characteristic (ROC) curve.

The classification performance evaluation is carried out for the classified results by the confusion matrix (*18*). *TN* is the number of correct predictions that an instance is negative, *FP* is the number of incorrect predictions that an instance is positive, *FN* is the number of incorrect predictions that an instance is negative, and *TP* is the

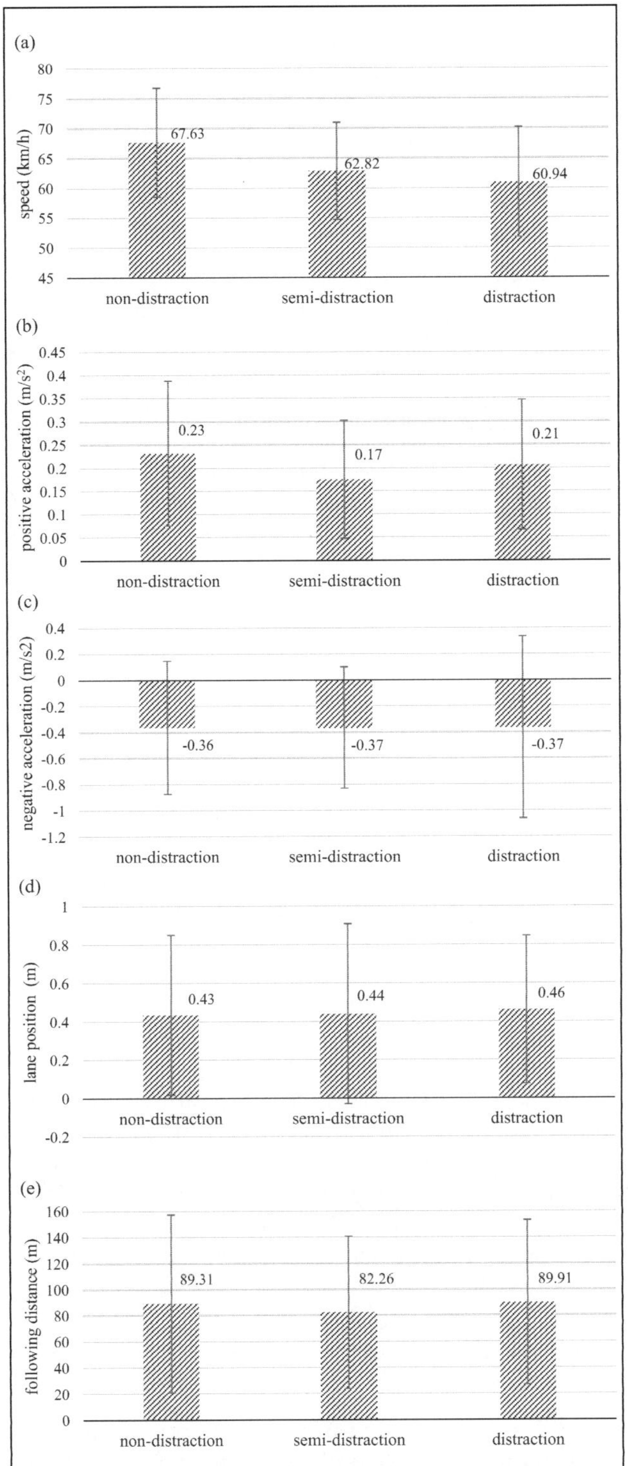

Figure 4. Distribution of driving behavior at three distraction levels for (*a*) speed, (*b*) positive acceleration, (*c*) negative acceleration, (*d*) LP, and (*e*) FD.

number of correct predictions that an instance is positive. The accuracy of a classifier is estimated by:

$$\text{accuracy} = \frac{\text{TP} + \text{TN}}{\text{TP} + \text{TN} + \text{FP} + \text{FN}} \tag{1}$$

However, in the process of evaluation, occasional inconsistency may happen. The accuracy of the whole test cannot exclude this inconsistency. Therefore, based on the confusion matrix, the kappa statistic is used to evaluate the consistency between predicted labels and test labels. The kappa coefficient (K) measures pairwise agreement among a set of coders making category judgments, thus correcting for expected chance agreement (*19*):

$$K = \frac{P(A) + P(E)}{1 - P(E)} \tag{2}$$

where $P(A)$ is the proportion of times that the coders agree and $P(E)$ is the proportion of times expected to authorize by chance.

The area under a ROC curve (AUC) is also an important evaluation indicator. AUC is statistically consistent and more discriminating than accuracy. AUC has increased sensitivity in Analysis of Variance tests, which is independent of the decision threshold, and is invariant to prior class probability distributions (*20*).

In order to use AUC to evaluate, it is necessary to convert multiclass learning problems into binary classification problems through the ECOC method. According to complete codes, the transition matrix is established. Table 3 illustrates the transition of three types of distraction to three binary classifiers *f0*, *f1*, and *f2*.

The binary classification results are applied to four kinds of distraction tasks, so the three modes of ROC curves can be obtained. When AUC is closer to 1, the model is more accurate.

Evaluation Results

Figure 7 presents the ROC curves under different distraction tasks. Results describe that each ROC curve of each task is close to 1, suggesting that the proposed method has a high identification accuracy.

The AUC value is the mean value of three binary classifiers (*21*). Word messages, voice messages, tuning the radio, navigation are represented by x_1, x_2, x_3, x_4 respectively. The evaluation results of RF model are:

- Accuracy: x_1 (94.79%), x_2 (94.04%), x_3 (93.46%), x_4 (93.72%)

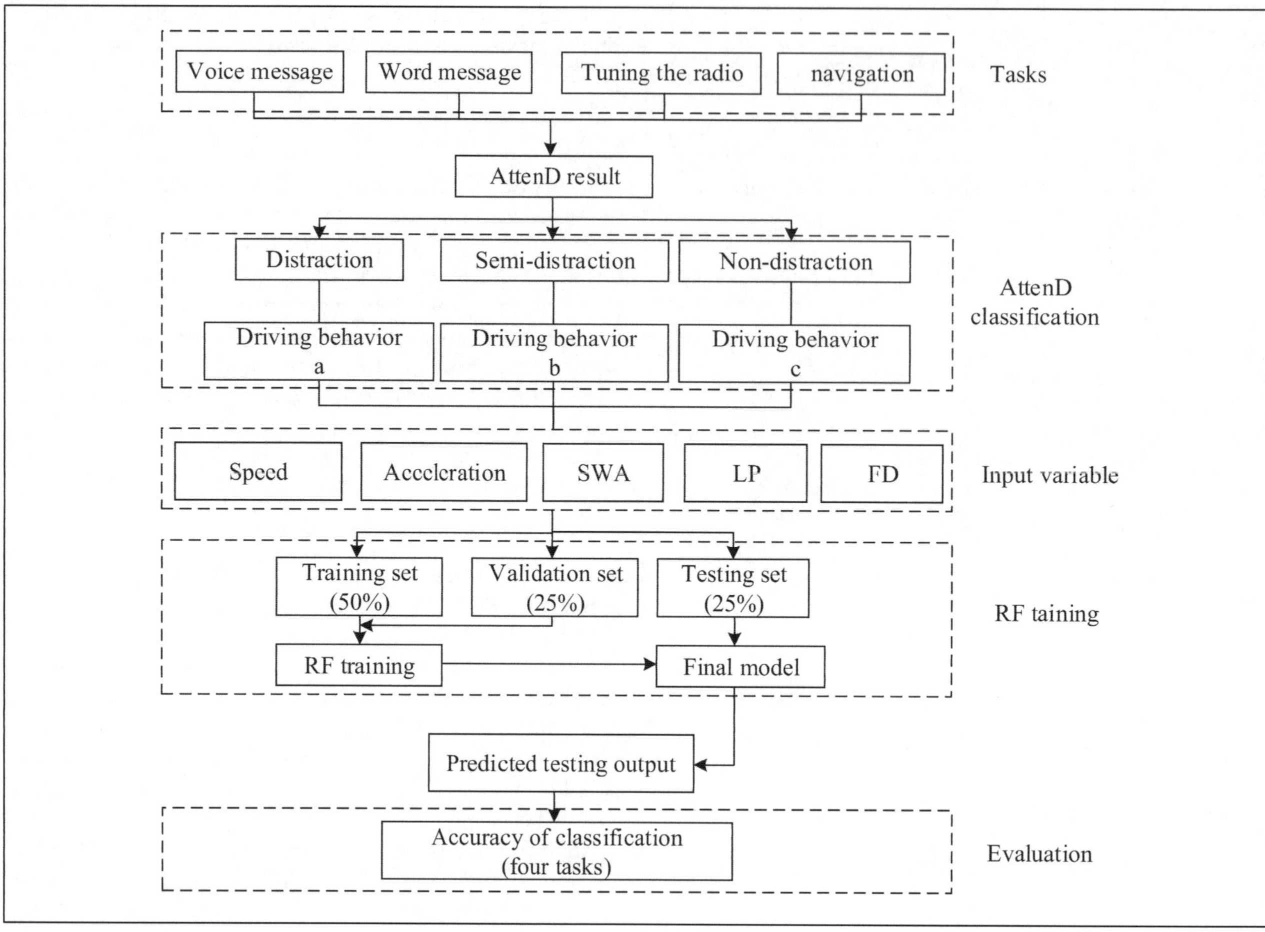

Figure 5. RF training process.

Table 3. Code Transition Matrix

	f0	*f1*	*f2*
Non-distraction	1	1	1
Semi-distraction	0	1	0
Distraction	0	0	1

- Kappa coefficient: x_1 (0.900), x_2 (0.887), x_3 (0.892), x_4 (0.837)
- AUC: x_1 (0.9417), x_2 (0.9364), x_3 (0.9436), x_4 (0.8962)

Distraction classification of each task has good reliability, since the Kappa coefficient is above 0.8, and its AUC is mainly greater than or border to 0.9. Based on the time of glancing away from the FRD, the complexity of four tasks could be sorted roughly. Sending word messages is most complex (average time = 14.52 s), sending voice messages is the second most complex (average time = 8.29 s); the next most complex is tuning the radio (average time = 6.99 s), and navigation is the simplest (average time = 3.19 s). It is obvious that when tasks are more complicated, the accuracy is higher. Simple tasks result in lower accuracy. Thus, it could be summarized that with the increasing of task complexity, the classification accuracy is higher. Therefore, this model is more applicable to seriously distracted driving.

Conclusion and Future Work

In this paper, an innovative classification method of distracted driving based on driving behavior combined with visual features is introduced. Experiments are conducted with simulated driving scenarios designed with four kinds of distraction tasks. The distraction classification method of visual data based on AttenD is presented. Relationships between drivers' visual features and driving behavior of distracted driving are explored after dividing the data into three distraction levels. Through

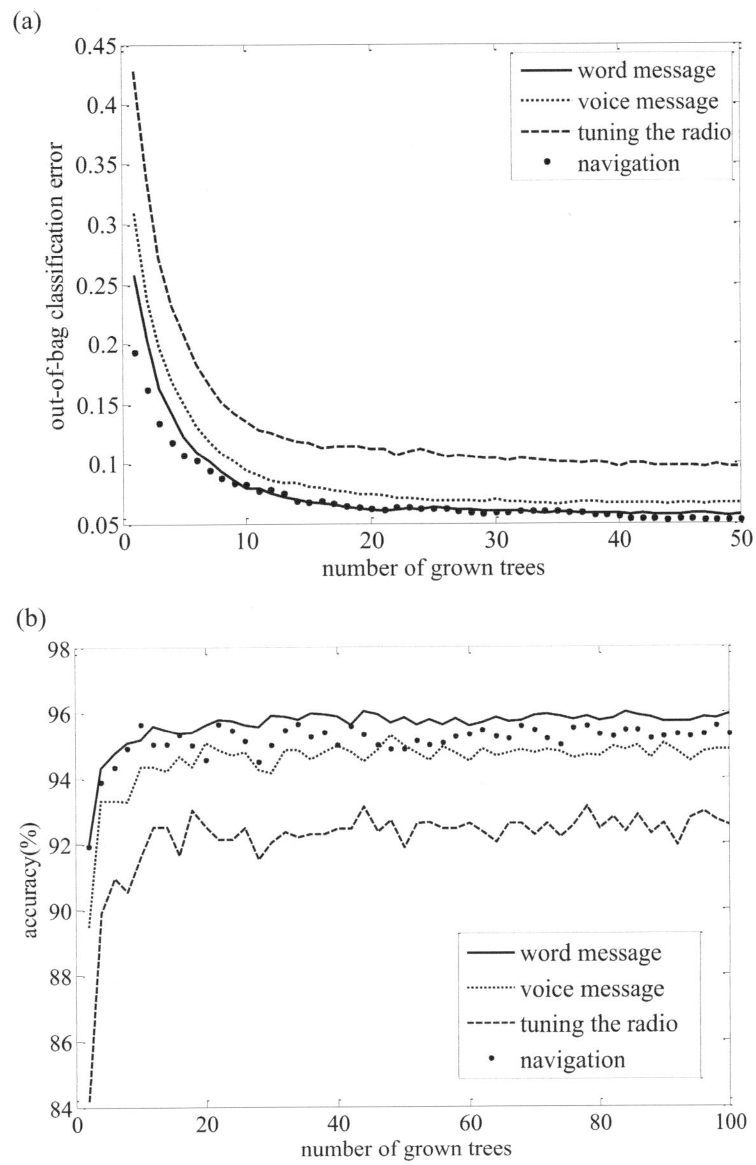

Figure 6. OOB error and cross validation.

the features described, the results demonstrate that driving behavior shows significant differences in different distraction tasks and distraction levels. A model is built by means of RF, and the accuracy of the model in each task is over 90% with high reliability (Kappa coefficient >0.8), with its AUC being mainly greater than or close to 0.9. In other words, this method can accurately predict and classify distraction levels, and the classification accuracy is higher for highly detailed or complex operations.

In this paper, both eye movement data and driving behavior data are used to build RF model. Eye movement data are used to classify distraction into different levels, while driving behavior data are calculated to explore the effects of different distraction levels on driving behavior. Eventually, the RF model connects between driving behavior characteristics and distraction levels. Therefore, the model is built based on drivers' visual features and behavior data, but can be applied in distraction detection in actual driving based on driving behavior alone.

There are other research areas for future studies. Firstly, some features obtained from driving simulator

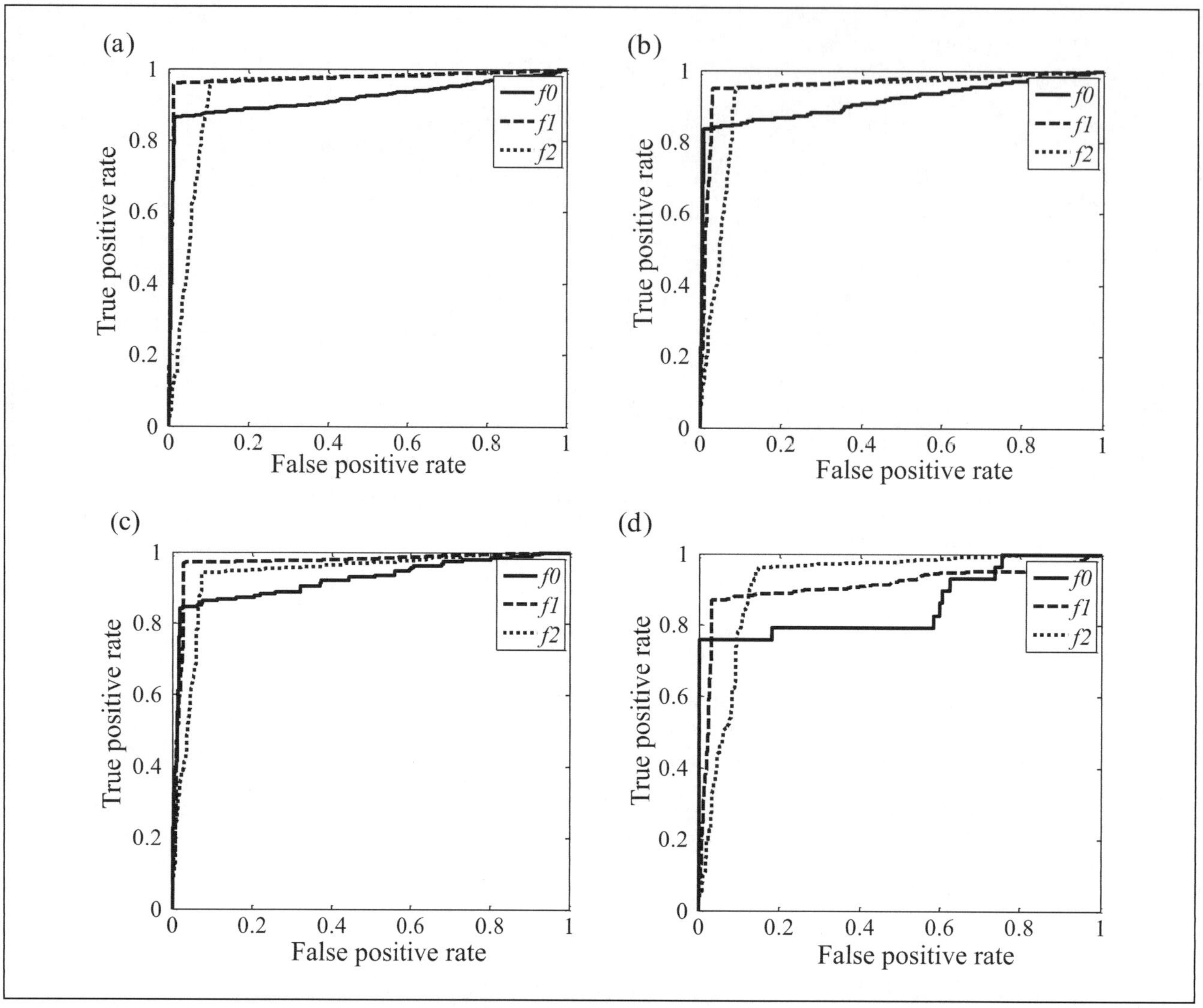

Figure 7. ROC curves of different distraction tasks: (*a*) voice message; (*b*) word message; (*c*) radio; (*d*) navigation.

may be difficult to collect in actual vehicles. Other indicators (e.g., speed, yaw angle etc.), which are easier to obtain directly from vehicles, will be considered. Nevertheless, this paper successfully connected driving behavior and visual data and established a foundation for practical applications in the future. Secondly, this research only studies four most common distraction tasks and does not include distraction tasks with high complexities, such as those involving multiple vehicles. In the future, more distraction tasks should be considered to expand the applicability of this method. In addition, more road alignment types and events can be added into virtual scenarios to develop models with higher robustness.

Acknowledgments

This work was supported by National Natural Science Foundation of China (Grant number 61672067), Open Project of Key Laboratory of Ministry of Public Security for Road Traffic Safety in China (Grant number 2016ZDSYSKFKT01), and the Beijing Engineering Research Center of Urban Transport Operation Guarantee, Beijing University of Technology.

Author Contributions

The authors confirm contribution to the paper as follows: study conception and design: YY and XZ; data collection: YY, HD, and JR; analysis and interpretation of results: YY and HD; draft manuscript preparation: YY and YZ. All authors reviewed the results and approved the final version of the manuscript.

References

1. National Highway Traffic Safety Administration. *Distracted Driving 2015*. Traffic Safety Facts Research Note,

2017. https://www.nhtsa.gov/sites/nhtsa.dot.gov/files/documents/812_381_distracteddriving2015.pdf. Accessed March 2017.
2. China Youth Daily. *Chinese Traffic Safety Day: Say "No" to Rude Driving and Distracted Driving*. http://news.xinhuanet.com/politics/2015-12/03/c_12849 279 6.htm. Accessed December 2015.
3. Stutts, J., J. Feaganes, D. Reinfurt, E. Rodgman, C. Hamlett, K. Gish, and L. Staplin. Driver's Exposure to Distractions in Their Natural Driving Environment. *Accident Analysis & Prevention*, Vol. 37, No. 6, 2005 pp. 1093–1101.
4. National Highway Traffic Safety Administration. *Visual-Manual NHTSA Driver Distraction Guidelines for In-Vehicle Electronic Devices*. National Highway Traffic Safety Administration, Department of Transportation, Washington, D.C., 2012.
5. Ranney, T. A. *Driver Distraction: A Review of the Current State-of-Knowledge*. No. HS-810 787. National Highway Traffic Safety Administration, Washington, D.C., 2008.
6. Horrey, W. J., and C. D. Wickens. Examining the Impact of Cell Phone Conversations on Driving using Meta-Analytic Techniques. *Human Factors the Journal of the Human Factors & Ergonomics Society*, Vol. 48, No. 1, 2006, pp. 196–205.
7. Wang, Y., C. Chen, J. Cao, and W. Zhang. Measurement of Degradation Effects of Mobile Phone Use on Driving Performance with Driving Simulation. *Logistics: The Emerging Frontiers of Transportation and Development in China*, 2009, pp. 2165–2170.
8. Victor, T. W., J. L. Harbluk, and J. A. Engström. Sensitivity of Eye-Movement Measures to In-Vehicle Task Difficulty. *Transportation Research Part F Traffic Psychology & Behaviour*, Vol. 8, No. 2, 2005, pp. 167–190.
9. Kutila, M. H., M. Jokela, T. Mäkinen, J. Viitanen, G. Markkula, and T. W. Victor. Driver Cognitive Distraction Detection: Feature Estimation and Implementation. *Proceedings of the Institution of Mechanical Engineers, Part D: Journal of Automobile Engineering*, Vol. 221, No. 9, 2007, pp. 1027–1040.
10. Yang, J., T. N. Chang, and E. Hou. Driver Distraction Detection for Vehicular Monitoring. *IECON 2010-36th Annual Conference on IEEE Industrial Electronics Society*. IEEE, 2010, pp. 108–113.
11. Lee, J. D., J. Moeckli, T. L. Brown, S. C. Roberts, C. Schwarz, L. Yekhshatyan, E. Nadler, Y. Liang, T. Victor, D. Marshall, and C. Davis. *Distraction Detection and Mitigation Through Driver Feedback*. No. DOT HS 811 547A. National Highway Traffic Safety Administration, Washington, D.C., 2013.
12. Kircher, K., and C. Ahlström. Issues Related to the Driver Distraction Detection Algorithm AttenD. *Proc., First International Conference on Driver Distraction and Inattention*. Gothenburg, Sweden, 2009.
13. Breiman, L. Random Forests. *Machine Learning*, Vol. 45, No. 1, 2001, pp. 5–32.
14. Dietterich, T. G., and G. Bakiri. Solving Multiclass Learning Problems via Error-Correcting Output Codes. *Journal of Artificial Intelligence Research*, Vol. 2, 1995, pp. 263–286.
15. Xinhuanet. *There Were 327 Million Motor Vehicle Drivers in 2015, and Women Accounted for 25.71%*. http://news.xinhuanet.com/local/2016-01/25/c_128667495.htm. Accessed January 2016.
16. Li, J., X. Zhao, S. Xu, J. Ma, and J. Rong. The Study of Driving Simulator Validation for Physiological Signal Measures. *Procedia - Social and Behavioral Sciences*, Vol. 96, 2013, pp. 2572–2583.
17. Kohavi, R. A Study of Cross-Validation and Bootstrap for Accuracy Estimation and Model Selection. *Ijcai*, Vol. 14. No. 2. 1995, pp. 1137–1143.
18. Fawcett, T. An Introduction to ROC Analysis. *Pattern Recognition Letters*, Vol. 27, No. 8, 2006, pp. 861–874.
19. Carletta, J. Assessing Agreement on Classification Tasks: The Kappa Statistic. *Computational Linguistics*, Vol. 22, No. 2, 1996, pp. 249–254.
20. Ling, C. X., J. Huang, and H. Zhang. AUC: A Statistically Consistent and More Discriminating Measure than Accuracy. *IJCAI*, Vol. 3, 2003, pp. 519–524.
21. Qin, F., H. Luo, Z. Cheng, S. Ren, and L. Chen. Evaluating Performance of Multiple Bayes Classifier Based on AUC Method. *Computer Engineering and Design*, Vol. 24, 2007, pp. 5919–5920.

The Standing Committee on Artificial Intelligence and Advanced Computing Applications (ABJ70) peer-reviewed this paper (18-04533).

TRR
JOURNAL OF THE TRANSPORTATION RESEARCH BOARD

Article

Transportation Research Record
2018, Vol. 2672(45) 222–231

DOI: 10.1177/0361198118777631
journals.sagepub.com/home/trr
SAGE

Traffic Congestion Detection from Camera Images using Deep Convolution Neural Networks

Pranamesh Chakraborty[1], Yaw Okyere Adu-Gyamfi[2], Subhadipto Poddar[1], Vesal Ahsani[1], Anuj Sharma[1], and Soumik Sarkar[3]

Abstract

Recent improvements in machine vision algorithms have led to closed-circuit television (CCTV) cameras emerging as an important data source for determining of the state of traffic congestion. In this study we used two different deep learning techniques, you only look once (YOLO) and deep convolution neural network (DCNN), to detect traffic congestion from camera images. The support vector machine (SVM), a shallow algorithm, was also used as a comparison to determine the improvements obtained using deep learning algorithms. Occupancy data from nearby radar sensors were used to label congested images in the dataset and for training the models. YOLO and DCCN achieved 91.5% and 90.2% accuracy, respectively, whereas SVM's accuracy was 85.2%. Receiver operating characteristic curves were used to determine the sensitivity of the models with regard to different camera configurations, light conditions, and so forth. Although poor camera conditions at night affected the accuracy of the models, the areas under the curve from the deep models were found to be greater than 0.9 for all conditions. This shows that the models can perform well in challenging conditions as well.

The U.S. Department of Transportation (DOT) defined traffic congestion as "one of the single largest threats" to the economic prosperity of the nation (*1*). The cost of congestion in 2014 was calculated to be $160 billion for the top 471 urban areas in the United States. This included 6.9 billion hours of wasted time and 3.1 billion gallons of wasted fuel (*2*). Undoubtedly, dissemination of real-time traffic information to road users can significantly improve the efficiency of traffic networks. Therefore, estimating real-time traffic states, and thereby detecting network anomalies such as congestion and incidents, have been of significant interest to researchers for some years.

Traditionally, traffic-state estimation is conducted using point-based sensors, including inductive loops, piezoelectric sensors, and magnetic loops (*3*). Recent advances in active infra-red/laser radar sensors have led to these devices gradually replacing the traditional point-based sensors (*4*). Also, with the increasing usage of navigation-based GPS devices, probe-based data are emerging as a cost-effective way to collect network-wide traffic data (*5*). Video monitoring and surveillance systems also are used for calculating real-time traffic data (*6*). Recent advances in image processing techniques have improved vision-based detection accuracy. Deep learning methods, such as convolution neural networks (CNNs), have been able to achieve human-level accuracy in image classification tasks (*7*). The basic advantage of these methods is that they do not require picking up hand-crafted features and can, therefore, do away with the painstaking calibration tasks needed when using camera images for traffic-state estimation (*8*).

Studies have also been performed fusing multiple sources of data for traffic state estimation (*9–11*). Van Lint and Hoogendoorn used an extended generalized Treiber-Helbing filter for fusing probe-based and sensor-based data (*9*). Choi and Chung used fuzzy regression and Bayesian pooling techniques for estimating link travel times from probe data and sensor data (*10*). Bachmann et al. investigated several multi-sensor data-fusion-based techniques to compare their ability to estimate freeway traffic speeds (*11*). State DOTs also traditionally use sensor data and probe vehicle data for

[1]Department of Civil, Construction and Environmental Engineering, Iowa State University, Ames, IA
[2]Department of Civil and Environmental Engineering, Columbia, MO
[3]Department of Mechanical Engineering, Iowa State University, Ames, IA

Corresponding Author:
Address correspondence to Pranamesh Chakraborty: pranames@iastate.edu

traffic-state estimation. However, they have also installed many roadside cameras on freeways and arterials for surveillance tasks such as incident detection. These cameras are used by traffic incident managers who can zoom, tilt, and pan the cameras according to their needs. The use of cameras for traffic-state estimation or congestion detection involves additional challenges due to frequent camera movements which can alter the default calibrations. However, algorithms should not rely on the exact placement of cameras and should be able to accurately detect traffic conditions for different placement scenarios.

In this study, we used camera images from different locations, orientations, and weather conditions to successfully detect traffic congestion. Three different models were used for congestion detection tasks. Two of these are deep neural networks: deep convolution neural networks (DCNNs) and you only look once (YOLO). Because these models require time-consuming and costly Graphical Processing Unit (GPU) training, the support vector machine (SVM), a shallow learning model, was used as a comparison to determine the advantages of using deep models.

The outline of this article is as follows: The present section provides a brief introduction and the importance of traffic congestion detection; the next section gives a review of previous work done on using cameras for traffic-state estimation; the third section gives an overview of the proposed models used for traffic-congestion determination; the fourth section provides a description of the data used in this study and the data preprocessing steps adopted for further analyses; the fifth section includes a discussion of the results obtained from the analyses; and the final section provides the conclusion and recommendations for future work.

Literature Review

During the last few decades, significant research efforts have been devoted to using closed-circuit television (CCTV) cameras to determine real-time traffic parameters such as volume, density, and speed (*12*, *13*). These methods can be broadly divided into three categories: (a) detection-based methods, (b) motion-based methods, and (c) holistic approaches.

Detection-based methods use individual video frames to identify and localize vehicles and thereby perform a counting task. Ozkurt and Camci used neural network methods to perform vehicle counting and classification tasks from video records (*6*). Kalman filter-based background estimation has also been used to estimate vehicle density (*14*). In addition, faster recurrent convolution neural networks (RCNNs) have been used for traffic density calculation (*15*). However, these were found to perform poorly for videos with low resolution and high occlusion. Recent achievements in deep learning methods in image recognition tasks have led to several such methods being used for traffic counting tasks. Adu-Gyamfi et al. used DCNNs for vehicle category classification (*16*). Oñoro-Rubio and López-Sastre used two variations of CNNs, namely counting CNN and hydra CNN, to conduct vehicle counting and predict traffic density (*17*). Recently, Zhang and Wu used both deep learning and optimization-based methods to perform vehicle counts from low frame-rate, high occlusion videos (*12*).

Several motion-based methods have been suggested in the literature to estimate traffic flow utilizing vehicle tracking information. Asmaa et al. used microscopic parameters extracted using motion detection in a video sequence (*18*). However, these methods tend to fail due to lack of motion information and low frame rates of videos; some vehicles appear only once in a video, making it difficult to estimate their trajectories.

Holistic approaches avoid the segmentation of each object. Rather, an analysis is performed on the whole image to estimate the overall traffic state. Gonclaves et al. classified traffic videos into different congestion types using spatiotemporal Gabor filters (*19*). Lempitsky and Zisserman performed a linear transformation on each pixel feature to estimate the object density in an image (*20*). This approach, however, was found to perform poorly in videos with a large perspective. Furthermore, both these methods require manual annotation of each object in the images to perform the training of the counting task.

Overall, significant studies have been conducted in the past using various deep and shallow learning models to implement vehicle counting tasks and thereby determine congestion states. In this study, we adopted the holistic approach to label an image as either congested or non-congested. We also did away with counting each vehicle to determine the congestion state. Rather, we assigned labels to the images based on nearby benchmark sensors and then conducted the classification task. The next section provides a detailed description of the methods used in our study.

Experimental Approach

Traffic congestion detection from camera images can be conducted in two broad ways. With the first approach, the input image can be fed into an object recognition model to determine the number of vehicles in the image and, when the number of vehicles exceeds a threshold, the image can be labeled as congested. With the second approach, the entire image can be classified as either congested or non-congested. In our study, we used the second approach, as it is much simpler and also does not require time-consuming manual annotation of individual vehicles.

We used three different algorithms for the traffic congestion detection task: two based on deep neural networks, which require time-consuming GPU training, and one from the shallow learning algorithm class, which does not require GPU training. The shallow algorithm was adopted primarily to determine the advantages, if any, of using GPU for this classification task. The three algorithms used in this study were:

1. Traditional Deep Convolution Neural Network (DCNN)
2. You Only Look Once (YOLO)
3. Support Vector Machine (SVM)

A detailed description of each of these algorithms is provided next.

DCCNs

Collectively, DCNNs are a state-of-the-art technique for object detection and image classification. We used a traditional ConvNet architecture consisting of convolution and pooling layers. The convolution architecture used in this study is shown in Table 1. Because images from different cameras were used in this study, the input images were of different sizes, the majority being 800 × 450 pixels. The images were then resized to 400 × 225 pixels to prevent memory allocation issues during the training of the model. Next, these images were fed into the model as two consecutive convolution layers 32 × 3 × 3 in size (i.e., kernel size 3 × 3) followed by a max pooling layer 2 × 2 in size. This was followed by two additional convolution layers 64 × 3 × 3 in size and then again max pooling with a 2 × 2 filter. Each max pooling layer was followed by dropout with a probability of 0.25 to prevent overfitting. Finally, two fully connected layers (dense) were used, the first one with 512 neurons and the final one with two neurons corresponding to the binary classes (congested and non-congested). A batch size of 32 was used throughout the model and Leaky-ReLU was used as an activation function.

DCNN models are computationally expensive and usually require millions of images to train the model to prevent overfitting. However, in our study, we had only 2,400 images, of which 1,400 were used for training. So, to prevent overfitting, along with dropout, we also used data augmentation, similar to Ahmed et al. (*21*). Here, we randomly flipped images horizontally with a probability of 0.5 and also performed horizontal and vertical shifts in the range of 10% of the total height and width. It took 25 minutes to train the model on a NVIDIA Tesla K20m GPU with 4 GB RAM memory. Keras (*22*), a deep learning library, was used to run the script in GPU.

Table 1. DCNN Model Architecture Used

Layer	Kernel	Stride	Output shape
Input			[400, 225, 3]
Convolution	3×3	1	[400, 225, 32]
Convolution	3×3	1	[398, 223, 32]
Max pooling	2×2	2	[199, 111, 32]
Dropout			[199, 111, 32]
Convolution	3×3	1	[199, 111, 64]
Convolution	3×3	1	[197, 109, 64]
Max pooling	2×2	2	[98, 54, 64]
Dropout			[98, 54, 64]
Dense			512
Dropout			512
Dense			2

YOLO

We adopted the YOLO model (*23*) for general purpose congestion detection and localization from CCTV video feeds. Current object detection systems repurpose powerful CNN classifiers to perform detection. For example, to detect an object, these systems take a classifier for that object and evaluate it at various locations and scales in the test image. YOLO reframes object detection; instead of looking at a single image 1,000 times to accomplish detection, it looks at an image only once (but in a clever way) to perform the full detection pipeline. A single convolutional network simultaneously predicts multiple bounding boxes and class probabilities for those boxes. This makes YOLO extremely fast and easy to generalize to difference scenes. YOLO is also a DCNN classifier, however in the rest of the analyses, we will denote it as YOLO and the traditional DCNN explained before as DCNN.

YOLO uses a simple CNN architecture shown in Table 2. This neural network uses only standard layer types: convolution with a 3 × 3 kernel and max pooling with a 2 × 2 kernel. The very last convolutional layer has a 1 × 1 kernel, which serves to reduce the data to the shape 13 × 13 × 125. This 13 × 13 shape is the size of the grid into which the image gets divided. There are 35 channels for every grid cell. These 35 numbers represent the data for the bounding boxes and the class predictions, as each grid cell predicts five bounding boxes and a bounding box is described by seven data elements:

- x, y, width, and height for the bounding box's rectangle;
- the confidence score; and
- the probability distribution over the two classes (congested and non-congested).

The key implementation steps for YOLO are as follows:

Table 2. YOLO Model Architecture Used

Layer	Kernel	Stride	Output shape
Input			[416, 416, 3]
Convolution	3×3	1	[416, 416, 16]
Max pooling	2×2	2	[208, 208, 16]
Convolution	3×3	1	[208, 208, 32]
Max pooling	2×2	2	[104, 104, 32]
Convolution	3×3	1	[104, 104, 64]
Max pooling	2×2	2	[52, 52, 64]
Convolution	3×3	1	[52, 52, 128]
Max pooling	2×2	2	[26, 26, 128]
Convolution	3×3	1	[26, 26, 256]
Max pooling	2×2	2	[13, 13, 256]
Convolution	3×3	1	[13, 13, 512]
Max pooling	2×2	1	[13, 13, 512]
Convolution	3×3	1	[13, 13, 1024]
Convolution	3×3	1	[13, 13, 1024]
Convolution	1×1	1	[13, 13, 35]

1. Resize the input image to 416 × 416 pixels.
2. Pass the image through a CNN in a single pass. The architecture of the CNN is described in the following section.
3. The CNN outputs a $13 \times 13 \times k$ tensor describing the bounding boxes for the grid cells. The value of k is related to the number of classes as follows: k = (number of classes + 5)*5.
4. Compute the confidence scores for all bounding boxes and reject all boxes that fall below a predefined threshold.

Because there are $13 \times 13 = 169$ grid cells and each cell predicts five bounding boxes, there are 845 bounding boxes in total. Ideally, the majority of these boxes would have very low confidence scores. In this study, a confidence threshold of 45% was used for congestion detection.

SVM

A SVM is one of the most widely used shallow algorithms for image classification tasks. It solves a constrained quadratic optimization problem to classify data into different categories. The resulting optimal hyperplane is determined by maximizing the largest minimum distance to training examples to make it least sensitive to noise. We utilized the Oriented FAST and Rotated BRIEF (ORB) (*24*) feature detector to detect the key points in each image, whereby the FAST (features from accelerated segment test) algorithm was used to extract the key points and the Harris corner distance was used to determine the top N points. The algorithm was run on the training dataset with 10-fold cross-validation to determine the optimal penalty parameter and kernel. This algorithm was run on Windows 7 with Intel Core i7-4790 CPU with 8 GB of RAM.

Data Description

Two different data sources were used in this study: camera images and radar-based Wavetronix sensors. Camera images were obtained from 121 cameras from the Iowa DOT CCTV camera database spread across the interstates and highways of Iowa. The database covered the major cities of Iowa: Des Moines, Sioux City, Cedar Rapids, Council Bluffs, Davenport, and Iowa City. Images were extracted from the cameras at five-minute intervals from October 2016 to March 2017, resulting in a total of 3.5 million images during the study period. The task of assigning a label to an image (congested or non-congested) consisted of four sub-tasks:

1. Associating each camera with a nearby wavetronix sensor pair.
2. Smoothing the wavetronix data.
3. Extracting the details (camera name, timestamp) of each image.
4. Assigning the label of the image based on sensor data.

Details of each of these tasks are discussed next.

Each camera was first associated with the two nearest Wavetronix sensor pairs covering both directions of the freeway on which the camera was placed. If the sensor pair was located more than 0.5 miles away from the camera, then the particular camera was removed from analysis.

The next step was to assign the traffic data from the sensor to each image. However, sensor data obtained from Wavetronix in 20-second intervals included too much noise; so, we used Wavelet smoothing to remove the noise. In this study, among the several families of wavelets that could be used, such as Haar, Daubechies, Biorthogonal, Symlets, Coiflets, Morlet, Mexican Hat, Meyer, and so forth, we used Daubechies extremal phase wavelets. Daubechies family wavelets are also known as "dbN," where N refers to the number of vanishing moments. The higher the value of N, the longer the wavelet filter and the smoother the wavelet. Based on our data, we used db2 with level 6 to achieve a smooth curve-like filter that followed most of the variations of the original signal. A sample of the original and smoothed data is shown in Figure 1.

The next step was to extract the details of each image. The top of each image showed the details of the image (direction, camera name, and timestamp). Optical character recognition (OCR) was used to extract the details

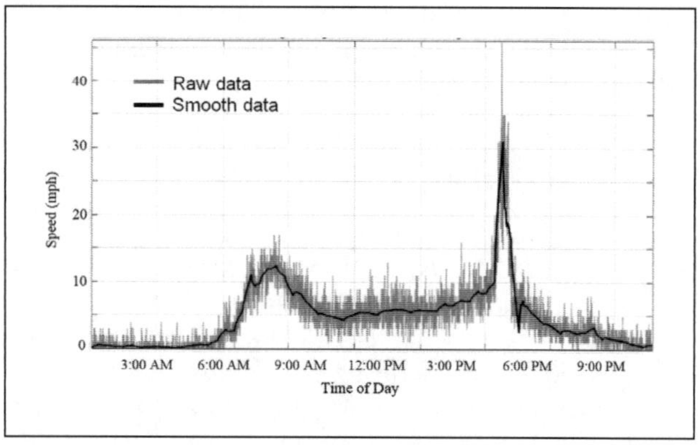

Figure 1. Original and smoothed occupancy using wavelet transform (db2 level 6).

from each image, which were then matched with the corresponding sensor data based on the camera's name and timestamp.

After obtaining the smoothed Wavetronix data, timestamp, and camera name for each image, we assigned the traffic data obtained from the sensor to the image. The traffic data comprised speed, volume, and occupancy observed at 20-second intervals. To assign the congested or non-congested label to the image, we used occupancy values, which are denoted by the percentage of the time the sensor is occupied by vehicles and have one-to-one mapping to traffic density or the number of vehicles in the unit distance. Persaud and Hall suggested that occupancy of 20% or more should be considered congested, whereas occupancy below that should be considered non-congested (*25*). Thus, if no congestion (occupancy <20%) was observed in either direction of the Wavetronix pair, then the image was classified as "non-congested"; if congestion was visible in any particular direction, or in both directions, then it was labeled as "congested." We adopted this approach to do away with manual labeling of congested and non-congested images and to follow a uniform methodology for assigning labels to the images.

Finally, we obtained 1,218 congested images and more than 3 million non-congested images. Due to class imbalance, we randomly chose 1,200 non-congested images out of the 3 million images. This dataset consisting of a total 2,418 images was then subdivided into a training set and a test set. The training set consisted of 1,400 images with equal proportions of congested and non-congested images. However, as will be discussed later, the YOLO approach of congestion detection requires manually annotating the region of congestion. For this purpose, 100 congested images were extracted from the training set and manually annotated with the congested region. The test set consisted of 1,018 images out of which 518 were congested and the rest were uncongested. Because sensor errors can occasionally cause misclassification of images, test set images were manually cross-checked and then the final labels were assigned. However, no manual cross-checking of labels was performed for the training set, as it was assumed that the algorithm itself should be able to determine the misclassifications, if any, in the training set.

Results

The performance of each of the three algorithms was trained on 1,400 images and tested on 1,018 test set images (518 congested and 500 non-congested). YOLO was trained and tested on NVIDIA GTX 1080 Ti 8 GB RAM GPU while for DCNN, NVIDIA Tesla K20m GPU with 4 GB RAM was used. Intel Core i7-4790 with 8 GB RAM CPU was used for training and testing of SVM. The training times for YOLO, DCNN and SVM were 22 hours, 26 minutes, and 50.4 seconds respectively. The testing times for the three algorithms were 0.01, 0.01, and 0.03 seconds/frame, respectively. The testing time does not include the time required for the model; rather, it includes only the time required to predict the class for each image. Because YOLO and DCNN are deep models, they had to be trained and tested using GPUs which involved time-consuming and costly training compared with its shallow counterpart, SVM. The testing times for DCNN and YOLO were lower, but they required GPU during testing time as well.

The performance of the algorithms was evaluated using the standard performance metrics of precision, recall, and accuracy (Equations 1–3, respectively). When a congested image was correctly labeled (i.e., the predicted label was also "congested"), it was classified as true positive (TP). Similarly, if a non-congested image was correctly labeled as "non-congested," then it was classified as true negative (TN). However, if the actual label was "congested" and the predicted label was "non-congested," it was classified as false negative (FN). And finally, if the actual label was "non-congested" and the predicted label was "congested," it was classified as false positive (FP).

$$Precision = \frac{TP}{TP + FP} \tag{1}$$

$$Recall = \frac{TP}{TP + FN} \tag{2}$$

$$Precision = \frac{TP + TN}{TP + FP + TN + FN} \tag{3}$$

Some examples of true classifications and misclassifications obtained from each algorithm (YOLO, DCNN, and SVM) are shown in Figure 2. Examples of true positives, for which YOLO, DCNN, and SVM correctly

Figure 2. Congestion detection classifcation examples: (*a–c*) true positives, (*d–f*) false positives, (*g–h*) false negatives, (*j–l*): true negatives from YOLO, DCNN, and SVM, respectively.

labeled congested images, are shown in Figure 2*a–c* respectively (YOLO also gives the bounding box for the congested region). On the other hand, examples of false positives, for which the algorithms misclassified non-congested images as congested, are shown in Figure 2*d–f*. It can be seen that YOLO misclassified an image as a congested region because of a group of vehicles located far away from the camera during nighttime (Figure 2*d*) and vehicles on a bridge led to misclassification by DCNN (Figure 2*e*). SVM, on the other hand, had misclassifications in adverse weather conditions (Figure 2*f*) because snow particles were detected as corners, which caused the image to be labeled as congested. Examples of false negatives, for which the algorithms failed to detect congested images correctly, are shown in Figure 2*g–i*. Congestion quite distant from the camera led to misclassification by YOLO (Figure 2*g*), whereas DCNN failed to detect congestion in a single lane when the other lane was closed and therefore empty (Figure 2*h*). Glare issues resulted in SVM misclassifications (Figure 2*i*). Finally, examples of true negatives, for which the algorithms correctly labeled non-congested images, are shown in Figure 2*j–l*.

The precision, recall, and accuracy values obtained from each algorithm are shown in Table 3. YOLO achieved the highest precision, recall, and accuracy followed closely by DCNN. Because YOLO achieved better accuracy compared with DCNN, we did not perform region-based CNN separately to determine the congested region of the results obtained from DCNN. YOLO, on the other hand, being a region-based classifier gives the congested region of the image by default (see Figure 2*a*,

Table 3. Precision, Recall, and Accuracy Values Obtained from the Three Algorithms

Method	Precision (%)	Recall (%)	Accuracy (%)
YOLO	88.6	94.3	91.4
DCNN	86.9	93.9	90.2
SVM	82.8	88.5	85.7

d). The accuracy obtained by SVM was comparatively lower (85.2%) than expected given the lower computation costs involved in such a shallow algorithm. In this context, it should be mentioned that a separate analysis, reported in a separate paper (*24*), using an ensemble of shallow learning algorithms (SVM with ORB, Shi-Tomasi, and Structured Edge Toolbox feature detector) gave an accuracy of 86.7% with the same dataset. Here, we used only SVM with ORB for comparison of deep learning models with a standard shallow model.

Sensitivity Analysis

Sensitivity analysis was also performed to determine which factors might affect the performance of the congestion detection system developed here. We evaluated two factors that could influence the classification task: first, the time of day the image was captured (daytime versus nighttime) and, second, camera resolution (blurring, rain, snow, and glare). The test database was then divided into four subgroups according to the combination of the two factors, as follows:

1. D-G: daytime, good resolution (436 images);
2. N-G: nighttime, good resolution (147 images);
3. D-P: daytime, poor resolution (190 images); and
4. N-P: nighttime, poor resolution (245 images).

Receiver operating characteristics (ROC) curves were then used to compare the performance of each algorithm for each subgroup based on the true positive and false positive rates (TPR and FPR, respectively), as defined in Equations 4 and 5, respectively:

$$TPR = \frac{TP}{TP + FN} \tag{4}$$

$$FPR = \frac{FP}{FP + TN} \tag{5}$$

For an efficient image classification model, the TPR should be higher than the corresponding FPR. On the other hand, for a poor-vision system model, when the sensitivity (TPR) increases, it loses the ability to discriminate between congested and non-congested images, which makes the TPR directly proportional to the FPR. The ROC curves for each subgroup obtained from the three models—YOLO, DCNN, and SVM—are shown in Figure 3*a–c*, respectively. The overall ROC curve for the three algorithms is shown in Figure 3*d*. Also, the area under each curve (AUC) is provided for each case. For all three algorithms, TPRs were higher than the corresponding FPRs irrespective of the prevailing conditions (daytime or nighttime, poor or good resolution). All of the algorithms performed well during the daytime, irrespective of the camera resolution. However, AUCs were found to be lowest for poor resolution images at night (N-P). Moreover, irrespective of the conditions, the AUCs from all algorithms for each subgroup were found to be mostly higher than 0.90, except for N-P conditions from SVM. This shows that the system works well even under challenging conditions.

In addition, ROC curves can be used by traffic management centers (TMCs) to choose an optimal threshold between TPR and FPR. Previous studies have shown that too many false calls is a major reason for limited integration of automatic incident detection algorithms in the TMC. Therefore, it is important for TMC personnel to know the accuracy that can be achieved given a particular FPR. For example, if a TMC wants to restrict the FPR to lower than 0.1, then the TPR obtained by YOLO, DCNN, and SVM will be 0.92, 0.96, and 0.82, respectively, during good daytime (D-G) conditions. Obviously, the accuracy would be lower with poor camera conditions at night. TMC personnel can, therefore, use the ROC curves to set the optimal threshold of TPR and FPR based on their specific needs.

Real-Time Implementation

In addition, the congestion detection algorithms can easily be implemented online. With a test time of 0.01 seconds per image, the algorithms can be adopted to detect traffic congestion of approximately 1,000 cameras in every 10-second interval using a single GPU. Figure 4 shows an example of congestion detection by DCNN algorithm on images extracted from a camera on a single day (27th October, 2017) at every 10-second interval. The congestion alarm occurrences from camera is shown in the background of Figure 4 along with the occupancy data obtained from nearest radar sensors (both directions). However, due to sensor issues, sensor data were missing from 8:51 a.m. to 12:57 p.m. So, a two-vehicle crash reported at around 10:30 a.m. was missed by sensor, but was detected successfully by the camera. Thus, this example also shows that using multiple data sources (cameras, sensors, etc.) can increase the reliability in traffic-state estimation. Callouts (i) and (ii) are also provided in Figure 4 to show samples of camera images when congestion alarms were triggered. To eliminate false alarms, alarms are triggered only when congestion

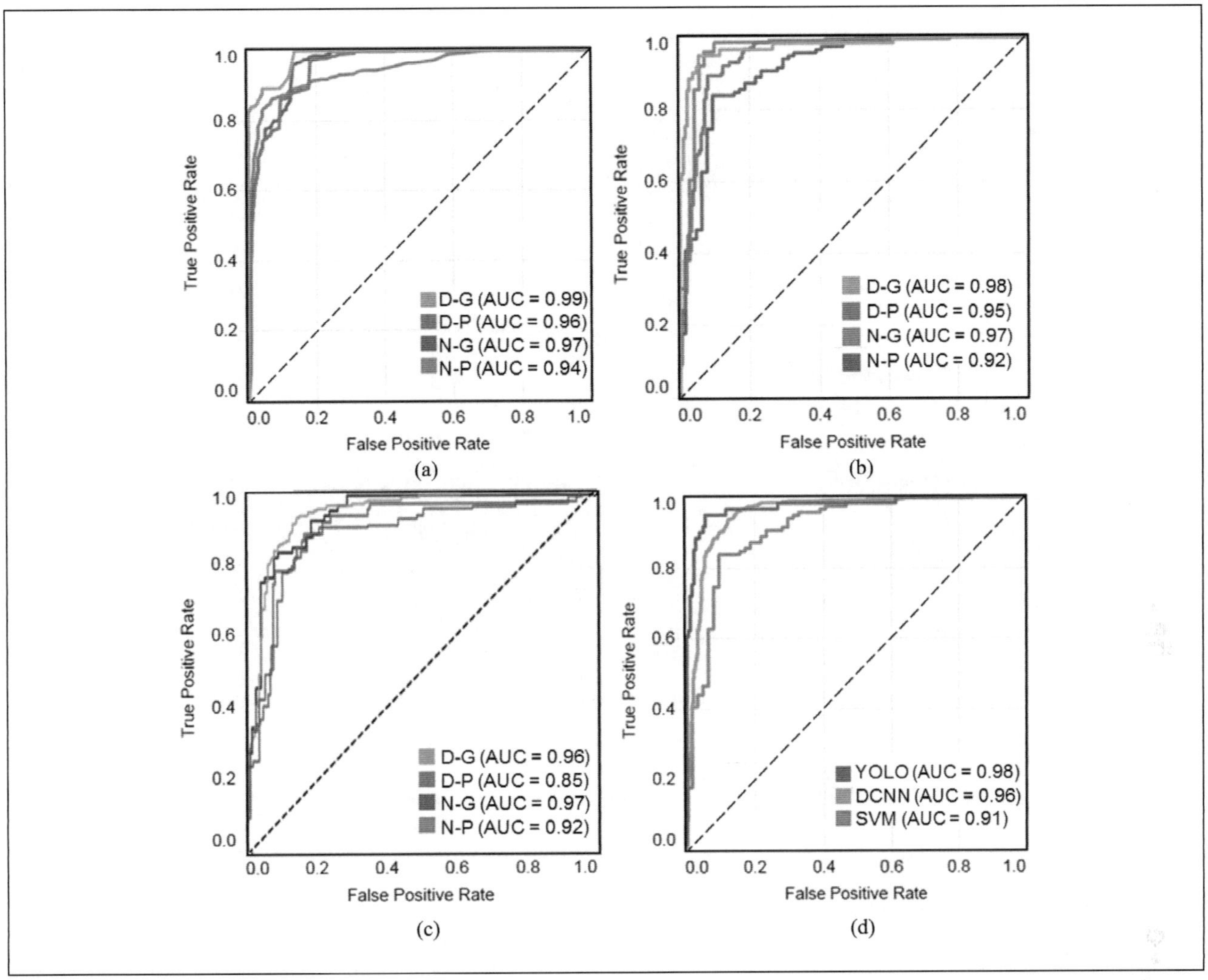

Figure 3. ROC curves under different prevalent conditions obtained from (*a*) YOLO, (*b*) DCNN, (*c*) SVM, and (*d*) all conditions combined for each algorithm.

is detected on three consecutive 10-second interval frames (persistency test). Also, alarms triggered within five-minute intervals are combined together to form a single continuous alarm. These "signal smoothing" techniques help in decreasing false alarm rates (FAR) and increasing detection rates (DR). Future studies can be done implementing better smoothing techniques like Fourier Transforms or Wavelet smoothing and determining the DR and FAR on a network of cameras.

Conclusion

Recent advances in machine-vision algorithms and high-performance computing have greatly improved image classification accuracy. In this study, two such deep learning techniques, the traditional DCNN and YOLO models, were used to detect traffic congestion from camera images. SVM also was used for comparison and to determine what improvements were obtained while using costly GPU techniques. To eliminate the time-consuming task of manual labeling and to maintain uniformity in congestion labeling, we used nearby Wavetronix sensors to correctly identify congested images. For testing purposes, we also labeled each image manually to remove misclassifications due to sensor errors.

The YOLO model achieved the highest accuracy of 91.2% followed by DCNN with an accuracy of 90.2%, while 85% of images were correctly classified by SVM. Congestion regions located far away from the camera, single-lane blockages, and glare issues were found to affect the accuracy of the models. To determine the sensitivity of the models to different camera configurations and light conditions, ROC curves were used. All the

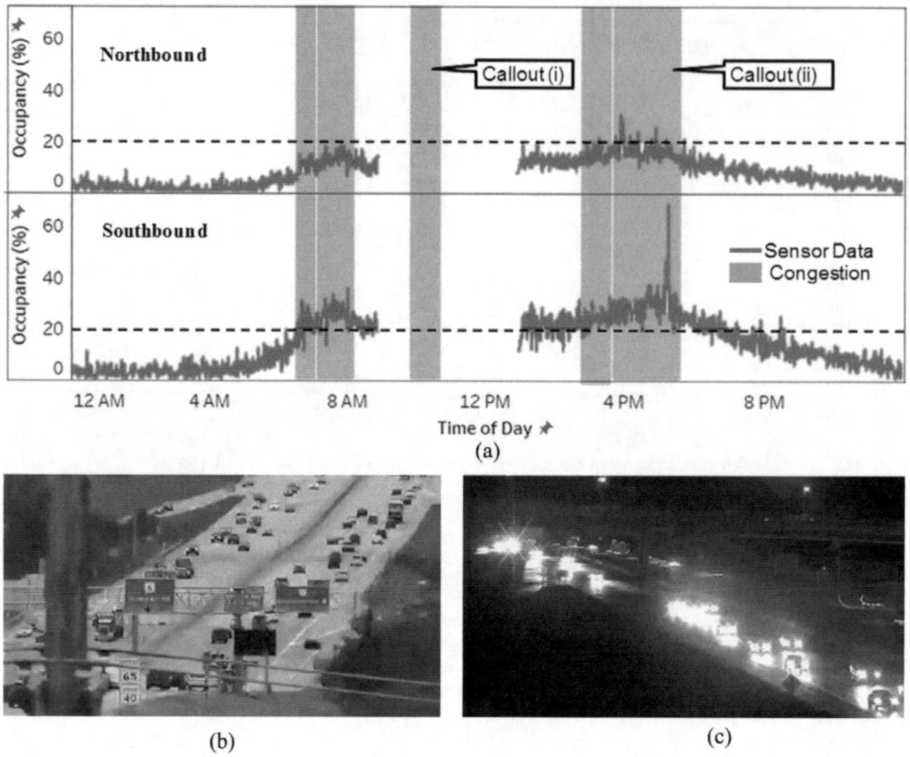

Figure 4. (*a*) Sensor occupancy data and congestion alarm from a camera on a particular date; (*b, c*) camera images of callouts (i–ii) shown in Part *a*.

algorithms were found to perform well in daytime conditions, but nighttime conditions were found to affect the accuracy of the vision system. However, for all conditions, the AUCs were found to be greater than 0.9 for the deep models. This shows that the models perform well in challenging conditions as well.

An example of the real-time implementation of congestion detection using DCNN algorithm is also performed using a continuous set of images extracted from a camera. Simple persistence test methods were applied to reduce the false alarms and smoothe the output signal. Future studies can look into different smoothing techniques (Fourier Transform, Wavelets) to denoise the output obtained from the algorithm and determine the overall detection rate and false alarm rates on a network of cameras. Future studies can also be done using different model architectural designs to improve detection accuracies. Such models can also be used for determining different levels of congestion (high, medium, or low) and also more accurate traffic state determination (speed, volume and occupancy).

Acknowledgments

Our research results are based on work jointly supported by the National Science Foundation Partnerships for Innovation: Building Innovation Capacity (PFI: BIC) program under Grant No. 1632116, National Science Foundation under Grant No. CNS-1464279, and Iowa DOT Office of Traffic Operations Support Grant.

Author Contributions

The authors confirm contribution to the paper as follows: study conception and design: Pranamesh Chakraborty, Yaw Okyere Adu-Gyamfi, Subhadipto Poddar, Anuj Sharma, Soumik Sarkar; data collection: Pranamesh Chakraborty, Subhadipto Poddar; analysis and interpretation of results: Pranamesh Chakraborty, Yaw Okyere Adu-Gyamfi, Subhadipto Poddar, Vesal Ahsani, Anuj Sharma, Soumik Sarkar; draft manuscript preparation: Pranamesh Chakraborty, Yaw Okyere Adu-Gyamfi, Subhadipto Poddar, Vesal Ahsani, Anuj Sharma, Soumik Sarkar. All authors reviewed the results and approved the final version of the manuscript.

References

1. Owens, N., A. Armstrong, P. Sullivan, C. Mitchell, D. Newton, R. Brewster, and T. Trego. Traffic Incident Management Handbook. *Public Roads*, Vol. 172, 2010, p. 116.
2. Schrank, D., B. Eisele, T. Lomax, and J. Bak. *2015 Urban Mobility Scorecard*. Texas A&M Transportation Institute, The Texas A&M University System, College Station, Texas, August 2015. https://static.tti.tamu.edu/tti.tamu.edu/documents/mobility-scorecard-2015.pdf. Accessed February 2, 2018.
3. Kotzenmacher, B. J., E. D. Minge, and B. Hao. *Evaluation of Portable Non-Intrusive Traffic Detection System*. Final Report MN-RC-2005-37. Mn/DOT, Saint Paul, MN, 2005.
4. Zhong, M., and G. Liu. Establishing and Managing Jurisdiction-Wide Traffic Monitoring Systems: North American Experiences. *Journal of Transportation Systems Engineering and Information Technology*, Vol. 7, No. 6, 2007, pp. 25–38.
5. Feng, Y., J. Hourdos, and G. A. Davis. Probe Vehicle Based Real-Time Traffic Monitoring on Urban Roadways. *Transportation Research Part C: Emerging Technologies*, Vol. 40, 2014, pp. 160–178.
6. Ozkurt, C., and F. Camci. Automatic Traffic Density Estimation and Vehicle Classification for Traffic Surveillance Systems Using Neural Networks. *Mathematical and Computational Applications*, Vol. 14, No. 3, 2009, pp. 187–196.
7. He, K., X. Zhang, S. Ren, and J. Sun. Deep Residual Learning for Image Recognition. *arXiv preprint arXiv:1512.03385v1*, Vol. 7, No. 3, 2015, pp. 171–180. http://arxiv.org/pdf/1512.03385v1.pdf.
8. Bauza, R., J. Gozalvez, and J. Sanchez-Soriano. Road Traffic Congestion Detection Through Cooperative Vehicle-to-Vehicle Communications. *Proc., Conference on Local Computer Networks*, LCN, 2010, pp. 606–612.
9. Van Lint, J. W. C., and S. P. Hoogendoorn. A Robust and Efficient Method for Fusing Heterogeneous Data from Traffic Sensors on Freeways. *Computer-Aided Civil and Infrastructure Engineering*, Vol. 25, No. 8, 2010, pp. 596–612.
10. Choi, K., and Y. Chung. A Data Fusion Algorithm for Estimating Link Travel Time. *Journal of Intelligent Transportation Systems*, Vol. 7, No. 3–4, 2010, pp. 235–260.
11. Bachmann, C., B. Abdulhai, M. J. Roorda, and B. Moshiri. A Comparative Assessment of Multi-Sensor Data Fusion Techniques for Freeway Traffic Speed Estimation Using Microsimulation Modeling. *Transportation Research Part C: Emerging Technologies*, Vol. 26, 2013, pp. 33–48.
12. Costeira, J. P., J. M. Moura, G. Wu, and S. Zhang. Understanding Traffic Density from Large-Scale Web Camera Data. *Proc., 2017 IEEE Conference on Computer Vision and Pattern Recognition (CVPR)*, 2017, pp. 4264–4273.
13. Darwish, T., and K. Abu Bakar. Traffic Density Estimation in Vehicular Ad Hoc Networks: A Review. *Ad Hoc Networks*, Vol. 24, 2015, pp. 337–351. http://dx.doi.org/10.1016/j.adhoc.2014.09.007
14. Balcilar, M., and A. C. Sönmez. Extracting Vehicle Density from Background Estimation Using Kalman Filter. *Proc., 2008 23rd International Symposium on Computer and Information Sciences*, Istanbul, 2008, pp. 1–5.
15. Ren, S., K. He, R. Girshick, and J. Sun. Faster R-CNN: Towards Real-Time Object Detection with Region Proposal Networks. *IEEE Transactions on Pattern Analysis and Machine Intelligence*, Vol. 39, No. 6, 2017, pp. 1137–1149.
16. Adu-Gyamfi, Y., S. Asare, A. Sharma, and T. Titus. Automated Vehicle Recognition with Deep Convolutional Neural Networks. *Transportation Research Record: Journal of the Transportation Research Board*, 2017. 2645: 113–122.
17. Oñoro-Rubio, D., and R. J. López-Sastre. Towards Perspective-Free Object Counting with Deep Learning. *Proc., European Conference on Computer Vision (EECV)*, Vol. 9911, 2016, Lecture Notes in Computer Science, pp. 615–629.
18. Asmaa, O., K. Mokhtar, and O. Abdelaziz. Road Traffic Density Estimation Using Microscopic and Macroscopic Parameters. *Image and Vision Computing*, Vol. 31, No. 11, 2013, pp. 887–894.
19. Gonçalves, W. N., B. B. Machado, and O. M. Bruno. Spatiotemporal Gabor Filters: A New Method for Dynamic Texture Recognition. *arXiv preprint arXiv:1201.3612*, 2012. http://arxiv.org/abs/1201.3612.
20. Lempitsky, V., and A. Zisserman. Learning to Count Objects in Images. *Advances in Neural Information Processing Systems*, Vol. 23, 2010, pp. 1324–1332. http://papers.nips.cc/paper/4043-learning-to-count-objects-in-images.pdf.
21. Ahmed, E., M. Jones, and T. K. Marks. An Improved Deep Learning Architecture for Person Re-Identification. *Proc., 2015 IEEE Conference on Computer Vision and Pattern Recognition (CVPR)*, 2015, pp. 3908–3916.
22. Chollet, F. et al. Keras: Deep Learning Library for Theano and TensorFlow. *GitHub Repository*, 2015. https://keras.io. Accessed December 15, 2017.
23. Redmon, J., S. Divvala, R. Girshick, and A. Farhadi. You Only Look Once: Unified, Real-Time Object Detection. *Proc., 2016 IEEE Conference on Computer Vision and Pattern Recognition (CVPR)*, 2016, pp. 779–788.
24. Rublee, E., V. Rabaud, K. Konolige, and G. Bradski. ORB: An Efficient Alternative to SIFT or SURF. *Proc., IEEE International Conference on Computer Vision*, 2011, pp. 2564–2571.
25. Persaud, B. N., and F. L. Hall. Catastrophe Theory and Patterns in 30-Second Freeway Traffic Data: Implications for Incident Detection. *Transportation Research Part A: General*, Vol. 23, No. 2, 1989, pp. 103–113.

The Standing Committee on Artificial Intelligence and Advanced Computing Applications (ABJ70) peer-reviewed this paper (18-04687).

Any opinions, findings, and conclusions or recommendations expressed in this material are those of the author(s) and do not necessarily reflect the views of the National Science Foundation.

TRR
JOURNAL OF THE TRANSPORTATION RESEARCH BOARD

Article

Transportation Research Record
2018, Vol. 2672(45) 232–241

DOI: 10.1177/0361198118782761
journals.sagepub.com/home/trr
SAGE

Performance Assessment of Urban Streets Addressing Improvement Issues for Automobile Mode of Transport

Suprava Jena[1], Abhishek Chakraborty[2], and Prasanta Kumar Bhuyan[1]

Abstract

This paper focusses on modeling automobile drivers' response patterns to assess urban-street service quality in developing countries. Several quality-of-service attributes affecting drivers' riding quality were investigated, from 102 urban street segments under widely varying geometric and traffic conditions. Traffic volume, effective road width, travel speed, pavement condition, on-street parking turnover, land use, hindrance due to public transits, non-motorized vehicles, and encounters are found to significantly affect drivers' comfort levels. Two novel artificial intelligence techniques, that is, artificial neural network (ANN) and functional linked artificial neural network (FLANN) were applied to predict automobile drivers' level of satisfaction scores (*ALOS_score*). The prediction performance of developed models is assessed in terms of various statistical parameters of a modified rank index. Bayesian regularization neural network has given the best fitted model in both training and testing data sets among the ANN models. However, application of the FLANN model shows better prediction performance in the present context, as no hidden layer exists. All input layer neurons are directly linked with output layer neurons with a lesser number of connections, which is advantageous over ANN in reducing accumulated error. The result shows 73% of studied segments are offering service category "C" or below. Sensitivity analyses reported that pavement condition is the most important variable, with relative importance of 26.78%, to influence drivers' riding quality. Similarly, other parameters were ranked in decreasing order of their relative importance, which will help highway authorities to prioritize budgets of future investments for improving service quality.

In preceding centuries, the range of mobility and travel was limited by geography. The advent of the automobile allowed drivers to travel distances that were previously unimaginable. The automobile is the most efficient mode of transportation, which gives incredible freedom of movement for traveling to and from work, for running errands, and so forth. In developing nations, the automobile is increasingly reshaping the pattern of living by increasing the accessibility and interconnectedness between different regions and cities and aiding the expansion of urban areas. At present, the planning, design, and management of any transportation system has become concerned with providing a comfortable ride for automobile drivers. In developing countries such as India, drivers hardly follow any lane discipline and share the through lanes mostly with non-motorized vehicles and heavy vehicles. This often leads to uncontrolled traffic growth and congestion during peak hours. Moreover, there are several other engineering factors, other than speed and delay, which influence drivers' satisfaction levels on urban streets in developing countries. Therefore, infrastructural development based on homogeneous traffic flow models will fail to quantify the performance of urban streets under heterogeneous traffic flow conditions. Transportation planners should focus on those attributes of the transportation system that are more influential to road users and how they affect drivers' satisfaction levels. Typically, the methods of evaluating operational performance of any transportation facility are being traditionally acknowledged as level of service (LOS) models. The most commonly used *Highway Capacity Manual* (HCM) is followed in almost all parts of the world, which defines LOS to be quantified in terms of users' satisfaction levels (*1*, *2*). However, the investment decisions in developing nations do not involve information about user perception. Therefore, the objective of this research is to develop an automobile drivers' level of satisfaction (ALOS) model to assess urban street service quality

[1]Department of Civil Engineering, National Institute of Technology Rourkela, India
[2]Department of Civil Engineering, Indian Institute of Technology Kharagpur, India

Corresponding Author:
Address correspondence to Suprava Jena: suprava728@gmail.com

by identifying the key influencing factors of driver satisfaction under mixed traffic flow conditions.

To achieve the objective of the research, the study framework is divided into four parts. The first identifies important road attributes that influence the comfort level of drivers under highly heterogeneous traffic flow conditions. The second develops a suitable model for the prediction of automobile drivers' level of satisfaction score (*ALOS_score*) using two artificial intelligence techniques, namely, artificial neural network (ANN) and functional linked artificial neural network (FLANN). The reason for choosing neural networks for model development is because of their ability to detect complex and nonlinear relationships between dependent and an expanded dimension of independent variables with reduced processing time and a high level of accuracy, which makes it a powerful alternative to statistical techniques. The third decides the most efficient model by checking the prediction performance of proposed models under prevailing conditions. The fourth defines the ranges of *ALOS_score* for six service categories (A–F) and performs a model sensitivity analysis, which will check the importance of input parameters and sort them in decreasing order of their relative importance. The intention is that it will support highway authorities in enhancing street infrastructure by investing budgets on a priority basis.

Literature Review

Some research has been carried out to define LOS criteria of urban street infrastructures in developing countries. Choocharukul et al. used an ordered probit model for highway LOS. The authors concluded that the HCM use of traffic density as the only measure of effectiveness may not accurately reflect the perceptions of road users (*3*). Flannery et al. recommended the incorporation of qualitative measures such as drivers' perceptions in the estimation of LOS (*4*). Dowling et al. proposed four LOS models to predict service quality on urban streets for four modes of transport (automobile, transit, bicycle, and pedestrian) (*5*). Flannery et al. developed a cumulative logit model to predict the service quality of automobile drivers' perceptions based on presence of median, landscaping, progression, and posted speed (*6*). Papadimitriou et al. identified and analyzed the perceived highway LOS of road users with respect to traffic conditions (*7*). Faezi et al. developed a method to forecast the overall performance level of motorcycles applying logistic regression. The most influential factors perceived by the rider were found to be volume, total lane width, pavement surface quality, and speed (*8*).

Some research has been carried out to determine urban street LOS under heterogeneous traffic flow conditions. Maitra et al. developed an empirical model to estimate congestion levels on urban road sections to determine LOS, with the help of a fundamental diagram of speed-flow, and operational and volume characteristics of mixed traffic (*9*). Marwah and Singh have investigated the behavior of heterogeneous traffic flow rate and classified LOS into four groups based on volume to capacity ratio (*10*). Shao and Sun characterized LOS into two parts: level of traffic facility provided and level of traffic operation (*11*). Mohapatra et al. applied a hybrid algorithm made of both genetic algorithm and Fuzzy C-Means clustering on speed data to define the ranges of six LOS categories. This research concluded the impact of physical and surrounding environmental characteristics for good LOS (*12*). Patnaik and Bhuyan used genetic programming clustering to classify street and average travel speeds into several LOS categories (*13*).

From the literature, it is concluded that the applicability of the above models in developing countries is apprehensive. This is because speed and travel time are not the only parameters influencing the riding quality of drivers. The influences of several important factors such as roadside commercial activities, parking turnover, interruptions by non-motorized vehicles and public transits, and so forth, are not addressed in these studies. Moreover, disregarding travel behavior, any approach would face a possibility of strong rejection. Therefore, this study has examined the impact of a wide range of road attributes and drivers' perceived satisfaction under a mixed traffic flow environment. The existing approaches are justified by exploring a new model to evaluate the service quality offered by urban road segments.

Selection of Study Area

Required quantitative as well as qualitative data sets were collected from 102 urban street segments of eight Indian cities, namely, Bhubaneswar (19 segments), Rourkela (15 segments), Vishakhapatnam (eight segments), Kolkata (27 segments), Durgapur (14 segments), Siliguri (six segments), Jalpaiguri (nine segments), and Coochbehar (four segments). Study locations were chosen in residential as well as commercial zones in urban communities. The behavior and composition of traffic along with the road infrastructures also vary between these eight cities. Therefore, these data sets can well represent the heterogeneity of traffic stream.

Data Collection

From the literature, basic information was gathered about the limitations of applying former methods to evaluate urban street service quality under highly heterogeneous traffic flow conditions. Therefore, both quantitative and qualitative parameters affecting drivers' comfort levels have been considered in the data collection process to achieve the objective of the research in the best possible way.

Quantitative Data

The required road geometric and traffic operational data sets affecting the comfort level of drivers on urban streets were collected with the help of a field survey. The

Table 1. QOS Parameters Collected from Urban Street Segments

	QOS parameters	Units			
1	No. of lanes in one direction of travel	Number			
2	Effective road width	Meters			
3	Pavement condition index	Rating scale: 1–5			
4	Average travel speed of vehicle	Km/h			
5	Number of stops per km	Number			
6	Motorized traffic volume	PCU/h			
7	Non-motorized traffic volume	PCU/h			
8	Composition of heavy vehicles in total traffic	Percentage			
9	Shy distance in meters	Meters			
10	Direction of travel	One way			1.0
		Both ways			0.5
11	Presence of median barrier	Absent			0.0
		Present	Width < 500 mm		0.5
			Raised & Width ≥ 500 mm		1.0
12	Presence of camber	Absent			0.0
		Present	Somewhat irregular		0.5
			Proper		1.0
13	Presence of grade separated sidewalk	Absent			0.0
		Present			1.0
14	Presence of separate bike lane	Absent			0.0
		Present			1.0
15	Land use pattern	Residential (no/low on-street commercial density)			0.0
		Mixed (medium on-street commercial density)			0.15
		Commercial (high on-street commercial density)			0.3
16	On-street parking turnover	Absent (< 5%)			0
		Low (5–25%)			0.13
		Moderate (25–45%)			0.27
		High (> 45%)			0.4
17	Hindrance due to frequently stopping public transits ahead	Presence of separate bus pull-out lane			0.0
		Frequency of interruptions by public transits (in absence of bus pull-out lanes)		Low	0.15
				Moderate	0.3
				High	0.45
				Very high	0.6
18	Average time interval to face an encounter	Minutes			

quality-of-service (QOS) attributes considered in this study are listed in Table 1. The road geometrical data sets, such as effective road width, width of median, shy distance, and so forth, were noted down with the help of measuring tape. The number of lanes per one direction of traffic varies from one to four, having two-way traffic flow, and in some cases traffic movements are restricted to one-way only.

The flow of traffic on a particular road segment was recorded with the help of a high-resolution video camera during both morning (8.00–11.00 a.m.) and evening peak hours (4.00–7.00 p.m.). These timings were chosen to reflect the worst traffic conditions encountered by any drivers in peak hours. On each road segment, the flow of traffic was recorded for a duration of three hours. Recorded video clips were projected onto a large screen in the laboratory to extract peak hourly vehicle counts for each category of vehicle present in the traffic mix on a particular road segment. Running average method was used for the estimation of peak hour traffic volume (*PHV*) from the three peak hours of traffic flow. Some video recordings were also replayed at random to cross-check the extracted vehicle count. Traffic flow data were not collected on weekends and holidays as the traffic volume is comparatively low on these days. The conversion factors mentioned for different types of vehicles in Indian Road Congress (IRC): 106 were multiplied with the classified traffic count to express the extracted peak hour traffic volume in terms of Passenger Car Units per hour (PCU/h) (*14*). Mathematically, total peak hourly traffic volume is the summation of peak hourly motorized traffic volume (*PMV*) and non-motorized traffic volume (*NV*), that is, $PHV = PMV + NV$. To estimate the total traffic volume per effective road width, a variable named PHV/W_e was calculated in this study. Effective road width (W_e) can be defined as the available road width for free movement of traffic. Mathematically, W_e

$= W - (W_P + W_v)$, Where, W = Total road width, W_P and W_v are the reduction in road width due to authorized or unauthorized on-street parking activity and vendor encroachment respectively. The average travel speed (S_{avg}) of vehicles was recorded with the help of a radar gun. The road surface quality was ranked in terms of pavement condition index (PCI), using a five-point scale in which 1 = worst and 5 = excellent. In addition to this, the percentage of heavy vehicles and non-motorized vehicles in the total traffic volume, on-street pedestrian volume, and so forth, were also extracted from the video clips.

This study incorporates several auxiliary variables, which vary significantly from place to place, such as land use pattern, on-street parking turnover, presence of bicycle lane, number of stops per kilometer, and so forth. Some attributes also cause a hindrance while driving on the road. These obstructions may originate from different sources, such as frequent stopping of public transits, slow movement of non-motorized vehicles, interaction with encounters (oppositely moving pedestrians or vehicles), and so forth. These attributes were inspected visually with reference to Table 1. For example: the obstruction in driving caused by frequently stopping public transits is quantified in terms of "O_{PT}" based on frequency of interruptions by public transits, that is, "O_{PT}" = 0 for the road segments passing through an area having separate bus pull-out lanes; the value of "O_{PT}" = 0.15, 0.3, 0.45, and 0.6 for the road segments passing through an area having low, moderate, high, and very high frequency of interruptions by public transits respectively. Similarly, the obstruction in driving caused by slow movement of non-motorized vehicles (bicycles, cycle-rickshaws, hand-driven carts, animal-driven carts, etc.) were quantified in terms of an adjustment factor "O_{NV}" based on the percentage of non-motorized traffic volume ($\%NV$) in total traffic. For example: $O_{NV} = 0$ (if $\%NV < 5$); 0.14 (if $\%NV$ = 5–10%); 0.28 (if $\%NV$ = 10–15%); 0.42 (if $\%NV$ = 15–20%); 0.56 (if $\%NV$ = 20–25%); 0.70 (if $\%NV$ >25%). The obstruction in driving caused by frequent interaction with encounters is quantified in terms of "O_E" based on the average time interval to face an encounter (pedestrians, bicycles, or vehicles coming from the opposite direction of through traffic) on a particular street segment. The minimum and maximum time interval to face an encounter (t_e), taken from the field survey was found to be 0.25 minutes and 5.5 minutes. To quantify the adverse effect of encounters, an O_E value between 0 and 0.2 was introduced for t_e of >5minutes and <0.25 minutes respectively. The intermediate ranges are determined by interpolation, that is, $O_E = 0$ (if $t_e > 5$ min); $O_E = 0.04$ (if t_e = 3–5 min); $O_E = 0.08$ (if t_e = 1–3 min); $O_E = 0.12$ (if t_e = 0.5–1 min); $O_E = 0.16$ (if t_e = 0.25–0.5 min); $O_E = 0.2$ (if $t_e < 0.25$ min).

Perception Survey

The road users' perception data were collected by a travelers' intercept survey to test how drivers' levels of satisfaction vary with respect to different road and traffic conditions. Around 4,047 drivers of varying socio-economic categories were interviewed in this survey just after driving a particular road segment. The drivers were requested to indicate their overall satisfaction (OS) level for the street segment on a Likert scale ranging from 1 = highly dissatisfied to 6 = highly satisfied, corresponding to the six levels (A–F) of the HCM. The strength of this survey is that it includes an extensive driving population, to both gather a huge sample size and make it cost-effective. However, drivers aged less than 18 years were excluded from this survey due to lack of enough experience to make a judgment on the road infrastructure and traffic conditions. The gender distributions of participants were roughly 43% and 57% for female and male. The age distributions of participants were around 37%, 31%, 29%, and 3% for age groups of 18–25, 26–35, 36–45, and ≥45 years respectively. Approximately, 45%, 37%, and 18% of the total number of participants drove the road section with two-wheelers, private cars, and light commercial vehicles respectively.

To identify a diverse set of issues related to service quality, road segments were selected from seven different mid-sized and large Indian cities, so that a good range of data could be obtained to incorporate different types of urbanized areas, influencing drivers' expectations and experience of roadway conditions under a mixed traffic stream. It has been observed that PHV of studied segments varies from 296 PCU/hour to 5,033 PCU/hour per one direction of travel. Some of the selected road segments of these cities carry urban residential settings with less traffic volume, which are separated from commercial zones and connected to high-density streets, whereas, medium-density or highly dense commercial developments are closed to roadways at some locations carrying medium or high volumes of traffic. The roadway cross-section begins with one lane and expands to four lanes or more at various points. Some road segments have medians and short stretches of sidewalk alongside the carriageway, whereas some segments do not have these facilities. Landscaping becomes more apparent for some segments with clean and smooth road surface quality. At some points, the road width narrows due to authorized or unauthorized on-street parking. Some road segments also contain separate bus pull-out lanes, whereas, some segments do not have these facilities. The average speed of vehicles varies from 15 to 50 km/h, according to the level of congestion and vehicular interaction. To represent the whole population, people from different socio-economic backgrounds were interviewed in the perception survey for exposure to a variety of driving environments and driver behavior. Considering the contextual diversity in the data sets in developing the ALOS model, it is anticipated that these study findings will be universally applicable to forecasting the service quality of urban street segments in developing countries.

Methodology

Before following the steps for model development, the significant independent variables affecting drivers' comfort levels on urban streets were selected with the help of Pearson's

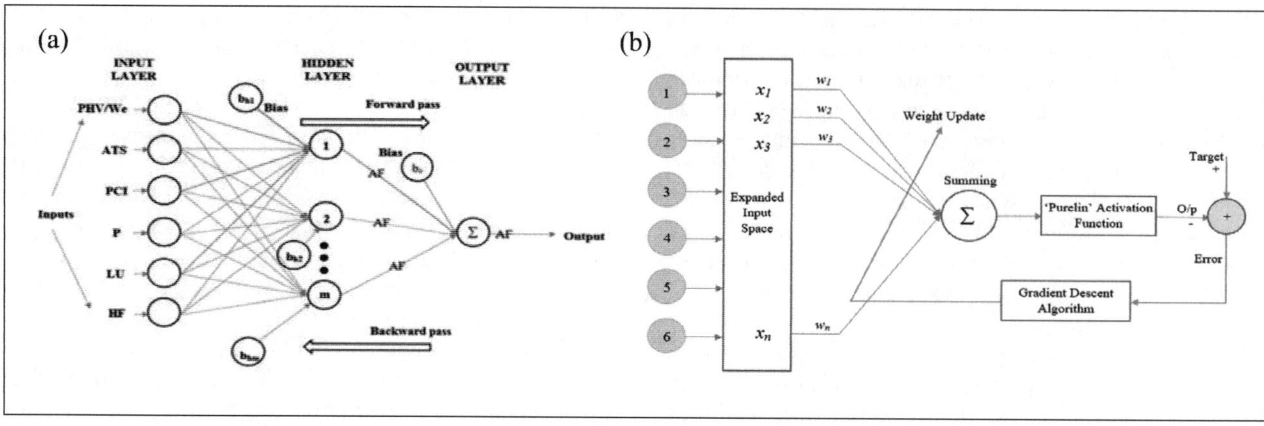

Figure 1. Schematic diagram of (*a*) ANN model architecture and (*b*) FLANN model architecture to predict *ALOS_score*.

correlation analysis. The influential variables were used as input variables in the model development process. As observed by many researchers, the information processing system using neural networks is easy to use and understand, for which neural networks are powerful alternatives to statistical techniques in solving prediction problems. ANN is a powerful, data-driven, self-adaptive, flexible computational tool, which can handle large amount of data sets. It has the ability to implicitly detect complex and nonlinear relationships between dependent and independent variables by taking into account all possible interactions between predictor variables. Apart from the previously mentioned advantages of ANN algorithms, FLANNs have a single-layer feed-forward network with an expanded dimension of inputs. That drastically reduces the processing time and achieves a high level of accuracy with low supervision compared with other algorithms. The idea behind choosing two different learning methods was to know the domain of competence for both the family of algorithms and to test the better performing one in the present context. The following sub-sections explain the principles of applying both ANN and FLANN techniques for the development of the ALOS model.

ANN

The ANN model architecture comprises three or more layers for information processing called neurons. The first layer has the input neurons, taking significant QOS attributes as input variables, while the last layer contains the output parameter, that is, the *ALOS_score*. In between them, there are one or more hidden layers to delineate and learn the patterns governing the network's data, which is shown in Figure 1*a*. Information is processed between neurons through the connecting links. Neural networks have to implement complex mapping functions. Therefore, they need a nonlinear activation function in order to bring the much-needed non-linearity property. The activation function enables the network to approximate any function and allow it to learn more complex functions than linear regression.

The main advantage of the ANN model is that it is a non-parametric model, while most statistical methods are parametric models requiring a higher background of statistics. Usually, neural networks are non-parametric, as they make judgments without assuming a particular family of distributions. The term "parametric" is used when a set of parameters (such as mean and variance of a normal distribution) fully specifies the distribution. ANNs do use parameters, but the aim is to find the parameter values that lead to a good engineering approximation of some functions. Furthermore, neural networks do not fix the number of parameters (weights) and there is no rule of thumb to pick the number of weights, number of nodes, and so forth, in a particular neural network architecture. The number of neurons and hidden layers is chosen by using a repetitive trial and error procedure during the training of a neural network to find a well-fitted model. Modification of neuron weights alone only serves to manipulate the shape/curvature of transfer function, whereas the bias neurons allow the transfer function curve to shift horizontally along the input axis, leaving the shape unaltered. Therefore, the input-to-output mapping can be customized to suit the particular requirements. These activation functions (*AF*) may be of different mathematical type depending on the nature of inputs and their characteristics. The output of the network is computed by multiplying the input (x) by the weight (w_{ij}) and passing the result through any of the three stated activation functions. So, the generalized formula can be written as shown in Equation 1:

$$ALOS_N = \varphi[b_0 + \Sigma_{k=1}^{k=m}\{w_k * \varphi(b_k + \Sigma_{i=1}^{i=n} w_{ik} * x_i)\}] \quad (1)$$

Where, $ALOS_N$ = normalized value in the range of [–1, 1]; φ = Activation function; b_o = bias at the output layer; w_k = connection weight between k^{th} neuron of hidden layer and the single output neuron; b_k = bias at the kth neuron of hidden layer; w_{ik} = connection weight between i^{th} input variable and k^{th} neuron of hidden layer; x_i = normalized input variable i in the range [–1, 1]; m = number of neurons in the hidden layer; and n = number of input variables.

Finally, $ALOS_N$ should be de-normalized to get the desired *ALOS_score* for each segment.

Functional Linked ANN

FLANN is a higher order neural network, which is a bit different from a conventional multilayer perceptron (MLP). It consists of a single-layer feed-forward type structure with an expanded dimension of inputs as shown in Figure 1*b*. The concept of FLANN is to some extent advanced when compared with traditional neural network models. ANN consists of one or more hidden layers. So, the total number of synapses or links is quite large due to the connection of each input, hidden and output layers' neurons. If a small error is associated with one link, sometimes it gets magnified. In this context, FLANN is quite advantageous because the number of connections is less and it generally helps to obtain a lesser error value as desired.

To train a FLANN network, back propagation algorithm is not generally practiced, as the optimal solution may be located in a region of local minima instead of global minima. The weights of FLANN network connection are updated by considering the error and a control parameter (mu) to control the error convergence in each step. The inputs are expanded into an expanded vector form following the Fourier expansion rule or trigonometric expansion (*15*).

To develop the FLANN input space matrix, different trials and iterations have been carried out to find the optimum expansion of the input variables. The weight updating process in each iteration followed a gradient descent technique. Therefore, the predicted *ALOS_score* can be expressed as:

$$ALOS_N = \varphi\{b + \Sigma_{i=1}^{n}(w_i * x_i)\} \tag{2}$$

Where, $ALOS_N$ = normalized in the range of [–1, 1]; b = bias of output layer; w_i = weight associated to i^{th} input variable; x_i = i^{th} input variable; n = number of input variables in expanded input space; and φ = activation function.

Here, all the inputs are normalized form in the range of [–1, 1]. Therefore, the output *ALOS_score* will be obtained by de-normalizing $ALOS_N$.

Result and Analysis

This section provides a detailed discussion to find out the significant road attributes and estimated model parameters to determine an *ALOS_score*. 70% of data sets (i.e., 71 randomly selected segments) were used to train the model, and the remaining 30% were used to test the model. The prediction performance of these models was assessed in terms of several statistical parameters to decide the better one in the present context.

Selection of Input Parameters

Pearson's correlation analysis was carried out on the list of QOS attributes with respect to *OS* scores to verify the statistical significance of the input variables at 99.99% confidence level. A total of eight variables were significantly identified, which were taken into account in the ALOS model development process, for example: peak hour traffic volume per effective road width (PHV/W_e), average travel speed (S_{avg}), PCI, on-street parking turnover (*P*), land use pattern (*LU*), hindrance due to frequent stopping of public transits ahead (O_{PT}), obstruction due to slow movement of non-motorized vehicles (O_{NV}), and encounters coming from the opposite direction of through traffic (O_E). To incorporate the operational effect of the last three variables (O_{PT}, O_{NV}, and O_E) on the comfort level of drivers, they were summed up and labeled as hindrance factor (*HF*). Mathematically, $HF = O_{PT} + O_{NV} + O_E$.

Prediction of ALOS_score with the Help of ANN Model Architecture

The development of the ANN model was carried out by taking six predictor variables (i.e., PHV/W_e, S_{avg}, *PCI, P, LU*, and *HF*) as input neurons and one response variable (i.e., *ALOS_score*) as output node. All predictor and output variables were normalized in the range of [–1, 1] to avoid their dimensional effects on model building. Several ANN models were developed using MATLAB 2015a, taking feed-forward perceptron with back propagation (FFBP) training algorithms, such as Bayesian regularization neural network (BRNN), Levenberg-Marquardt neural network (LMNN), scaled conjugate gradient neural network (SCGNN), and gradient descent with momentum and adaptive learning rate back propagation neural network (GDXNN) with three most commonly used activation functions such as "hyperbolic tangent-sigmoid" (tan-sig), "logistic sigmoid" (log-sig), and "linear" (pure-lin) (*16*). The number of hidden layers is fixed to 1. The general architecture of developing the ALOS model using ANN is shown in Figure 1*a*. Prediction performances of the four training algorithms, along with the three activation functions, were tested separately in the present context. A best-fit model has been developed using each algorithm by varying the number of neurons in the hidden layers. A total of 84 models have been developed, taking a combination of four learning algorithms; three activation functions with a varying number of neurons from 3 to 15 in the hidden layer.

Comparing the performance of all models in terms of coefficient of correlation (R), coefficient of determination (R^2), root-mean-square error (RMSE), mean absolute error (MAE), and over-fitting ratio (OR) for training and testing data sets individually, BRNN is found to be the better

Table 2. FLANN Model Parameters for Different Types of Expansion of Six Input Variables

Sl. no.	Input variable	Weight	Sl. no.	Input variable	Weight	Sl. no.	Input variable	Weight
1	*a*	0.001233	9	*c*	0.013293	17	*exp(f)*	0.007029
2	*2a*	−0.02798	10	*2c*	0.020177	18	*exp(-f)*	0.036933
3	*3a*	−0.04233	11	*3c*	0.034201	19	*sin(d)*	−0.02308
4	*4a*	−0.05547	12	*4c*	0.037293	20	*sin(2d)*	−0.03923
5	*b*	0.008938	13	*exp(d)*	−0.04549	21	*sin(e)*	−0.00408
6	*2b*	0.02591	14	*exp(-d)*	0.009545	22	*sin(2e)*	−0.00899
7	*3b*	0.034773	15	*exp(e)*	−0.03924	23	*sin(f)*	−0.03401
8	*4b*	0.038675	16	*exp(-e)*	−0.00261	24	*sin(2f)*	−0.1229

performing one in this regard compared with other models. The main aim of the Bayesian regularizer is to penalize the weight functions, which are much more complex, in favor of simpler functions. In addition, it reduces the need of cross-validating the data. The value of OR nearer to one indicates a good-fit model. It is also observed that for 'tansig' and 'purelin' transfer functions, 10 neurons in the hidden layer is the better performing one, whereas in the case of 'logsig' function, 15 neurons in the hidden layer is found to be best performing. Therefore, these three models are compared in Table 3 in terms of average absolute error (AAE) to find the competent BRNN model.

Prediction of ALOS_scores with the Help of FLANN Model Architecture

In the process of developing the FLANN model, different trials were carried out for the expansion of six input variables using Fourier expansion, Trigonometric, algebraic, or a combination of them. Each input was associated with some weights. Before starting the analysis, a fixed bias and weights were chosen randomly. Based on the output obtained, the weights were updated at the end of every iteration by a gradient descent algorithm. Since no hidden layer exists in this case, the expanded input variables have a direct link with the output. The general architecture of developing the ALOS model using FLANN is shown in Figure 1*b*.

After several trials and combinations, the likelihood of achieving the best result was observed with a mixture of algebraic, exponential, and trigonometric expansion functions. After many trials with different combinations, the six input variables were expanded to generate a prolonged input space up to 24 transformations. Therefore, the six variables *PHV/*W_e, S_{avg}, *PCI, P, LU*, and *HF* are represented as *a, b, c, d, e*, and *f* respectively, to reduce the space and complexity in showing the details of 24 parameters along with their associated connection weights in Table 2. These weights were obtained by training the FLANN model in MatLab. Within 30–50 iterations, the error becomes minimized as per the desired value. However, ANN took a greater number of iterations. Two types of FLANN models were found as good models by comparing RMSE and OR values. It was observed that FLANN models converge faster than the BRNN model. The details of FLANN model expansions are mentioned in Table 2. As there is no hidden layer in between the input and output layer, the sum of the products of these parameters with their corresponding weight coefficient gave rise to the model output as shown in Equation 4.

Model Validation and Comparison of Different Model Parameters

All these models are validated for both training and testing data sets to examine the requirement of a good validation. Figure 2 shows some graphs, which are plotted keeping predicted output (*ALOS_score*) on the *y*-axis and actual output on the *x*-axis.

To determine which model is the best one among the developed models, rank index (RI) was proposed by Cai et al. and later a modified rank index (MRI) parameter was proposed by Beura and Bhuyan (*17*, *18*). The performance measuring parameters were chosen from both.

Mathematically, the expression of MRI is shown in Equation 3:

$$MRI = R_1 + R_2 + R_3 + R_4 + R_5 \tag{3}$$

Where, R_1 = rank of a model on the basis of slope of the best-fit line of the predicted to observed *ALOS_scores* plot, R and efficiency coefficient "E" as proposed by Nash and Sutcliffe (*19*).

R_2 = rank of a model on the basis of error terms like RMSE, AAE, and MAE.

R_3 = rank of a model on the basis of the arithmetic calculations of the ratio of predicted to observed *ALOS_scores*, that is, mean and standard deviation.

R_4 = rank of a model on the basis of P_{50} and P_{90} 50% and 90% cumulative probability values and obtained by plotting cumulative probability values of w.r.t to the ratio of "predicted to observed" *ALOS_scores*.

R_5 = rank of a model on the basis of *ALOS_scores* within ±20% accuracy level.

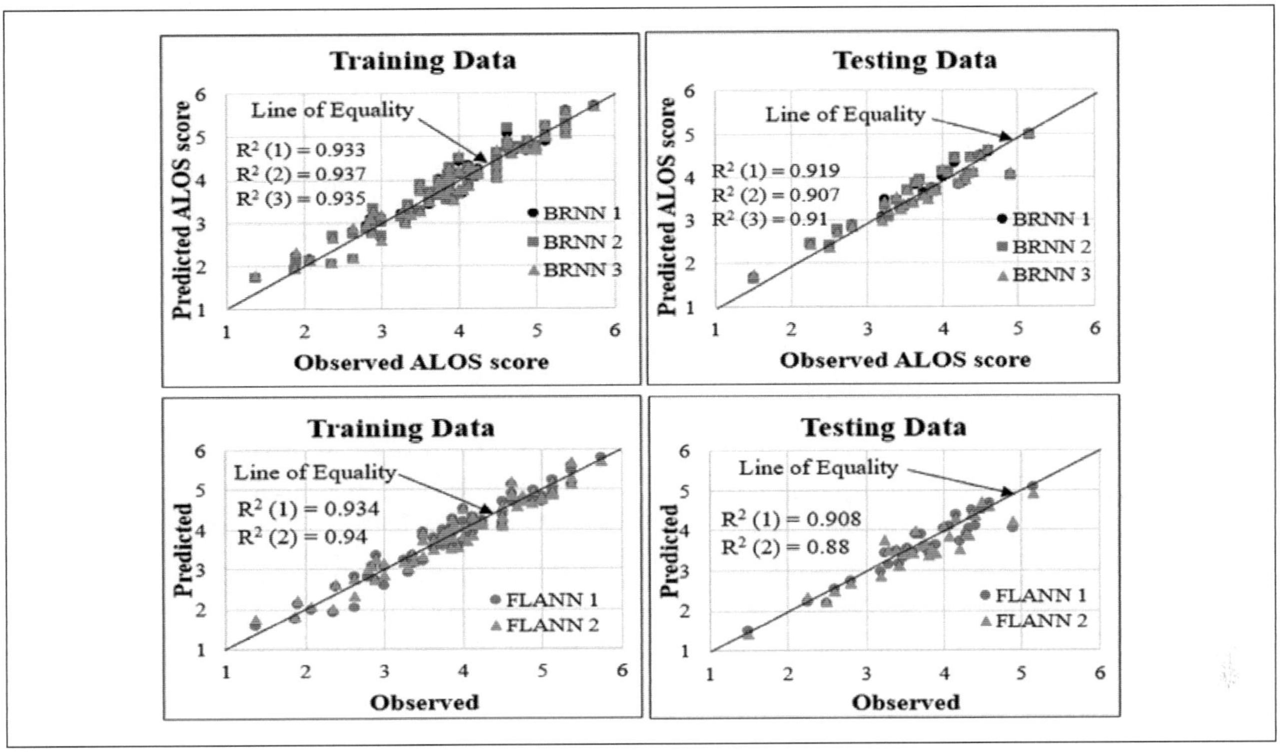

Figure 2. Validation plots of ANN and FLANN models for training and testing data sets.

Table 3. Model Comparison using MRI

Model			BRNN 1		BRNN 2		BRNN 3		FLANN 1		FLANN 2	
Data type			Training	Testing	Training	Testing	Training	Testing	Training	Testing	Training	Testing
Best-fit calculation	$ALOS_{Pr}/ALOS_{Act}$		0.93	0.85	0.95	0.87	0.94	1.00	0.99	0.94	0.95	0.92
	R^2		0.933	0.919	0.938	0.908	0.935	0.910	0.934	0.908	0.940	0.880
	E		0.933	0.904	0.937	0.898	0.935	0.903	0.931	0.884	0.940	0.850
	R_1		5		3		1		2		4	
Error measuring parameters	RMSE		0.108	0.110	0.105	0.113	0.107	0.111	0.121	0.121	0.106	0.137
	AAE		0.090	0.083	0.085	0.090	0.088	0.090	0.197	0.199	0.085	0.109
	MAE		0.215	0.389	0.268	0.397	0.251	0.362	0.269	0.395	0.253	0.313
	R_2		1		3		2		5		4	
$ALOS_{Pr}/ALOS_{Act}$	mu		1.001	0.987	1.004	0.988	1.009	0.993	1.000	0.970	1.005	0.967
	sigma		0.075	0.063	0.071	0.064	0.077	0.068	0.073	0.059	0.070	0.073
	R_3		2		3		4		1		5	
Cumulative probability	Probability	0.995	0.977	0.995	0.970	1.003	0.982	0.997	0.975	0.997	0.957	
		1.095	1.069	1.095	1.071	1.102	1.073	1.100	1.057	1.108	1.056	
	R_4		1		4		3		2		4	
20% accuracy	Log-normal		99	100	99	100	98	100	100	100	100	100
	Histogram		98.59	100	98.59	100	97.18	100	100	100	98.59	100
	R_5		3		3		5		1		2	
Overall rank	MRI		12		16		15		11		19	
	Final rank		3		4		3		1		5	

The details of calculating MRI for the five developed models are shown in Table 3. It was observed that "FLANN 1" model with gradient descent algorithm and linear activation function, showing MRI value of 11, was the best among five models. Therefore, FLANN-1 model architecture was applied for the evaluation of the service quality of urban streets.

The expansion form of the input space along with their weights are shown in Table 2. Mathematically, predicted $ALOS_N$ can be expressed as shown in Equation 4:

$$\begin{aligned} ALOS_N &= 0.001233*a - 0.2798*2a - 0.04233 \\ &*3a - 0.05547*4a + 0.008938*b + 0.02591 \\ &*2b + 0.034773*3b + 0.038675*4b + 0.013293 \\ &*c + 0.020177*2c + 0.034201*3c + 0.037293 \\ &*4c - 0.04549*\exp(d) + 0.009545*\exp(-d)) \\ &-0.03924*\exp(e) - 0.00261*\exp(-e) + 0.007029 \\ &*\exp(f) + 0.036933**\exp(-f) - 0.02308*\sin(d) \\ &-0.03923*\sin(2d) - 0.00408*\sin(e) - 0.00899 \\ &*\sin(2e) - 0.03401*\sin(f) - 0.1229*\sin(2f) \end{aligned} \tag{4}$$

Finally, $ALOS_N$ is de-normalized to get the predicted *ALOS_scores* for each street segment.

Defining Ranges of ALOS Classes (A–F)

The model-predicted *ALOS_scores* obtained for street segments were classified into six classes (A–F) to specify the range of *ALOS_scores* for six LOS categories. The mean value of perceived *ALOS_scores* was approximately 3.5, which corresponds to the boundary between LOS C and D. Considering the symmetry of this boundary point, the ranges of *ALOS_scores* for six service categories (A–F) are determined manually as follows.

If the ranges of *ALOS_scores* were >5.1, >4.33 to ≤5.1, >3.5 to ≤4.33, >2.67 to ≤3.5, >1.84 to ≤2.67, and ≤1.84, then the corresponding ALOS categories would be A (excellent), B (good), C (average), D (below average), E (poor), and F (very poor) respectively.

Predicted ALOS categories for all the street segments were estimated, which showed that the predicted service categories have 95.1% matchings with the observed *OS* scores. Therefore, the proposed model is appropriate for assessing urban street service levels under mixed traffic flow conditions.

Sensitivity Analysis

The variation in drivers' satisfaction levels with respect to percentage change in the input parameters was tested using a model sensitivity analysis (*20*). PCI was found to have the highest relative importance of 26.78% on the comfort level of drivers than other road attributes. PHV, W_e, S_{avg}, O_{NV}, O_{PT}, P, LU, and O_E are ranked second, third, fourth, fifth, and sixth most important parameters with relative importance of 24.93%, 14.65%, 13.89%, 6.28%, and 5.39% respectively. This suggests that the lane width of urban streets in India is shared by automobiles, non-motorized vehicles, public transits, pedestrians, and so forth, without any lane discipline. This leads to congestion and delay in through movement of automobiles, as well as a decrease in drivers' satisfaction levels. Therefore, there should be the provision of separate bicycle lanes and bus pull-out lanes. The remaining parameters, such as P, LU, and O_E, have relative importance of 3.59%, 2.69%, and 1.79% respectively. These variables have considerable negative impacts by decreasing the degree of driver satisfaction on urban streets.

Conclusion

A new approach is proposed in this paper to model automobile users' response patterns by taking both quantitative as well as qualitative parameters together under mixed traffic flow conditions. Based on the MRI value for both training and testing data sets, it was concluded that the FLANN model is the most reliable technique for the evaluation of urban street service quality.

The result of this study shows that only 9% of the studied segments are labeled under ALOS category "A". More than 73% of the street segments are designated under ALOS "C" or below. Based on model sensitivity analysis, the relative importance of nine input variables influencing drivers' satisfaction levels was checked. PCI has the highest impact on drivers' satisfaction levels. Therefore, pavement surfaces require regular maintenance to achieve a higher *ALOS_score*. Whereas, on-street parking turnover and commercial activities, shared non-motorized traffic, public transits, oppositely moving encounters, and so forth, have considerable negative impacts on the through movement of automobiles. Separate bus pull-out lanes and bicycle lanes are essential to increase driver satisfaction. Therefore, the proposed model will help traffic engineers to measure the service quality offered by existing road infrastructures and to design future implements on a priority basis.

This model highlights drivers' expectations of service quality under the influence of unauthorized on-street parking and commercial activities; interruptions by non-motorized traffic, public transits and oppositely moving encounters; and so forth, which have not been addressed previously. The drivers' satisfaction score is poorer under such driving environments. Therefore, the proposed model offers some new insights of traveler satisfaction levels under highly heterogeneous traffic flow conditions and has scope for wide application in developing countries such as India. However, traffic characteristics are somewhat different for developed countries and users' perceived satisfaction scores may vary accordingly. Therefore, this research needs further investigation related to service attributes affecting comfort level of drivers in homogenous traffic flow conditions. To overcome this limitation, the methodology needs some modifications by iteratively changing the input parameters in varying road and traffic conditions for effective assessment of the service quality of urban street infrastructures in a global scenario.

Author Contributions

The authors confirm contributing to the paper as follows: study conception and design: Suprava Jena, Abhishek Chakraborty, Prasanta Kumar Bhuyan; data collection: Suprava Jena, Abhishek Chakraborty; analysis and interpretation of results: Suprava Jena, Abhishek Chakraborty; draft manuscript preparation: Suprava Jena, Abhishek Chakraborty, Prasanta Kumar Bhuyan. All authors reviewed the results and approved the final version of the manuscript.

References

1. *Highway Capacity Manual*. TRB, National Research Council, Washington, D.C., 2000.
2. *Highway Capacity Manual 2010*. Transportation Research Board of the National Academies, Washington, D.C., 2010.
3. Choocharukul, K., K. C. Sinha, and F. L. Maneering. User Perceptions and Engineering Definitions of Highway Level of Service: An Exploratory Statistical Comparison. *Transportation Research Part A*, Vol. 38, No. 9–10, 2004, pp. 677–689. http://dx.doi.org/10.1016/j.tra.2004.08.001.
4. Flannery, A., K. Wochinger, and A. Martin. Driver Assessment of Service Quality on Urban Streets. *Transportation Research Record: Journal of the Transportation Research Board*, 2005. 1920: 25–31.
5. Dowling, R., A. Flannery, B. Landis, T. Petritsch, N. Rouphail, and P. Ryus. Multimodal Level of Service for Urban Streets. *Transportation Research Record: Journal of the Transportation Research Board*, 2008. 2071: 1–7.
6. Flannery, A., N. Rouphail, and D. Reinke. Analysis and Modeling of Automobile Users' Perceptions of Quality of Service on Urban Streets. *Transportation Research Record: Journal of the Transportation Research Board*, 2008. 2071: 26–34.
7. Papadimitriou, E., V. Mylona, and J. Golias. Perceived Level of Service, Driver, and Traffic Characteristics: Piecewise Linear Model. *Journal of Transportation Engineering*, Vol. 136, No. 10, 2010, pp. 887–894. http://dx.doi.org/10.1061/(ASCE)TE.1943-5436.0000154.
8. Faezi, S. F., H. Hamid, and H. K. Sanij. Development of Service Performance Model for Exclusive Motorcycle Lanes. *Science International*, Vol. 25, No. 3, 2013, pp. 461–468.
9. Maitra, B., P. K. Sikdar, and S. L. Dhingra. Modelling Congestion on Urban Roads and Assessing Level of Service. *Journal of Transportation Engineering*, Vol. 125, No. 6, 1999, pp. 508–514. http://dx.doi.org/10.1061/(ASCE)0733-947X(1999)125:6(508).
10. Marwah, B. R., and B. Singh. Level of Service Classification for Urban Heterogeneous Traffic: A Case Study of Kanpur Metropolis. *Proc., 4th International Symposium on Highway Capacity*, Hawaii, 2000, pp. 271–286.
11. Shao, M., and L. Sun. United Evaluation Model of Traffic Operation Level for Different Types of Urban Road. *Journal of Tongji University (Natural Science)*, Vol. 38, No. 11, 2010, pp. 1593–1598. http://dx.doi.org/10.3969/j.issn.0253-374x.2010.11.006.
12. Mohapatra, S. S., P. K. Bhuyan, and K. V. K. Rao. Genetic Algorithm Fuzzy Clustering Using GPS Data for Defining Level of Service of Criteria Urban Streets. *European Transport \ Trasporti Europei*, Vol. 49, No. 52, 2012, pp. 1–19.
13. Patnaik, A. K., and P. K. Bhuyan. Application of Genetic Programming Clustering in Defining LOS Criteria of Urban Street in Indian Context. *Travel Behaviour and Society*, Vol. 3, 2016, pp. 38–50. http://dx.doi.org/10.1016/j.tbs.2015.08.003.
14. Indian Road Congress. *Guidelines for Capacity of Urban Roads in Plain Areas*. Indian Road Congress, New Delhi, 1990.
15. Jiang, L. L., D. L. Maskell, and J. C. Patra. A FLANN-Based Controller for Maximum Power Point Tracking in PV Systems under Rapidly Changing Conditions. *Proc., 2012 IEEE International Conference on Acoustics, Speech and Signal Processing (ICASSP)*, Kyoto, Japan, IEEE, New York, 2012, pp. 2141–2144.
16. Dorofki, M., A. H. Elshafie, O. Jaafar, O. A. Karim, and S. Mastura. Comparison of Artificial Neural Network Transfer Functions Abilities to Simulate Extreme Runoff Data. *International Proceedings of Chemical, Biological and Environmental Engineering*, Vol. 33, 2012, pp. 39–44.
17. Cai, G., S. Liu, L. Tong, and G. Du. Assessment of Direct CPT and CPTU Methods for Predicting the Ultimate Bearing Capacity of Single Piles. *Engineering Geology*, Vol. 104, No. 3–4, 2009, pp. 211–222.
18. Beura, S. K., and P. K. Bhuyan. Modeling Quality of Bicycle Accommodations on Urban Road Segments using Functional Networks and Multivariate Adaptive Regression Splines Techniques. Presented at 96th Annual Meeting of the Transportation Research Board, Washington, D.C., 2017.
19. Nash, J. E., and J. V. Sutcliffe. River Flow Forecasting through Conceptual Models Part 1 – A Discussion of Principles. *Journal of Hydrology*, Vol. 10, No. 3, 1970, pp. 282–290.
20. Gandomi, A. H., G. J. Yun, and A. H. Alavi. An Evolutionary Approach for Modeling of Shear Strength of RC Deep Beams. *Materials and Structures*, Vol. 46, No. 12, 2013, pp. 2109–2119. http://dx.doi.org/10.1617/s11527-013-0039-z.

The Standing Committee on Artificial Intelligence and Advanced Computing Applications (ABJ70) peer-reviewed this paper (18-04773).

Article

TRR
JOURNAL OF THE TRANSPORTATION RESEARCH BOARD

Transportation Research Record
2018, Vol. 2672(45) 242–252
© National Academy of Sciences: Transportation Research Board 2018
Article reuse guidelines:
sagepub.com/journals-permissions
DOI: 10.1177/0361198118791365
journals.sagepub.com/home/trr
SAGE

Evaluation of Clustering Algorithms for the Prediction of Trends in Bus Travel Time

Hima Elsa Shaji[1], Arun K. Tangirala[2], and Lelitha Vanajakshi[1]

Abstract
Providing accurate and reliable travel time information to travellers is essential to improve the quality of public transit systems. With the availability of the latest technologies, it has become possible to collect a large amount of traffic data to analyze and understand these systems better. Traffic in India is characterized by lack of lane discipline and the presence of vehicles of varying static and dynamic characteristics, which makes prediction of bus travel time especially challenging. The aim of this study is to identify both a prediction algorithm that can handle high variability and suitable inputs or regressors to be used. Earlier studies performed offline manual grouping considering the patterns observed, which leads to limitations for automated field implementations. The present study explores the use of data-driven approaches, primarily clustering, to address the challenges for the prediction of bus travel time trends. Discrete wavelet transform (DWT) was used to extract trends from the travel time measurements. Three popular clustering algorithms—*k*-means, hierarchical, and self organizing maps (SOM)—were used to identify patterns. Travel time trends were then predicted by searching for similar cluster patterns within the historical database using pattern sequence-based forecasting (PSF). A comparison of the performance of these algorithms was carried out based on prediction errors. The clustering + prediction framework developed was also compared with the case when no clustering was done on the regressor dataset.

Travel time is one of the most popular measures of traffic system performance, as it can be easily perceived by users as well as by operators, planners, and researchers. Most of the private and public vehicles operating in metropolitan cities are now equipped with global positioning system (GPS) devices. Due to the low operation and maintenance costs associated with GPS devices, they are widely used for real-time tracking of vehicles such as public buses. In this process of tracking vehicles' trajectories, a huge amount of trajectory data is generated which serves as a rich historical database. Many studies have reported on travel time prediction using GPS data from buses (*1–3*). In all these studies, however, a systematic method to identify the best inputs for the prediction process is lacking.

The accuracy of any prediction algorithm depends on the input provided to it. Hence, there is a need to identify correct inputs for the prediction algorithm by considering the natural groupings in the data and also the high variability in the system. Travel time, in general, follows certain patterns such as trip-wise, daily, weekly, monthly, and yearly, and they influence the selection of the correct regressor (*4–7*). In the majority of the earlier studies on bus travel time prediction, suitable regressors were selected manually offline, based on these patterns. However, for real-time implementations, there is a need for an automated method to analyze the data in order to identify the correct inputs, referred to as regressors here, to be used to predict bus travel times. For example, Kumar et al. (*7*) predicted travel times by identifying significant regressors using statistical analysis by considering each day of the week separately. The patterns were identified offline and were not dynamic in nature. However, in the real world these patterns may not be static and may vary depending on the day, time, location, etc. The system being very dynamic, a technique needs to be developed to group the data automatically rather than separating them manually using known patterns such as

[1]Department of Civil Engineering, Indian Institute of Technology Madras, Chennai, India
[2]Department of Chemical Engineering, Indian Institute of Technology Madras, Chennai, India

Corresponding Author:
Address correspondence to Lelitha Vanajakshi: lelitha@iitm.ac.in

daily and/or weekly trajectories. Thus, there is a need to identify the correct regressors for the prediction algorithm by considering the natural groupings in the data and also the high variability in the system, ideally with an automated approach. Dynamic techniques need to be identified to do this grouping of data automatically rather than separating it manually using daily or weekly patterns. Clustering is one such data-driven, dynamic technique which can be used to identify the natural groups within a data set. This approach is possible if a large historical database is available. Once such groups are identified, the characteristics of the group can be used for better travel time prediction. For example, Van Der Voort et al. (*8*) predicted traffic flow on a motorway by dividing the data set manually into weekdays and weekends and also by clustering it using Kohonen maps. It was shown that the latter worked better than the case when the regressor dataset was divided manually.

The choice of the type of clustering technique to be used and the number of clusters into which the data should be divided are two major questions that should be answered. According to Jain et al. (*9*), there is no clustering technique that can be applied universally. Hence, the identification of the clustering technique that is best suited to the data is a major concern. The important clustering algorithms include: partitional, hierarchical, model-based, density-based, grid-based, and self organizing maps (SOM). Partitioning clustering algorithms, like k-means clustering, divide the dataset containing n unlabeled objects into k clusters such that each of the clusters formed has at least one object in it (*10*, *11*). However, the optimum value of k should be known. Hierarchical clustering (*12*) is another popular clustering algorithm which uses either agglomerative or divisive techniques to create a hierarchy of clusters. In a model-based clustering approach, the data is assumed to belong to a mixture of probability distributions, each representing a different cluster. It then finds the best fit of data to that model (*13*). SOM, or Kohonen maps (*14*), is a type of artificial neural network which projects higher dimensional data into a lower dimension (typically 2-D). Other popular clustering algorithms include density-based clustering (*15*) and grid-based clustering (*16*). However, parameters such as density thresholds (*17*) need to be defined and are not used for time series clustering due to high complexity (*18*).

Clustering techniques have been reported for prediction of traffic parameters. Chung (*19*) used a clustering algorithm to group historical travel time data. The computation time reduced significantly after the data was clustered, as only similar segments of the historical database need to be searched to identify the patterns. Ladino et al. (*20*) used a k-means clustering algorithm to organize a historical database into clusters and then used Kalman filtering technique to predict travel time for each cluster. Park et al. (*21*) used fuzzy c-means clustering and Kohonen SOM to cluster historical link travel times and calibrated individual artificial neural network (ANN) for each class. However, all these studies were limited to homogenous traffic conditions.

A detailed literature review on bus travel time prediction was carried out. The techniques used for travel time prediction can be categorized into naive models, regression models, linear and non-linear time series models, state space models, and ANN, as summarized in Table 1.

A prediction methodology was developed based on a pattern sequence-based forecasting (PSF) algorithm. The two main steps in PSF include the use of a suitable clustering technique to obtain cluster labels and prediction of future values by searching for the previous cluster labels within the database. Alvarez et al. (*36*) used PSF to predict future energy demand by clustering the data set using k-means. It was reported that this technique showed remarkable improvement compared with predictions made using autoregressive integrated moving average (ARIMA) or ANN. Jin et al. (*37*) extended energy

Table 1. Travel Time Prediction Techniques

Authors	Method	Description
Weigang et al. (*22*); Jeong et al. (*23*)	Naive models	Simple or weighted average of the historical travel times
Bertini et al. (*24*); Patnaik et al. (*25*); Abdelfattah et al. (*26*)	Regression models	Function of a set of regressor variables
Bhandari (*27*); Suwardo et al. (*28*); Kumar et al. (*29*)	Linear and non-linear time series models	Used AR, ARIMA, seasonal ARIMA, and non-linear time series models
Vanajakshi et al. (*30*); Padmanabhan et al. (*31*); Kumar et al. (*3*); Kumar et al., (*32*)	State space models	Used Kalman Filter to estimate state space models
Chien et al. (*33*); Yu et al. (*34*); Fan et al. (*35*)	Artificial neural networks	Used ANN to predict bus travel times

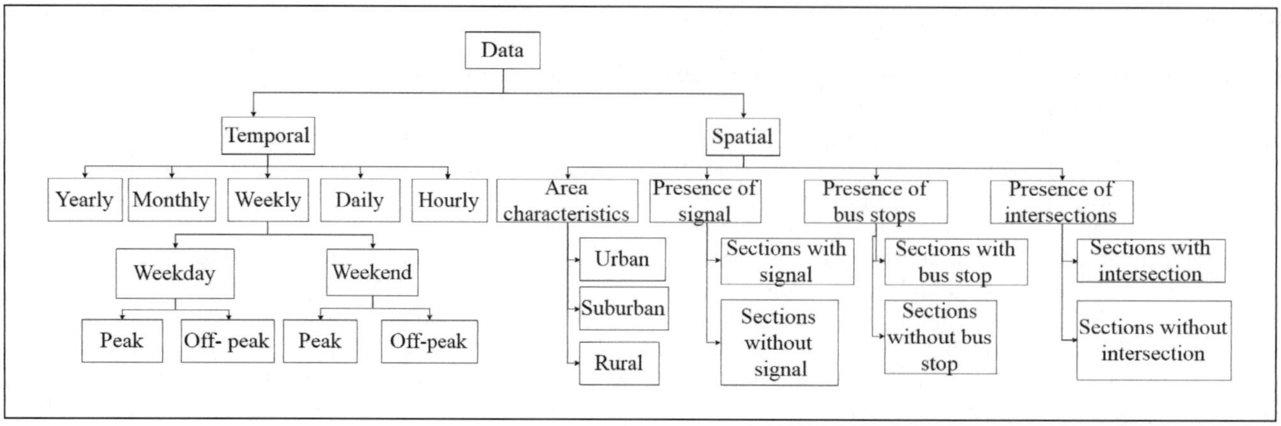

Figure 1. Data hierarchy.

demand forecasting using PSF technique, with SOM as the clustering technique. It was shown that PSF which used SOM as initial classifier worked better when compared with k-means as the initial classifier.

From the review of the literature, it was seen that not many studies which used clustering as a pre-processing tool for prediction involved mixed traffic conditions. In addition, in the majority of these studies, patterns in travel time were identified manually using chronological factors. There is a lack of an automated method for travel time prediction based on bus GPS data by identifying the trends in travel times. Considering these limitations, the present study aims to:

1. Identify natural groupings in the data using popular clustering techniques such as k-means, hierarchical, and SOM clustering.
2. Carry out the prediction of trends in travel time using PSF technique and compare the prediction performance using each of the clustering techniques.
3. Compare the performance of the clustering + prediction framework with the no clustering case.

Data Collection

The data used for the study were collected using GPS devices fixed to Metropolitan Transport Corporation buses in Chennai, the capital city of the state of Tamil Nadu, India. The 19B bus route was chosen as the study stretch. The route spans a length of 29.4 km and connects Kelambakkam, a suburban area of the city, to Saidapet, a major commercial area of the city. The northbound 19B route was chosen for the analysis. The GPS data were collected every 5 s from the GPS units fitted to the buses. Data were collected from a total of 1,024 trips during a period of 45 days. The obtained GPS data includes date, timestamp, and latitude and longitude of the bus location. The distance between two consecutive GPS points was calculated using the Haversine formula (*38*). For the purpose of analysis, the route was divided into 500-m sections, resulting in 56 sections. The travel times for each 500-m section were calculated using interpolation.

Preliminary Analysis

Travel times vary both spatially and temporally. In addition, travel times may follow yearly, monthly, weekly, daily, and hourly patterns. Many of these patterns are assumed to be recurrent in nature. Within a week, travel time patterns may differ between weekdays and weekends. Travel time also varies spatially and is affected by the characteristics of the area and the presence of signals, bus stops, and intersections. Based on all these, a data hierarchy structure of the travel time is prepared first and is shown in Figure 1. It can be seen from Figure 1 that the travel time data is high dimensional and varies with many factors.

To understand the variation of travel time in each section, the variance of travel time in each section was calculated using available data. Figure 2 shows the variance in travel times calculated for 500-m sections. It can be seen that the travel time along the study stretch is highly variable. The high peaks in the plot correspond to sections with bus stops and/or intersections. For example, the highest variance can be observed for section 47, a section near TIDEL Park, which is a major information technology park in Chennai. This section has a bus stop and two major intersections within 100 m distance. This causes delays, thereby increasing the travel times to a high value. It can also be seen that initial sections, such as section 10, have low travel time variance. The reason

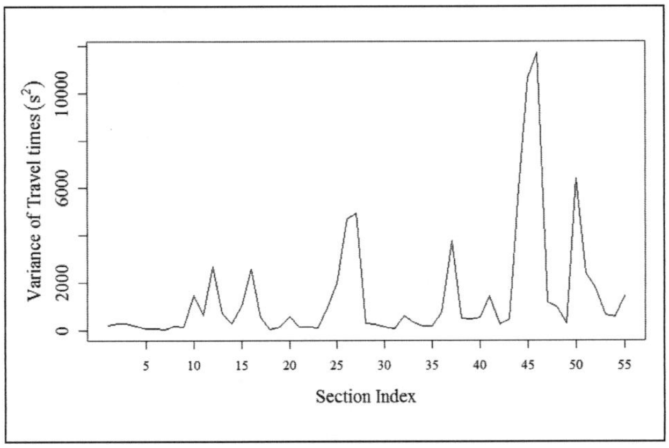

Figure 2. Variance in travel times for each section of route 19B.

could be that these sections are located in the suburban area before the bus enters the city limits. Section 33 was taken as a representative sample section from within the city limits, but with no bus stops and/or intersections. It can be observed that the variance of travel time in this section is not as high as section 47, nor as low as section 10. These three sections were chosen as sample sections for further analyses, namely, section 10 for low travel time, section 33 for urban section with no bus stops or intersections, and section 47 for urban section with a bus stop and intersection.

Proposed Methodology

The proposed prediction methodology uses a PSF algorithm. This algorithm involves two main steps: clustering and prediction. The data is initially clustered and cluster labels are obtained. Further, the prediction is carried out based on the cluster labels. This method can be applied to predict travel times in a time series, either one-dimensional or multi-dimensional. Figure 3 shows the prediction framework used in this study. A detailed description of all the steps in the algorithm is provided in the following sections.

Extraction of Travel Time Trends

A signal can be considered to be a combination of a trend component and a high-frequency component. The definition of trend is subjective based on the application. The trend, in general, can be defined as the predictable

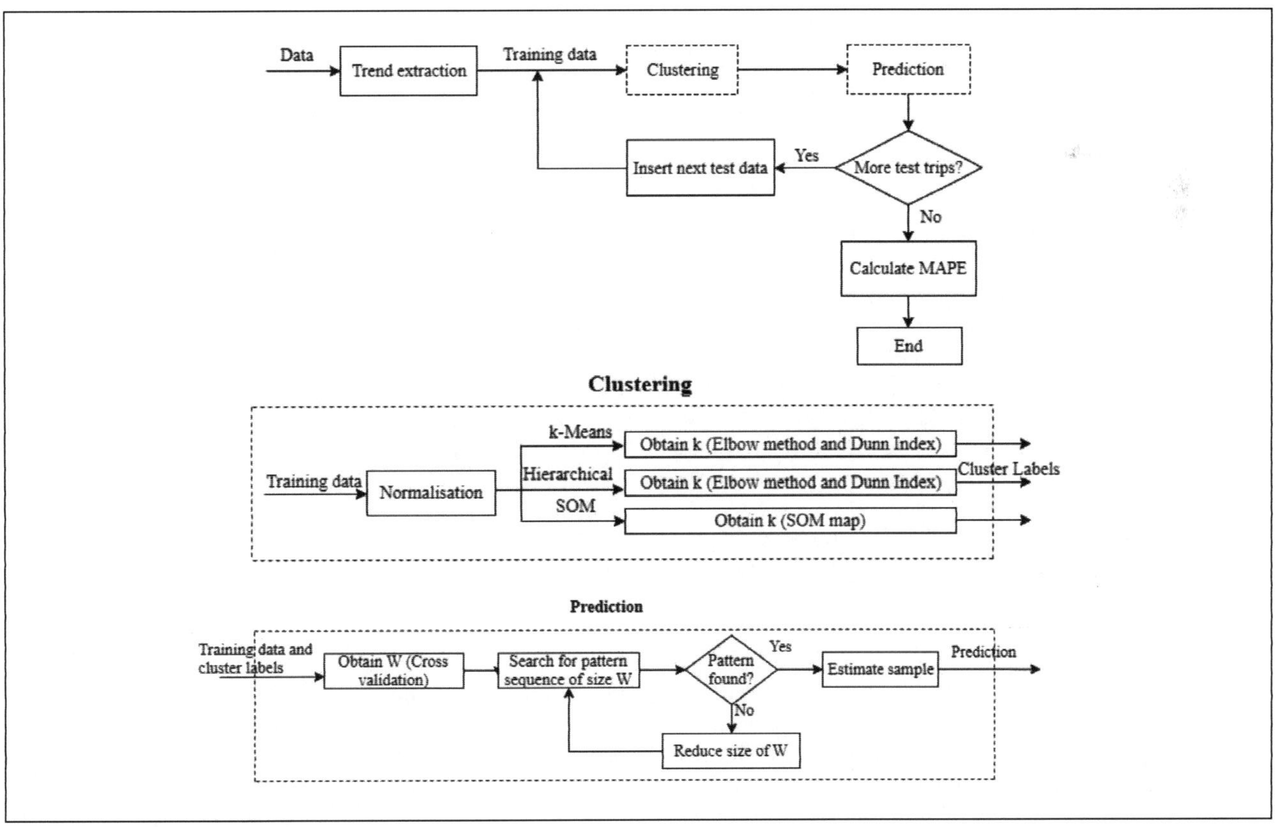

Figure 3. Proposed methodology of PSF.

component of a signal, which will be predominantly the low-frequency part. It is the smoothened part of the signal and holds significant information contained in the signal. The travel time data under consideration is highly variable, consisting of both trends and high-frequency variations. In this study, the travel time trends extracted from the travel time measurements were used as regressors for prediction, while the high-frequency variations will be handled by a more sophisticated algorithm later.

Trends can be extracted from time series data using various techniques such as moving average filtering (*39*), median filtering (*40*), and discrete wavelet transforms (DWT) (*41*). Wavelet-based trend extraction was reported to be more efficient as it can reconstruct the signal without loss of much information (*41*). Hence, in this study, the trends in travel time were extracted using DWT. For this, the travel time data is first decomposed into wavelets at different levels of resolution. The wavelet coefficients at the lower levels of resolution were then thresholded using soft policy. In soft policy, the wavelet coefficients less than the threshold value are set to zero, whereas the threshold value is subtracted from those wavelet coefficients which are greater than the threshold. The threshold levels were chosen such that the trend extracted captures at least 70% of the signal. After thresholding, the signal is reconstructed. This reconstructed signal now contains the trends extracted.

Clustering

Clustering is a data-driven technique which can be used to find k clusters from the group of data such that objects belonging to the same cluster are similar and dissimilar to the objects in the other clusters. The type of clustering technique to be used and the number of clusters into which the data should be divided into are the two major questions that should be answered. From the literature review, it was seen that k-means, hierarchical, and SOM are three popular clustering techniques used to group highly varying data. Hence, these three clustering algorithms were chosen for further analysis. The travel time trends of each of the selected sections were clustered using k-means, hierarchical, and SOM separately and were used as regressors to the prediction. Based on the predictions made, performance analysis of the three clustering algorithms was done to identify the clustering technique suitable for each scenario.

The travel time trend data used for clustering was scaled to improve the convergence properties of the clustering algorithms (*42*). The optimum number of clusters (k) for each of the clustering algorithms needs to be determined. For k-means clustering and hierarchical clustering, many validation indices are available to find the optimum number of clusters. Out of these, two popular methods were chosen: elbow method (*43*), and Dunn index (*44*). For SOM, the number of optimum clusters (k) was identified from the U-Matrix (*45*). The optimum number of clusters k using k-means, hierarchical clustering, and SOM, respectively are: Section 10: {9, 5, 3}; Section 33: {9, 6, 4} and Section 47: {6, 6, 5}.

Pattern Sequence-Based Forecasting (PSF)

Given the travel time trends of a section up to trip d, our aim is to predict the travel time trend of the next trip $d+1$ for the same section. Let $Y(i)$ be the travel time trend of the section for the ith trip and $C(i)$ €{1,…,k} be the cluster label for the ith trip. Let $S_w^{\ i}$ denote the sequence of cluster labels of the section travel time trend for W consecutive trips, from day i backward.

$$S_W^i = [C_{i-W+1}, C_{i-W+2}, \ldots, C_i] \tag{1}$$

The prediction algorithm then searches for the sequence of labels of length $S_W{}^d$ in the database to obtain the subsequence set ES_{d+1}.

$$ES_{d+1} = \{j \text{ such that } S_W^j = S_W^d\}. \tag{2}$$

In case no matching pattern is found, the value of W is reduced by 1 and the pattern search is again initiated. Once the pattern is identified, the predicted value of travel time trend for the next trip $d+1$ is obtained by averaging the travel time trends of the members in the set ES_{d+1} as:

$$\hat{Y}(d+1) = \frac{1}{size(ES_{d+1})} \sum_{j \in ES_{d+1}} Y(j+1), \tag{3}$$

where $size(ES_{d+1})$ is the number of elements in the set ES_{d+1}. The process is repeated until all the test trips are predicted.

Determining the Number of Previous Trip Labels.

The number of previous trip labels to be considered (W) for the PSF algorithm is another parameter that needs to be determined. The PSF algorithm searches the label sequence of W previous trips within the training data set in order to make a prediction. The value of W should be chosen such that the forecasting errors, as given in Equation 4, are minimized.

$$\sum_{d \in ts} \left\| \hat{Y}(d+1) - Y(d+1) \right\|, \tag{4}$$

where $\hat{Y}(d+1)$ is the predicted travel time trend of the$(d+1)^{th}$ trip and $Y(d+1)$ is the measured travel time trend of the $(d+1)^{th}$ trip and ts denotes the training dataset. However, to find the value of $\hat{Y}(d+1)$, the members in the set ES_{d+1} should be known, which again depends on the value of W. Hence the value of W was found using cross-validation.

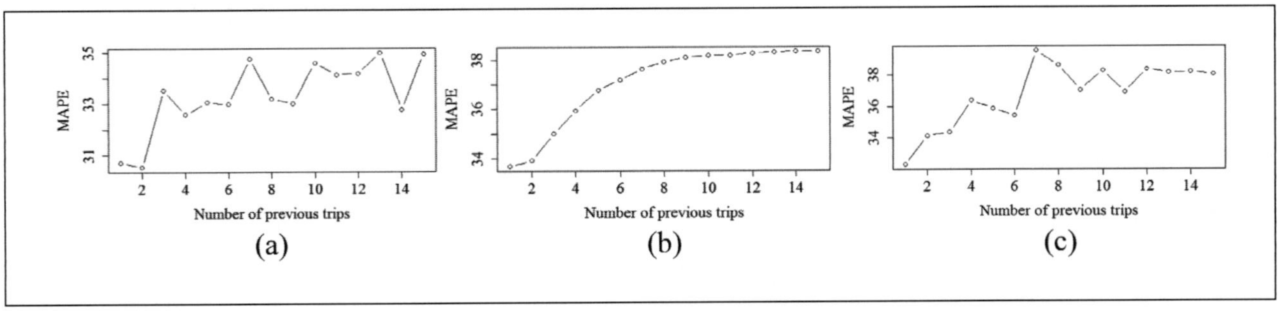

Figure 4. *W* for section 47 using (*a*) *k*-means, (*b*) hierarchical, and (*c*) SOM clustering.

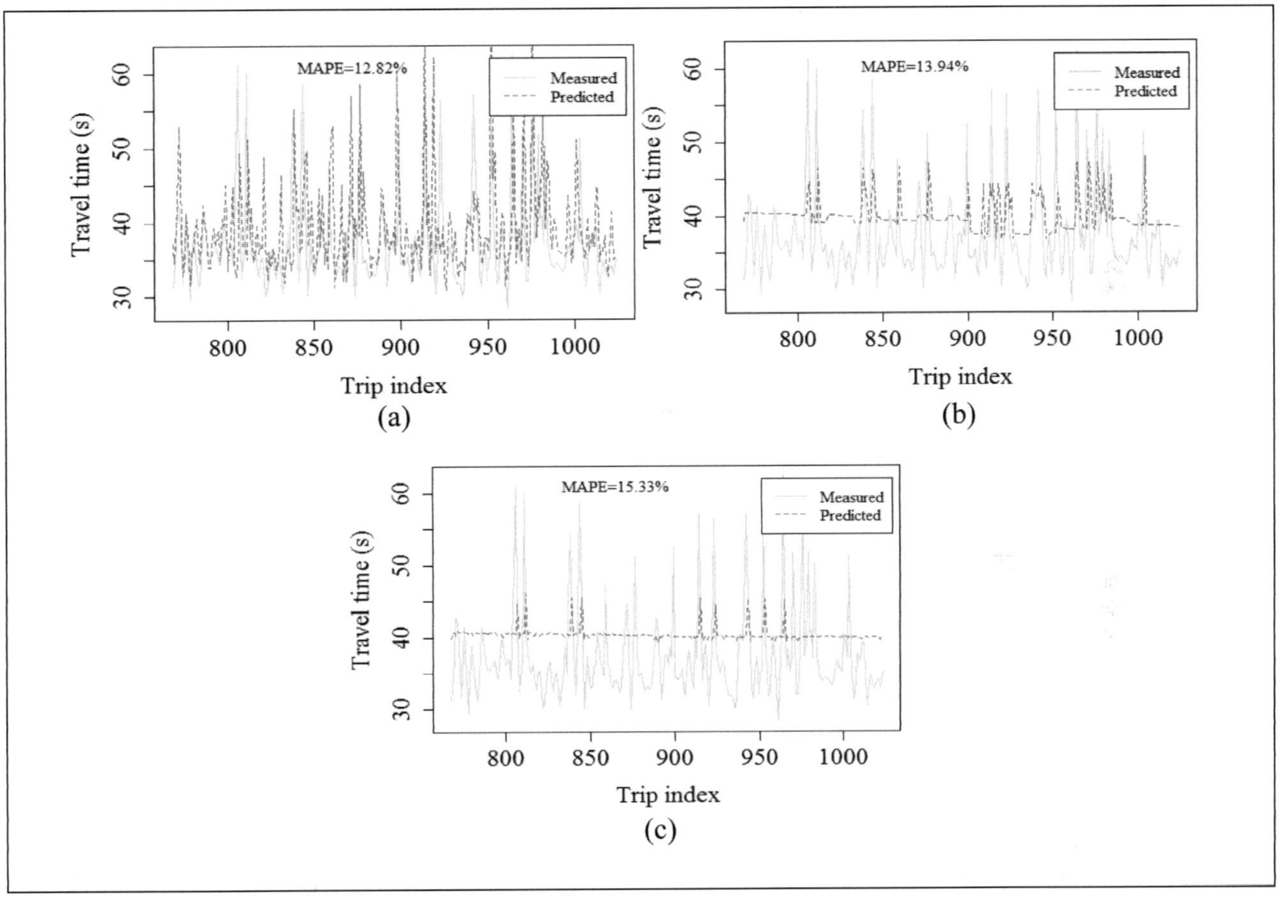

Figure 5. Measured and predicted travel time trends of section 10 using (*a*) *k*-means, (*b*) hierarchical, and (*c*) SOM clustering.

Cross-validation is the process of dividing a data set into a training data set for training the model and a test data set to evaluate the performance of the model (*46*). The traditional cross-validation can be applied only to data sets where the observations are independent. Hence, cross-validation is done on a rolling basis in this study (*47*). A 10-fold cross-validation is used for our study. The data set is divided into 10 partitions. In fold 1: partition [1] was used as training and partition [2] as the test set. In fold 2: partitions [1] and [2] were used as training and partition [3] was used as the test set. This process is repeated for all the folds. The mean absolute percentage error for cross-validation ($MAPE_{cv}$) was calculated for every fold, varying W. The average value of $MAPE_{cv}$ (e_j) is then calculated for each window size by averaging across all 10 folds, as shown in Equation 5. W is selected as that value which minimizes the average value of $MAPE_{cv}$ as shown in Equation 6.

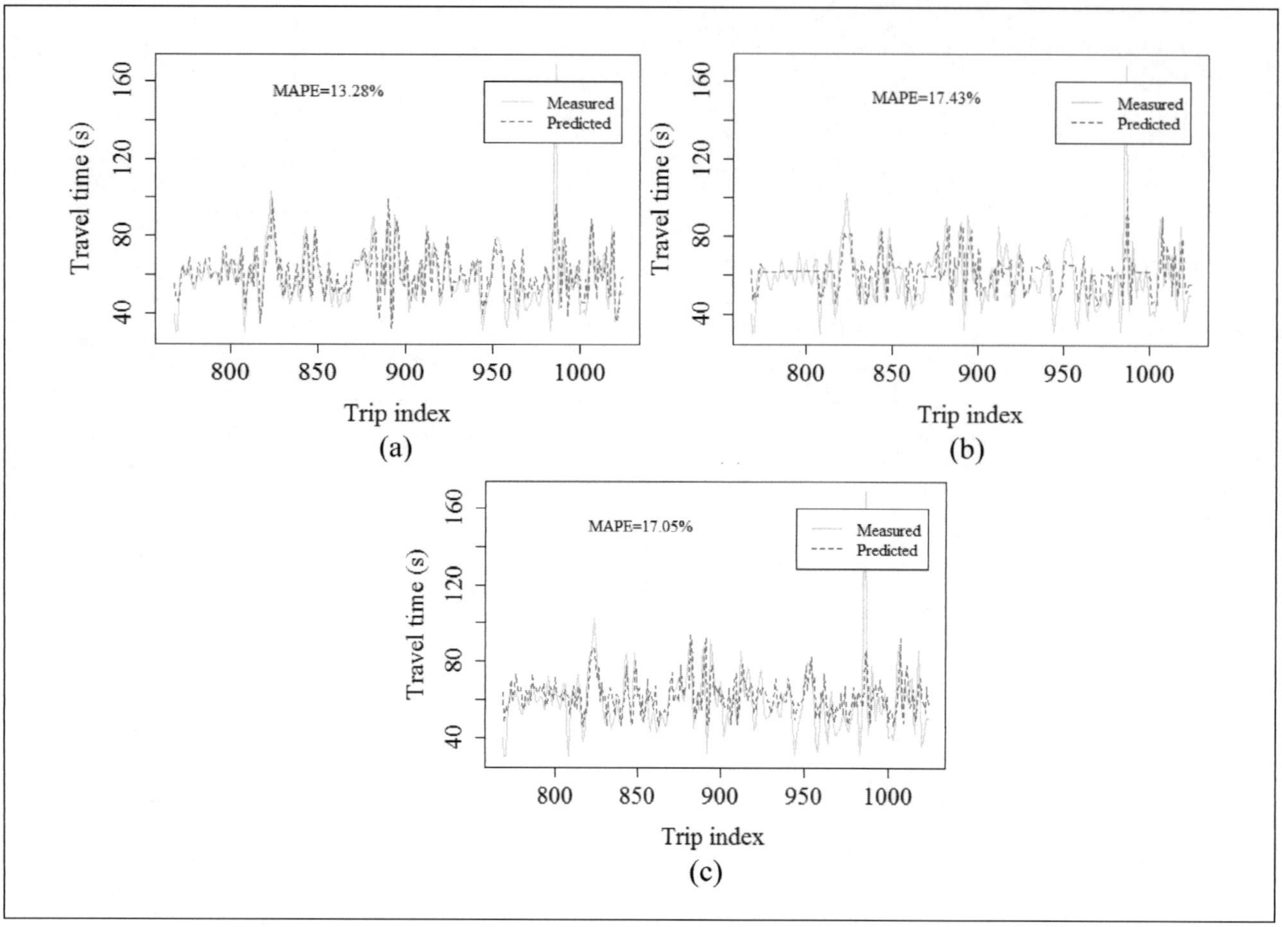

Figure 6. Measured and predicted travel time trends of section 33 using (*a*) *k*-means, (*b*) hierarchical, and (c) SOM clustering.

$$e_j = \frac{1}{10}\sum_{i=1}^{10} MAPE_{cv}\{W = j\}, \quad (5)$$

$$W = \arg\min\{e_j\}, \quad (6)$$

where j varies from 1 to W_{max}. Figure 4*a, b* and *c* below shows the sample plot of rolling cross-validation performed for travel time trends of section 47 using *k*-means, hierarchical, and SOM clustering, respectively. The values of W are chosen as 2, 1, and 1 for section 47 using *k*-means, hierarchical, and SOM clustering, respectively. A similar analysis was done for sections 10 and 33. The value of W obtained using *k*-means, hierarchical, and SOM clustering, respectively, are: Section 10: {3, 1, 1}; Section 33: {1, 2, 1} and Section 47: {2, 1, 1}.

Performance Evaluation

The PSF algorithm was implemented for each of the sections using the optimum number of clusters (k) and number of previous trip labels (W) obtained, as discussed above. The performance accuracy of each of the clustering algorithms for each section was quantified using mean absolute percentage error (MAPE) as follows:

$$MAPE = \frac{1}{n}\sum_{i=1}^{n}\frac{|\hat{Y} - Y_m|}{Y_m}100, \quad (7)$$

where $\hat{Y}$ is the predicted value of travel time trend, Y_m is the corresponding measured travel time trend, and n is the number of observations in the test data set. Figure 5*a*, *b*, and *c* show the measured versus the predicted travel time trends of section 10 using *k*-means, hierarchical, and SOM clustering. It can be seen that *k*-means works better for this section, which had low variance in travel times. Both hierarchical and SOM fail to capture the variations for the low variance section.

Figure 6 shows the measured versus the predicted travel time trends for section 33, which is an urban section without any signals or intersections. It can be seen that *k*-means also works best in this case, capturing the variations in travel time trends better than the other

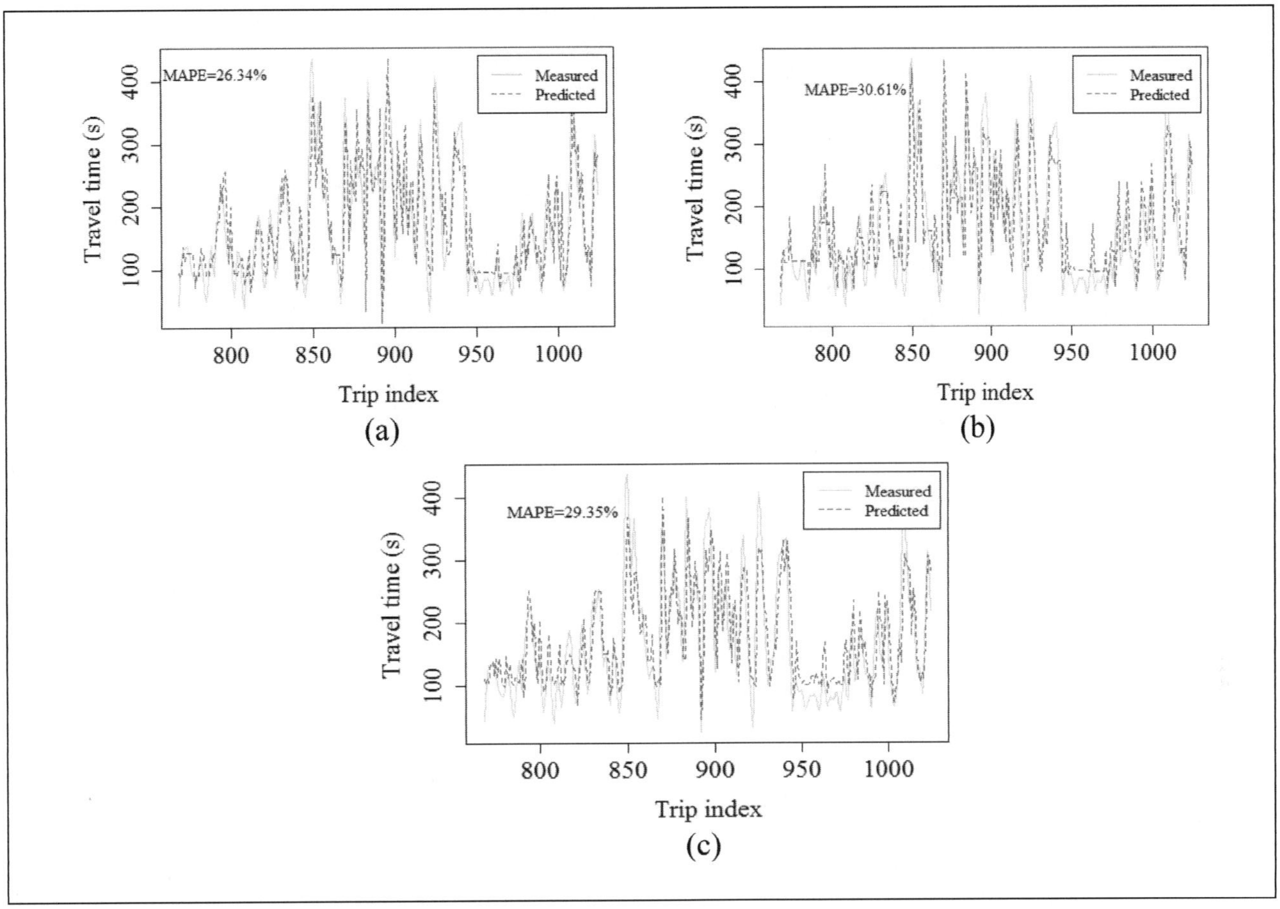

Figure 7. Measured and predicted travel time trends of section 47 using (*a*) *k*-means, (*b*) hierarchical, and (*c*) SOM clustering.

methods. Hierarchical clustering gives poor performance in this case too, whereas the performance of SOM has improved.

Figure 7 shows the results for section 47, which is the section with the highest variance in travel times along the selected route. It can be seen that the predictions based on *k*-means clustering work best, yielding a low value of MAPE and also capturing the variations in travel time trends. The predictions based on hierarchical clustering remain poor, whereas those based on SOM have improved. Hence it was concluded that *k*-means works best for all the sections under study. The trend predictions based on hierarchical clustering remain poor in all three cases, whereas the performance of SOM improves as the section variance increases.

Clustering prior to prediction helps to improve the prediction accuracy. The results of the clustering + prediction framework obtained using the PSF algorithm were compared with the case when no clustering was done. In the no clustering case, predictions were obtained by taking the average of the previous W travel time trends of the same section. Figure 8 shows the comparison between the predictions made using the PSF algorithm and the no clustering case for a sample section on the seven days in the test dataset. It can be seen that, in all the cases, the MAPE obtained using clustering + PSF algorithm was lower compared with the no clustering case, indicating that the clustering helped to improve the prediction accuracy.

Summary and Conclusion

Travel time is a highly variable quantity, which depends on many factors, such as road conditions, weather conditions, driver behavior, etc. Addressing the high variability in travel time is a challenging task. Travel time prediction becomes even more complex in mixed traffic conditions, such as those prevailing in India. The present study identifies the proper regressors to be used for prediction of travel time trends, by comparing the performance of three popular clustering techniques—*k*-means, hierarchical, and SOM—for road sections with different

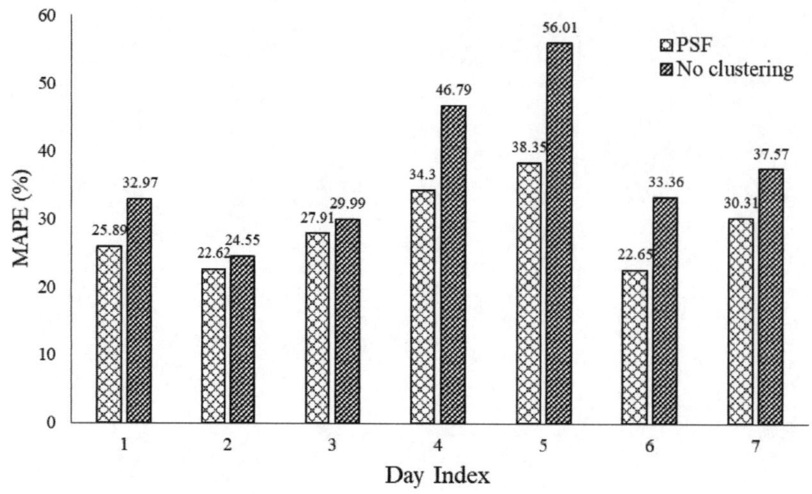

Figure 8. MAPE obtained for all days using PSF algorithm and no clustering case.

characteristics. The following conclusions were drawn from the study:

1. Due to the high variability in travel time, manually grouping travel times into weekdays, weekends, or day-wise is not a very efficient method. Data-driven techniques such as clustering can be used to identify the natural groupings in the data automatically. This becomes possible with the availability of huge amounts of historical data from automated sensors.
2. A clustering + prediction framework was developed using a PSF algorithm to predict travel time trends. It was seen that, for all the three sections with different characteristics which were analyzed, PSF based on *k*-means clustering performed well. The performance of hierarchical clustering remained poor in all the cases, whereas PSF based on SOM was able to capture the variations more effectively as the variation in travel time trends increased. In the later stages of the study, when the dimensionality of the regressor data increases, predictions based on *k*-means may not be efficient. In this case, clustering techniques which work efficiently with high dimensional data such as SOM might be required.
3. The clustering + prediction framework developed was compared with the case when no clustering was done. It was seen that predictions based on PSF yielded low MAPE value when compared with the case when no clustering was performed on the regressor dataset. Hence it was concluded that clustering prior to prediction helps to improve the accuracy of the prediction.

Acknowledgments

The authors acknowledge the Centre of Excellence in Urban Transport, IIT Madras, for the data collection.

Author Contributions

The authors confirm contribution to the paper as follows: study conception and design: Hima Elsa Shaji, Arun K. Tangirala, and Lelitha Vanajakshi; data collection: Shaji and Vanajakshi; analysis and interpretation of results: Shaji, Tangirala, and Vanajakshi; draft manuscript preparation: Shaji, Tangirala, and Vanajakshi. All authors reviewed the results and approved the final version of the manuscript.

References

1. Zhang, M., F. Xiao, and D. Chen. Bus Arrival Time Prediction Based on GPS Data. *Proc., 4th International Conference on Transportation Engineering, ICTE 2013*, China, American Society of Civil Engineers, 2013, pp. 1470–1475.
2. Bin, Y., Y. Zhongzhen, and Y. Baozhen. Bus Arrival Time Prediction Using Support Vector Machines. *Journal of Intelligent Transportation Systems*, Vol. 10, No. 4, 2006, pp. 151–158.
3. Kumar, S. V., L. Vanajakshi, and S. C. Subramanian. A Model Based Approach to Predict Stream Travel Time Using Public Transit as Probes. *Proc., 2011 IEEE Intelligent Vehicles Symposium (IV)*, Baden-Baden, Germany, IEEE, New York, 2011, pp. 101–106.
4. Kwon, J., B. Coifman, and P. Bickel. Day-to-Day Travel-Time Trends and Travel-Time Prediction from Loop-Detector Data. *Transportation Research Record: Journal of the Transportation Research Board*, 2000. 1717: 120–129.
5. Vlahogianni, E. I., M. G. Karlaftis, J. Golias, and N. Kourbelis. Pattern-Based Short-Term Urban Traffic Predictor. *Proc., 2006 IEEE Intelligent Transportation Systems*

Conference, Toronto, Ontario, Canada, IEEE, New York, 2006, pp. 389–393.
6. Kumar, S. V., and L. Vanajakshi. Pattern Identification Based Bus Arrival Time Prediction. *Proc., Institution of Civil Engineers-Transport*, Vol. 167, No. 3, Thomas Telford Ltd., 2014, pp. 194–203.
7. Kumar, B. A., L. Vanajakshi, and S. C. Subramanian. Pattern-Based Bus Travel Time Prediction under Heterogeneous Traffic Conditions. Presented at 93rd Annual Meeting of the Transportation Research Board, Washington, D.C., 2014.
8. Van Der Voort, M., M. Dougherty, and S. Watson. Combining Kohonen Maps with ARIMA Time Series Models To Forecast Traffic Flow. *Transportation Research*, Vol. 4, No. 5, 1996, pp. 307–318.
9. Jain, A. K., M. N. Murty, and P. J. Flynn. Data Clustering: A Review. *ACM Computing Surveys (CSUR)*, Vol. 31, No. 3, 1999, pp. 264–323.
10. Macqueen, J. Some Methods for Classification and Analysis of Multivariate Observations. *Proceedings of the Fifth Berkeley Symposium on Mathematical Statistics and Probability*, Vol. 1, No. 233, 1967, pp. 281–297.
11. Bradley, P. P. S., U. Fayyad, and C. Reina. Scaling Clustering Algorithms to Large Databases. *Proc., 4th International Conference on Knowledge Discovery and Data Mining*. AAAI Press, New York, 1998, pp. 9–15.
12. Kaufman, L., and P. J. Rousseeuw. *Finding Groups in Data: An Introduction to Cluster Analysis*. John Wiley & Sons, Inc., Vol. 344, New Jersey, 2009. http://dx.doi.org./10.1002/9780470316801
13. Shavlik, J. W., and T. G. Dietterich. *Readings in Machine Learning*. Morgan Kaufmann, Inc., San Mateo, California, 1990.
14. Kohonen, T. The Self-Organizing Map. *Proceedings of the IEEE*, Vol. 78, No. 9, 1990, pp. 1464–1480.
15. Ester, M., H. P. Kriegel, J. Sander, and X. Xu. A Density-Based Algorithm for Discovering Clusters in Large Spatial Databases with Noise. *Knowledge Discovery and Data Mining*, Vol. 96, No. 34, 1996, pp. 226–231.
16. Grabusts, P., and A. Borisov. Using Grid-Clustering Methods in Data Classification. *Proc., International Conference on Parallel Computing in Electrical Engineering, PARELEC 2002*, Warsaw, Poland, IEEE, New York, 2002, pp. 425–426.
17. Gan, G., C. Ma, and J. Wu. *Data Clustering: Theory, Algorithms, and Applications*. ASA-SIAM Series on Statistics and Applied Probability, Vol. 20, Philadelphia, 2007.
18. Aghabozorgi, S., A. Seyed Shirkhorshidi, and T. Ying Wah. Time-Series Clustering – A Decade Review. *Information Systems*, Vol. 53, 2015, pp. 16–38.
19. Chung, E. Classification of Traffic Pattern. *Proc., 11th World Congress on ITS, Chicago*, 2003, pp. 687–694.
20. Ladino, A., A. Kibangou, H. Fourati, and C. C. De Wit. Travel Time Forecasting from Clustered Time Series via Optimal Fusion Strategy. *Proc., 2016 European Control Conference, ECC 2016*, Aalborg, Denmark, IEEE, New York, 2017, pp. 2234–2239.
21. Park, D., and L. Rilett. Forecasting Multiple-Period Freeway Link Travel Times Using Modular Neural Networks. *Transportation Research Record: Journal of the Transportation Research Board*, 1998. 1617: 163–170.
22. Weigang, L., W. Koendjbiharie, R. C. de M Juca, Y. Yamashita, and A. MacIver. Algorithms for Estimating Bus Arrival Times Using GPS Data. *Proc., IEEE 5th International Conference on Intelligent Transportation Systems*, Singapore, IEEE, New York, 2002, pp. 868–873.
23. Jeong, R., and L. Rilett. The Prediction of Bus Arrival Time Using AVL Data. Presented at 83rd Annual Meeting of the Transportation Research Board, Washington, D.C., 2004.
24. Bertini, R. L., and A. M. El-Geneidy. Modeling Transit Trip Time Using Archived Bus Dispatch System Data. *Journal of Transportation Engineering*, Vol. 130, No. 1, 2004, pp. 56–67.
25. Patnaik, J., S. Chien, and A. Bladikas. Estimation of Bus Arrival Times Using APC Data. *Journal of Public Transportation*, Vol. 7, No. 1, 2004, p. 1.
26. Abdelfattah, A. M., and A. M. Khan. Models for Predicting Bus Delays. *Transportation Research Record, Journal of the Transportation Research Board*, 1998. 1623: 8–15.
27. Rajbhandari, R. *Bus Arrival Time Prediction Using Stochastic Time Series and Markov Chains*. Ph.D. thesis, New Jersey Institute of Technology, Newark, New Jersey, 2004.
28. Suwardo, W., M. Napiah, and I. Kamaruddin. ARIMA Models for Bus Travel Time Prediction. *Journal of the Institute of Engineers Malaysia*, Vol. 71, No. 2, 2010, pp. 49–58.
29. Kumar, S. V., and L. Vanajakshi. Short-Term Traffic Flow Prediction Using Seasonal ARIMA Model with Limited Input Data. *European Transport Research Review*, Vol. 7, No. 3, 2015, pp. 1–9.
30. Vanajakshi, L., S. C. Subramanian, and R. Sivanandan, R. Travel Time Prediction Under Heterogeneous Traffic Conditions Using Global Positioning System Data from Buses. *IET Intelligent Transport Systems*, Vol. 3, No. 1, 2009, pp. 1–9.
31. Padmanaban, R. P. S., L. Vanajakshi, and S. C. Subramanian. Estimation of Bus Travel Time Incorporating Dwell Time for APTS Applications. *Proc., 2009 IEEE Intelligent Vehicles Symposium*, Xi'an, China, IEEE, New York, 2009, pp. 955–959.
32. Kumar, B. A., L. Vanajakshi, and S. C. Subramanian. Bus Travel Time Prediction Using a Time-Space Discretization Approach. *Transportation Research Part C: Emerging Technologies*, Vol. 79, 2017, pp. 308–332.
33. Chien, S. I. J., Y. Ding, and C. Wei. Dynamic Bus Arrival Time Prediction with Artificial Neural Networks. *Journal of Transportation Engineering*, Vol. 128, No. 5, 2002, pp. 429–438.
34. Yu, B., Z. Z. Yang, K. Chen, and B. Yu. Hybrid Model for Prediction of Bus Arrival Times at Next Station. *Journal of Advanced Transportation*, Vol. 44, No. 3, 2010, pp. 193–204.
35. Fan, W., and Z. Gurmu. Dynamic Travel Time Prediction Models for Buses Using only GPS Data. *International Journal of Transportation Science and Technology*, Vol. 4, No. 4, 2015, pp. 353–366.

36. Alvarez, F. M., A. Troncoso, J. C. Riquelme, and J. S. A. Ruiz. Energy Time Series Forecasting Based on Pattern Sequence Similarity. *IEEE Transactions on Knowledge and Data Engineering*, Vol. 23, No. 8, 2011, pp. 1230–1243.
37. Jin, C. H., G. Pok, Y. Lee, H. W. Park, K. D. Kim, U. Yun, and K. H. Ryu. A SOM Clustering Pattern Sequence-Based next Symbol Prediction Method for Day-Ahead Direct Electricity Load and Price Forecasting. *Energy Conversion and Management*, Vol. 90, 2015, pp. 84–92.
38. Chamberlain, R. *Great Circle Distance between Two Points*. 2014. http://www.movable-type.co.uk/scripts/gis-faq-5.1.html.
39. Alvarez-Ramirez, J., E. Rodriguez, and J. C. Echeverría. Detrending Fluctuation Analysis Based on Moving Average Filtering. *Physica A: Statistical Mechanics and its Applications*, Vol. 354, 2005, pp. 199–219.
40. Fried, R., J. Einbeck, and U. Gather. Weighted Repeated Median Smoothing and Filtering. *Journal of the American Statistical Association*, Vol. 102, No. 480, 2007, pp. 1300–1308.
41. Dash, S., M. R. Maurya, V. Venkatasubramanian, and R. Rengaswamy. A Novel Interval-Halving Framework for Automated Identification of Process Trends. *American Institute of Chemical Engineers Journal*, Vol. 50, No. 1, 2004, pp. 149–162.
42. Neuristique, L. B., and Y. Bengio. Convergence Properties of the *K*-Means Algorithms. *Proc., 7th International Conference on Neural Information Processing Systems*, Denver, Colo., MIT Press, Cambridge, Mass., 1995, pp. 585–592.
43. Thorndike, R. L. Who Belongs in the Family? *Psychometrika*, Vol. 18, No. 4, 1953, pp. 267–276.
44. Dunn, J. C. Well-Separated Clusters and Optimal Fuzzy Partitions. *Journal of Cybernetics*, Vol. 4, No. 1, 1974, pp. 95–104.
45. Ultsch, A. Self-Organizing Neural Networks for Visualization and Classification. In *Information and Classification* (Opitz, O., B. Lausen, and R. Klar, eds.), Springer-Verlag, Berlin, 1993, pp. 307–313.
46. Kohavi, R. A Study of Cross-Validation and Bootstrap for Accuracy Estimation and Model Selection. *Proc., International Joint Conference on Artificial Intelligence (IJCAI)*, Canada, Morgan Kaufmann, Inc., California, Vol. 5, 1995, pp. 1–7.
47. Arlot, S., and A. Celisse. A Survey of Cross-Validation Procedures for Model Selection. *Statistics Surveys*, Vol. 4, 2010, pp. 40–79.

The Standing Committee on Artificial Intelligence and Advanced Computing Applications (ABJ70) peer-reviewed this paper (18-05451).

TRR
JOURNAL OF THE TRANSPORTATION RESEARCH BOARD

Article

Transportation Research Record
2018, Vol. 2672(45) 253–264

DOI: 10.1177/0361198118798734
journals.sagepub.com/home/trr
SAGE

A Kalman Filter-Based Queue Length Estimation Method with Low-Penetration Mobile Sensor Data at Signalized Intersections

Juyuan Yin[1], Jian Sun[1], and Keshuang Tang[2]

Abstract

Queue length estimation is of great importance for signal performance measures and signal optimization. With the development of connected vehicle technology and mobile internet technology, using mobile sensor data instead of fixed detector data to estimate queue length has become a significant research topic. This study proposes a queue length estimation method using low-penetration mobile sensor data as the only input. The proposed method is based on the combination of Kalman Filtering and shockwave theory. The critical points are identified from raw spatiotemporal points and allocated to different cycles for subsequent estimation. To apply the Kalman Filter, a state-space model with two state variables and the system noise determined by queue-forming acceleration is established, which can characterize the stochastic property of queue forming. The Kalman Filter with joining points as measurement input recursively estimates real-time queue lengths; on the other hand, queue-discharging waves are estimated with a line fitted to leaving points. By calculating the crossing point of the queue-forming wave and the queue-discharging wave of a cycle, the maximum queue length is also estimated. A case study with DiDi mobile sensor data and ground truth maximum queue lengths at Huanggang-Fuzhong intersection, Shenzhen, China, shows that the mean absolute percentage error is only 11.2%. Moreover, the sensitivity analysis shows that the proposed estimation method achieves much better performance than the classical linear regression method, especially in extremely low penetration rates.

Vehicle queues have been recognized as a key factor for traffic flow dynamics at signalized intersections. An accurate and high-resolution estimation method of queue length is of great importance for both performance measures in terms of delay, stops and travel time (*1*), and signal optimization (*2–6*). The methods for queue length estimation can be generally categorized into two classes: the first is based on the analysis of cumulative traffic input–output of a single link (*2*, *7*, *8*), and the other is based on Lighthill–Whitham–Richards (LWR) shockwave theory (*9–13*). Compared with the first one, the methods of the second approach can provide the temporal-spatial dynamics of the queuing process with input data from loop detectors or probe vehicles. Existing methods of queue length estimation are mostly based on data from fixed detectors, either aggregate level or event based (*11*, *13*). However, most of these methods have limitations in that the accuracy of estimation depends on locations of detectors; more specifically, when queues exceed the position of detector, the accuracy of queue length estimation degrades. And detector errors caused by a high rate of damaged detectors in real-world situations also have impacts on the performance of estimation.

With the latest developments in connected vehicle technology and mobile internet technology, there is a growing trend for the application of mobile sensors. Previous studies involved with mobile sensor data have focused on road traffic state estimation. Seo et al. proposed a method that simply aggregates data from ADAS-equipped probe vehicles to estimate flow and density with traffic flow models given by a fundamental

[1]Department of Transportation Engineering, Key Laboratory of Road and Traffic Engineering, Ministry of Education, Tongji University, Shanghai, China
[2]Department of Transportation Information and Control Engineering, Key Laboratory of Road and Traffic Engineering, Ministry of Education, Tongji University, Shanghai, China

Corresponding Author:
Address correspondence to Jian Sun: sunjian@tongji.edu.cn

diagram (FD) and they improved their method later without an assumed FD (*14*, *15*). Hofleitner et al. used sparse probe data to estimate arterial traffic state based on a probabilistic framework (*16*). Utilizing probe data, Ramezani and Geroliminis developed a method based on the link travel times of probe vehicles to estimate the arterial travel time distribution by considering spatiotemporal non-linear correlations (*17*).

As for queue length estimation based on mobile sensor data, some studies show a preference to shockwave theory. Ban et al. used sample travel times of probe vehicles to estimate the queue rear no-delay times, then constructed a queue length curve (*10*). Sun and Ban proposed a vehicle trajectory reconstruction method to estimate the missing parts of trajectories, and estimated long queue length (*18*). Cheng et al. identified critical points from high-resolution vehicle trajectory data, and estimated queue length with fitting curves to critical points for queue-forming and discharging waves (*19*). Ramezani and Geroliminis proposed a queue length estimation method with no information about signal settings and arrival patterns, and modeled the queue-forming and discharging wave by piecewise linear function and achieved better performance than the uniform arrival procedure (*20*). There are still some major challenges:

- Some of the studies rely on assumptions such as known arrival patterns and known signal timings, which may be unrealistic in practical applications;
- Most of the studies based on shockwaves estimate the queue-forming wave without sufficient consideration of inherent dynamics of queue forming, for example, methods of linear or non-linear fitting within a single cycle consider neither data errors nor systematic correlations with other cycles;
- Accuracy of the estimation methods in low-penetration situations (e.g., $<10\%$) which may be usual in practice, is not satisfying.

To address the challenges, this study proposes a queue length estimation method with low-penetration mobile sensor data as the only input, which combines Kalman Filtering with shockwave theory. More specifically, the main contributions of this study are: (1) to introduce the processing method of dynamic mobile phone GPS data (low sampling interval and low penetration) collected from the DiDi Chuxing App; (2) to establish a state-space model which characterizes the queue-forming dynamics; (3) to apply the Kalman Filter in estimation of queue-forming waves, simultaneously considering data measurement errors and system errors, which facilitates the accuracy in low-penetration situations; (4) to realize the estimation of queue lengths as well as maximum queue lengths; and (5) to test the performance of the proposed method in both real-world situation and simulation scenarios with different penetration rates and intersection geometrics.

Background

In this section, basic definitions are first given, and then several assumptions made in this study are explained. Two main theory foundations, the shockwave model for signalized intersections and the standard discrete Kalman Filter, are briefly introduced.

Definitions

Queuing Status and Moving Status. There are two defined types of motion status, that is, queuing and moving, during the process of vehicles' passing through the intersection. Queuing status refers to vehicles approaching with a speed very close to zero or fully stopped at the back of queue. Moving status refers to vehicles moving at a required speed.

Back of Queue, Queue Length and Maximum Queue Length. The back of queue is defined as the position of the last one of all vehicles in queuing status at a certain timestamp. Queue length is defined as the total length (or number) of vehicles in queuing status of a lane at a certain timestamp. Queue length is also understood as the distance from the back of a queue to the stop bar. Maximum queue length is defined the maximum value of queue lengths within a cycle. As for multilane, the above three take the average of all lanes in the same lane group.

Assumptions

- Signal settings such as cycle length and splits are assumed to be unknown, while the bounds of the settings are assumed known. That is, signal settings may change unpredictably, but the bounds of the settings such as minimum cycle length and minimum green, are normally pre-determined, thus available.
- Volumes and arrival patterns will not change sharply within a short period, and moving speed among vehicles approximate a desired value.

Shockwave Model for Signalized Intersection

Shockwaves are usually used to model the process of formation and dissipation of queue at signalized intersections, based on LWR theory (*9*). Shockwaves are boundaries dividing different traffic states determined by volume and density which are correlated in a FD. Figure 1*a* shows a classic triangular FD. The boundaries

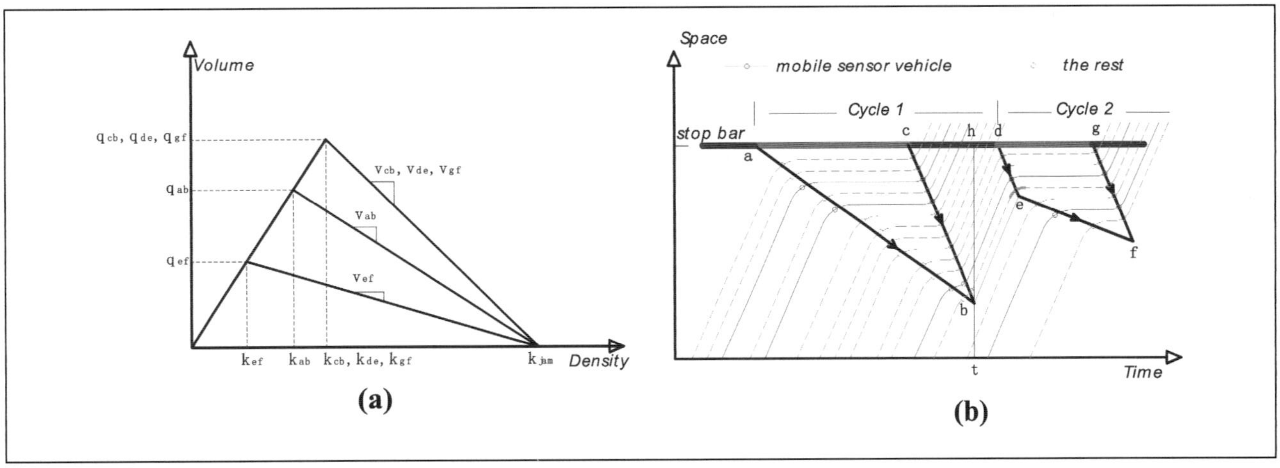

Figure 1. Illustrations of fundamental diagrams and shockwaves at signalized intersections. (*a*) Fundamental diagrams and (*b*) shockwaves.

formed by time-space points, of which speed changes abruptly, characterize the features of interrupted traffic flow with an expression of the queue formation and dissipation. These points, called critical points in the following sections, are elements for constructing shockwaves.

Figure 1*b* depicts a regular process of queue formation and dissipation at signalized intersections with a shockwave model. In a normal cycle, when vehicles reach the intersection during the red phase they are stopped to join a queue successively, and this state of queue formation propagates backward as a mode of wave (see line *ab* and the speed is *Vab*), called a queue-forming wave. The distance from point on the wave to the stop bar is the queue length, thus it increases. Once the green phase starts, those queuing vehicles orderly leave the queue, and this state of queue dissipation also propagates backward, and is called the queue-discharging wave (see line *cb* and *gf*). During a cycle, if we regard the start of the red phase as the cycle start, the queue-discharging wave chases the queue-forming wave with a higher speed, so they finally meet at a point which indicates the maximum queue length of a cycle. For example, *cb* and *ab* meet at point *b*, so *bh* is not only the queue length at time *t*, but also the maximum queue length of *Cycle 1*. When a cycle is oversaturated, a residual queue from this cycle will form the initial queue in the next cycle (see line *de* in *Cycle 2*). This part of the queue-forming wave, with a speed approximating the queue-discharging speed, should be distinguished, because it cannot characterize the feature of the naturally arrived traffic flow.

The Standard Discrete Kalman Filter

The Kalman Filter (*21*) is simply an optimal recursive data-processing algorithm, which can be used widely in processes such as signal noise filtering and estimation of parameters. For a linear system with measurement and system noises drawn from a zero-mean Gaussian distribution, the Kalman Filter has been proved the optimal estimator because of its minimum mean squared errors. Consider a discrete-time linear system based on the state-space model described by state equation (Equation 1) and measurement equation (Equation 2):

$$\mathbf{X}_k = \mathbf{\Phi}_{k|k-1}\mathbf{X}_{k-1} + \mathbf{w}_k \tag{1}$$

$$\mathbf{Z}_k = \mathbf{H}_k\mathbf{X}_k + \mathbf{v}_k \tag{2}$$

with $\mathbf{w}_k$ and $\mathbf{v}_k$ satisfying

$$E[\mathbf{w}_k] = 0 \tag{3}$$

$$E[\mathbf{v}_k] = 0 \tag{4}$$

$$E[\mathbf{w}_k\mathbf{w}_{k'}{}^T] = \begin{cases} \mathbf{Q}_k, k = k' \\ 0, k \neq k' \end{cases} \tag{5}$$

$$E[\mathbf{v}_k\mathbf{v}_{k'}{}^T] = \begin{cases} \mathbf{R}_k, k = k' \\ 0, k \neq k' \end{cases} \tag{6}$$

where, $\mathbf{X}_k$ is $n \times 1$ state vector, $\mathbf{Z}_k$ is $m \times 1$ measurement vector, $\mathbf{\Phi}_{k|k-1}$ is $n \times n$ state transition matrix, $\mathbf{H}_k$ is $m \times n$ measurement matrix, $\mathbf{w}_k$ is $n \times 1$ system noise vector, $\mathbf{v}_k$ is $m \times 1$ measurement noise vector, $\mathbf{Q}_k$ is $n \times n$ covariance matrix of $\mathbf{w}_k$, and $\mathbf{R}_k$ is the $m \times m$ covariance matrix of $\mathbf{v}_k$.

The standard Kalman Filter updates the process with two stages: state update (Equations 7, 8) and measurement update (Equations 9–11) to recursively calculate the estimated state vector $\hat{\mathbf{X}}_t$:

$$\hat{\mathbf{X}}_{k|k-1} = \mathbf{\Phi}_{k|k-1}\hat{\mathbf{X}}_{k-1} \tag{7}$$

$$\mathbf{P}_{k|k-1} = \mathbf{\Phi}_{k|k-1}\mathbf{P}_{k-1}\mathbf{\Phi}_{k|k-1}{}^T + \mathbf{Q}_{k-1} \tag{8}$$

$$\mathbf{K}_k = \mathbf{P}_{k|k-1}\mathbf{H}_k{}^T(\mathbf{H}_k\mathbf{P}_{k|k-1}\mathbf{H}_k{}^T + \mathbf{R}_k)^{-1} \quad (9)$$

$$\hat{\mathbf{X}}_k = \hat{\mathbf{X}}_{k|k-1} + \mathbf{K}_k(\mathbf{Z}_k - \mathbf{H}_k\hat{\mathbf{X}}_{k|k-1}) \quad (10)$$

$$\mathbf{P}_k = (\mathbf{I} - \mathbf{K}_k\mathbf{H}_k)\mathbf{P}_{k|k-1}(\mathbf{I} - \mathbf{K}_k\mathbf{H}_k)^T + \mathbf{K}_k\mathbf{R}_k\mathbf{K}_k{}^T \quad (11)$$

where, the $n \times n$ matrix $\mathbf{P}_k$ is the error covariance of $\hat{\mathbf{X}}_k$, $\hat{\mathbf{X}}_{k|k-1}$ and $\mathbf{P}_{k|k-1}$ respectively denotes the one-step prediction from $\hat{\mathbf{X}}_{k-1}$ and $\mathbf{P}_{k-1}$, and the $n \times m$ matrix $\mathbf{K}_k$ is the Kalman gain.

Methodology

This study aims to utilize mobile sensor data consisting of positions, timestamps, and vehicle IDs as the only input; that is, without any other information such as signal timings, traffic demands, and vehicle arrival patterns. Mobile sensor data can be described as a series of spatiotemporal points (t_i^k, x_i^k), where: x_i^k is the distance from current position to the stop bar, t_i^k is the timestamp, i is the vehicle ID and k is time step. The sampling intervals are assumed to be in a range of 3–6 s, which is reckoned to be a low sampling interval and is available from civilian technology (e.g., DiDi Chuxing App data collected from mobile phone GPS).

The basic idea of the proposed method includes these key steps: (1) to process the data, with the critical points identified and allocated to different cycles, (2) to estimate the queue-discharging wave in each cycle with the method of linear regression, (3) to establish the state-space model for queue forming and use the Kalman Filter to estimate the queue-forming waves; the queue length and maximum queue length are then estimated. The principal steps of the method are explained in detail in the following subsections.

Step 1: Processing of the Critical Points

In order to construct traffic shockwaves, the critical points, including joining points and leaving points, should be identified and allocated to respective cycles.

Step 1.1: Classification of Vehicles' Motion Status. Vehicles' motion status could be generally classified to queuing status and moving status at signalized intersections, according to the first definition, and the two statuses could be easily determined by a speed threshold. For any (t_i^k, x_i^k), the corresponding motion status M_i^k is calculated by Equations 12 and 13:

$$M_i^k = \begin{cases} 1, if\ v_i^k \leq \bar{v} \\ 0, if\ v_i^k > \bar{v} \end{cases} \quad (12)$$

$$v_i^k = \frac{x_i^{k+\Delta_v} - x_i^{k-\Delta_v}}{t_i^{k+\Delta_v} - t_i^{k-\Delta_v}} \quad (13)$$

where, $M_i^k = 1$ or 0 represents queuing or moving status, v_i^k denotes the point speed calculated by the Δ_v th adjacent points, and, $\bar{v}$ denotes the speed threshold.

Step 1.2: Identification of the Critical Points. As the vehicle moves forward and stops to join the queue, its motion status transfers from moving into queuing at a certain spatiotemporal point, and such points are called joining points. Accordingly, points at which a transition from queuing status to moving status occurs are called leaving points. The critical points are identified by the correlation of adjacent motion statuses, as shown in Equations 14–17. For any (t_i^k, x_i^k) before the stop bar, $x_i^k < x_{stop}$, IP_i^k indicates which class the point belongs to: NCP for non-critical points, PJP and SJP for joining points, and LP for leaving points. Δ_c refers to the length of consideration for consistent motion statuses.

$$IP_i^k = \begin{cases} LP, \text{if} \sum_{j=k+1}^{k+\Delta_c} M_i^j = 0 \text{ and } TAG_i^k = 1 \\ PJP, \text{if} \prod_{j=k+1}^{k+\Delta_c} M_i^j = 1 \text{ and } TAG_i^k = 0 \\ SJP, \text{if} \prod_{j=k+1}^{k+\Delta_c} M_i^j = 1 \text{ and } TAG_i^k = 2 \\ NCP, \text{otherwise} \end{cases} \quad (14)$$

$$TAG_i^k = \begin{cases} 0, \text{if } TAG_i^{k-1} = 0 \text{ and } IP_i^{k-1} = NCP \\ 1, if\ IP_i^{k-1} = PJP \text{ or } IP_i^{k-1} = SJP \text{ or } TAG_i^{k-1} = 1, k>1 \\ 2, if\ IP_i^{k-1} = LP \text{ or } TAG_i^{k-1} = 2 \end{cases} \quad (15)$$

$$TAG_i^1 = 0 \quad (16)$$

$$IP_i^1 = NCP \quad (17)$$

Normally, for a trajectory, if any joining point exists, it ought to be followed by a leaving point, because vehicles stop then leave all the time. TAG is such a variable for alternately identifying joining and leaving points. Note that joining points are further classified: PJP denotes the premier joining points produced by natural arrivals, whereas SJP denotes the secondary joining points produced by a residual queue. This further classification of joining points may contribute to estimation of queue length, which will be discussed later.

Step 1.3: Allocation for Critical Points to Cycles. As shockwaves are associated with cycles, critical points should be allocated to various cycles so that shockwaves can be estimated. Combining the known information of the bounds of signal settings, the leaving points were allocated to different cycles and then the joining points associated to corresponding cycles.

Allocation for Leaving Points. Compared with arrival patterns, vehicles' leaving patterns are relatively steady, because queued vehicles leave the queue one by one when the green phase starts, and they mostly pass through the intersection at a saturated flow. As a result, the speeds of queue-discharging waves vary slightly over the cycles (see Vcb, Vgf in Figure 1*a*). Therefore, given the default speed of the queue-discharging wave VD_{default} and the leaving point $(t_i^k, x_i^k)_{CP_i^k=2}$, then the corresponding timestamp of green start GS_i^k could be estimated by Equation 18:

$$GS_i^k = t_i^k + \frac{x_{\text{stopbar}} - x_i^k}{VD_{\text{default}}} \tag{18}$$

If two time-adjacent leaving points $(t_i^k, x_i^k)_{CP_i^k=2}$ and $(t_{i'}^{k'}, x_{i'}^{k'})_{CP_{i'}^{k'}=2}$ belong to the same cycle, the difference between the two estimated green-start timestamps are supposed to be negligible, that is, $|GS_i^k - GS_{i'}^{k'}| < \varepsilon$. With the known information of the bounds of signal settings, we can derive the minimum time interval between green starts, I_{GS}. Let $\varepsilon = I_{GS}$, the principle for deciding that two time-adjacent leaving points belong to the same cycle is given in Equation 19:

$$|GS_i^k - GS_{i'}^{k'}| < I_{\text{GS}} \tag{19}$$

The procedure of the allocation method is detailed as follows:

- *Substep 0*
  Sort all the leaving points $lp \in LP : \{lp_j : (t_j, x_j) | j = 1, 2, ..., N\}$ in time-ascending order $lp \in LP_{sort} : \{lp^{(s)} : (t^{(s)}, x^{(s)}) | s = 1, 2, ..., N\}$.
- *Substep 1*
  1. Initialize the group number, $Gnum = 1$, and the point number, $Pnum = 1$;
  2. Initialize the leaving points' set of the first cycle, Ω_{LP}^1, with the first leaving point $lp^{(1)}$, that is $\Omega_{LP}^1 : \{lp^{(1)}\}$;
  3. Calculate $GS_{lp^{(1)}}$ by Equation 18;
  4. Set $Pnum = Pnum + 1$.
- *Substep 2*
  1. Calculate $GS_{lp^{(Pnum)}}$ by Equation 18;
  2. Check the Equation 19 and if it holds, add $lp^{(Pnum)}$ to Ω_{LP}^{Cnum}, otherwise set $Gnum = Gnum + 1$, and create Ω_{LP}^{Gnum} with $lp^{(Pnum)}$, that is $\Omega_{LP}^{Gnum} : \{lp^{(Pnum)}\}$;
  3. Set $Pnum = Pnum + 1$;
  4. If $Pnum \leq N$, repeat Substep 2, otherwise, end the procedure.

The proposed allocation method ensures each group Ω_{LP}^n $\forall n = 1, 2, ...$, with at least one leaving point, whereas those cycles with no leaving point will be inevitably neglected because of a lack of signal settings, that is, only the cycle with leaving points will be recognized.

Allocation for Joining Points. As each joining point is paired with a leaving point after the identification, joining points could be simply associated with corresponding leaving points to their respective cycles.

Figure 2 illustrates the allocation of the critical points; leaving points near a certain magenta arrow with their GS_i^k very close to each other are allocated to the same cycle, represented by a green dashed area; the red dashed areas, which denote different cycles of joining points, are associated to the cycles of green dashed areas.

Step 2: Estimation of Queue-Discharging Waves

Based on LWR theory, as the queued vehicles basically pass through the intersection at a saturated flow, the queue-discharging waves have strong linearity. Therefore, the queue-discharging wave for each cycle could be represented by a single line fitted to the leaving points.

Consider cycle n (which actually means the group in which the number of leaving points are non-zero because cycles with no leaving points are not discussed in this section) with $\Omega_{LP}^n : \{(t^{(1)}, x^{(1)}), (t^{(2)}, x^{(2)}), ..., (t^{(N)}, x^{(N)})\}$ denoting the set of leaving points in cycle n, and N denoting the number of leaving points, there are conditions: (a) $N \geqslant 2$ and timestamps of all leaving points are not simultaneously equal, (b) $N \geqslant 2$ and timestamps of all leaving points are simultaneously equal, and (c) $N = 1$.

- For condition (a), $\hat{\beta}_0^n$ and $\hat{\beta}_1^n$ are derived from Equation 20, which denotes a least squares problem:

$$\min \sum_{s=1}^{N} [x^{(s)} - (\hat{\beta}_0^n + \hat{\beta}_1^n \cdot t^{(s)})]^2 \tag{20}$$

- For condition (b) and (c), use the default speed of queue-discharging wave VD_{default} instead of the slope for such cycles because no line can be fitted, that is:

$$\hat{\beta}_1^n = VD_{\text{default}} \tag{21}$$

Then use Equation 22 to calculate the intercept:

$$\hat{\beta}_0^n = \frac{1}{N}\sum_{s=1}^{N} x^{(s)} - \hat{\beta}_1^n \cdot \frac{1}{N}\sum_{s=1}^{N} t^{(s)} \tag{22}$$

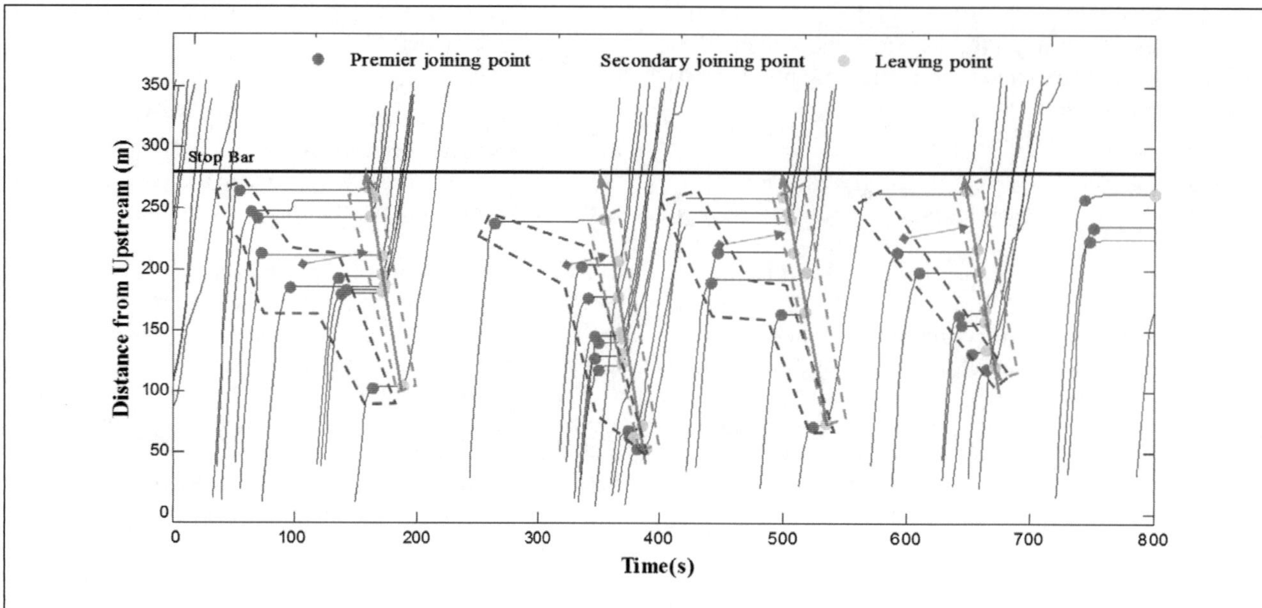

Figure 2. An example of the processing of the critical points.

Step 3: Estimation of Queue-Forming Waves and Queue Lengths

Step 3.1: State-Space Model for a Discrete-Time Queue-Forming System. An appropriate state-space model is needed for application of the Kalman Filter. As the vehicles arrive at intersections discretely and stochastically, the queue forming is considered as a discrete stochastic system with state variables of the back of queue QX_k and the queue-forming speed VF_k at time step k. Let us assume the gradient of VF_k, that is the acceleration of queue forming, a_k, is a white Gaussian noise with:

$$E[a_k] = 0 \tag{23}$$

$$E[a_k a_{k'}{}^T] = \begin{cases} \frac{q}{T}, k = k' \\ 0, k \neq k' \end{cases} \tag{24}$$

where q is covariance intensity of a_k and T refers time interval between time steps k and $k\text{-}1$. Then, the state equation (Equation 1) of the system could be established as:

$$\begin{bmatrix} QX_k \\ VF_k \end{bmatrix} = \begin{bmatrix} 1 & T \\ 0 & 1 \end{bmatrix} \begin{bmatrix} QX_{k-1} \\ VF_{k-1} \end{bmatrix} + \begin{bmatrix} T^2/2 \\ T \end{bmatrix} a_k \tag{25}$$

where, the state vector $\mathbf{X}_k = \begin{bmatrix} QX_k \\ VF_k \end{bmatrix}$, $\mathbf{\Phi}_{k|k-1} = \begin{bmatrix} 1 & T \\ 0 & 1 \end{bmatrix}$, $\mathbf{w}_k = \begin{bmatrix} T^2/2 \\ T \end{bmatrix} a_k$.

With mobile sensor data, queue length is observable. So the measurement matrix from measurement equation (Equation 2) is given $\mathbf{H}_k = [1 \quad 0]$.

Step 3.2: Joining Points' Sub-Grouping for Measurement Input. After the above processing, there are probably two problems for immediate use of joining points as measurement input: (a) the secondary joining points, which denote an over-saturated situation with residual queued vehicles, cannot reflect real information of natural arrivals because of the impacts of previous signal control; (b) some exceptional changes of queue lengths can be observed from individual points and may lead to the Kalman Filter diverging.

Using only premier joining points can simply solve the first problem despite some losses in data. Let us analysis the second problem. Theoretically, as the queue-forming wave propagates backward, queue length is supposed to increase with the time before the queue is discharged. However, a decrease of queue length may be occasionally observed from real-world mobile sensor data because of either positioning errors or sampling uncertainties. Particularly for multilane links, queue lengths vary randomly over lanes, and a decrease of queue length may be frequently observed from mobile sensor data that is probably caused by difference of queue lengths on diverse sampled lanes. Meanwhile, exceptional increases of queue lengths are also considerable. Therefore, premier joining points are sub-grouped for reducing those exceptional changes in order to improve the accuracy and robustness of the estimation.

The principles of sub-grouping for premier joining points can be elaborated as follows:

Given all the premier joining points sorted to time-ascending order, $pjp \in PJP_{\text{sort}} : \{pjp^{(s)} : (t^{(s)}, x^{(s)}) | s = 1, 2, ..., N\}$, and the first sub-group Ω_{SG}^{1} initialized with $pjp^{(1)}$, any two time-adjacent premier joining points $pjp^{(s-1)} : (t^{(s-1)}, x^{(s-1)})$, $pjp^{(s)} : (t^{(s)}, x^{(s)})$, with $pjp^{(s-1)}$ grouped in Ω_{SG}^{SGnum}, can be determined as following:

- If $pjp^{(s)}$ and $pjp^{(s-1)}$ are from different cycles, group $pjp^{(s)}$ to $\Omega_{SG}^{SGnum+1}$.
- If $pjp^{(s)}$ and $pjp^{(s-1)}$ are within the same cycle, and if the Equation 26 holds, then group $pjp^{(s)}$ to $\Omega_{SG}^{SGnum+1}$; otherwise, group $pjp^{(s)}$ to Ω_{SG}^{SGnum}.

$$\begin{cases} x^{(s)} - \min\limits_{x \in \Omega_{SG}^{SGnum}} \{\mathbf{x}\} < \bar{x} \\ t^{(s)} - t^{(s-1)} > \bar{t} \end{cases} \tag{26}$$

where, $\min\limits_{x \in \Omega_{SG}^{SGnum}} \{\mathbf{x}\}$ is the minimum x in Ω_{SG}^{SGnum}, $\bar{x}$ and $\bar{t}$ refer to a spatial and temporal threshold respectively for the sub-grouping:

After the sub-grouping, each group Ω_{SG}^{k} will be represented by a new point (τ^k, ξ^k) on its centroid:

$$\begin{cases} \tau^k = \operatorname{mean}\limits_{t \in \Omega_{SG}^{k}}\{\mathbf{t}\} \\ \xi^k = \operatorname{mean}\limits_{x \in \Omega_{SG}^{k}}\{\mathbf{x}\} \end{cases} \tag{27}$$

These points are used as measurement inputs for queue length estimation based on Kalman Filter, that is, $\mathbf{Z}_k = [\xi^k]$ at timestamp τ^k.

Step 3.3: Kalman Filter-Based Estimation of Queue length. Based on the proposed state-space model, the Kalman Filter recursively estimates the back of queue QX_k and the queue-forming speed VF_k at discrete-time step k, with measurement update and state update. At each time step, current queue length will be estimated, while additionally for some certain time steps (see following condition 2), the maximum queue length is estimated. Note that the process of Kalman Filter is independent of the queue length estimation whereas the latter is dependent on the former.

Given the initial information, i.e. VF_1, $\mathbf{P}_1$, $\mathbf{R}$ and q, Kalman Filter-based estimation of queue length is implemented as follows:

For $k = 1$, with the initial measurement input $\mathbf{Z}_1 = [\xi^1]$ at τ^1, set $QX_1 = \mathbf{Z}_1$, then we have $\hat{\mathbf{X}}_1 = \begin{bmatrix} QX_1 \\ VF_1 \end{bmatrix}$ at the first time step. For $k>1$, there are two conditions:

- *Condition 1* current measurement input (τ^k, ξ^k) belongs to the same cycle with(τ^{k-1}, ξ^{k-1}).

For this condition, maximum queue length would not be estimated and standard measurement update and state update are implied. That is, for state update, set $T = \tau^k - \tau^{k-1}$, $\hat{\mathbf{X}}_{k|k-1}$ and $\mathbf{P}_{k|k-1}$ can be calculated by Equations 7 and 8; then, for measurement update, with $\mathbf{Z}_k = [\xi^k]$ at τ^k, the current estimated state vector $\hat{\mathbf{X}}_k$ and error covariance matrix $\mathbf{P}_k$ are calculated by Equations 9–11. So the queue length at time step k is:

$$QL^k = x_{\text{stopbar}} - QX_k \tag{28}$$

- *Condition 2* current measurement input (τ^k, ξ^k) belongs to the cycle n and (τ^{k-1}, ξ^{k-1}) belongs to the cycle *n–1*.

This condition means that the queue of cycle *n-1* has been discharged within $[\tau^{k-1}, \tau^k]$, so the maximum queue length of cycle *n–1* would certainly occur within this interval. Therefore, the maximum queue length of the cycle *n–1* could be estimated using $\hat{\mathbf{X}}_{k-1}$ and the queue-discharging wave of cycle *n–1*. The queue-forming wave from (τ^{k-1}, ξ^{k-1}) and queue-discharging wave are, respectively, represented by linear functions:

$$x = VF_{k-1}(t - \tau^{k-1}) + QX_{t-1} \tag{29}$$

$$x = \hat{\beta}_0^{n-1} + \hat{\beta}_1^{n-1} t \tag{30}$$

The $(t_{n-1}{}^*, x_{n-1}{}^*)$ simultaneously satisfying two equations is the point at which the queue-forming wave and queue-discharging wave meet in cycle n-1, so the estimated maximum queue length for the cycle is:

$$MQL_{n-1} = x_{\text{stopbar}} - x_{n-1}{}^* \tag{31}$$

In addition, state vector at current time step k, $\mathbf{X}_k$ should also be estimated. However, for condition 2, a standard measurement update and state update cannot be applied because the queue-forming wave has been interrupted. Instead, current back of queue QX_k is given the value of current measurement input, $\mathbf{Z}_k$, and queue-forming speed VF_k is given the same value as the previous queue-forming speed, VF_{k-1}, that is, with $\mathbf{Z}_k = [\xi^k]$ at τ^k,$\hat{\mathbf{X}}_k = \begin{bmatrix} QX_k \\ VF_k \end{bmatrix} = \begin{bmatrix} \mathbf{Z}_k \\ VF_{k-1} \end{bmatrix}$ and $\mathbf{P}_k = \mathbf{P}_{k-1}$.

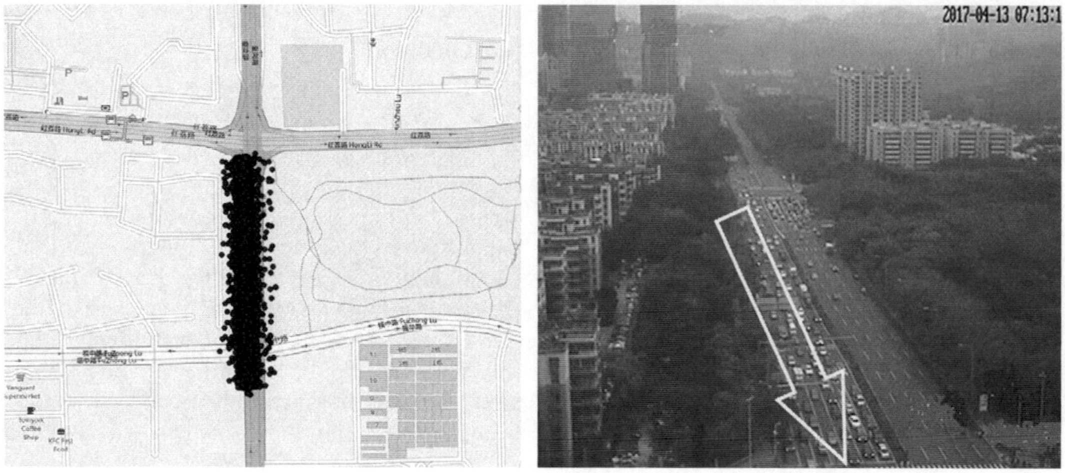

(a) (b)

Figure 3. Illustrations of data for the case study in Huanggang-Fuzhong intersection. (*a*) DiDi GPS scatter points; and (*b*) Monitoring video.

Evaluations and Results

In this section, a case study using real-world GPS data and a simulation test with various penetration rates and intersection geometrics is made for the evaluation of the proposed queue length estimation method.

For the case study, the method is tested on GPS data from DiDi Chuxing, which is a major ride-sharing company providing transportation services including taxi hailing, private car hailing, and so forth, to users in China via a smartphone application. The data contain comprehensive individual vehicle information such as instantaneous position, links, and time stamps at the resolution of 3–6 s. The data are for a 4-lane through-link on the northbound approach of Huanggang-Fuzhong intersection, Shenzhen, from 7:00 to 12:00. The ground truth queue length is surveyed from field monitoring video. For the entire study period, the total number of cycles is 75 and the average volume is 2728 veh/h (682 veh/h/lane); the average penetration rate of DiDi vehicles is 7.4%. Figure 3 shows the DiDi data and monitoring video, and the yellow arrow in Figure 3*b* indicates the through-link to be considered.

In order to quantitatively evaluate the proposed method, the cycle-by-cycle maximum queue lengths are estimated and compared using two methods: the proposed Kalman Filter-based method and the linear regression method. The linear regression method produces a single line fitting all the joining points or leaving points within a cycle with least squares, so the queue-forming and discharging wave is assumed to be linear, and for any cycle with only one joining point, the queue-forming speed is substituted by that of the last cycle. In the linear regression method, estimation of queue-forming waves is similar to that of queue-discharging waves in the proposed method which is presented in Step 2. Note that the processing of raw data and critical points is identical for the two methods. Given critical points allocated into cycles, $\hat{\beta}_0^n$ and $\hat{\beta}_1^n$, which determine the linear function of queue-forming or discharging waves, can be derived from Equation 20. Then, maximum queue length can be derived from the two linear functions. Further insight into the linear regression method can be found elsewhere (*20, 22*). The necessary parameters of the proposed method in the case study are listed below:

L_1: 10 m, η: 0.5, $\bar{v}$: 1 m/s, Δ_v: 1, Δ_c: 1, $\bar{x}$: 0 m, $\bar{t}$: 5 s, VD_{default}: –5 m/s, I_{GS}: 100 s, q: 0.0075, $\mathbf{R}$: 237.5, VF_1: –1.5 m/s, $\mathbf{P}_1$: $\begin{bmatrix} 15000 & 0 \\ 0 & 1.5 \end{bmatrix}$. Some of the parameters and initial conditions for the Kalman Filter, that is, VD_{default}, q, $\mathbf{R}$ and VF_1 in this case, are extracted from field data (monitoring video), and $\mathbf{P}_1$, the initial error covariance, which is supposed to be sufficiently large to ensure a steady-state filtering process (*21*), is set $diag(Q_{11}, Q_{22})$ where Q_{11} and Q_{22} are diagonal elements of $\mathbf{Q}$ calculated by Equations 5 and 24. Some studies have analyzed the influences and means of tuning a Kalman Filter (*23–25*), but it is not further discussed in this study.

For a general evaluation on the performance of both methods, the mean absolute error (MAE) and mean absolute percentage error (MAPE) of maximum queue lengths were calculated over the entire study period. MAE and MAPE are formulated in Equations 32 and 33.

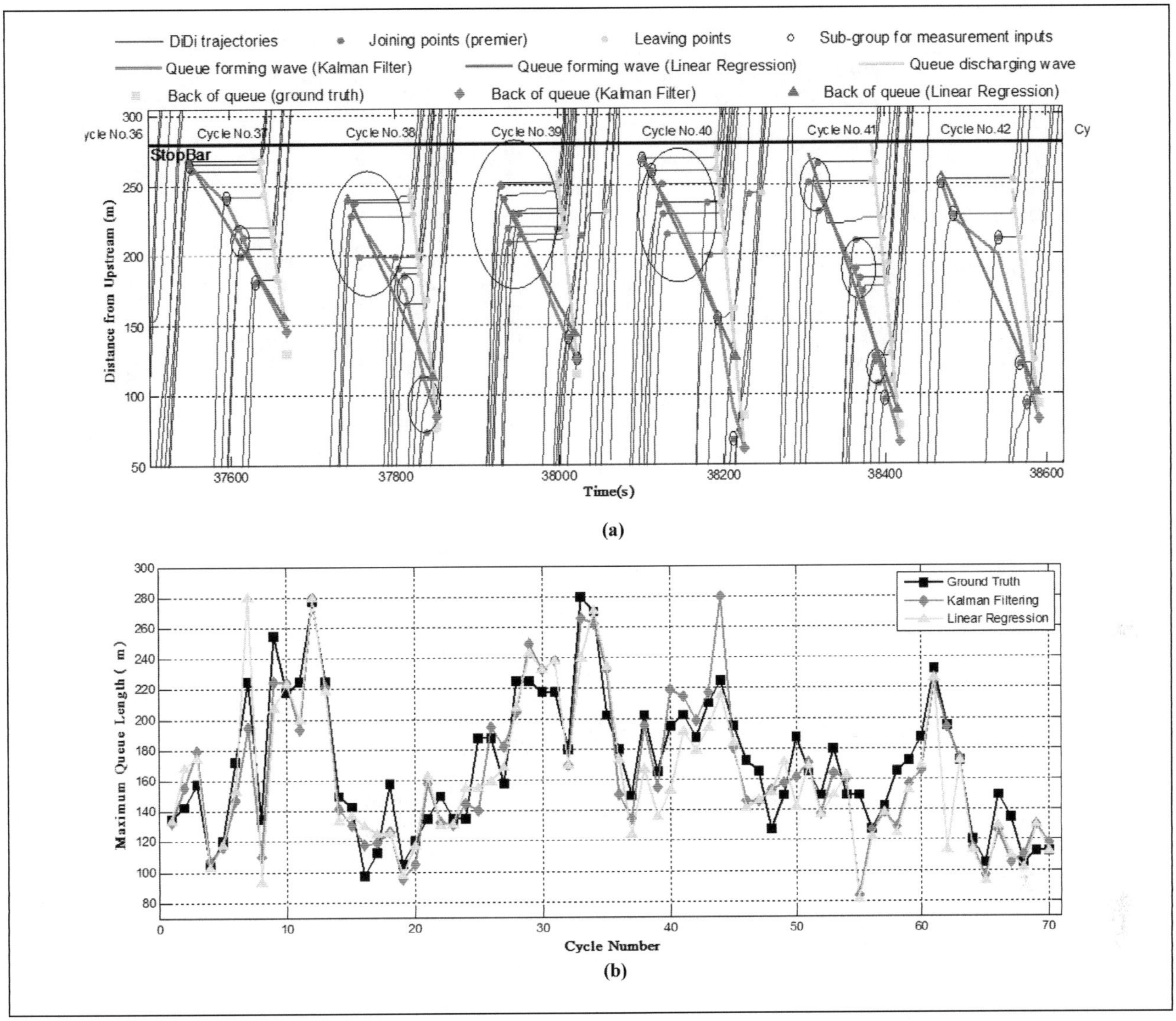

Figure 4. Results for the case study. (*a*) Illustrations for the estimation in six cycles; and (*b*) Cycle-by-cycle results of estimation.

$$MAE = \frac{1}{n}\sum_{i}^{n} |\hat{Y} - Y| \tag{32}$$

$$MAPE = \frac{1}{n}\sum_{i}^{n} \left|\frac{\hat{Y} - Y}{Y}\right| \times 100\% \tag{33}$$

where, n is the total number of cycles, that is, $n = 75$, $\hat{Y}$ is estimated maximum queue length and Y is ground truth. Note that ground true maximum queue lengths are calculated by averaging for multilane.

Figure 4*a* illustrates the estimation of shock waves and queue lengths for six cycles from the study period based on the proposed method (magenta) and linear regression method (gray). There are no evident differences on the estimated back of queue (cyan squares, magenta diamonds, and gray triangles, respectively, represent the back of queue for ground truth, Kalman Filter-based method, and linear regression method) for the both methods in the 37th and 41st cycle, because of its intrinsic linearity of queue forming. However, the arrival patterns are not always uniform in real-world networks and queue-forming waves are not consistently linear. Exactly as the 38th, 39th, 40th and 42nd cycle shows, when the queue-forming pattern becomes more complex and no longer strongly linear, the accuracy of the linear regression method drops obviously whereas the Kalman Filter-based method maintains a much better accuracy. It is also seen in the 38th, 39th, and 40th cycle that joining points' sub-grouping for measurement input reduces the impact of exceptional distribution of some individual points to avoid unrealistic results.

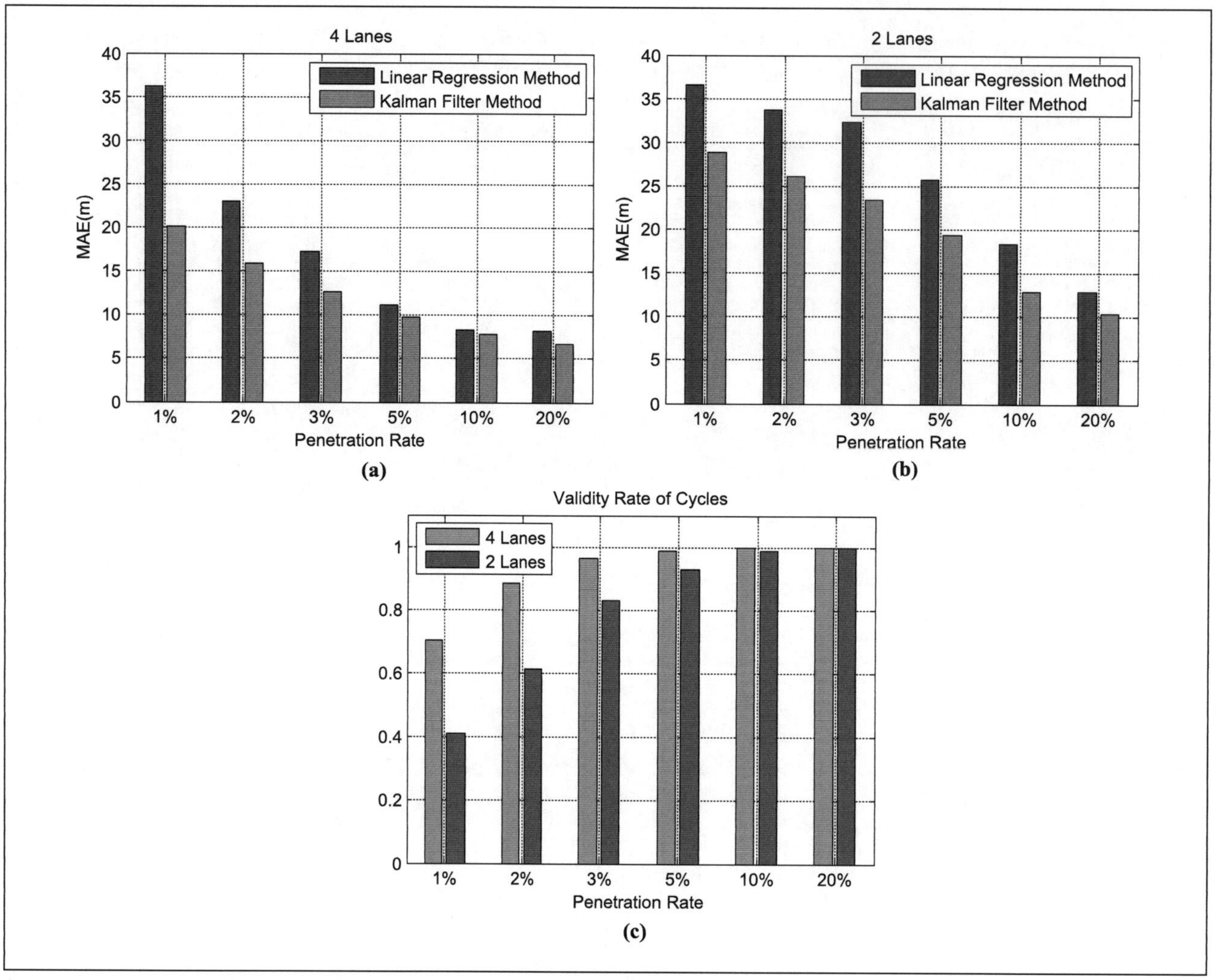

Figure 5. Simulation results of different scenarios. (*a*) MAE for the four-lane approach; (*b*) MAE for the two-lane approach; (*c*) Validity rate of cycles.

Among all 75 cycles in the study period five are invalid, which means there is no queued DiDi vehicle in the cycle. A cycle with at least one queued DiDi vehicle so that the estimation method can be applied is called a valid cycle. Therefore, the validity rate of cycles in this case is 93.3%. Figure 4*b* shows the maximum queue lengths for all valid cycles. For all 75 cycles, MAE of the proposed method and the linear regression method is, respectively, 18.3 m and 23.2 m; MAPE is, respectively, 11.2% and 15.8%. Note that the impact of invalid cycles has been taken into account, as the maximum queue lengths of invalid cycles are substituted equally by the mean of all the valid cycles.

The result of the case study show that the proposed Kalman Filter-based method achieves a MAE/MAPE of 18.3 m (2 or 3 vehicles)/11.2% in cycle-by-cycle maximum queue length estimation, and achieves better performance compared with the linear regression method (20% decrease in MAE).

Scenarios with different penetration rates and intersection geometrics (number of lanes) are tested in simulations for sensitivity analysis. The first simulated approach is modeled on that in the case study (Huanggang-Fuzhong intersection), in which the number of lanes is four, the volume per lane is 700 veh/h, the green duration is 100 s, and the cycle time is 200 s. Only the number of lanes is changed to two for the second simulated approach, with other parameters remaining unchanged. The simulation models are also calibrated for desired speed, stopping space, and saturated flow rate, so that their parameters are set equal to the case study's. The two approaches are respectively simulated at penetration rates of 1%, 2%, 3%, 5%, 10%, and 20%. Figure 5 shows the simulation results.

The result of simulation shows that: (a) with the decrease of penetration rate, the superiority of the Kalman Filter-based method becomes more significant, especially in extremely low penetration situations (e.g., the Kalman Filter-based method achieves 15 m and 40% decrease in MAE against the linear regression method at 1% penetration rate, Figure 5*a*); (b) even with the same penetration rate, an increase in the number of lanes significantly improves the accuracy of the estimation and validity rate of cycles (e.g., with the same penetration rate of 5%, the Kalman Filter-based method achieves 10 m and 50% decrease in MAE with lane number increasing from 2 to 4).

Conclusion

This study proposes a queue length estimation method based on the combination of Kalman Filtering and shockwave theory with low-penetration mobile sensor data at signalized intersections. Unlike most of the existing methods, the proposed method uses mobile sensor data with spatiotemporal information as the only input for the estimation, without knowing signal timings, traffic demands, and vehicle arrival patterns. The conclusions of this study can be highlighted as follows:

1. A state-space model characterizing the queue-forming dynamics is established for the Kalman Filter, so that the estimation is more reasonable in low-penetration situations because it depends on not only current measurement but correlations of states.
2. With the application of Kalman Filter, queue length, which can be further used as the input for adaptive control algorithms, is recursively estimated.
3. The proposed method is tested in both real-world case and simulation scenarios, and the MAE/MAPE of the maximum queue length is 18.3 m/11.2% in real-world situation at a penetration rate of 7.4%.
4. The simulation results show that the proposed Kalman Filter-based estimation method achieves a much better performance than the linear regression method at extremely low penetration rates.

Nevertheless, this study also has some limitations:

1. The method may lose effectiveness for those situations with complex vehicle interactions (e.g., lane-changing behaviors among approach lanes), because the queue formation and queue dissipation may not be well modeled.
2. If the system noise or measurement noise of the state-space model is set very biased, the errors of estimation may not be satisfactory.

A future study is to validate the performance of the proposed method in various real-world scenarios. In addition, the Kalman Filter may be improved so that it adapts itself to system noises and measurement noises in real time in order to stabilize the estimation.

Acknowledgments

This research was supported by the Natural Science Foundation of China [51422812], the Shanghai Science and Technology Project (16DZ1203602), the Fundamental Research Funds for the Central Universities and DiDi Chuxing.

Author Contributions

The authors confirm contribution to the paper as follows: study conception and design: Jian Sun, Juyuan Yin and Keshuang Tang; data collection: Juyuan Yin; analysis and interpretation of results: Jian Sun and Juyuan Yin; draft manuscript preparation: Jian Sun and Juyuan Yin. All authors reviewed the results and approved the final version of the manuscript.

References

1. Balke, K. N., and H. Charara. *Development of a Traffic Signal Performance Measurement System (TSPMS)*. No. FHWA/TX-05/0-4422-2. Texas Transportation Institute College Station, TX, 2005.
2. Webster, F. V. *Traffic Signal Settings*. Road Research Technical Paper 39. H. M. Stationery Office, London, United Kingdom, 1958.
3. Newell, G. F. Approximation Methods for Queues with Application to the Fixed-Cycle Traffic Light. *SIAM Review*, Vol. 7, No. 2, 1965, pp. 223–240.
4. Chang, T., and J. Lin. Optimal Signal Timing for an Oversaturated Intersection. *Transportation Research Part B: Methodological*, Vol. 34, No. 6, 2000, pp. 471–491.
5. Mirchandani, P. B., and N. Zou. Queuing Models for Analysis of Traffic Adaptive Signal Control. *IEEE Transactions on Intelligent Transportation Systems*, Vol. 8, No. 1, 2007, pp. 50–59.
6. Varaiya, P. Max Pressure Control of a Network of Signalized Intersections. *Transportation Research Part C: Emerging Technologies*, Vol. 36, No. 22, 2013, pp. 177–195.
7. Viti, F., and H. J. V. Zuylen. Probabilistic Models for Queues at Fixed Control Signals. *Transportation Research Part B: Methodological*, Vol. 44, No. 1, 2010, pp. 120–135.
8. Sharma, A., D. M. Bullock, and J. A. Bonneson. Input-Output and Hybrid Techniques for Real-Time Prediction of Delay and Maximum Queue Length at Signalized Intersections. *Transportation Research Record: Journal of the Transportation Research Board*, 2007. 2035: 69–80.
9. Lighthill, M. J., and G. B. Whitham. On Kinematic Waves. II. A Theory of Traffic Flow on Long Crowded

Roads. *Proceedings of the Royal Society: A Mathematical Physical & Engineering Sciences*, Vol. 229, No. 1178, 1955, pp. 317–345.
10. Ban, X., P. Hao, and Z. Sun. Real Time Queue Length Estimation for Signalized Intersections Using Travel Times from Mobile Sensors. *Transportation Research Part C: Emerging Technologies*, Vol. 19, No. 6, 2011, pp. 1133–1156.
11. Wu, X., and H. X. Liu. A Shockwave Profile Model for Traffic Flow on Congested Urban Arterials. *Transportation Research Part B: Methodological*, Vol. 45, No. 10, 2011, pp. 1768–1786.
12. Skabardonis, A., and N. Geroliminis. Real-time Estimation of Travel Times on Signalized Arterials. Presented at 16th International Symposium on Transportation and Traffic Theory, University of Maryland, College Park, MD, 2005.
13. Skabardonis, A., and N. Geroliminis. Real-Time Monitoring and Control on Signalized Arterials. *Journal of Intelligent Transportation Systems*, Vol. 12, No. 2, 2008, pp. 64–74.
14. Seo, T., T. Kusakabe, and Y. Asakura. Estimation of Flow and Density Using Probe Vehicles with Spacing Measurement Equipment. *Transportation Research Part C: Emerging Technologies*, Vol. 53, No. 6, 2015, pp. 134–150.
15. Seo, T., and T. Kusakabe. Probe Vehicle-based Traffic State Estimation Method with Spacing Information and Conservation Law. *Transportation Research Part C: Emerging Technologies*, Vol. 59, No. 3, 2015, pp. 391–403.
16. Hofleitner, A., R. Herring, P. Abbeel, and A. Bayen. Learning the Dynamics of Arterial Traffic from Probe Data Using a Dynamic Bayesian Network. *IEEE Transactions on Intelligent Transportation Systems*, Vol. 13, No. 4, 2012, pp. 1679–1693.
17. Ramezani, M., and N. Geroliminis. On the Estimation of Arterial Route Travel Time Distribution with Markov Chains. *Transportation Research Part B: Methodological*, Vol. 46, No. 10, 2012, pp. 1576–1590.
18. Sun, Z., and X. Ban. Vehicle Trajectory Reconstruction for Signalized Intersections Using Mobile Traffic Sensors. *Transportation Research Part C: Emerging Technologies*, Vol. 36, No. 11, 2013, pp. 268–283.
19. Cheng, Y., X. Qin, J. Jin, and B. Ran. An Exploratory Shockwave Approach to Estimating Queue Length Using Probe Trajectories. *Journal of Intelligent Transportation Systems*, Vol. 16, No. 1, 2011, pp. 12–23.
20. Ramezani, M., and N. Geroliminis. Queue Profile Estimation in Congested Urban Networks with Probe Data. *Computer-Aided Civil and Infrastructure Engineering*, Vol. 30, No. 6, 2015, pp. 414–432.
21. Kalman, R. E. A New Approach to Linear Filtering and Prediction Problems. *Journal of Basic Engineering*, Vol. 82, No.1, 1960, pp. 35 45.
22. Izadpanah, P., B. Hellinga, and L. Fu. Automatic Traffic Shockwave Identification Using Vehicles' Trajectories. Presented at 88th Annual Meeting of the Transportation Research Board, Washington, D.C., 2009.
23. Maybeck, P. S. Optimal Filtering with Linear System Models. In *Stochastic Models, Estimation, and Control Volume I*, Academic Press, New York, 1979, pp. 203–279.
24. Cinquemani, E. A Superposition Principle for The Kalman Filter. *Systems & Control Letters*, Vol. 55, No. 1, 2006, pp. 38–44.
25. Sangsuk-Iam, S., and T. E. Bullock. Analysis of Discrete-time Kalman Filtering under Incorrect Noise Covariances. *IEEE Transactions on Automatic Control*, Vol. 35, No. 12, 1990, p. 13041309.

The Standing Committee on Artificial Intelligence and Advanced Computing Applications (ABJ70) peer-reviewed this paper (18-06204).

TRR
JOURNAL OF THE TRANSPORTATION RESEARCH BOARD

Article

Transportation Research Record
2018, Vol. 2672(45) 265–273

DOI: 10.1177/0361198118796071
journals.sagepub.com/home/trr
SAGE

Dijkstra's-DBSCAN: Fast, Accurate, and Routable Density Based Clustering of Traffic Incidents on Large Road Network

Yang Zhang[1], Lee D. Han[2], and Hyun Kim[3]

Abstract

Incident hotspots are used as a direct indicator of the needs for road maintenance and infrastructure upgrade, and an important reference for investment location decisions. Previous incident hotspot identification methods are all region based, ignoring the underlying road network constraints. We first demonstrate how region based hotspot detection may be inaccurate. We then present Dijkstra's-DBSCAN, a new network based density clustering algorithm specifically for traffic incidents which combines a modified Dijkstra's shortest path algorithm with DBSCAN (density based spatial clustering of applications with noise). The modified Dijkstra's algorithm, instead of returning the shortest path from a source to a target as the original algorithm does, returns a set of nodes (incidents) that are within a requested distance when traveling from the source. By retrieving the directly reachable neighbors using this modified Dijkstra's algorithm, DBSCAN gains its awareness of network connections and measures distance more practically. It avoids clustering incidents that are close but not connected. The new approach extracts hazardous lanes instead of regions, and so is a much more precise approach for incident management purposes; it reduces the $O(n^2)$ computational cost to $O(n)$, and can process the entire U.S. network in seconds; it has routing flexibility and can extract clusters of any shape and connections; it is parallellable and can utilize distributed computing resources. Our experiments verified the new methodology's capability of supporting safety management on a complicated surface street configuration. It also works for customized lane configuration, such as freeways, freeway junctions, interchanges, roundabouts, and other complex combinations.

Spatial hotspot identification benefits transportation management in many ways. Identification of traffic incident hotspots is a powerful tool for lane safety management. It exposes dangerous road segments and serves as a direct indicator for the needs of road maintenance and infrastructure updates (*1–3*). To choose the best location for a new project with budgetary constraints, departments of transportation rely on a comprehensive cluster ranking method to evaluate the priority of the candidate sites. In addition to safety applications, the growing availability of real-time vehicle positional data from various sources, such as GPS, smart phones, RFID devices, and Bluetooth, is creating new opportunities for using spatial clustering techniques to study traffic flows at both local and inter-regional scales. Vehicle trajectory clusters uncover "hot routes" and enable urban planners to optimize the allocation of traffic control devices (*4*), better understand congestion (*5–7*), build and train models to predict dynamically traffic demands and reliable travel times (*8–12*). In a big data era, massive traffic data imposes remarkable challenges for traditional hotspot detection and spatial clustering methods. Conventional region based methods cannot provide the detailed roadway level information we want for further in-depth analysis; an algorithm with high computational complexity will easily fail to handle larger amounts of data; capability of parallelization becomes a requirement in order to utilize distributed computing resources; and compatibility with large-scale networks allows a method to be applicable on a broader spatial scale. Hence it is worthwhile to rethink such aspects of classical approaches.

Two major hotspot detection methods currently in use are kernel density estimation (KDE) and clustering analysis (*13*). KDE method (*14, 15*) divides the space into

[1]Apple Inc., Cupertino, CA
[2]Department of Civil and Environmental Engineering, University of Tennessee, Knoxville, TN
[3]Department of Geography, University of Tennessee, Knoxville, TN

Corresponding Author:
Address correspondence to Lee D. Han: lhan@utk.edu

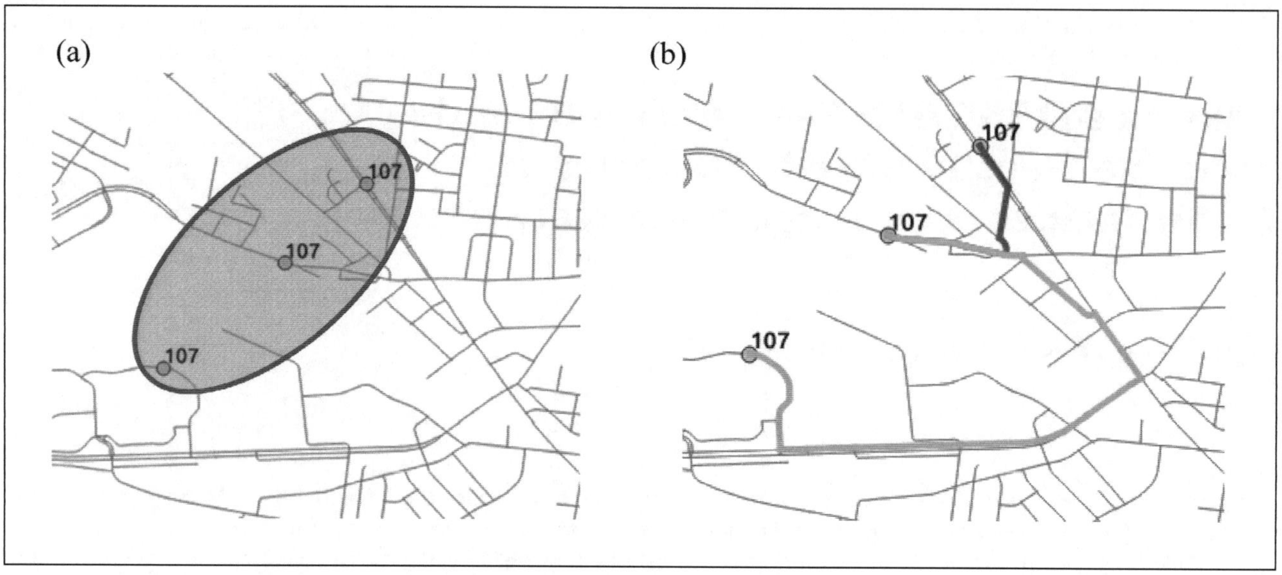

Figure 1. Clustering and underlying network. (*a*) Three incidents formulate an incident hotspot because they are close in Euclidean space. (*b*) The incidents actually have no close road connections and so the cluster in (*a*) is invalid.

cells given a user-defined cell size. The density value of each cell is calculated by summing up the influences from other cells. Kernel function defines how the influence varies with distance. There are different types of kernel function, such as Gaussian kernel, quartic kernel, triangular kernel, minimum variance kernel, etc. Hotspots are identified as the regions with high accumulative density values. A variety of non-kernel clustering algorithms have also been applied to incident hotspot analysis: hierarchical (*16*, *17*), K-means (*14*, *18*, *19*), DBSCAN (*20*–*22*), etc. Both hierarchical and K-means clustering require users to specify the number of clusters, which makes them impractical for a large dataset because researchers sometimes can barely visualize a large dataset (*23*), not to mention provide a reasonable estimation of number of clusters. DBSCAN (density based spatial clustering of applications with noise) has been reported to have better performance and so is adopted more often than the others because of its simple parameters (*23*).

One important fact that most previous related studies did not address is that these methods were all designed for general spatial analysis, where the space is at least two dimensional. A spatial hotspot refers to, for example, a neighborhood with high crime rates, a town experiencing disease epidemics, an area where biodiversity is endangered, etc. The fundamental difference between traffic related hotspots and general spatial hotspots is that the space of traffic events is neither two dimensional nor higher dimensional, but the underlying road network. For one thing, a two-dimensional region cluster is not a direct measure of the danger level of road segments. An urban region with dense road connections is not as dangerous as a rural region with the same area and incident counts, but which has only one road across it, even though they are equivalent in region based approaches. On the other hand, the Euclidean distance measure is not accurate and even misleading to decision makers when applying region based methods to traffic incident clustering. Figure 1 gives an example.

To present our new algorithm, Dijkstra's-DBSCAN, which precisely clusters traffic incidents and further supports lane management and decision making effectively, the rest of this paper is organized as follows. The second section briefly introduces the DBSCAN algorithm. The third section modifies Dijkstra's shortest distance algorithm for our purpose, proves its correctness, and finally presents Dijkstra's-DBSCAN algorithm. The fourth section applies the algorithm to a case study using data on five years of fatalities on the entire U.S. road network. The fifth section analyzes the computational complexity and the parallelization of the algorithm. The sixth and final section discusses the results and concludes the paper with possible extensions.

DBSCAN Algorithm

The principle of DBSCAN was published in 1996 (*24*) and has been extensively applied in spatial analysis and hot spot detection. A core point o has at least minpts, or minimum number of points, neighbors in its $\varepsilon-$ *neighborhood*, that is, the $\varepsilon-$ *neighborhood* of o is a set $N_o = \{x_i | \ d(o, x_j) < \mathrm{eps}\}$, where $|N| \geq \mathrm{minpts}$. Here $d(\cdot)$ denotes the distance measure (metric) of two points, $|\cdot|$ denotes the size (cardinality) of a set. A core is a direct neighbor to itself. Two points, p and q, are density-connected if there exists a path $o_1o_2...o_n$, where $p \in N_{o_1}, q \in N_{o_n}$, and $o_i \in N_{o_{i+1}}, i = 1, 2, ..., n-1$. A density cluster is defined as a core and all its density-connected neighbors.

There are two important properties from the definition of a DBSCAN cluster.

1. A cluster contains at least minpts points. A cluster contains at least one core, which has at least minptsdirectly connected neighbors.
2. Any element in a cluster can reach at least one other point within a ε distance. Suppose an element $p$does not reach any other point within ε, then $\forall o_i,\ p \notin N_{o_i}$. Further, for an arbitrary core point o_j, there exists no path connecting p and o_j, by the definition of a cluster, p is not an element of the cluster containing core o_j. Contradiction.

Property (a) constrains the minimum counts of points to constitute a valid cluster, and property (b) guarantees closeness of the points. The higher minptsand the lower ε, the more dense the cluster. These two properties make DBSCAN an intuitive density clustering algorithm.

The pseudo code of DBSCAN is shown in Figure 2. It starts with obtaining the $\varepsilon - neighborhood$ of each data point. If $\varepsilon - neighborhood$ set has at least minpts elements, this point becomes a core point, indicating that it is qualified for starting up a cluster and expanding it later. The cluster expansion is a recursive depth first search (DFS) process: it starts from an arbitrary core point, traverses its directly connected neighbors, and marks them as "visited." If any neighbor point is also a core, the above steps will be repeated on its $\varepsilon - neighborhood$. The recursion stops after all the expandable cores are fully expanded. The recursion procedure is actually searching for all the path connected neighbors for the first core point. After each cluster expansion, we update the cluster identity to start a new cluster until all the elements of the dataset are labeled as visited. Figure 3 illustrates the cluster expansion procedure with an example.

```
Dijkstra_DBSCAN(D, ε, minpts){
    id ← 0
    for each point p in dataset D:
    N(p) ← Modified_Dijkstra(G, p, ε)
        if |N(p)| ≥ minpts :
            mark p as core
    for each unvisited core point o :
            mark o as visited
            C_id = {o}
            id++
            ExpandCluster(o)
}
ExpandCluster(o){
    for each unvisited p' in N(o) :
        mark p' as visited
        C_id = C_id ∪ {p'}
        if p' is core
            ExpandCluster (p')
}
```

Figure 2. DBSCAN implementation.

Modified Dijkstra's Shortest Path Algorithm

As a classic algorithm, Dijkstra's algorithm finds the shortest path from one source s to one or multiple targets t in a graph $G = (V, E)$, where V is the set of vertices and E of edges. Without repeating the basics of this well-known algorithm, we simply recap the essence of the algorithm, based on which we will modify it to fit our purpose. First, two disjoint sets of vertices are used: an explored set S, where the shortest distances $d(s, v)$ from vertex v to source s were determined, and an unexplored set $V - S$, where the Dijkstra's distances $d(s, u)$ are as yet undetermined. Initially, $S = \{s\}$. Second, in each step, a new element r_i is moved from unexplored set to explored set, where r_i and v_i are determined by: $\min_{v_i \in S,\ r_i \in V-S} [d(s, v_i) + e(v_i, r_i)]$, and $d(s, r_i) = d(s, v_i) + e(v_i, r_i)$, where $e(v_i, r_i)$ represents the length of the edge connecting r_i and an element v_i from the unexplored set. In the kth step, the size of the explored set is $|S| = k + 1$ (including s), while the size of the unexplored set is $|V - S| = |V| - k - 1$.

Theorem: For a non-negatively weighted graph, the vertex determined in step $k + 1$has a longer Dijkstra's distance than that of the vertex determined in step k, that is, $d(s, r_{k+1}) \geq d(s, r_k)$.

Proof: We use induction method.

(1) When $k = 0$, $d(s, r_1) \geq 0$, $d(s, r_0) = d(s, s) = 0$. The inequality holds.

(2) When $k = n - 1$, suppose the inequality is satisfied. $d(s, r_n) \geq d(s, r_{n-1})$.

(3) We need to prove that when $k = n$, $d(s, r_{n+1}) \geq d(s, r_n)$.

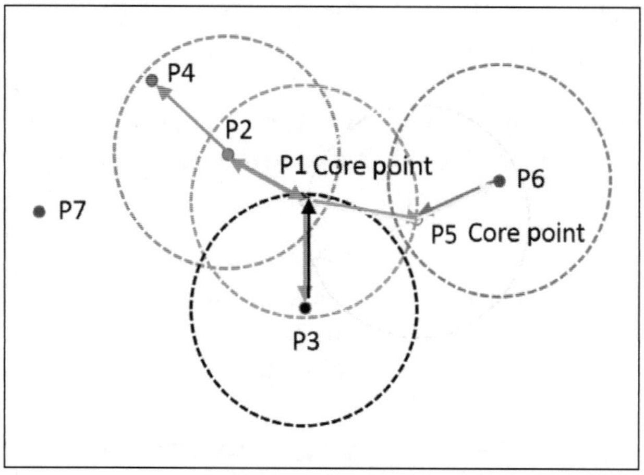

Figure 3. Illustration of cluster expansion. min*pts* = 3. P1 and P5 are core points because they both have three directly reachable neighbors within ε (the radius of the circles). P1 is selected as a root core point, the algorithm expands the cluster started by P1 recursively. $N(P1) = \{P1, P2, P3, P5\}$ $C_0 = \{P1, P2, P3, P5\}$. In the directly accessible neighbor set of P1, only P5 is core. The unvisited neighbor of P5 is $N(P5) = \{P6\}$. The cluster is then expanded $C_0 = \{P1, P2, P3, P5\} \cup \{P6\}$. The cluster is completely expandable and all points were visited. The algorithm stops and the cluster involves five points.

In step 2, r_n is determined by satisfying $\min_{v_i \in S_{n-1}, r_i \in V - S_{n-1}} [d(s, v_i) + e(v_i, r_i)]$. Since $d(s, r_n)$ is the minimum for all v_i in S_{n-1} and all r_i in $V - S_{n-1}$, we have: $d(s, r_n) \leq d(s, v_i) + e(v_i, r_{n+1})$, $v_i \in S_{n-1}$. To determine the next vertex r_{n+1}, we have $d(s, r_{n+1}) = \min_{v_i \in S_n, r_i \in V - S_n} [d(s, v_i) + e(v_i, r_i)]$. Notice that $S_n = S_{n-1} \cup \{r_n\}$. There are two possibilities for v_i:

if $v_i = r_n$, $d(s, r_{n+1}) = d(s, r_n) + e(r_n, r_{n+1}) \geq d(s, r_n)$;

else, $v_i \in S_{n-1}$, $d(s, r_{n+1}) = d(s, v_i) + e(v_i, r_{n+1}) \geq d(s, r_n)$.

So $d(s, r_{k+1}) \geq d(s, r_k)$ holds for $k = n$ as well.

By the principle of induction, the inequality is true for all steps. QED.

Our modification of Dijkstra's algorithm (Figure 4) is based on what has been proved above, that the shortest distance ordered by steps is a monotonically increasing sequence. Instead of terminating the routing when the target is found, as the original Dijkstra's does, we control the algorithm by comparing the most lately determined shortest distance with the maximum distance ε in DBSCAN. When a vertex becomes a larger distance than ε, there is no need to evaluate any other nodes because their distance to the source will be even greater. The modified Dijkstra's algorithm returns the vertices in the explored set, which are exactly the directly reachable neighbors that we need in network based DBSCAN.

The complexity analysis is consistent with original Dijkstra's. The difference is that, instead of knowing the numbers of edges and vertices, the routing here is on a local neighborhood, the scale of which is determined by ε. The larger ε is, the larger the local network will be, and so the more edges and vertices will be visited. We denote the number of edges and vertices as $|E(\varepsilon)|$ and $|V(\varepsilon)|$ respectively. The time complexity for obtaining the neighborhood of one incident point will be $O(|E(\varepsilon)| + |V(\varepsilon)|\log(|V(\varepsilon)|))$. Density based clustering

```
Modified_Dijkstra(G, s, ε){
    d[s] ← 0
    Q = {s}
    neighbor_list = {}
    while Q is not empty:
        u ←Q[0]
        Q←Q- {u}
        for each neighbor v of u:
            if v is not visited and d[u]+e(u,v) < d[v]:
                d[v] = d[u]+e(u,v)
                Q←Q ∪ {v}
        sort Q in ascending order based on the distance values of the elements
        if d[Q[0]] > ε :
                return neighbor_list
        else:
                neighbor_list←neighbor_list ∪ Q[0]
                mark Q[0] as visited
}
```

Figure 4. Implementation of modified Dijkstra's algorithm to fetch ε – *neighborhood*.

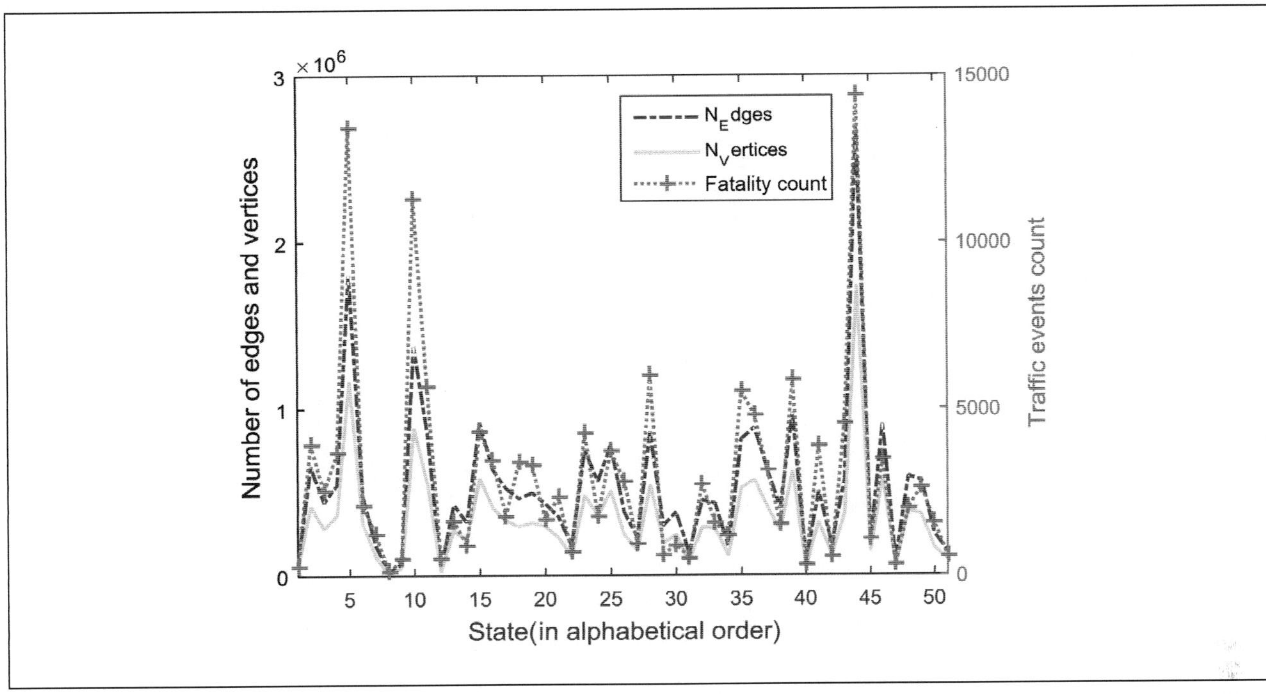

Figure 5. Numbers of links, nodes, and five-year fatality counts in the U.S. road network.

usually uses a small ε to capture the most significant clusters, so the network is small and obtaining the neighborhood for one event point takes little time. We should also emphasize one fact—that the time complexity of the neighborhood retrieval step is determined by the number of traffic event points and ε, but not the scale of the road network. This fact is the basis for the scalability of the Dijkstra's-DBSCAN algorithm.

Dijkstra's-DBSCAN for Incident Location Hotspots Identification: A Case Study

We test Dijkstra's-DBSCAN on the road network of the entire United States. The road network topology includes 27,145,945 edges and 17,464,790 vertices. The traffic incident data we used are the 152,089 traffic fatalities over a five-year period (2009 to 2013) provided by the Fatality Analysis Reporting System (FARS). We partition the U.S. road network and event data into 51 groups labelled by state. Figure 5 shows the basic statistics of the road network and fatality counts of the 51 states. The fatality count of each state is positively related to the scale of the road network.

Figure 6 presents some sample clusters generated with $\varepsilon = 1,000m$ and $\min pts = 3$. Each row of Figure 6 shows one of the new features of Dijkstra's-DBSCAN compared with previous incident hotspot identification algorithms. Row 1 shows that Dijkstra's-DBSCAN provides more accurate clusters than Euclidean based DBSCAN. Even though some points are close to others by Euclidean distance, they are not included in the cluster because of the lack of actual short road connections. Row 2 shows the routing flexibility of the algorithm. A cluster generated by Dijkstra's-DBSCAN is not interrupted by intersections, roundabouts, ramps, etc. It can freely cross the corners and connects to other incidents. Row 3 shows the capability of mining clusters of irregular and complicated shapes. With the awareness of road network connections, Dijkstra's-DBSCAN can capture density clusters of any shape.

Computational Cost Analysis and Parallelization

We have analyzed the time complexity of modified Dijkstra's as $O(|E(\varepsilon)| + |V(\varepsilon)|\log(|V(\varepsilon)|))$, where E and V represent the number of edges and vertices of the local graph around an incident point, and ε determines the scale of the local graph. Suppose the dataset contains N traffic incidents, then the neighbor retrieval step has a time complexity of $O(N \cdot (|E(\varepsilon)| + |V(\varepsilon)|\log(|V(\varepsilon)|)))$. The other step in Dijkstra's-DBSCAN is cluster labeling and expanding, which is essentially a DFS. Every vertex is visited once, so the complexity is $O(N)$, which is ignorable compared with the complexity of neighbor retrieval. So the overall complexity is $O(N \cdot (|E(\varepsilon)| + |V(\varepsilon)|\log(|V(\varepsilon)|)))$.

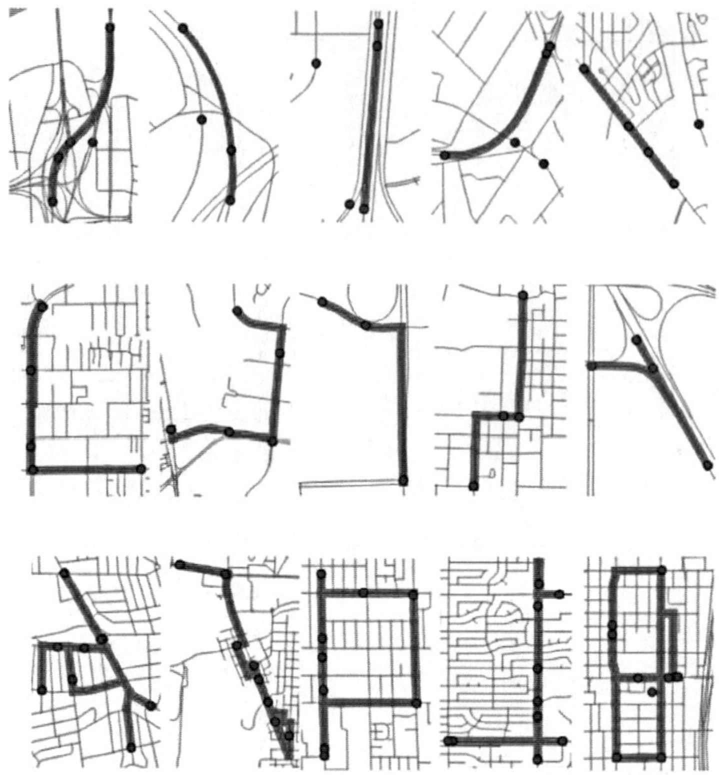

Figure 6. Sample incident clusters.

We consider a relatively large traffic event dataset. Since the local network is somewhat small for retrieving the neighboring points, $|E(\varepsilon)| + |V(\varepsilon)|\log(|V(\varepsilon)|$ will also be a relatively small number compared with N, so the overall complexity of Dijkstra's-DBSCAN is near linear. Compared with the Euclidean distance based DBSCAN, which has a $O(N^2)$ complexity, the network-aware Dijkstra's-DBSCAN performs much faster. When we use a larger ε, the local graphs will be bigger and so it will definitely take longer time to complete the clustering process, but this does not change the order of magnitude of the overall time complexity.

The parallelization of DBSCAN (Figure 7) has been a challenging problem, mainly because data access in both steps is sequential in nature (*25*, *26*), and partitioning data will cause heavy communication overheads. Road network and traffic events data, however, have something special in common that makes data partitioning easy: an administrative entity tag is available. Any road network database has an attribute showing to which state or county a road segment belongs. For traffic events data, even if the administrative area tag is not directly available, we can still quickly determine the administrative entity in which it is situated from GPS coordinates. Hence, we can divide the data by their territorial areas.

```
Job Scheduling(S,V[S],P)
{       L[P] ← {}
        J[P] ← {}
        for i = 0: S-1
                p ←argmin(L)    find the processor that has the minimum load
                J[p] ← J[p] ∪ {i}
                L[p] ← L[p] + V[i]
        end for
        return J
}
```

Figure 7. Job scheduling for parallelization.

Now suppose both the road network and event data are split into disjoint territorial groups. Different groups contain different volumes of data. As stated above, the time cost of the algorithm is near linearly related to the number of event points, which then becomes an appropriate measure of the computational loads of each data group. Suppose we have S groups, and their event data

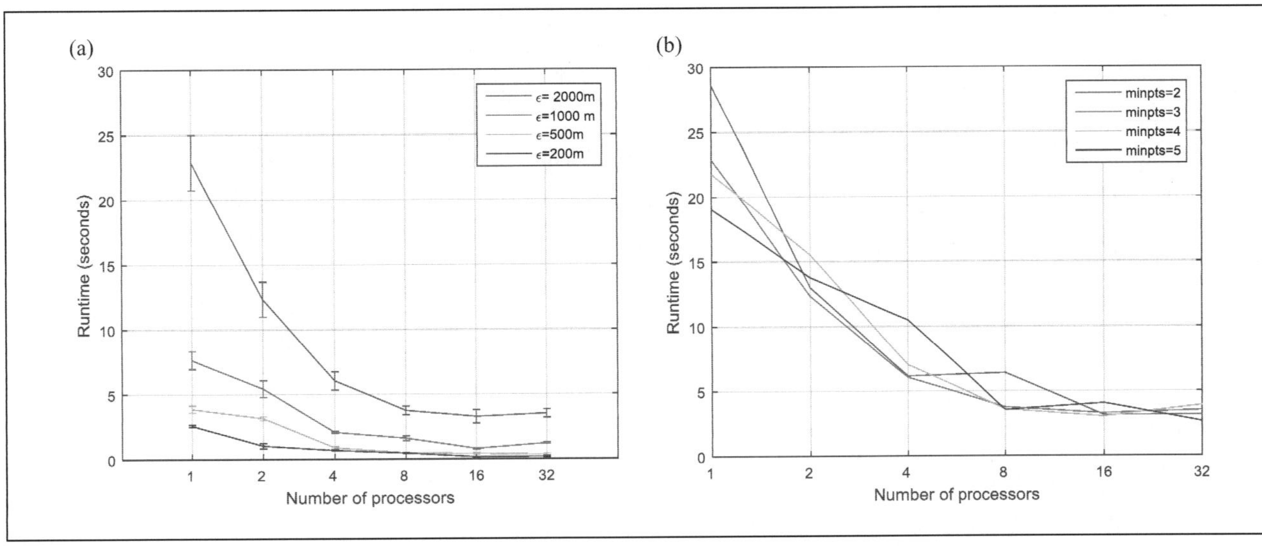

Figure 8. Benchmarks of parallel Dijkstra's-DBSCAN: (*a*) with different ε values; (*b*) with different minpts values.

volumes are in $V[0, 1, ..., S-1]$, we have P distributed processors. $P \leq S$. Allocating the data to the processors can find its solution from a well-studied list scheduling problem (*27*, *28*), the heuristic solution to which is to iteratively assign an unassigned job to the processor that currently owns the least load until all jobs have been allocated.

We now prove that the differences between the loads of different processors are bounded. The loads of two processors p_1, p_2 are $L[p_1]$, $L[p_2]$ respectively, then $|L[p_1] - L[p_2]| \leq \max(V)$.

Proof: The two loads must be determined in two different steps. Without loss of generality, we assume $L[p_2]$ is determined as step k, which is later than $L[p_1]$. Then, at the beginning of step k, $L_k[p_2] \leq L[p_1]$ and $L[p_2] = L_k[p_2] + V[k] \leq L[p_1] + V[k]$, so $L[p_2] - L[p_1] \leq V[k] \leq \max(V)$. This tells us that the processors have generally balanced loads and can finish the computations within similar lengths of time.

We benchmarked the algorithm on distributed computer clusters. We also changed the parameters to see their influences on the performance. Figure 8 show the results. Each machine in this experiment is a one-core, 4GB-memory machine and the programming environment is C + + . Here is what we observe:

1. Dijkstra's-DBSCAN is fast. Even on a single one-core computer, it can finish clustering on the entire U.S. network in less than 25 seconds. We also tried Euclidean based DBSCAN, and even for Tennessee state, which has 573,444 edges and 366,940 vertices, and a fatality count of 4,580, it takes more than 20 minutes to complete the clustering process. The reason for the differences in time complexity is that Dijkstra's-DBSCAN has near linear complexity while the original DBSCAN has square order complexity. In the neighborhood retrieval step, one incident point has to compare with all other points, so Euclidean DBSCAN is a global approach; However, Dijkstra's-DBSCAN does a local search; the underlying road network enables a point to search just within a shortest distance neighborhood without going beyond that range.
2. Dijkstra's-DBSCAN is scalable. The more processors used, the faster the algorithm runs. When using 16 processors, it takes only three to four seconds to find the clusters that have a minimum distance of two kilometers. While Dijkstra's-DBSCAN is parallelization friendly, Euclidean based DBSCAN is hard to parallelize, as explained above.
3. The two parameters have different effects on the parallel performance of the algorithm. ε is more significant than min*pts*. This is because the time cost of Dijkstra's-DBSCAN is mainly determined by the neighborhood retrieval step. When ε is big, a point has to search in a broader range, which then will take more time. min*pts* determines how many points will be labeled as cores. The cluster expanding step is essentially a DFS on the cores. We have analyzed that the time complexity of cluster expanding step equals the number of cores, which can be ignored compared with the neighborhood retrieval step. A smaller min*pts* will identify more core points and indeed increase the time cost, as we can see from Figure 8*b*. But the effect of min*pts* becomes less and less significant as more processors are used.

Conclusion and Future Work

We have introduced and implemented Dijkstra's-DBSCAN, a density based clustering algorithm specifically for traffic incident hotspot identification, lane safety management, and investment location decision support. Compared with region based methods such as KDE, DBSCAN, K-means, or hierarchical clustering, the new algorithm has the following advantages. It is based on network space and uses routing distance as distance measure, and so the clusters are more accurate than region based approaches for traffic management purpose. It has routing flexibility and can extract clusters of irregular shapes. It has a time complexity of O(n), which is a large increase in speed from the O(n^2) complexity of all other algorithms. It is parallelable and can utilize distributed computing resources to handle large-scale networks.

There are still numerous topics to investigate on Dijkstra's-DBSCAN to fully support incident hotspot analysis and investment decision making. One important issue is that all algorithms, including our new algorithm, will obtain a number of clusters, but there has been no research published looking into cluster ranking to find the few most significant clusters. This problem becomes essential when investment budgets are so limited that only a few locations will be selected for investment in improvements. The advantage is that Dijkstra's-DBSCAN gives clusters together with additional information on the incidents' distance relations. We will be developing a measure of the overall "density" of a cluster to support practical and reasonable incident cluster ranking, which hopefully will provide a valuable reference to investment decision making.

Although we have only used physical distance to develop the algorithm, this "distance" should be extended as a broad "cost function," for example, travel time, or other inclident exposure function in safety analysis. By substituting the distance with these metrics, Dijkstra's-DBSCAN will be able to extract clusters following the new cost functions.

This article focuses on the computational side of the algorithm; we will further conduct in-depth case studies using this algorithm to show how it supports transportation analysis. For example, the algorithm can help identify dangerous routes for cycling, where bicycle involved incidents occur frequently. This is what region based methods cannot achieve. Another possible application is to identify congested routes using floating car data, a common method to determine traffic speed using cellular data.

Author Contributions

The authors confirm contribution to the paper as follows: study conception and design: Lee Han and Yang Zhang; data collection: Yang Zhang; analysis and interpretation of results: Yang Zhang, Hyun Kim, and Lee Han; draft manuscript preparation: Yang Zhang. All authors reviewed the results and approved the final version of the manuscript.

References

1. Prasannakumar, V., H. Vijith, R. Charutha, and N. Geetha. Spatio-temporal Clustering of Road Accidents: GIS Based Analysis and Assessment. *Procedia-Social and Behavioral Sciences*, Vol. 21, 2011, pp. 317–325.
2. Truong, L. T., and S. V. Somenahalli. Using GIS to Identify Pedestrian–Vehicle Crash Hot Spots and Unsafe Bus Stops. *Journal of Public Transportation*, Vol. 14, No. 1, 2011, p. 6.
3. Bills, S. D. *Examining Hotspots of Traffic Collisions in Redlands*. California, 2009.
4. Li, X., J. Han, J.-G. Lee, and H. Gonzalez. Traffic Density-based Discovery of Hot Routes in Road Networks. *Advances in Spatial and Temporal Databases*, 2007, pp. 441–459.
5. Wen, H., J. Sun, and X. Zhang. Study on Traffic Congestion Patterns of Large City in China Taking Beijing as an Example. *Procedia-Social and Behavioral Sciences*, Vol. 138, 2014, pp. 482–491.
6. Anbaroglu, B., B. Heydecker, and T. Cheng. Spatio-temporal Clustering for Non-recurrent Traffic Congestion Detection on Urban Road Networks. *Transportation Research Part C: Emerging Technologies*, Vol. 48, 2014, pp. 47–65.
7. Montazeri-Gh, M., and A. Fotouhi. Traffic Condition Recognition Using the K-means Clustering Method. *Scientia Iranica*, Vol. 18, No. 4, 2011, pp. 930–937.
8. Zhang, B., K. Xing, X. Cheng, L. Huang, and R. Bie. Traffic Clustering and Online Traffic Prediction in Vehicle Networks: A Social Influence Perspective. *2012 Proceedings IEEE Infocom*, 2012. IEEE.
9. Won, J.-I., S.-W. Kim, J.-H. Baek, and J. Lee. Trajectory Clustering in Road Network Environment. *IEEE Symposium on CIDM'09*. 2009, pp. 299–305. IEEE.
10. Roh, G.-P., and S.-W. Hwang. Nncluster: An Efficient Clustering Algorithm for Road Network Trajectories. *Database Systems for Advanced Applications*. 2010.
11. Kim, J., and H. S. Mahmassani. Spatial and Temporal Characterization of Travel Patterns in a Traffic Network Using Vehicle Trajectories. *Transportation Research Part C: Emerging Technologies*, Vol. 59, 2015, pp. 375–390.
12. Bartin, B., K. Ozbay, and C. Iyigun. Clustering-based Methodology for Determining Optimal Roadway Configuration of Detectors for Travel Time Estimation. *Transportation Research Record: Journal of the Transportation Research Board*, 2007. 2000: 98–105.
13. Lawson, A. B. Hotspot Detection and Clustering: Ways and Means. *Environmental and Ecological Statistics*, Vol. 17, No. 2, 2010, pp. 231–245.
14. Anderson, T. K. Kernel Density Estimation and K-means Clustering to Profile Road Accident Hotspots. *Accident Analysis & Prevention*, Vol. 41, No. 3, 2009, pp. 359–364.
15. Yang, C., R. Duraiswami, N. A. Gumerov, and L. Davis. Improved Fast Gauss Transform and Efficient Kernel Density Estimation. *Proc., 9th IEEE International Conference on Computer Vision, 2003*. IEEE, 2003.

16. Kidane, P., and A. Bonds. *An Analysis of Hierarchical Clustering Algorithms for Hotspot Detection in Geographical Request Maps*, KTH Royal Institute of Technology, Stockholm, Sweden, 2015.
17. Phillips, P., and I. Lee. Mining Positive Associations of Urban Criminal Activities Using Hierarchical Crime Hot Spots. *Intelligence and Security Informatics*, Vol. 3917, 2006, pp. 127–132.
18. Di Martino_affanb, F., V. Loia, and S. Sessa. Extended Fuzzy C-means Clustering Algorithm for Hotspot Events in Spatial Analysis. *International Journal of Hybrid Intelligent Systems*, Vol. 5, No. 1, 2008, pp. 31–44.
19. Faria, A. M., M. E. Cintra, A. F. de Castro, and D. C. Lopes. Criminal Hot Spot Detection using Formal Concept Analysis and Clustering Algorithms. *Encontro Nacional de Inteligência Artificial, ENIAC*, Universidade Federal Rural do Semi-Árido, Brazil, pp. 85–90.
20. Shirai, M., M. Hirota, H. Ishikawa, and S. Yokoyama. A Method of Area of Interest and Shooting Spot Detection using Geo-tagged Photographs. *ACM SIGSPATIAL Workshop on Computational Models of Place*. Association for Computing Machinery, New York City, 2013.
21. Lopez, X. M., O. Debeir, C. Maris, S. Rorive, I. Roland, M. Saerens, I. Salmon, and C. Decaestecker. Clustering Methods Applied in the Detection of Ki67 Hot-spots in Whole Tumor Slide Images: An Efficient Way to Characterize Heterogeneous Tissue-based Biomarkers. *Cytometry Part A*, Vol. 81, No. 9, 2012, pp. 765–775.
22. Birant, D., and A. Kut. ST-DBSCAN: An Algorithm for Clustering Spatial-temporal Data. *Data & Knowledge Engineering*, Vol. 60, No. 1, 2007, pp. 208–221.
23. Divya, G., and K. Selvan. Suitability of Clustering Algorithms for Crime Hotspot Analysis. *International Journal of Science, Engineering and Computer Technology*, Vol. 4, No. 7, 2014, 231.
24. Ester, M., H.-P. Kriegel, J. Sander, and X. Xu. A Density-based Algorithm for Discovering Clusters in Large Spatial Databases with Noise. *KDD'96 Proc., 2nd International Conference on Knowledge Discovery and Data Mining*, Menlo Park, CA, 1996.
25. Patwary, M., D. Palsetia, A. Agrawal, W.-k. Liao, F. Manne, and A. Choudhary. A New Scalable Parallel DBSCAN Algorithm Using the Disjoint-set Data Structure. *Proc., 2012 International Conference for High Performance Computing, Networking, Storage and Analysis (SC)*, IEEE, 2012.
26. Brecheisen, S., H.-P. Kriegel, and M. Pfeifle. Parallel Density-based Clustering of Complex Objects. *Advances in Knowledge Discovery and Data Mining*, 2006, pp. 179–188.
27. Mokotoff, E., J. L. Jimeno, and A. I. Gutiérrez. List Scheduling Algorithms to Minimize the Makespan on Identical Parallel Machines. *TOP, An Official Journal of the Spanish Society of Statistical and Operations Research*, Vol. 9, No. 2, 2001, pp. 243–269.
28. Kleinberg, J., and É. Tardos. *Algorithm Design*, Pearson, London, 2005.

The Standing Committee on Artificial Intelligence and Advanced Computing Applications (ABJ70) peer-reviewed this paper (18-06780).

Article

TRR
JOURNAL OF THE TRANSPORTATION RESEARCH BOARD

Transportation Research Record
2018, Vol. 2672(45) 274–283

DOI: 10.1177/0361198118787657
journals.sagepub.com/home/trr
SAGE

A Voxel-Based Method for Automated Detection and Mapping of Light Poles on Rural Highways using LiDAR Data

Suliman A. Gargoum[1], James C. Koch[1], and Karim El-Basyouny[1]

Abstract
The number of light poles and their position (in terms of density and offset off the roadside) have significant impacts on the safe operation of highways. In current practice, inventory of such information is performed in periodic site visits, which are tedious and time consuming. This makes inventory and health monitoring of poles at a network level extremely challenging. To relieve the burden associated with manual inventory of poles, this paper proposes a novel algorithm which can automatically obtain such information from remotely sensing data. The proposed algorithm works by first tiling point cloud data collected using light detection and ranging (LiDAR) technology into manageable data tiles of fixed dimensions. The data are voxelized and attributes for each data voxel are calculated to classify them into ground and nonground points. Connected components labeling is then used to perform 3D clustering of the data voxels. Further clustering is performed using a density-based clustering to combine connected components of the same object. The final step involves classifying different objects into poles and non-poles based on a set of decision rules related to the geometric properties of the clusters. The proposed algorithm was tested on a 4 km rural highway segment in Alberta, Canada, which had substantial variation in its vertical alignment. The algorithm was accurate in detecting nonground objects, including poles. Moreover, the results also highlight the importance of considering the length of the highway and its terrain when detecting nonground objects from LiDAR.

Periodic assessment of road infrastructure is essential for safe and efficient operation of roads. Whether it is streetlight poles, traffic signs, bridges, or pavement surface, all these assets must be routinely assessed to ensure they meet the necessary standards throughout the road's service life. The density of light poles, their dimensions, and placement have significant impact on the safety of a highway. In fact, the relationship between the presence of light poles on a highway and safety is a complex one. For instance, while increasing the light pole density may help reduce some types of night-time crashes as a result of the improved visibility, the greater density could also result in more fixed object collisions.

Many studies have found a significant relationship between the offset of poles from the road and the safety of highways (*1*, *2*). Owing to the significant damage poles could cause in fixed object collisions, design guides require that they are placed at a certain offset from the roadside, depending on their sizes and dimensions. This helps in the efforts to design more forgiving highways where a driver is given the best chance of recovery in the case of a collision.

Developing an inventory of light poles and their locations on a highway corridor is also essential to increase the efficiency of maintenance and street aesthetics. Unfortunately, in current practice, inventory information is obtained in manual inspections which are error prone and require several man-hours of work. Obtaining such information on a network level is therefore expensive, and constantly updating existing records is often infeasible. In recent years, light detection and ranging (LiDAR) technology has been considered as an alternative for the inventory and assessment of many elements of a road's infrastructure (*3–7*).

LiDAR technology combines laser scanners, global navigation satellite systems (GNSS) and inertial measurement units (IMU) into one system to collect a 3D point cloud of the road's environment. In mobile laser

[1]Department of Civil and Environmental Engineering, University of Alberta, Edmonton, AB, Canada

Corresponding Author:
Address correspondence to Suliman A. Gargoum: gargoum@ualberta.ca

scanning (MLS), the scanning system is mounted on a truck that travels down a highway, creating a 360° virtual image of the highway consisting of closely spaced points of known positional coordinates and intensity information. Although not useful for pole extraction because of the low-reflective surface texture, intensity information is commonly used to extract highly reflective objects such as traffic signs and lane markings. The high speeds at which data are collected enables efficient and safe mapping of extensive areas.

Although a fair number of studies have attempted the extraction of poles from LiDAR, the majority of those algorithms have been developed for short, flat segments in an urban setting. Consequently, there is a lack of studies in which pole detection is attempted on high-speed (>80 km/h) rural highway with rolling terrain. Unlike flat urban roads, rural highways often experience significant variation in vertical alignments. This creates huge challenges when extracting the nonground surface from the point cloud while maintaining points representing roadside poles which is an integral step to most of the pole extraction algorithms. The high speeds at which LiDAR data is collected in rural environment also impacts the properties of the point cloud, including point density. Such conditions, although challenging, are commonly present, and so there is a need for research to explore the efficiency of detecting pole-like features in such an environment.

This paper develops an algorithm using data on a 4 km rural highway corridor where significant variation in vertical profile exists. The extraction revealed the importance of tiling LiDAR data before attempting the extraction of poles on such a large scale. Testing also revealed that, even in such a challenging environment, pole-like objects could be extracted and mapped at a decent level of accuracy. The study also points out the importance of further filtering in the post-segmentation stage when extracting poles in a rural high-speed environment.

Literature Review

Pole-like object extraction from LiDAR in the literature generally covers two areas: tree detection for forestry applications and roadside furniture extraction for transportation applications. The methods used in previous studies range from simple scanline segmentation and shape-based clustering to more complex voxel-based approaches.

Early pole detection attempts from LiDAR used circular cross-sections (*8*). These methods involved taking a horizontal slice of the point cloud, creating a corresponding raster image, clustering pixels with a higher number of points, and circle fitting. Using such approaches, Bienert et al. reported 97.4% detection rates (*8*).

Golovinskiy et al. proposed an object detection method that involved locating, segmenting, characterizing, and classifying point clusters (*9*). Iterative plane fitting is first used to filter out points close to the ground. Locating candidate objects then occurs using a k-nearest neighbors (kNN) graph and a clustering algorithm. Segmentation was attempted by combining the kNN graph used to connect points into foreground objects by distance thresholds with information on the degree to which the foreground object are connected to the background. After characterizing foreground objects, objects are classified using random forests and support vector machine classifiers. Testing revealed that the algorithm was effective in locating and segmenting objects with 92% and 93% accuracy, respectively. Recognition rates were significantly lower at 58% precision and 65% recall.

Lehtomäki et al. proposed a scanline-based algorithm to extract pole-like objects from mobile LiDAR data (*10*). In scanline LiDAR data, poles will exist as sweeps (i.e., point groups) in each scanline. Point groups which are on top of each other in adjacent scanlines are then clustered. Clusters that constitute the same pole are merged using principal component analysis (PCA). A cluster is defined as part of a pole-like object if it meets specific geometric properties. The algorithm was tested on a 450 m straight and flat section, and the authors reported a 77.7% detection rate and 81.0% correctness rate. False-positives included pillars in buildings and different wall structures. Lamp posts were found to be the easiest to detect, with a detection rate of 93% as compared to traffic signs and tree trunks at 73.3% and 76.1%, respectively.

Pu et al. presented a method to classify MLS point clouds into three categories: ground surface, objects on the ground, and objects off the ground (*11*). Additionally, objects on the ground are classified into detailed groups such as traffic signs, trees, and utility poles. A surface growing algorithm taken from Vosselman et al. (*12*) is employed to determine the points representing the ground surface. To detect poles, the common feature of a vertical principal axis are used. Objects are divided into quartiles based on their height, and the third quartile (measured from the lowest elevation point of a given cluster) is used for further analysis. This helps omit objects such as bushes and trees when classifying ground objects into poles and non-poles. The authors reported an 87% success rate for detection of pole-like objects using their procedure.

The procedure proposed by El-Halawany and Lichti to extract poles starts by organizing the point cloud using a *KD* tree data structure (*13*). A 2D density-based segmentation is performed using a density-based clustering algorithm (DBSCAN) which finds clusters of high density in local neighborhoods. The proximity threshold in the DBSCAN search is defined based on a utility pole radius of 25 cm. The output of the clustering is then used in a vertical region-growing procedure to extract upright objects starting from the lowest elevation object detected in the previous step as the seed for the vertical regions. To merge different vertical segments that are close enough to be considered part of the same object, segment merging is based on the horizontal distance between centroids of the vertical regions grown in the previous step. Objects are then classified using several criteria including object height range, the surface normal direction, and the largest normalized eigenvalue. The algorithm was tested using data collected on three urban streets ranging in length from 103 to 768 m, with a reported processing time of 4–6 hours. The average detection rate was 86% for the three segments and the accuracy was 97%.

Yan et al. proposed a four-step procedure to extract poles and towers from LiDAR (*14*). The method consists of ground filtering, unsupervised clustering, classification, and data cleaning. Filtering the ground surface from the LiDAR point cloud is done based on the statistical distribution of the points (assuming normality of ground points). This allows for a statistical skewness balancing algorithm applied to the height attribute to differentiate ground and nonground points. The paper then uses DBSCAN to cluster the height-normalized nonground points. Each cluster is classified into one of five types of poles based on a set of defined decision rules. The final stage involves using least square circle fitting algorithms on the lower 10–20% portion of the pole structure to eliminate ground points from the extracted pole object. The proposed algorithm was tested on an urban site in Toronto, Ontario, resulting in a 91% detection rate for five types of light poles and towers.

Wu et al. proposed a voxel-based method for identification of street trees from LiDAR (*15*). The method consists of voxelization, calculating values of voxels, searching and marking neighborhoods, extracting potential trees, and using morphological parameters to eliminating pole-like objects other than trees. It is worth noting that the voxel layer that fell 1.2–1.4 m above the ground was used to begin the neighborhood marking and searching to extract trees. The proposed algorithm was tested on two 300 m long flat urban street segments with less than 1 m difference in elevation—therefore, height normalization with respect to a ground surface was not required. This resulted in a completeness and correctness of over 98% in detection.

Cabo et al. also proposed an automatic voxel-based extraction of pole-like objects from MLS LiDAR (*16*). The data are first voxelized to reduce data size for processing. Each horizontal layer of the voxelization is analyzed and segmented separately and segments are then merged to form the selected 3D features. The 2D analysis is carried out to identify pole-like candidates in three stages: segmentation of connected horizontal elements; selection of elements greater than the maximum area criteria; and selection of elements by isolation criteria. Both the second and third stages are based on the assumptions that poles have a relatively small cross-sectional area and are isolated. The results provide a set of segments associated with a Z coordinate of a candidate part of a pole. The third step involves connecting all voxel elements that share a face, edge, or vertex among all elevation layers. A minimum vertical height is set for connected groups to differentiate pole-like objects. The algorithm was successfully tested on four sites, with an average completeness of 92.3% and a correctness of 83.8%.

Lehtomäki et al. proposed an automated voxel-based method to process large MLS LiDAR point clouds and create road environment infrastructure maps (*17*). The proposed method consists of the following workflow: isolating nonground points from the point cloud, object segmentation, segment classification, and object location estimation. The authors use connected component labeling to perform object segmentation. Feature descriptors calculated from voxels making up a segmented object include local descriptor histograms, spin images, and general shape/point distribution attributes to apply machine learning techniques for object classification. The paper was successful in extracting and classifying trees, lamp posts, traffic signs, cars, pedestrians, and advertising boards. The authors tested their algorithm on a 900 m stretch of road in a suburban area in Espoo, Finland. In general, the authors report between 66.7% and 94.3% recall for the six defined object classes.

As evident in the review, there has been some progress in efforts to extract pole-like objects from LiDAR. The advanced techniques recently adopted have helped increase the accuracy of the results; however, in most existing studies testing was conducted on relatively short and flat urban streets. Accordingly, the feasibility of extraction on rural segments where the terrain and the length of the highway segments pose a challenge remains unknown. To address the aforementioned gaps, this paper builds on research in existing studies to develop a more robust method that can extract poles in a more challenging rural environment with rolling terrain. Specifically, the paper builds on the voxel-based approach used to perform nonground extraction since it was found to be the most effective method when dealing with significant variations in vertical alignment common in rural environments. To increase the efficiency of the voxel-based nonground extraction and reduce processing time, which is typically a challenge when dealing with long segments, the paper introduces the concept of data tiling, in which the point cloud is split into tiles or quadrants. Moreover, when segmenting the pole-like candidates, the paper performs density-based clustering on voxels as opposed to points. This helps address the issue of the low density of points on some pole structures, which is expected in rural environments. In addition, this also helps reduce processing time while maintaining the same level of accuracy.

Proposed Method

Pre-processing

Pre-processing was done in three different steps. First, points of very low intensity were removed from the point cloud. These low-intensity points are generally isolated particles within the atmosphere that do not represent the road environment. The second step of pre-processing involves filtering out points within the point cloud that have elevations lower than the mean elevation of the point cloud by six standard deviations. These points are typically a result of multipath errors in which the laser beam is refracted by a surface and thus the range measurement from source to destination is severely degraded.

The final step of pre-processing involves segmenting the LiDAR point cloud into square "tiles" of data. Since the proposed procedure is intended for long LiDAR segments, it is essential that the data be segmented into manageable portions. Each "tile" of data is chosen to represent a 50 m^2 subset of the entire LiDAR dataset. The dataset is divided into these tiles by defining a grid of 50 m cells based on the dataset's coordinate extents. Points are then grouped into each "tile" using a 3D histogram counting algorithm (*18*). This algorithm takes three vectors delineating each direction into N point groups and returns the i, j, k indices that correspond to a specific data "tile" for each point.

Stage A: Voxelization

The first stage of the processing pipeline involves voxelization of each data tile in the point cloud. Voxelization is the process of discretizing the LiDAR point cloud into three-dimensional voxels of a certain size similar to two-dimensional pixels in a normal image.

Let $v(i, j, k)$ denote a voxel and $v(I, J, k)$ denote all voxels in layer k. If V represents the voxel grid system consisting of K layers, then V can be defined as the union of all voxel layers:

$$V = \bigcup_{k=0}^{K} k \tag{1}$$

Unlike data tiling discussed above, the voxelization of each point in the point cloud is performed based on the spatial coordinates of the point as illustrated in Figure 1.

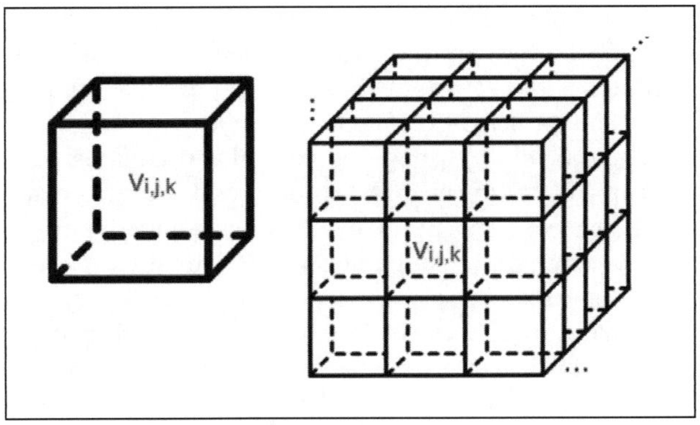

Figure 1. Voxel representation (*19*).

A point $P(x, y, z)$ is assigned to a voxel $v(i, j, k)$ as follows. If Δx, Δy, and Δz denote the dimensions of a single voxel cell (v) in the x, y, and z directions and x_0, y_0, and z_0 denote the origin of the voxel grid (V), then the ID of the voxel, $v(i, j, k)$, in which the point $P(x, y, z)$ falls can be computed as follows:

$$i = \frac{\text{int}(x - x_0)}{\Delta_x} \tag{2}$$

$$j = \frac{\text{int}(y - y_0)}{\Delta_y} \tag{3}$$

$$k = \frac{\text{int}(z - z_0)}{\Delta_z} \tag{4}$$

The actual voxelization of each data tile is based on the same three-dimensional histogram counting algorithm (*18*). The algorithm works by classifying points into bins based on the preset dimensions for individual voxel cells (v) and information on the size of the entire voxel grid (V).

The dimensions of the voxel are user-defined. For best overall results, it is recommended that voxel dimensions be defined based on the laser scanner properties. Since the data scanned in this study were done in scanlines that are approximately 20 cm apart, a 20 cm cell size for voxelization was used. This also ensures that points scanned from adjacent scanlines would fall in neighboring voxels, which is important when performing the connected components labeling.

Stage B: Ground vs Nonground Filtering

After voxelization occurs, all unique foreground voxels (i.e., voxels containing points) in a certain voxel layer (k) are found with their corresponding related points and a set of scalar and distributional geometric shape descriptors are generated. The attributes assigned to each voxel consist of a scalar geometric center and principal components based on the distribution of points within a local neighborhood of the voxel. The local neighborhood is defined as the 26 neighbors of a voxel in three dimensions. The principal components are calculated using PCA, specifically by using singular value decomposition (SVD) to find the principal directions and their eigenvalues. Since the SVD approach requires a large enough sample, only voxels containing more than three points were considered in this step.

The PCA attributes computed for each voxel include measures of linearity, flatness, and angular measures with respect to the Cartesian directions. Voxel linearity and flatness can be surmised from the eigenvalues corresponding to the principal directions (i.e., eigenvectors) since the eigenvalues can be seen as a measure of the variance along each principal direction. Linearity can be defined by dividing the largest principal eigenvalue by the sum of the eigenvalue of the principal directions, as in Equation 5. Similarly, flatness can be defined by dividing the smallest principal eigenvalue (λ) by the sum of the eigenvalues of all principal directions, as in Equation 6:

$$\text{Linearity} = \frac{\lambda_1}{\lambda_1 + \lambda_2 + \lambda_3} \tag{5}$$

$$\text{Flatness} = \frac{\lambda_3}{\lambda_1 + \lambda_2 + \lambda_3} \tag{6}$$

The angle of the component can then be computed as follows. If V denotes an eigenvector and U_Z denotes the Cartesian Z unit vector, then the following equations are used to compute the angular measures:

$$\text{Angle of first principal component w.r.t } \mathrm{U_Z} = \arctan(\|V_1 \times U_Z\|, V_1 \cdot U_Z) \tag{7}$$

$$\text{Angle of third principal component w.r.t } \mathrm{U_Z} = \arctan(\|V_3 \times U_Z\|, V_3 \cdot U_Z) \tag{8}$$

Both attributes of flatness and angle of the third eigenvector with the unit z vector help filter voxels that are expected to have high flatness and close to vertical third principal component (i.e., voxels representing the ground surface).

Once ground surface voxels are filtered out, all remaining voxels will then represent nonground objects. Data points must be voxelized into a small enough voxel size for the third principal component of potential road segments to be close to vertical, especially considering many rural highways have large vertical curves. Therefore, the choice of a 20 cm voxel size.

Stage C: Connected Component Labeling

After voxelization and ground filtering, a binary 3D image of the nonground voxels within each data tile is generated with foreground voxels represented as value "1" and background voxels represented as value "0". Connected component labeling (CCL) is then performed. This involves placing voxels within close proximity under a single label (see Figure 2). In this paper, CCL is done based on 26-connectivity, which means that, for each central voxel, adjacent voxels in the 3D space (up to the 26 neighbors) are considered when grouping voxels.

Once connected components are found, they are reported as potential objects in each voxel layer. These potential clusters of objects are typically only part of a larger object. For instance, different parts of tall objects such as poles could fall in voxel layers that are not adjacent and thus segment merging is required. Two-dimensional DBSCAN clustering is used to perform segment merging on cluster centroids. This filter ensures that changes in the 2D distance between cluster centroids in

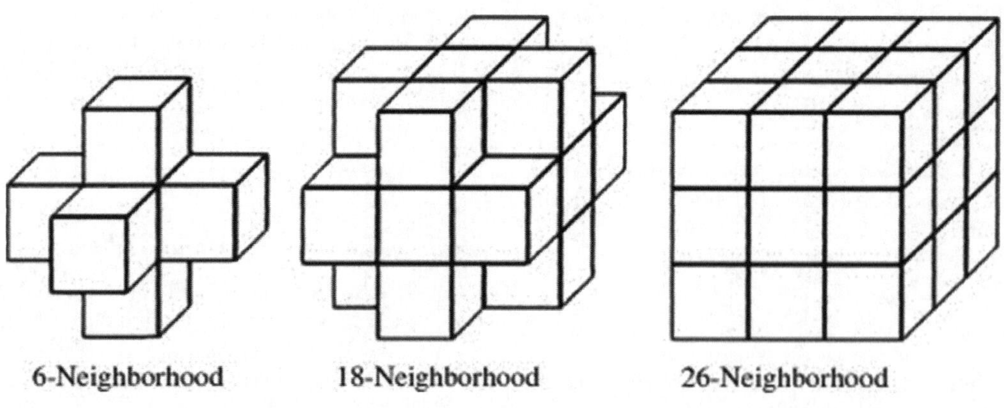

Figure 2. Central voxel neighborhood (in this paper, CCL is based on 26-neighborhood connectivity) (*20*).

adjacent voxel layers are not significant, while at the same time merging clusters in different voxel layers with insignificant changes in 2D distance. Employing these two criteria, segment merging is performed which enables merged segments to be representative of meaningful physical objects in the road environment.

Stage D: Pole Identification and Classification

This stage involves the computation of shape descriptors for each of the merged segments (i.e., objects). These descriptors are then used to further classify detected objects into poles and non-poles. In addition to the general geometric descriptors such as number of points and the height of objects, PCA is used to compute the same attributes as for the individual voxels but now on segmented objects. This includes linearity, flatness, and two angle measures. Streetlight and utility poles are thin vertical objects with a distinct range in the first principal component (i.e., high linearity) and first and third principal component angles with respect to the Cartesian z direction. Specifically, the following decision rules were used in the classification:

For east two-thirds of the test segment:

1. linearity between 0.7 and 0.8
2. flatness greater than 0.975
3. surface normal angle between 85° and 95°
4. first principal component angle less than 75°

For west one-third of the test segments:

1. linearity between 0.8 and 0.9
2. flatness greater than 0.9
3. surface normal angle between 65° and 95°

It is worth noting here that different filters for different portions of the 4 km test segment were used since parts of the segment included poles with incomplete structure (this is discussed further in the Results and Discussion section).

Case Study

The algorithm developed in this paper was tested on LiDAR data collected on a 4 km segment along Highway 20 in Alberta, Canada. The test segment in this paper was much longer (4 km) than in previous studies, and included high variation in vertical alignment (Δz = 796 m). A longer segment length with varying vertical alignment was purposely chosen to test the algorithm since it represents a realistic rural setting. The next two paragraphs provide more details on LiDAR data collection procedure and the test segment.

LiDAR Data

LiDAR data were collected using REIGL's VMX 450 Laser Scanning System. The RIEGL VMX-450 system uses two VQ-450 scanners along with the GNSS/IMU units to collect the data. The laser scanners are symmetrically configured on the left and right sides, pointing toward the rear of the vehicle at a heading angle of approximately 145°. The VQ-450 scanner has a scan rate reaching 1.1 million points per second and a speed of 400 lines per second (*13*). The density of the points on a scanned object depends on the range, and the speed of the data collection truck; provincial surveys conducted at 90 km/h result in point densities on the pavement surface of 150–1000 points/m^2 (*14*).

Test Segment

The analyzed portion of Highway 20 is an undivided highway located close to Sylvan Lake (west of Red Deer). A portion of the test segment is shown in Figure 3. The speed limit on the segment is 100 km/h; however, the speed does drop for a 1 km portion of the segment that runs close to the town of Sylvan Lake. Vegetation along the sides of the road is moderate. In addition to the drastic change in vertical alignment along the 4 km section, there is also a reverse horizontal curve on the segment and a rural intersection. Ground truth data used in the validation of the extracted information was collected through manual exploration of the test segment.

Results and Discussion

Before presenting the results, it is worth emphasizing that the test segment used in this paper was 4 km long with rolling terrain. The length of the segment increases the number of nonground objects from which poles are filtered. On the segment tested in this study, the number of objects detected before classification was 24,767 nonground objects. This includes signs, trees, building facades, vegetation, and poles. Out of those 24,767, only 41 poles existed; as a result, the likelihood of other objects having the same properties as poles is much higher when compared to analyzing a short 100 m segment where fewer objects exist. Moreover, the rolling terrain on the test segment causes a difference in elevation of 796 m. This makes extracting the nonground surface using voxelization of the point cloud extremely challenging. Despite that, nonground classification was highly accurate.

Figure 3. Highway 20.

Figure 4. LiDAR points representing detected poles.

Figure 4 shows points in the LiDAR point cloud representing the detected poles. Figure 5*a* shows a map of all the detected poles along the 4 km segment with the ground truth shown in Figure 5*b*. Similarly, Figures 5*c* and *d* show the detected poles and the ground truth for a 1 km test segment that runs adjacent to Sylvan Lake. It is noticeable from the figures that the algorithm was more effective in classifying poles in the section closer to Sylvan Lake town.

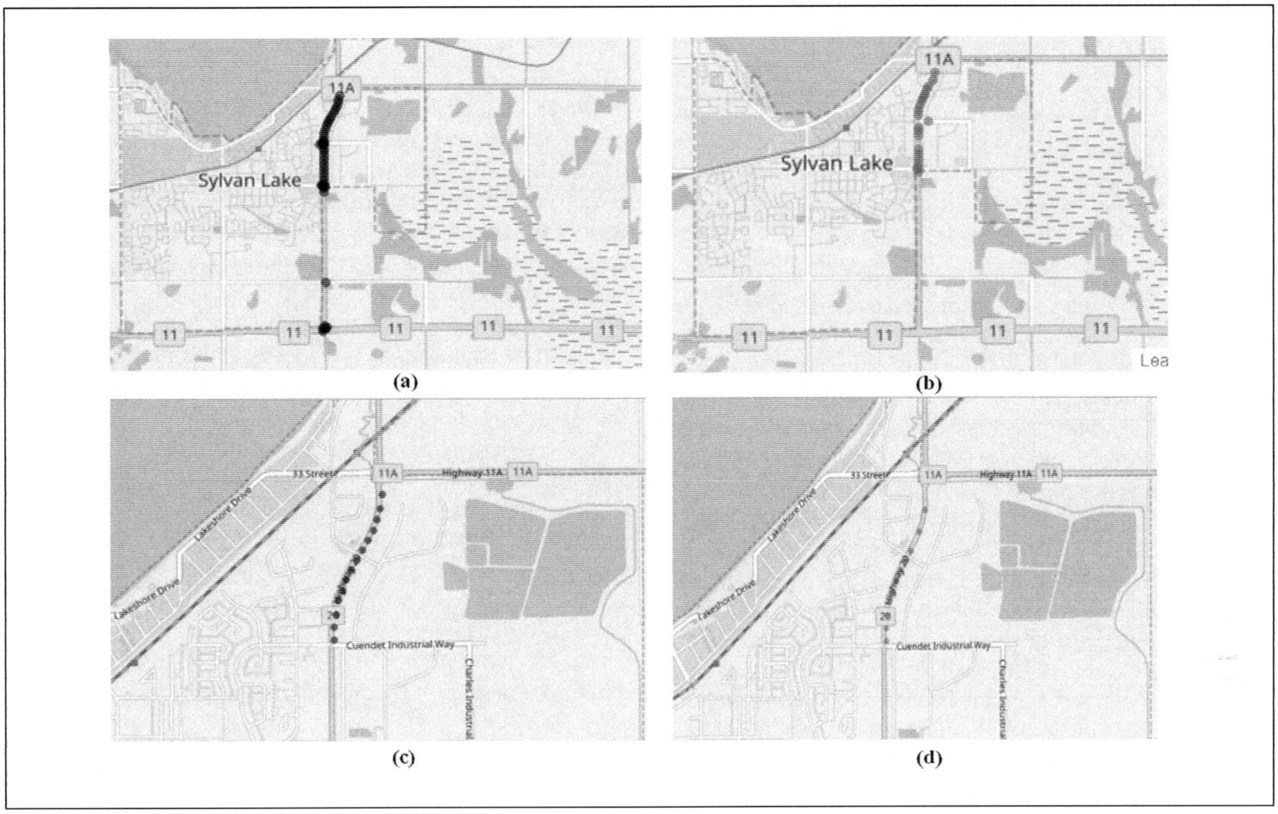

Figure 5. Pole distribution along the test segment: (a) ground truth (4 km segment); (b) true-positives (4 km segment); (c) ground truth (1 km segment—low speed); (d) true-positives (1 km segment—low speed).

The extracted information can be mapped on the highway segment with pinpoint accuracy, which facilitates measuring the density of light poles on a road segment along with their offsets from the road. These two critical performance measures were found to be the most significant factors in studies that analyzed the effects of lighting standards on safety.

To numerically assess the validity of the results, three metrics were calculated. The metrics were precision, detection rate, and accuracy. These were calculated as follows:

$$\text{Precision} = \frac{\text{TP}}{\text{TP} + \text{FP}} \tag{9}$$

$$\text{Detection rate} = \frac{\text{TP}}{\text{TP} + \text{FN}} \tag{10}$$

$$\text{Accuracy} = \frac{\text{TP} + \text{TN}}{\text{TP} + \text{FN} + \text{FP} + \text{TN}} \tag{11}$$

where TP and TN denote the number of true-positives (actual poles detected) and the true-negatives (actual non-poles detected). FP denotes the number of false-positives (i.e., the number of objects detected as poles which were not actually poles) and FN denotes the number of false-negatives (i.e., the number of objects that were actually poles but that were not detected by the algorithm as poles).

Accuracy measures how effective the algorithm is in the valid classification of both true-positives and true-negatives. This measure is also known as quality and provides a compound performance metric that balances detection rate and precision (*21*). Detection rate, also known as completeness, measures how effective the algorithm is in the valid classification of true-positives only. Finally, precision, also known as correctness, measures how successful the algorithm is in applying the classification filters.

As evident in the table, both precision and accuracy are relatively high for both the entire segment and the section closer to Sylvan Lake, which points to the robustness of the proposed algorithm in extracting over-ground objects. The detection rate, however, is high for the portion close to Sylvan Lake but relatively low for the entire segment. This indicates that the classification procedure is not as efficient as the extraction (i.e., poles are being detected but some poles are getting misclassified as other objects). This is mainly caused by two things, both of which have been reported in previous research (*22*). First, there are some cases in which the poles are too close to

Table 1. Result Validity Assessment

Metric	4 km segment	Sylvan Lake (1 km segment)
Precision (%)	68	78.6
Detection rate (%)	49	78.6
Accuracy (%)	98	95.4

other objects such as vegetation (bushes) or signs. This results in mixed pixels between the two objects (the pole and other objects) and, as a result, the objects get segmented into one object in the CCL process.

The other reason some poles were misclassified is because portions of the pole structure were missing (i.e., the pole structure was incomplete). The incomplete structure of poles causes the CCL to segment the pole into two objects. The CCL works by segmenting adjacent foreground voxels into a single object; however, when one of the adjacent voxels does not contain points, this leads to the voxel groups being classified as two different objects. This causes an issue when classifying pole structures since a pole that is divided between two segments or only has the top portion detected does not have the same geometric properties as a complete pole that belongs to a single cluster.

Although DBSCAN clustering was used in an attempt to regroup those unconnected segments, this was not always possible. This issue was more prevalent on the portion of the highway further away from the town of Sylvan Lake, as evident in the detection rates shown in Table 1. Therefore, it could be a matter of the data collection truck traveling at higher speeds away from the city bounds causing a reduction in point density. Object occlusion could also be a reason why some pole structures were incomplete. To minimize the effects of this, different filtering thresholds were used to classify poles in different regions of the test segment. The incomplete structure of a pole might also affect the ability to measure dimensions of the detected pole, although this is beyond the scope of this paper.

Conclusions and Future Research

This paper proposes an algorithm to automatically extract and classify pole-like objects using LiDAR point cloud data on rural highways with rolling terrain. The algorithm involves breaking the data down into manageable data tiles. Each data tile is then voxelized and voxel attributes are used to classify the point cloud into ground and nonground points. After voxelization, CCL is used to segment different parts of the nonground surface into potential objects. Further clustering of objects is then done and the objects are classified into poles and nonpoles based on their geometric attributes. The proposed algorithm is tested on a 4 km rural highway segment in Alberta, Canada. In general, the results were accurate; however, misclassification of poles did exist owing to low point density and pole occlusion by other objects. This paper illustrates the importance of accounting for the road type and its terrain when extracting information from LiDAR. The results also indicate that more robust segmentation may be required when extracting poles in a rural high-speed environment to account for the lower point density caused by the high data collection speeds. To address this issue, future research might consider employing machine learning techniques in the segmentation stage. Similarly, ellipsoidal region-growing algorithms could also help overcome the limitations associated with the CCL when dealing with discontinuity in pole structure. The proposed procedure could also be replicated on a segment where LiDAR is collected using a phase-based scanner instead of the time-of-flight scanner used in this paper. Phase-based scanners often result in a more dense point cloud, albeit the range of the data collection is shorter (*23*). This could help increase the density of points on the pole-like objects.

Acknowledgments

The authors would like to thank Alberta Transportation for sponsoring and providing data for this study. Acknowledgments are also extended to Alberta Innovates and Alberta Advanced Education for financial support.

Author Contributions

The authors confirm contribution to the paper as follows: All authors contributed to the study conception and design; algorithm development—Suliman A. Gargoum and James C. Koch; analysis, testing, and interpretation of results—Suliman A. Gargoum and James C. Koch; all authors contributed to the drafting manuscript preparation. All authors reviewed the results and approved the final version of the manuscript.

References

1. Zegeer, C. V., and M. R. Parker, Jr. Effect of Traffic and Roadway Features on Utility Pole Accidents. *Transportation Research Record: Journal of the Transportation Research Board*, 1984. 970: 65–76.
2. El Esawey, M., and T. Sayed. Evaluating Safety Risk of Locating Above Ground Utility Structures in the Highway Right-of-Way. *Accident Analysis & Prevention*, Vol. 49, 2012, pp. 419–428.
3. Gargoum, S. A., K. El-Basyouny, J. Sabbagh, and K. Froese. Automated Highway Sign Extraction Using Lidar Data. *Transportation Research Record: Journal of the Transportation Research Board*, 2017. 2643: 1–8.
4. Guan, H., J. Li, Y. Yu, C. Wang, M. Chapman, and B. Yang. Using Mobile Laser Scanning Data for Automated

Extraction of Road Markings. *ISPRS Journal of Photogrammetry and Remote Sensing*, Vol. 87, 2014, pp. 93–107.
5. Holgado-Barco, A., B. Riveiro, D. González-Aguilera, and P. Arias. Automatic Inventory of Road Cross-Sections from Mobile Laser Scanning System. *Computer-Aided Civil and Infrastructure Engineering*, Vol. 32, No. 1, 2017, pp. 3–17.
6. Kumar, P., C. P. McElhinney, P. Lewis, and T. McCarthy. An Automated Algorithm for Extracting Road Edges from Terrestrial Mobile LiDAR Data. *ISPRS Journal of Photogrammetry and Remote Sensing*, Vol. 85, 2013, pp. 44–55.
7. Castro, M., L. Iglesias, J. A. Sánchez, and L. Ambrosio. Sight Distance Analysis of Highways Using GIS Tools. *Transportation Research Part C: Emerging Technologies*, Vol. 19, No. 6, 2011, pp. 997–1005.
8. Bienert, A., S. Scheller, E. Keane, F. Mohan, and C. Nugent. Tree Detection and Diameter Estimations by Analysis of Forest Terrestrial Laserscanner Point Clouds. *Proc., ISPRS Workshop on Laser Scanning 2007 and SilviLaser 2007*, Espoo, 2007, pp. 50–55.
9. Golovinskiy, A., V. G. Kim, and T. Funkhouser. Shape-Based Recognition of 3D Point Clouds in Urban Environments. *Proc., 2009 IEEE 12th International Conference on Computer Vision*, Kyoto, Japan, IEEE, New York, 2009. pp. 2154–2161.
10. Lehtomäki, M., A. Jaakkola, J. Hyyppä, A. Kukko, and H. Kaartinen. Detection of Vertical Pole-Like Objects in a Road Environment Using Vehicle-Based Laser Scanning Data. *Remote Sensing*, Vol. 2, No. 3, 2010, pp. 641–664.
11. Pu, S., M. Rutzinger, G. Vosselman, and S. O. Elberink. Recognizing Basic Structures from Mobile Laser Scanning Data for Road Inventory Studies. *ISPRS Journal of Photogrammetry and Remote Sensing*, Vol. 66, No. 6, 2011, pp. S28–S39.
12. Vosselman, G., B. G. Gorte, G. Sithole, and T. Rabbani. Recognising Structure in Laser Scanner Point Clouds. *International Archives of Photogrammetry, Remote Sensing and Spatial Information Sciences*, Vol. 46, No. 8, 2004, pp. 33–38.
13. El-Halawany, S. I., and D. D. Lichti. Detection of Road Poles from Mobile Terrestrial Laser Scanner Point Cloud. *Proc., 2011 International Workshop on Multi-Platform/Multi-Sensor Remote Sensing and Mapping (M2RSM)*, Xiamen, China, IEEE, New York, 2011. pp. 1–6.
14. Yan, L., H. Liu, J. Tan, Z. Li, H. Xie, and C. Chen. Scan Line Based Road Marking Extraction from Mobile LiDAR Point Clouds. *Sensors*, Vol. 16, No. 6, 2016, p. 903.
15. Wu, B., B. Yu, W. Yue, S. Shu, W. Tan, C. Hu, Y. Huang, J. Wu, and H. Liu. A Voxel-Based Method for Automated Identification and Morphological Parameters Estimation of Individual Street Trees from Mobile Laser Scanning Data. *Remote Sensing*, Vol. 5, No. 2, 2013, pp. 584–611.
16. Cabo, C., C. Ordoñez, S. García-Cortés, and J. Martínez. An Algorithm for Automatic Detection of Pole-Like Street Furniture Objects from Mobile Laser Scanner Point Clouds. *ISPRS Journal of Photogrammetry and Remote Sensing*, Vol. 87, 2014, pp. 47–56.
17. Lehtomäki, M., A. Jaakkola, J. Hyyppä, J. Lampinen, H. Kaartinen, A. Kukko, E. Puttonen, and H. Hyyppä. Object Classification and Recognition from Mobile Laser Scanning Point Clouds in a Road Environment. *IEEE Transactions on Geoscience and Remote Sensing*, Vol. 54, No. 2, 2016, pp. 1226–1239.
18. Luong, B. N-Dimensional Histogram. MATLAB Central File Exchange, 2009. Updated August 2011. https://www.mathworks.com/matlabcentral/fileexchange/23897-n-dimensional-histogram. Accessed May 2017.
19. Zirbes, R. Scientific Visualization: Volume Surface Rendering. http://johnrichie.com/V2/richie/isosurface/volume.html. Accessed October 2017.
20. Kinkingnéhun, S., E. Volle, M. Pélégrini-Issac, J. L. Golmard, S. Lehéricy, F. du Boisguéheneuc, S. Zhang-Nunes, D. Sosson, H. Duffau, and Y. Samson. A Novel Approach to Clinical–Radiological Correlations: Anatomo-Clinical Overlapping Maps (AnaCOM)—Method and Validation. *Neuroimage*, Vol. 37, No. 4, 2007, pp. 1237–1249.
21. Rutzinger, M., F. Rottensteiner, and N. Pfeifer. A Comparison of Evaluation Techniques for Building Extraction from Airborne Laser Scanning. *IEEE Journal of Selected Topics in Applied Earth Observations and Remote Sensing*, Vol. 2, No. 1, 2009, pp. 11–20.
22. El-Halawany, S. I., and D. D. Lichti. Detecting Road Poles from Mobile Terrestrial Laser Scanning Data. *GIScience & Remote Sensing*, Vol. 50, No. 6, 2013, pp. 704–722.
23. Guan, H., J. Li, S. Cao, and Y. Yu. Use of Mobile LiDAR in Road Information Inventory: A Review. *International Journal of Image and Data Fusion*, Vol. 7, No. 3, 2016, pp. 219–242.

The Standing Committee on Geographic Information Science and Applications (ABJ60) peer-reviewed this paper (18-04591).

Article

TRR
JOURNAL OF THE TRANSPORTATION RESEARCH BOARD

Transportation Research Record
2018, Vol. 2672(45) 284–293
© National Academy of Sciences:
Transportation Research Board 2018
Article reuse guidelines:
sagepub.com/journals-permissions
DOI: 10.1177/0361198118793000
journals.sagepub.com/home/trr
SAGE

A Conflation Methodology for Two GIS Roadway Networks and Its Application in Performance Measurements

Farzad Daneshgar[1], **Kaveh Farokhi Sadabadi**[2], **and Ali Haghani**[1]

Abstract

In the current environment, in which private sector organizations play an ever increasing role in providing traffic data and tend to report data based on proprietary maps, it is inevitable that public sector agencies will want to adapt the spatial data and to use it in their legacy systems. Thus, for the same transportation infrastructures, a variety of GIS roadway datasets based on different highway reference location systems and segmentations are developed. This paper presents a method and a workflow to conflate such networks. The matching process usually involves projecting one network onto another. The first step in this process is to match the typically proprietary private sector network with the public sector legacy network. Different public and private entities may use different location referencing systems for analysis and reporting of data. So, in most cases, it is necessary to transform the spatial data from one location referencing system to another. Geographic information systems (GIS) offer a powerful tool for this purpose. GIS uses standard referencing systems to locate the features in a transportation network. This study proposes a methodology that takes advantage of GIS capabilities in the projection of transportation networks. A workflow is introduced based on the proposed methodology. A tool is created to perform the projection in the ESRI ArcGIS environment. Sanity checks and some reporting options are presented. Results are reported on projecting a subset of the Highway Performance Measurement System (HPMS) used by Maryland State Highway Administration onto the Traffic Message Channel (TMC) network in Maryland.

Applying the concept of geographic information systems (GIS) to transportation (GIS-T) (*1*), transportation communities have developed a variety of GIS roadway datasets. These networks serve many purposes in the analysis, reporting, and visualization of transport related geospatial data. In a given jurisdiction, transportation networks provide a representation of the transportation infrastructure. However, private and public entities create and maintain different networks which are not necessarily similar regarding coverage, segmentation, or resolution as they reflect the different needs, capabilities, and most importantly priorities of these entities. In the current environment, the private sector is assuming an ever increasing role as a data provider in most areas related to transportation operations and planning. To maintain their strategic business and market advantage, private sector entities tend to report their data based on their proprietary maps. The public sector, on the other hand, relies heavily on networks developed in-house to run legacy systems. Any use of external spatial data that is based on a different segmentation scheme requires matching between the two underlying networks and conflating them.

Networks conflation is the process of merging transportation data associated with networks (*2*). The first step in this process is to identify corresponding spatial features in both networks. While the minimum number of common attributes in the feature tables may prove to be very useful for this purpose, matching based on feature attributes alone usually is not sufficient. This is mainly due to the use of different segmentation, different naming conventions, lack of standard reporting, and mismatch in reporting precision between the two networks. Another possibility for conflating networks and matching corresponding features is to take advantage of their geometry. Due to their sheer size, manual matching of transportation networks is prohibitively time consuming

[1]Civil and Environmental Engineering Department, University of Maryland, College Park, MD
[2]Center for Advanced Transportation Technology (CATT), Civil and Environmental Engineering Department, University of Maryland, College Park, MD

Corresponding Author:
Address correspondence to Farzad Daneshgar: farzadd@umd.edu

and expensive. Moreover, networks tend to be frequently updated to reflect the latest changes and to enable new capabilities. Thus, an automated network projection tool serves a crucial role in the process of integrating outsourced traffic data into each entity's own mapping and reporting environment.

To tackle this problem, this study proposes a methodology based on matching both attributes data and feature geometries of networks. The proposed method focuses on finding appropriate matches and projections for a single feature in the target network at every iteration. The step-by-step workflow using standard tools in the ESRI ArcGIS is presented. The developed tool is employed in projecting a subset comprising freeway and expressway facilities in the Highway Performance Measurement System (HPMS) network in Maryland. The features in the HPMS network are generated based on linear referencing system (LRS) standards. HPMS networks are widely used by transportation agencies around the country in their data collection, data reporting, asset management, and visualization functions. The second network considered in this projection is a Traffic Message Channels (TMC) network. TMC is a widely adopted industry standard for roadway segmentation and spatial traffic data reporting. Criteria for performing sanity checks are presented to ensure precision of the projections and to identify areas where additional attention may be needed.

In general, the introduction of the GIS concept into transportation and especially GIS applications has made it possible to tackle problems which previously were not easy to solve . NCHRP Report 359 (*1*) introduces the definition and applications of GIS-T and provides a management guide to assist transportation professionals and transportation agencies in the implementation of GIS-T.

The rest of this paper is organized as follows. First, the background to highway location referencing systems and road segmentations used by public agencies and private sector organizations, and also network conflation, is presented. The problem statement and methodology to tackle the problem are presented in the next sections. The methodology section includes the steps to create the developed tool to match and project features. The proposed methodology and developed tool are applied to a subset of the transportation highway network in Maryland. In the final sections, the case study results and conclusion are presented.

Linear Referencing System

Situations frequently occur in which there is a need to reference locations on a transportation network in a standard solid format. A method of spatial referencing has been developed, LRS, to express a location as a distance from a specific point in an indicated direction (*3*). In this referencing system, the locations of elements are indicated by either a point or a line and are described in terms of distances along a linear section from a specific point (*4*).

Linearly-referenced data comprises most of the data managed by transportation agencies (*5*). The features in HPMS networks are generated based on LRS standards. "Highway Performance Monitoring System (HPMS) is a national level highway information system database, which is the official source of data on the performance, condition, use, extent, and operational characteristics of all public roads in the United States" (*6*). All public roads must be included in HPMS. All states are required to report data, as specified in the HPMS Field Manual (*7*), in an ESRI shapefile or ESRI geodatabase format which includes their LRS. States must provide their LRS, which links the attribute and geospatial data of their roadway network. This allows the analysis of HPMS data in a GIS environment (*7*). National Highway System (NHS) (*8*) and All Roads Network of Linear referenced Data (ARNOLD) (*9*) are the other two databases that reference and include the characteristics of the public roads.

Linear referencing systems have been critical to the operations of transportation agencies for years, but there are significant issues to developing effective, comprehensive LRS-based analysis systems. One of the major challenges that large transportation agencies face is the integration of dispersed databases (*10*). LRS data models vary across transportation agencies, and often within organizations as well, and this has made the integration of data impossible (*5*).

Industry Spatial Segmentation and Reporting Standards

In recent years, integration of GIS with intelligent transport systems (ITS) has elevated the GIS-T capability because by integrating GIS into ITS improved territorial representation is possible in a very legible manner (*11*). Some private sector traffic data providers such as INRIX, TomTom, and HERE use newer highway location referencing systems other than systems used by public state and federal agencies. In their reporting, they describe the locations along a transportation network in static systems such as TMC or dynamic systems such as OpenLR, which are generally based on newer technologies. These newer referencing systems are widely used. Different entities in the industry may adopt variations of the same standard. For example, National Performance Management Research Data Set (NPMRDS), which is freely available to all states, is based on slightly different TMC definitions (*12*).

The Traveler Information Services Association (TISA) is the organization which preserves the TMC standard. As defined by TISA, "Traffic Message Channel (TMC) is a specific application of FM radio data system used for broadcasting real-time traffic, travel, and weather information to motor vehicle drivers. Data messages are received silently and decoded by a TMC-equipped navigation system, and delivered to the driver, typically, by offering dynamic route guidance." (*13*). INRIX reports traffic data by the distance between successive TMC locations codes, or TMC segments.

A well-known dynamic location referencing system is OpenLR (Open, Compact and Royalty-free Dynamic Location Referencing). OpenLR segments are based on a system of dynamic encoding and decoding of location referencing. OpenLR is an open source system promoted by TomTom that enables "data exchange and cross-referencing using digital maps of different vendors" (*14*). AGORA-C is also another type of dynamic location referencing system (*15*).

Networks Conflation

As discussed, public agencies and private sector entities may segment and reference highway networks based on different highway location referencing systems and segmentation approaches. Therefore, the conflation of the different files is critical to GIS-T performance. Conflation is the merging or fusion of spatial and attribute data from different files. Networks conflation is the process of merging transportation data associated with networks (*2*). Conflation is important as the source file may be rich in terms of attribute data but poorly positioned, while the target file may be perfectly positioned but lacking attribute data (*16*). The development of a highly automated conflation tool is a major step forward in data integration and data sharing from multiple data sources and is strongly demanded by agencies (*17*).

There are a few studies dealing with networks conflation. Green et al. (*18*) explored an approach to use private sector speed data to create performance measures for the HPMS network in Kentucky. Their approach was to transform one network to a point shapefile and then project it to the other network. Li and Liu (*19*) proposed two approaches for conflating GIS roadway datasets. These two approaches are linear referencing approach and spatial join approach. Lee et al. (*17*) also developed a conflation tool for the ESRI ArcGIS using linear feature matching techniques to determine matching and non-matching features. Another study, by Hackeloeer et al. (*20*), reviewed georeferencing methods and different types of information (geometric, topological, and semantic information), which can be used to reference geographical objects, and how they could be used for conflating maps. Yang and Zhang (*21*) proposed a pattern-mining method to conflate crowdsourcing road networks with points of interest by using geometric and semantic information. Zhang et al. (*22*) developed an automatic road-network conflation approach to transfer pedestrian ways from one map to another.

The literature on network conflation still lacks a profound and clear methodology accompanied by an automated tool which could be applied to actual GIS roadway datasets and deals with imperfections associated with them. Unfortunately, the current GIS roadway datasets typically are not perfect regarding positioning, resolution, and accuracy, or attribute data. This imperfection in networks needs to be taken into account for network conflation. Our study proposes a comprehensive methodology for the network conflation, and a workflow is introduced based on the proposed methodology to handle networks conflation for large networks. A tool is created to perform the projection in the ESRI ArcGIS environment and, to tackle the imperfection typically associated with the GIS datasets, a set of sanity check steps are presented. The proposed methodology provides a solution that could use tools in popular commercial software such as ESRI ArcGIS which is accessible to users.

Problem Statement

Consider the very common scenario where an analysis is performed by a private entity, and the results are reported based on that entity's own roadway segmentation method. However, the results may need to be transformed to segments based on another highway location referencing system described in the HPMS Field Manual (*7*) before they can be digested by public agencies. Therefore, for each feature in the first network, it is necessary to determine the overlapping segments from the second network and to determine the fraction of each overlapping segment to the subject feature.

In general, let us consider a case where, for the same highway network, there are two separate shapefiles, one based on highway location reference system 1, shapefile P, and another one based on location reference system 2, shapefile Q. As shown in Figure 1, for each segment of shapefile P denoted by P(i), the problem is to determine the overlapping segments from network Q (in this figure segments Q1, Q2, and Q3), and to determine the approximate percentages of the segment P(i) that overlap with each of these segments from network Q. For example, for segment P(i) in Figure 2, approximately 25% of the segment is projected into segment Q1, 40% of the segment is projected into segment Q2, and 35% of the segment is projected into segment Q3. Therefore, relevant attributes for segment P(i) could be estimated using

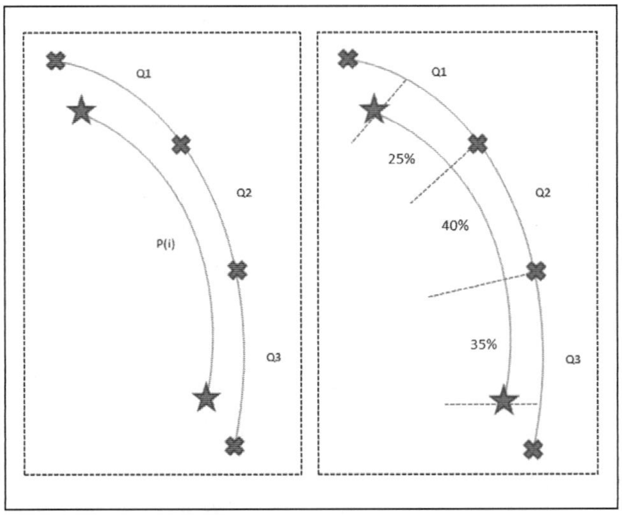

Figure 1. Matching segments from network Q for a target segment of network P.

attributes of segments Q1, Q2, and Q3 using the above rates. For example, assume the number of crashes is reported for the shapefile Q and segments Q1, Q2, and Q3. The number of crashes is reported as 60, 50, and 80. Then the number of crashes for the target segment P(i) is estimated to be $0.25*60 + 0.4*50 + 0.35*80 = 63$. In general, the problem is to determine overlapping segments from the matching network Q for each segment from the target network P and determine the percentage of the selected target segment covered by each matching segment.

Methodology

A geospatial-oriented algorithm is developed to use a set of tools and features in the GIS environment to build the model to solve the problem. In particular, existing standard tools in the ESRI ArcGIS software are utilized to develop the solution method. The main idea is to calibrate one network based on the geometric measures of the other network. The proposed methodology is implemented in Model Builder to create a specific customized tool for this purpose. The created tool needs two input shapefiles. Assume the following as inputs:

- Shapefile based on highway location reference system 1, target shapefile S1 (type: Polyline M)
- Shapefile based on highway location reference system 2, matching shapefile S2 (type: Polyline)

These shapefiles may include attributes on the geometry and other characteristics of the road such as road name, travel direction, location, traffic characteristics, length, and functional classification. A matching field between two shapefiles such as area location or road name is desired to remove irrelevant matching segments, and as a result speed up the process, for each target segment.

As stated above, the problem is to determine overlapping S2 segments for each subject S1 segment and to determine the percentage of each S2 segment on the selected S1 segment. To accomplish this task a step-by-step procedure needs to be followed, as presented below:

Step 1: Project shapefiles to the same projection (if already not in the same projection).
Step 2: Select S1 segments one by one.
Step 3: Select S2 segments by matching relevant attributes (such as location, road name, and direction) with the selected S1 segment. This step reduces the number of potential S2 matching segments that may match the subject S1 target segment. As a result, the accuracy of the results improves while the running time decreases.
Step 4: Convert shapefile S2 (selected in step 2) to M aware layer (mile post assigned to features) so it could be calibrated based on shapefile S1.
Step 5: Using S1 poly line shapefile, create S1 point shapefile based on all vertices.
Step 6: Locate all point features of S1 shapefile along S2 route feature and use the frequency tool to find unique overlapping segments from shapefile S2. Then, all corresponding S2 segments for each S1 segment are determined.
Step 7: Calibrate S2 selected M aware layer (created in step 3) by S1 point layer.
Step 8: Calculate mile post values of all overlapping S2 segments for each subject S1 segment.
Step 9: Calculate the percentage of the length of S1 segment overlapping with its matching S2 segment, to the length of the target S1 segment, as shown in Figure 2. The calculation is based on the position of S2 segment compared to the S1 segment, as follow:

Assume S1, and S2 segments' start and end mile posts are L_S, L_E, T_S, and T_E, respectively:

$$\text{If } T_S < L_S \text{ and } T_E < L_E, \text{ the percentage is}: (T_E - L_S)/(L_E - L_S) \quad (1)$$

$$\text{If } T_S > L_S \text{ and } T_E > L_E, \text{ the percentage is}: (L_E - T_S)/(L_E - L_S) \quad (2)$$

$$\text{If } T_S > L_S \text{ and } T_E < L_E, \text{ the percentage is}: (T_E - T_S)/(L_E - L_S) \quad (3)$$

$$\text{If } T_S < L_S \text{ and } T_E > L_E, \text{ the percentage is}: 100\% \quad (4)$$

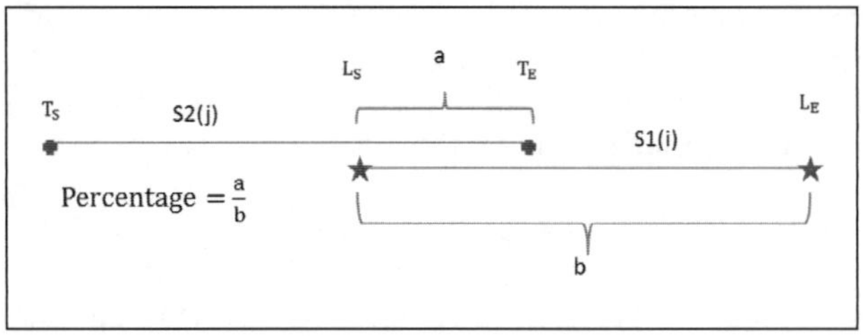

Figure 2. Percentage calculation.

Sanity Check

As mentioned, current GIS roadway datasets typically are not perfect regarding positioning, resolution, accuracy, and attribute data. Such imperfections may cause some undesirable or erroneous results which may need further attention. To resolve this issue, our model flags suspect features using pre-defined binary codes to determine if a sanity check may be needed for a specific segment. Depending on the imperfection associated with datasets different types of flags may be faced. There are a few reasons why it may be necessary to perform sanity checks for some results, among which are:

- When the calibration estimation is not accurate or matching segments are not calibrated at all. In this case, there may be a percentage value in an unacceptable range. This may occur when the overlapping segments only barely touch one another, and as a result, there are not enough overlapping points based on which to calibrate. Usually, hese segments can be removed from the results because in reality there is not a significant amount of overlap between segments.
- Once there are locations which are not covered by either shapefiles, for a few target segments, we may get a null result which is generally due to the non-existence of an overlapping matching segment at the subject location. This may also occur because the matching field between two shapefiles is labeled differently. It may then be necessary to go back and modify the matching fields.
- A manual check is needed for segments when the total percentage of their overlapping matching segments does not sum up to about 100%.

Data Description

Maryland State Highway Administration (MD-SHA) uses an LRS network internally for geospatial analysis. MD-SHA has licensed the use of third-party probe speed data through I-95 Corridor Coalition's Vehicle Probe Project (VPP). MD-SHA performs analysis on the VPP data (mainly INRIX data) to generate annual performance measures while the VPP data is reported for TMC segments, and as a result, the performance measures are also in TMC format. Now, the challenge is to transform these results into the MD-SHA-defined LRS format so that they can be readily used for reporting and evaluation purposes.

Therefore, in this case study two shapefiles for the same highway network in Maryland are considered. In Figure 3, INRIX Mobility TMC segments are indicated by blue polylines, and Maryland SHA HPMS network is highlighted in red. Note that these are only subsets of the complete TMC and MD HPMS network for the highway system in the state of Maryland. There are a few locations in these networks that are not referenced by both networks. These non-overlapping segments are highlighted only as red or blue. Generally, the HPMS network should be the more inclusive as all state agencies are required by FHWA to include all public roads in their HPMS report. The proposed methodology is applied to conflating these two networks.

Results

Detailed attributes for all LRS segments are determined. For each LRS segment, the result indicates the matching TMC segments, mile posts of the start and end points of the LRS segment (L_S and L_E) in question, and overlapping TMC segments (T_S and T_E). The fractions of the subject LRS segment overlapping with TMC segments are also presented. As an example, as presented in Table 1, LRS segment 144, which has start and end mile points of 2.16 and 3.86, is matched with three TMC segments: 122P04845, 122 + 04846, and 122P04846. These three TMC segments cover 11%, 75%, and 14% of the length of LRS segment 144, respectively.

The model was run for an LRS shapefile with 586 LRS segments and a TMC shapefile with 1,994 TMC segments. On a computer with 8GB RAM, the running time was less than 10 seconds for each LRS segment. Out of a total of 586 LRS segments, this method produced reliable results for 501 segments, while the results for the remaining 85 segments were returned as missing (null). A manual check shows that no overlapping TMC segment exists for these 85 LRS segments, and, therefore, confirms the initial results. Initial results for all non-missing segments are prepared, and results for the first 20 segments are presented in Table 2. The results for which a sanity check is desired are flagged in four different categories.

As discussed above, for each LRS segment, the methodology first selects the matching TMC segments by matching relevant attributes with the LRS segment. This step reduces the number of potential matching segments. Therefore, the running time decreases significantly and,

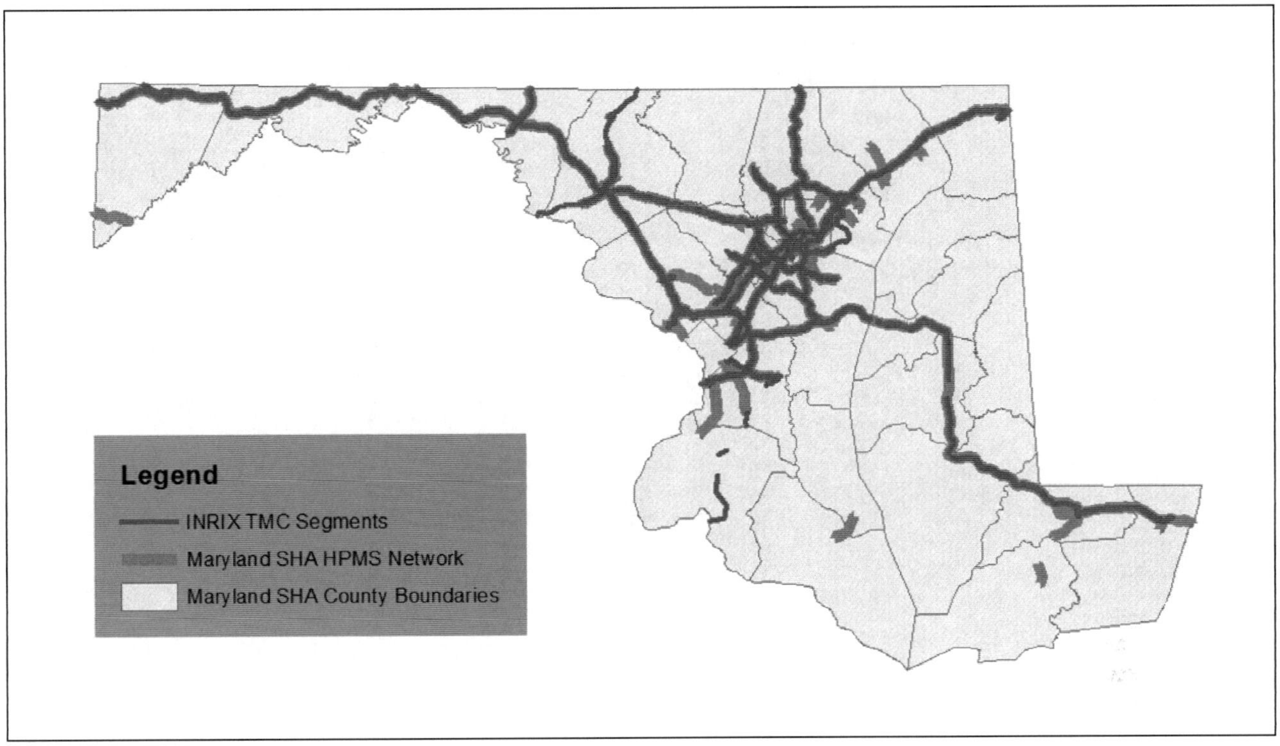

Figure 3. Maryland SHA HPMS and INRIX TMC segments network map (partial network).

Table 1. Single LRS Segment and Matching TMC Segments

LRS	TMC	L_S	L_E	T_S	T_E	Percentage
144	122P04845	2.16	3.86	1.94	2.34	11%
	122 + 04846			2.34	3.61	75%
	122P04846			3.61	4.13	14%

upon availability of sufficient matching attributes, becomes less dependent on the number of TMS segments (matching shapefile) and is more a matter of the number of LRS segments (target shapefile). Fortunately, as the results indicate, the developed tool is quite fast and accurate in determining the result for each segment. The developed tool can be applied to networks with large numbers of matching and target segments and provides accurate results in a timely manner.

Flag 1 indicates TMC segments that are not properly calibrated. This issue generally happens in three different scenarios. First, where a TMC segment is very small; since there is not enough overlap with the LRS segment, it may not be calibrated properly. In this case, we need to check it manually and correct the mile posts and percentages. Second, as shown in Figure 4*a* and *b*, where the TMC segment is in a different alignment or direction but has the same matching attribute data as the LRS segment, the TMC segment may be selected as matching the LRS segment. In this case, we should remove this TMC segment from the result. In the developed tool we filter the files only based on road names as additional matching data was not available. Adding a filter, based on the direction, will eliminate these unrelated segments from the result. Third, as shown in Figure 4*c*, where the TMC segment only touches the LRS segment but does not overlap, again, the TMC segment may be identified as matching with the LRS segment. In this case, similar to the second scenario, we may remove the TMC segment from the result.

Flag 2 determines TMC segments with negative percentage values. This may occur once TMC segments are not calibrated, or percentage values are not calculated properly because of similar reasons explained above.

Table 2. Initial Results

LRS	TMC	LS	LE	TS	TE	Percentage	Flag 1	Flag 2	Flag 3	Flag 4
0	122P04854	12.69	13.03	12.65	12.97	83%	0	0	0	0
	122 + 04855			12.97	13.02	15%	0	0	0	0
	122P04855			13.02	13.06	2%	0	0	0	0
1	110N04483	12.77	13.61	0.00	3752.18	0%	1	0	0	0
	110P05252			12.75	13.49	86%	0	0	0	1
3	110 + 04619	2.37	3.95	0.00	7867.67	0%	1	0	0	0
	110P04617			2.34	2.70	21%	0	0	0	0
	110 + 04618			2.70	3.76	67%	0	0	0	1
4	110P04528	3.04	3.41	2.41	3.24	53%	0	0	0	0
	110 + 04529			3.23	4.29	47%	0	0	0	0
6	110N06363	10.63	11.82	0.00	1273.69	0%	1	0	0	0
	110P05597			10.51	10.83	17%	0	0	0	0
	110 + 05598			10.83	11.63	67%	0	0	0	1
8	110N04568	0.20	1.42	–0.44	2.56	100%	0	0	0	0
	110P04503			0.13	0.25	4%	0	0	0	0
	110 + 04504			0.25	2.28	96%	0	0	1	0
10	110N04633	18.54	19.59	18.28	19.00	44%	0	0	0	0
	110-04632			19.00	19.47	44%	0	0	0	0
	110N04632			19.47	19.86	12%	0	0	0	0
11	110 + 04459	0.81	1.85	0.00	310.70	0%	1	0	0	0
	110P04457			0.64	0.97	15%	0	0	0	0
	110 + 04458			0.97	1.29	31%	0	0	0	0
	110P04458			1.29	1.40	10%	0	0	0	0
	110P04459			1.47	1.51	4%	0	0	0	0
	110 + 04460			1.51	1.87	33%	0	0	0	0
12	110P04428	0.00	0.53	–0.31	0.36	67%	0	0	0	0
	110N04428			–0.30	0.42	79%	0	0	1	0
14	110P04267	1.97	3.63	1.75	2.14	10%	0	0	0	0
	110 + 04268			2.14	3.44	79%	0	0	0	0
	110P04268			3.44	3.79	11%	0	0	0	0
15	110P04500	0.54	1.61	0.00	3854.33	0%	1	0	0	0
	110P04580			0.35	0.72	16%	0	0	0	0
	110 + 04581			0.71	1.02	28%	0	0	0	0
	110P04581			1.01	1.91	56%	0	0	0	0
16	110P04624	11.88	12.52	11.70	12.06	28%	0	0	0	0
	110 + 04625			12.06	12.38	50%	0	0	0	0
	110P04625			12.38	12.75	21%	0	0	0	0
17	110 + 04511	0.21	1.13	–0.16	0.22	1%	0	0	0	0
	110P04511			0.22	0.33	13%	0	0	0	0
	110 + 04512			0.35	0.45	11%	0	0	0	0
	110P04512			0.48	0.61	15%	0	0	0	0
	110 + 04513			0.61	0.87	29%	0	0	0	0
	110P04513			0.87	1.13	28%	0	0	0	0
19	110-04489	0.00	3.81	0.01	1.00	26%	0	0	0	0
	110-04488			0.17	3.46	86%	0	0	1	0
	110N04488			3.46	4.16	9%	0	0	1	0
20	110P04471	10.48	14.27	10.08	10.83	9%	0	0	0	0
	110 + 04472			10.83	14.03	85%	0	0	0	0
	110P04472			14.03	14.26	6%	0	0	0	0

Note: Values in columns L_S, L_E, T_S, and T_E are mile posts.

These segments need to be checked manually and fixed for mile posts or percentage calculations. We do not expect all segments to have the total percentage value of exactly one. Then, Flag 3 and 4 specify LRS segments with percentage values of more than 100% and less than 90%, respectively. Total percentages more than 100% may occur if a segment is not calibrated; the percentage is not calculated properly; an unrelated segment is matched with the subject LRS segment, or there are more than one TMC segment covering the same section

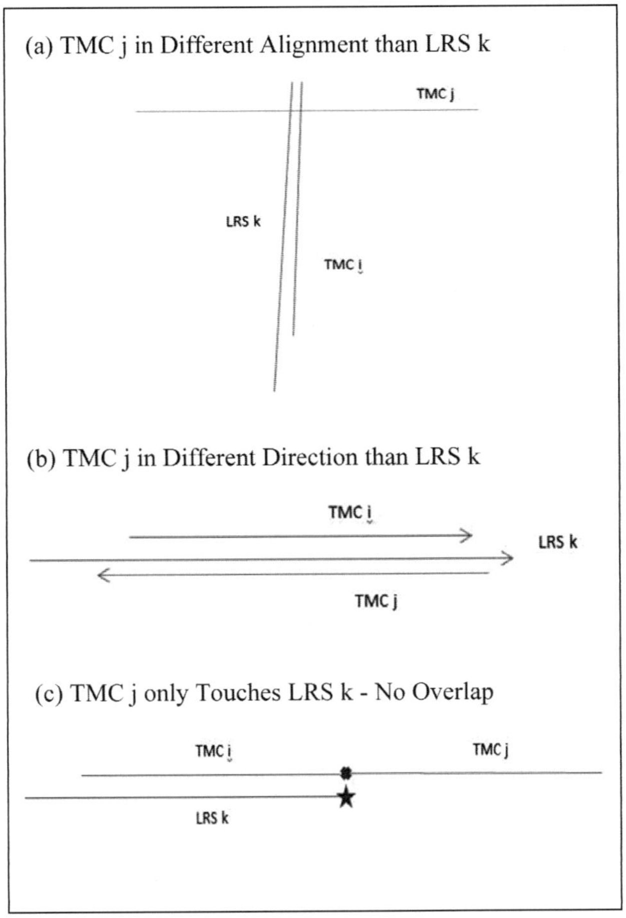

Figure 4. Schematic diagram of possible flags.

of the road (TMC segments overlap with each other). These segments need to be evaluated manually and fixed or removed, if necessary. Total percentage less than one could be acceptable as no TMC segment may exist at the subject LRS segment. However, a manual check may be needed to confirm the initial results.

The result after sanity check is prepared for all LRS segments and a subset of the final table after sanity check is presented in Table 3. This table is presented to review a few examples and see how they are addressed. For this study, all missing LRS segments refer to the locations that are not covered by TMC segments. Therefore, there are no TMC segments to match LRS segments 2, 5, 7, 9, 12, 13, and 18. Also, some flagged TMC segments such as TMC segments 110N04483 and 110P04500 are removed because they have different alignment or direction compared with the subject LRS segment. Preprocessing of the files and adding more matching attribute data such as road directions will remove these results from the analysis. Also, a few flagged segments such as TMC segments 110 + 04619 and 110N06363 are removed because they only touch the subject LRS segment and do not have a substantial overlap. Also, for a few cases, reported percentages add up to more than one. Typically, this may occur where TMC segments overlap with each other or where one TMC segment is part of another TMC segment. So, in these cases, preprocessing of the shapefiles can significantly reduce undesired results. Finally, we need to confirm that total percentages for all LRS segments add up to around one and determine if any matching segment may be missing from the result. However, in many cases when the total percentage is less than one, it is because TMC segments do not fully cover LRS segments such as LRS segments 1, 3, and 6.

Summary and Conclusion

For analysis and reporting of data in GIS format, public agencies and private sector organizations may use different highway location referencing systems and spatial segmentations for the same transportation infrastructures. Therefore, it is essential to be able to conflate the spatial data in different networks. This study proposes a methodology and workflow to conflate such networks. A customized tool based on the proposed methodology is created to perform the projection in the GIS environment. The methodology is to calibrate one network based on geometry measurements of the target network. This makes it possible to compare locations on different maps as they are on the same network and based on the same geometry.

The proposed methodology is accompanied by a tool which could be applied to actual GIS roadway datasets and deals with the imperfection associated with them. The developed tool provides a solution by using tools in popular commercial software such as ESRI ArcGIS which is accessible to users. This tool eliminates the need for additional commercial packages.

The developed tool is applied to project a subset of the Maryland HPMS onto the INRIX TMC highway mobility network. The results confirm the capability of the methodology and the created tool. The proposed tool can create results for large networks in a timely fashion with high accuracy. This tool is essentially useful for reporting purposes as federal and state highway agencies need to report based on their own location referencing systems. This tool may also be applied to project the different networks of different private sector organizations onto each other, when results based on a different system used by an external agent need to be used as an input for internal analysis.

The imperfection associated with networks needs to be considered. A sanity check is required as there could be cases in which undesirable or erroneous results are produced, which are mainly because of the inconsistencies between shapefiles' coverages or imperfections associated

Table 3. Final Results

LRS	TMC	LS	LE	TS	TE	Percentage	Total percentage
0	122P04854	12.69	13.03	12.65	12.97	83%	100%
	122 + 04855			12.97	13.02	15%	
	122P04855			13.02	13.06	2%	
1	110P05252	12.77	13.61	12.75	13.49	86%	86%
3	110P04617	2.37	3.95	2.34	2.70	21%	88%
	110 + 04618			2.70	3.76	67%	
4	110P04528	3.04	3.41	2.41	3.24	53%	100%
	110 + 04529			3.23	4.29	47%	
6	110P05597	10.63	11.82	10.51	10.83	17%	84%
	110 + 05598			10.83	11.63	67%	
8	110N04568	0.20	1.42	−0.44	2.56	0%	100%
	110P04503			0.13	0.25	4%	
	110 + 04504			0.25	2.28	96%	
10	110N04633	18.54	19.59	18.28	19.00	44%	100%
	110-04632			19.00	19.47	44%	
	110N04632			19.47	19.86	12%	
11	110 + 04459	0.81	1.85	1.40	1.47	6%	100%
	110P04457			0.64	0.97	15%	
	110 + 04458			0.97	1.29	31%	
	110P04458			1.29	1.40	10%	
	110P04459			1.47	1.51	4%	
	110 + 04460			1.51	1.87	33%	
14	110P04267	1.97	3.63	1.75	2.14	10%	100%
	110 + 04268			2.14	3.44	79%	
	110P04268			3.44	3.79	11%	
15	110P04580	0.54	1.61	0.35	0.72	16%	100%
	110 + 04581			0.71	1.02	28%	
	110P04581			1.01	1.91	56%	
16	110P04624	11.88	12.52	11.70	12.06	28%	100%
	110 + 04625			12.06	12.38	50%	
	110P04625			12.38	12.75	21%	
17	110 + 04511	0.21	1.13	−0.16	0.22	1%	95%
	110P04511			0.22	0.33	13%	
	110 + 04512			0.35	0.45	11%	
	110P04512			0.48	0.61	15%	
	110 + 04513			0.61	0.87	29%	
	110P04513			0.87	1.13	28%	
19	110-04489	0.00	3.81	0.01	0.17	4%	100%
	110-04488			0.17	3.46	86%	
	110N04488			3.46	4.16	9%	
20	110P04471	10.48	14.27	10.08	10.83	9%	100%
	110 + 04472			10.83	14.03	85%	
	110P04472			14.03	14.26	6%	

Note: Values in columns L_S, L_E, T_S, and T_E are mile posts.

with each shapefile. Preprocessing of shapefiles is essential and will improve the accuracy of the tool. This helps to remove undesired results, and as a result, reduces the number of segments which may need a sanity check.

Newly introduced systems such as OpenLR are dynamic location reference systems. Although they convey more detailed information on flow and traffic characteristics, they do not follow a persistent solid geometry in defining each code. Segments implied by these systems may change based on traffic characteristics. As a result, another useful approach for using our tool would be to use the dynamic reference systems in the first place and transform the results into systems with persistent segments subsequently. This approach helps to keep the segments the same while taking advantage of more detailed information such as speed, travel time, and flow information provided by these systems.

Author Contributions

The authors confirm contribution to the paper as follows: study conception and design: Daneshgar, Sadabadi; data collection: Daneshgar, Sadabadi; analysis and interpretation of results:

Daneshgar; draft manuscript preparation: Daneshgar, Sadabadi, Haghani. All authors reviewed the results and approved the final version of the manuscript.

References

1. Vonderohe, A. P. *Adaptation of Geographic Information Systems for Transportation*. National Academies Press, National Research Council, Washington, D.C., 1993.
2. Brown, J. N., A. Rao, and J. Baran. Automated GIS Conflation: Coverage Update Problems and Solutions. *Proc., Geographic Information Systems for Transportation Symposium (Gis-T)*, Sparks, NV, 1995, pp. 220–229.
3. *Linear Referencing and Other Forms of Location Expression for Transportation*. Final Report. California Department of Transportation, Sacramento, CA, 2002.
4. *What Is Linear Referencing?* ESRI. http://resources.arcgis.com/en/help/main/10.1/index.html#//003900000001000000. Accessed July 2017.
5. Adams, T. M., N. Koncz, and A. P. Vonderohe. Functional Requirements for a Comprehensive Transportation Location Referencing System. *Proc., North American Travel Monitoring Exhibition and Conference* (NATMEC), Wisconsin Department of Transportation, Madison, WI, 2000, p. 13.
6. *Highway Performance Monitoring Systems*. FHWA Office of Highway Policy Information. http://www.fhwa.dot.gov/policyinformation/hpms.cfm. Accessed July 2017.
7. *Highway Performance Monitoring Systems Field Manual*. Office of Highway Policy Information, March 2014.
8. *National Highway System*. FHWA Office of Planning, Environment, and Reality (HEP) Planning. http://www.fhwa.dot.gov/planning/national_highway_system/nhs_maps/. Accessed July 2017.
9. *Certified Public Road Mileage Data and ARNOLD*. FHWA Office of Highway Policy Information. http://www.fhwa.dot.gov/policyinformation/hpms/arnold.cfm. Accessed July 2017.
10. Kiel, D., J. Pollock, and M. Wang. Issues in Adapting Linear Referencing Systems for Transportation Applications. *Journal of Computing in Civil Engineering*, Vol. 12, No. 2, 1998, pp. 60–61.
11. Pons, J. M., and M. R. Pérez. Geographic Information Systems and Intelligent Transport Systems: Technologies Used to Form New Communication Networks. *Networks and Communication Studies*, Vol. 17, No. 1–2, 2003, pp. 53–70.
12. *National Performance Management Research Data Set*. FHWA Freight Management and Operations. http://ops.fhwa.dot.gov/Freight/freight_analysis/perform_meas/vpds/npmrdsfaqs.htm. Accessed July 2017.
13. *The Traffic Message Channel*. Traveler Information Services Association. http://tisa.org/technologies/tmc/. Accessed July 2017.
14. TomTom International B.V. OpenLR™ White Paper (Version 1.5, Revision 2). http://www.openlr.org/data/docs/OpenLR-Whitepaper_v1.5.pdf. Accessed July 2018.
15. AGORA-C. *Technology and Licensing Overview, ISO 17572-3 Standard for Location Referencing for Geographic Databases Part 3: Dynamic Location References*. http://www.via-corp.com/uploadedFiles/US/Licensing_Programs/AGORA-C/AGORA-C%20Technology%20and%20Licensing%20Overview.pdf. Accessed July 2018.
16. Sutton, J. Data Attribution and Network Representation Issues in GIS and Transportation. *Transportation Planning and Technology*, Vol. 21, No. 1–2, 1998, pp. 25–41.
17. Lee, D., W. Yang, and N. Ahmed. Conflation in Geoprocessing Framework-Case Studies. *Proc., 6th International Conference on Advanced Geographic Information Systems, Applications, and Services*, Barcelona, Spain, 2014.
18. Green, E., J. Ripy, M. Chen, and X. Zhang. Conflation Methodologies to Incorporate Consumer Travel Data into State HPMS Datasets. Presented at 92nd Annual Meeting of the Transportation Research Board, Washington, D.C., 2013.
19. Li, Y., and C. Liu. Spatial Approaches for Conflating GIS Roadway Datasets. *Proc., Ninth Asia Pacific Transportation Developement Conference, Chongqing, China*, 2012, pp. 290–298.
20. Hackeloeer, A., K. Klasing, J. M. Krisp, and L. Meng. Georeferencing: A Review of Methods and Applications. *Annals of GIS*, Vol. 20, No. 1, 2014, pp. 61–69.
21. Yang, B., and Y. Zhang. Pattern-Mining Approach for Conflating Crowdsourcing Road Networks with POIs. *International Journal of Geographical Information Science*, Vol. 29, No. 5, 2015, pp. 786–805.
22. Zhang, M., W. Yao, and L. Meng. Automatic and Accurate Conflation of Different Road-Network Vector Data towards Multi-Modal Navigation. *ISPRS International Journal of Geo-Information*, Vol. 5, No. 5, 2016, p. 68.

The Standing Committee on Geographic Information Science and Applications (ABJ60) peer-reviewed this paper (18-05373).

Article

TRR
JOURNAL OF THE TRANSPORTATION RESEARCH BOARD

Transportation Research Record
2018, Vol. 2672(45) 294–304

DOI: 10.1177/0361198118798295
journals.sagepub.com/home/trr
SAGE

Snow and Ice Removal Route Optimization in Kentucky

Benjamin Blandford[1], Erin Lammers[1], and Eric Green[1]

Abstract

This research describes a geographic information system-based methodology developed for the Kentucky Transportation Cabinet (KYTC) to improve the efficiency of Kentucky's snow and ice removal program. The state's program uses a system of snowplow routes that must factor in the location and attributes of trucks, facilities, materials, and roadways. Although functionally effective, the program had never been evaluated for efficiency. Researchers at the Kentucky Transportation Center used ArcGIS to assess the performance of KYTC's current snowplow routing procedures and identify strategies for optimizing them. The research established optimized routings for four counties located in two of the state's highway districts. Esri's Network Analyst and the vehicle routing problem toolset were used for this process. Early models were county-based, meaning that trucks were assigned to routes within counties based on truck type and route priority. These early models achieved some optimization but it was limited and routes were overly complex. In subsequent models, researchers devised more optimal results by modeling routes at the multi-county or district level. These successful models eliminate nine snowplow trucks across four counties, resulting in savings of US$225,000 per year. In addition to cost savings, optimizing the routing system can improve efficiency, increase safety, and reduce the amount of time needed to treat roadways. This report documents the challenges, successes, and lessons learned from this research. Specific examples are provided to demonstrate the advantages of route optimization. Subsequent research will expand upon the four successful counties and attempt to optimize snow and ice removal throughout the state.

The Kentucky Transportation Cabinet (KYTC) manages a system of trucks and facilities used for snow and ice removal throughout the state of Kentucky. During a snow event, trucks follow established routes that have been used for decades. Currently, routes are largely county-based, with each county having a specific number of trucks and facilities available to address its needs. Routes in each county are divided into categories and prioritized based on several factors, with average annual daily traffic being the most influential. Routes are cleared and treated within given time frames based on their assigned priority.

Although the current snow and ice treatment system is functionally effective, it has not been tested for efficiency. Geographic information system (GIS) analytical tools have been used in many KYTC applications and have the potential to further optimize the snow and ice treatment system by offering an in-depth examination of the network.

To this end, KYTC contracted the Kentucky Transportation Center (KTC) to perform a data-driven analysis of GIS-based methods to determine the feasibility of new methods. The purpose was to examine the current system and optimize truck routing throughout the network for a typical snow event. Key measures of optimization include time, location of salt storage facilities, number of trucks, and truck characteristics. The primary goals of KYTC were to improve efficiency, increase safety, and reduce the amount of time and funding needed to clear the roadways during winter storms.

Kentucky's Snow and Ice Program

KYTC typically spends between US$40 million and US$80 million per year on snow and ice removal and subsequent road treatment. This sum includes materials, contract equipment, funded state labor, and state equipment. KYTC is responsible for more than 63,000 lane miles of roads, with maintenance crews working to keep roadways open in the winter. The goals of KYTC's snow and ice control program are to:

[1]Kentucky Transportation Center, University of Kentucky, Lexington, KY

Corresponding Author:
Address correspondence to Erin Lammers: erin.lammers@uky.edu

- Provide adequate traction on road surfaces
- Promote safe and timely driving conditions
- Provide uniformity of pavement conditions within the route priority system
- Account for economic and environmental factors.

Future goals for KYTC's program include updating geospatial abilities and pursuing real-time tracking options. GIS specialists are enhancing the visualization of real-time data into dashboards that can be used for fleet management in order to streamline routing and material distribution. Work is also being done to develop performance measures to minimize roadway treatment and ensure salt distribution is necessary and efficient. Researchers kept these future goals in mind throughout the process of route analysis in order to ensure the tool's usefulness in the future.

The current system remains straightforward: roads throughout the state are categorized into routes and receive a priority ranking before the onset of the winter season. Route maps are created at the highway district level (Kentucky is divided into 12 highway districts) and submitted to the KYTC Division of Maintenance. Designating route priority takes into account the following:

- Priority A: Includes interstates, parkways, federal-aid primary routes, and road segments in the counties with average daily traffic (ADT) higher than that of the federal-aid primary route in that county. Additionally, any non-federal-aid secondary routes with an ADT greater than 1,000 may be included. Total Priority A mileage within a county should be between 20% and 50% of the two-lane road mileage within the county.
- Priority B: Includes federal-aid secondary routes not designated as Priority A and routes having an ADT greater than 500. Total Priority B mileage within a county should be between 20% and 50% of the two-lane road mileage within the county.
- Priority C: Includes all state-maintained routes not designated as Priority A or Priority B. Total Priority C mileage within a county should be between 20% and 50% of the two-lane road mileage within the county.

The total miles of a priority should be between 20% and 50% of the two-lane road mileage within the county. This ensures an even distribution of priority ranking throughout the state. For each priority level, KYTC has established guidelines regarding how quickly and often routes must be treated:

- Interstates and parkways—including all mainline, distributor, and collector lanes and ramps—should be treated every hour, as needed
- All other A routes should be treated every 2 h, as needed
- B routes should be treated every 4 h, as needed
- C routes should be treated every 8 h, as needed. Treatment of C routes can be deferred beyond the 8-h time window if storm conditions dictate that A and B routes require more extensive attention.

Each district has a designated number of state-owned and operated trucks to use for snow and ice removal. Although the features of these trucks vary, they can be divided into two categories: tandem trucks and single axle trucks. Tandem trucks are larger and have a greater salt carrying capacity than single axle trucks. Because of their larger size, tandem trucks are not able to treat some rural C routes, particularly those with narrow lanes or on roadways exposed to high winds. Only single axle trucks can treat these smaller rural routes. In addition to the state-owned trucks, KYTC also contracts with private operators to treat roads during snow and ice events. Contract trucks are the largest and have the greatest salt carrying capacity. As a result, contract trucks are used solely on the largest highways with the widest lanes, including all interstates, parkways, and other major highways.

KYTC's vehicle and equipment fleet includes 980 snowplows. In addition, the Cabinet has recourse to 436 contracted snowplow trucks to assist with snow and ice removal. Initial research focused on two districts: District 6 has 41 single axle trucks, 19 tandem trucks, and 67 contract trucks; District 7 has 40 single axle trucks, 17 tandem trucks, and 93 contract trucks.

Trucks generally operate at set speed limits to ensure safe and efficient clearing and treating. Speeds are set according to road type:

- For interstates, parkways, and major highways, truck speed is set at 30 mph.
- For all other highways, including ramps, turning lanes, and rural state routes, truck speed is set at 15 mph.

All control and new route calculations are based on the set speeds above. However, it is important to note that although plows attempt to stay at these speeds, there is sometimes a discrepancy between planned speed and actual speed. Speed is often a function of the snowfall intensity, time of day, and traffic. Traveling too slowly or too rapidly can adversely affect the distribution of salt.

Salt and brine storage locations are an important aspect of route development. KYTC has a total statewide storage capacity for 314,000 tons of salt; salt is held at KYTC district and county highway maintenance facilities. The Mega Cavern, an underground facility in

Louisville, contains the largest salt supply. It currently houses 180,000 tons of salt and serves as a reserve for districts across the state. When salt supplies run low, KYTC replenishes them from multiple contracted vendors. Brine is made by mixing water with salt and is applied to roadways before most winter weather events. In low temperatures, a calcium chloride solution is added to salt and brine application to help melt snow and ice.

Truck capacity influences how routes are organized. Salt is distributed from the trucks at an average rate of 350 lb per two-lane mile. The application rate may vary according to road type and size. Trucks often are limited to treating 20 to 25 mi before returning to a facility for more salt. These facilities are located at central areas within each district and are used as the home base for trucks. Routes begin and end at these facilities. The time it takes for a truck to drive from the facility to a starting point of a route is called dead head time. Driving without treating is inefficient, therefore well-organized routes minimize dead head time. Moreover, driving on already treated routes or routes that are not ready to be treated (in cases where salt is needed for higher priority routes) also contributes to inefficiencies. Mid-route stops at a facility usually take 15 to 30 min. When necessary and when arrangements have been made, trucks may refuel/re-salt at a different facility across county lines.

Literature Review

Researchers began this project by reviewing methods for snowplow routing used nationwide and beyond. Many transportation agencies have attempted to streamline their snow-removal process through a variety of methods. Most involved a heuristic approach, often with additional mathematical models. Heuristic refers to the fact that models were interactive, hands-on, and reality-influenced, as this produced the most reliable results.

In 2005, Iowa Department of Transportation implemented a web-based Winter Maintenance Decision Support System that integrates ArcIMC ActiveX Connector with ArcIMS RouteServer Extension and various other web technologies. By integrating geospatial analytical techniques and the existing asset management systems, the tool enhanced the ability to evaluate different snow-removal procedures. The system is based on the capacity to manually create, manipulate, and delete routes. It manages resources and provides expert advice to assist with complex decision-making such as routing, optimal resource allocation, and monitoring live weather information. Expert knowledge is required to determine the estimated snowplowing time, how many snowplow runs are needed, and total snowplowing time (*1*).

Quebec, Canada, improved snowplow vehicle routes using a combination of methods in 2008. The main approach used a model comprised on a multi-commodity network flow structure. In addition, they examined two heuristic solution methods based on mathematical optimization for vehicle routing. Service to each route satisfied a set of operational constraints, real-life restrictions, and a time objective. Although the model was useful for small areas, larger networks could not be solved. In addition, route priorities were not addressed (*2*).

A more recent (2013) study in Alberta, Canada, created a set of routes that would improve safety and mobility for road users while satisfying agency-directed operational constraints. The Ministry of Transportation began by abandoning the network optimization method, which can be limiting for multiple areas with unique conditions and restraints. It instead used a mathematical optimization model based on the capacitated arc routing problem to minimize total travel distance for snowplows. A metaheuristic algorithm was used to solve this model. It was tested in one city and proved sensitive to the depot location and number of routes (*3*).

FHWA performed a study in 2011 to provide state Departments of Transportation (DOTs) with the knowledge and tools to address road weather conditions in three ways: public advisory, traffic control, and snow-removal treatment. Researchers analyzed traffic density, speed, and capacity; they also examined how precipitation affects those numbers. For snowplow operations, routing could be manipulated to treat heavily trafficked roads first. This resulted in better decision-making by DOTs and motorists, and improved reliability, safety, and mobility on the roadway system during adverse weather (*4*).

In 2011, Missouri's DOT (MoDOT) developed procedures for snow-removal operations involving fleet location, route decision, and material selection. The development was based on a heuristic approach, with minimal repeatable methodology. Optimizing truck allocation and revising route decision policies let MoDOT close some maintenance buildings without sacrificing efficiency or high levels of service (*5*).

A 2013 study approached the snowplow problem from two directions: routing and application rate of salt/chemicals. A mixed integer linear problem was used to minimize the total operation time of all snowplows and reduce the longest individual truck time. The methods developed were applied to an empirical case study and computational results indicated that it solves the problem efficiently. This study is being further developed to create software that will assist in planning routes (*6*).

Vermont Agency of Transportation (VTrans) developed storm-specific routes for its snowplows in 2013 to improve service efficiency while minimizing labor hours and fuel. The project optimized the service areas for each maintenance garage based on the travel time between each garage and the surrounding road network. Vehicles were reassigned to the garages based on this method and routed according to the combined service time/fuel consumption metric. The model performance varied depending on the input parameters, but provided information that will influence future snowplow-related operations (7).

Also in 2013, the city of Centennial, Colorado, revised snowplow allocation based on research showing a significant variability among trucks both in miles traveled and in coverage of priority-level roads. As part of the reallocation, routes were divided into two priority levels. Optimization was based on time to complete the route and involved clearing all Priority 1 roads before starting on Priority 2 roads. Routes were developed using a mathematical version of the "postman problem," in which deliveries are made via an optimized route. The reallocation reduced miles traveled, decreased fuel costs, cut labor hours, and lowered CO_2 emissions (8).

Most efforts to improve and optimize snowplow routing use a heuristic interactive approach, often with supplemental mathematical models. Several agencies have experienced improved snowplow operations after implementing their chosen method; others undertook computational experiments but stopped short of implementation for various reasons. In addition, most of this methodology was relatively old, of questionable accuracy, or not available to third parties. There is no precedent for implementing a GIS-based method. The lack of promising research on other methods combined with KTC researchers' extensive experience with GIS prompted the development and evaluation of an entirely new GIS-based method for routing snowplows.

Methodology

KTC researchers met with district officials to evaluate the current system for routing snowplows. District officials articulated the specific needs for new methods and suggested that KTC focus on optimizing routes for a typical snow event, which is defined as a 1-in. snowfall. With this goal in mind, KTC defined work parameters, evaluated the current routes as a control, and created new routes using the Network Analyst feature of ArcGIS.

Control Routes

Each district has a snow and ice book that outlines the currently used control routes. Researchers used these books to verify route prioritization and evaluate control routes. In some cases, districts had additional digital files they shared with KTC. A combination of books, digital files, and other notes were used to compile current routes in Network Analyst. Typical routes are 20 to 25 mi. Time required for these routes was calculated through Network Analyst, and total time tallied for all routes. Current route times are compared with new route times to serve as the baseline for analysis.

New Routes

New routes were created for each district in Kentucky with the goal of improving the efficiency and simplicity of truck routes. Throughout the process, researchers adhered to level of service figures and other expectations for each roadway. For routing analysis purposes, it was assumed that a truck uses 80% of its capacity before returning to the facility. Each truck was limited to one repeatable route. All other conditions adhered to the parameters originally set up in Network Analyst according to KYTC expectations.

Vehicle Routing Problem

All routes, including the existing control routes and the newly created routes, were developed using esri's Network Analyst. Routes were set up as a vehicle routing problem (VRP), which finds the best routes for a fleet of vehicles to service many orders or deliveries of a product. In this case, it allowed researchers to treat salt as a product being delivered to specific sites along roadways.

VRP is a generalization of the traveling salesman problem, the goal of which is to find the optimal solution for a fleet of vehicles to service a set of orders. In the context of this research project, the set of orders referred to deliveries of salt on the roadways. This process was identified as the most suitable to optimize snow and ice clearance routes. The road network dataset was prepared to enable the performance of several different types of analysis using the ArcGIS Network Analyst extension.

Setting Up the Network. Researchers used the centerline roadway network from KYTC and an intersection database (9), but the data required some manipulation before it could be used for this project. The road layer was overlaid with the required roadway attributes: number of lanes, traffic volume, one-way/two-way, route type, speed limit. Using the intersect tool, points were created at all intersections on the new layer. This created single point features from multipoint features. The split tool was used to dissolve road layers at intersection points so that they could be treated as individual entities during the VRP analysis. This gave a complete map of all roads and intersections requiring treatment. Subsequent steps included assigning road network directionality, geoprocessing to minimize the number of line segments, and creating parallel lines for multi-lane roads.

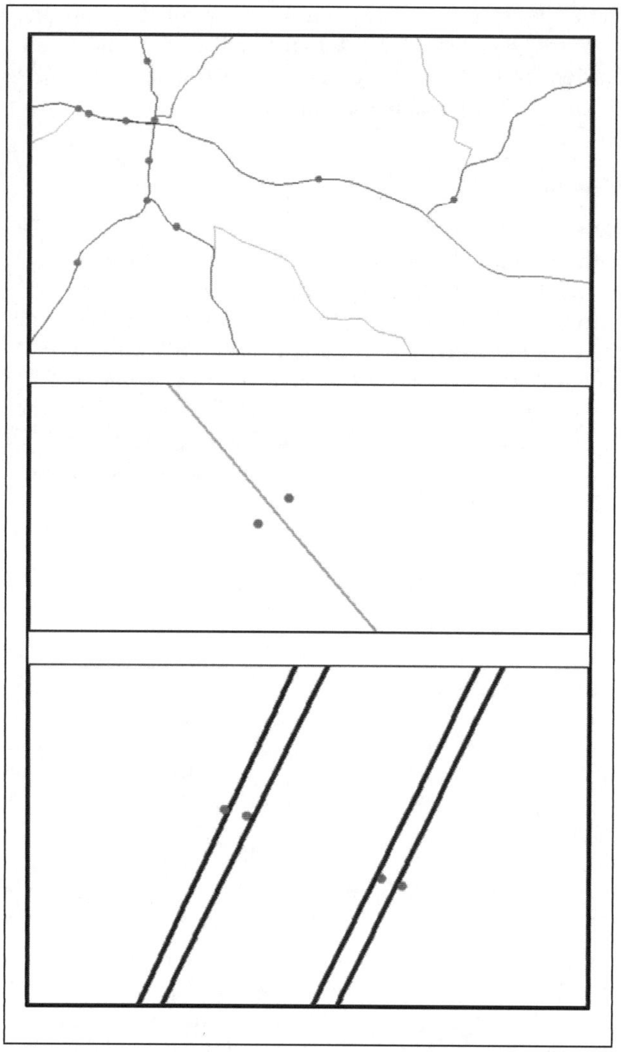

Figure 1. "Delivery" points for salt on either side of roadway.

After the main network was set up, it was manipulated such that it could be effectively used for the snowplow routing problem. To account for salt deliveries, the road segments were converted to delivery points that included fields indicating the amount of salt needed for application on each road segment. To create salt delivery points, midpoints were created for all the road segments with the ET GeoWizards tool "Points Along Polylines." The midpoints were offset to the right of the driving direction (Figure 1). For lines representing two-way roads, a point was created and offset from the line on both sides. Constraints were set that limited deliveries to the right side of the vehicle.

These geometric and data-based parameters ensured that snowplows travel in the correct direction and that a single truck does not cover two points of delivery in the same pass. An additional factor specifies the length of the road segment as a delivery quantity. As a truck treats a road, the length factor ensures that no truck exceeds its capacity in one trip.

An average salt distribution rate must be set to determine the amount of salt needed. For this project, an average salt distribution rate of 350 lb of per lane mile was chosen. Using this number, rate calculations were made to estimate the amount of salt needed for each road segment based on segment length. For VRP, this field was labeled *DeliveryQuantities*.

The problem of time windows for salt delivery was also addressed in the *Salt Delivery Points* layer. Time windows refers to KYTC's Snow and Ice Clearance (SNIC) priority system, in which routes are designated as A, B, or C routes. For each priority level, trucks are assigned to treat the routes at specified intervals. Time windows were used to designate within their specified interval when a truck needs to deliver salt to each point. This research was designed to model truck movements during a "typical" winter storm, so a "typical" 8-h shift was assumed for truck operators. Within this timeframe interstates and parkways require eight treatments (once every hour), A routes four treatments (every 2 h), B routes two treatments (every 4 h), and C routes one treatment. To account for multiple passes, salt delivery points were duplicated based on the number of passes needed. Within the attribute table, multiple time window fields were created to account for duplicate points. Table 1 displays an example of how this was created.

Several other fields were added to the salt delivery points to prepare for VRP. A field titled *CurbApproach* was created, and the Field Calculator used to populate all entries with the number 1. This indicates to the VRP solver that salt deliveries are only allowed on the right side of the vehicle. Thus, on two-way roads with points on either side, the truck must traverse the road in both directions to plow and deliver salt to both sides.

A final field added was titled *SpecialtyNames*. Creating this field enabled greater customization and control over how the trucks solve VRP. The *SpecialtyNames* field is simply an expression of the SNIC priority level, in which each point in the field is calculated as either A, B, C, or I, with I representing interstates and parkways.

Setting Up the Vehicle Routing Problem. After the network was established, VRP parameters were defined and implemented. The following list details the definitions used for the step.

- Orders: In the context of VRP, *orders* refer to the deliveries that need to be made by the fleet of trucks. Orders data are sorted into fields for quantities, service time, curb approach (indicates from

Table 1. Time Parameters with Regard to Route Priority Level

SNIC priority	Time window start 1	Time window end 1	Time window start 2	Time window end 2	Time window start 3	Time window end 3	Time window start 4	Time sindow end 4
A	8:00 AM	10:00	10:01 AM	12:00PM	12:01 PM	2:00 PM	2:01 PM	4:00 PM
B	8:00 AM	12:00	12:01	4:00 PM	na	na	na	na
C	8:00 AM	4:00 PM	na	na	na	na	na	na

Note: na = an irrelevant field. All time windows fit within an 8-h frame; after 8 h, the cycle is repeated.

Table 2. Minutes to Treat All Routes per County (District 7)

County	Current routes	Modeled routes	Difference	% improvement
Anderson	1,358	1,296	62	4.6%
Bourbon	1,529	1,287	242	15.8%
Boyle	926	892	34	3.7%
Clark	2,651	2,394	257	9.7%
Fayette[a]	2,004	1,376	628	31.3%
Garrard	997	961	36	3.6%
Jessamine	1,050	925	125	11.9%
Madison[a]	957	824	133	13.9%
Mercer	1,114	1,114	0	0.0%
Montgomery	1,210	1,210	0	0.0%
Scott[a]	3,177	3,177	0	0.0%
Woodford	1,089	1,089	0	0.0%
TOTAL	18,062	16,545	1,517	8.4%

[a]Does not include interstate and parkway routes.

which side of a truck the delivery can be made), and delivery time windows

- Depot
- Specialties
- Routes
- Route renewals
- VRP properties
  - Time and distance attributes
  - U-turns set to "Allowed only at intersections and dead ends."

VRP parameters correlated with SNIC parameters according to the following:

- Orders: Salt delivery points (i.e., midpoint of road segment), and the amount of salt that must be delivered to each road segment
- Depot: County maintenance facility containing equipment and salt
- Specialties: Restricts certain types of trucks to certain types of routes
- Routes: Number and type of truck available for the county.

Results

Researchers developed new routing solutions for each county. These routes were then combined to show all the county-level routes within a district. Initially, KYTC officials discouraged intercounty travel because organization over such a large area is difficult; however, this thinking was revised as discussed later. KYTC officials told researchers that it was not imperative to treat all A routes before B or C routes; however, this was an ancillary goal of the analysis since optimization is likely to correlate with route prioritization. Multiple iterations, combined with regular input from district officials, produced satisfactory results. These results demonstrate an improvement in the network that saves time and money.

First Iteration

To determine the performance of the new optimized routes, they were compared with control routes (i.e., routes currently used for the KYTC SNIC program). Differences in treatment times were quantified in minutes and percent improvement. Table 2 summarizes these differences for each county in District 7.

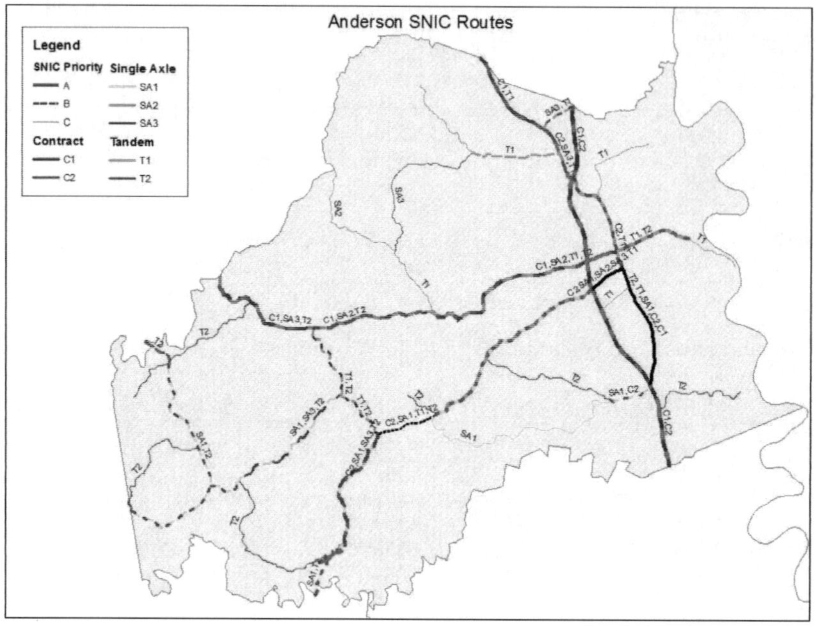

Figure 2. Example of first iteration of route optimization, as shown by Anderson County.

The initial results held promise, with an overall 8.4% improvement across the district in treatment time. Fayette County stood to benefit the most, with a 31.3% reduction in treatment time. Four counties—Mercer, Montgomery, Scott, and Woodford—showed no potential for improvement.

In some cases, the model demonstrated that trucks could not complete their routes as described within the time window. However, KYTC officials did not view this as a major problem for most routes. Occasional violation times are acceptable. If violation times were consistently high, it would indicate a need for more trucks. KYTC officials did, however, specify violation times of zero for interstates, as it is critical for interstates to receive treatment in a timely manner.

Closer examination of the routes by KYTC personnel revealed problems with the routes produced in the initial run. Most significantly, the routes produced by the model were overly complicated and not feasible if attempted. In many cases, the model would send multiple trucks to treat the same stretch of roadway at different times during the day. For example, the model might assign an A route to be treated by one truck the first pass of the day, whereas on the second, third, and fourth pass, different trucks would be assigned to the route. This meant that trucks would travel all over a county at different times during the day and all the trucks would have to coordinate precisely to achieve proper timing. This convoluted method is mapped in Figure 2.

A second problem with the results was that the complexity of the routes would limit the ability of truck drivers to become familiar with the routes they run. On this point, the participation of truck operators on the KYTC Study Advisory Committee was critical. Drivers stated that their familiarity with the routes was important for promoting safety and effectiveness. They described the importance of knowing even minute details of roads because they are covered in snow—and may not be visible—during operations. Knowing the locations of culverts, guardrails, mailboxes, nearby trees, or other obstacles of the road is critical as they can affect the effectiveness of snow and ice removal. As a result, overly complex routes were met with hesitation from snowplow operators.

A third problem with the routes was that it was unclear whether the model's results were mathematically optimized. There seemed to be some issue with the Network Analyst VRP solving algorithm: the discrepancy between the model's results and a mathematically optimized solution resulted when too many variables were added to the algorithm. Close inspection of the model results revealed instances of routes that were neither optimized nor optimal. To address these problems, further investigation of the problems with the VRP solver was set aside so researchers could adjust the methodology.

Modifications

To address the problems found with the results from the model's first iteration, the methodology was revised considerably. This called for redefining the project's research goals. In other words, rather than searching for the mathematically optimized solution to the routing problem, the goal became to identify an optimal and realistically applicable solution. This meant no longer viewing the Network Analyst VRP as the ultimate producer of the solution, repositioning it instead as a tool to guide development of a proper solution. This required more hands-on scrutiny of each county and each route, and understanding how their interconnections affect the ability of trucks to treat each route on schedule.

Most significantly, the Study Advisory Committee expressed a preference for truck routes that centered on specific A routes and then built outward to address the B and C routes. This became known as the route tree solution. Where possible, each truck route for a county began with a center trunk (A route) and branched out to treat the adjoining branches (B and C routes). KYTC truck operators preferred this approach, as it will allow them to quickly gain familiarity with their routes. A secondary benefit of a route tree system is that is better suited for adaptation in a real-world setting, when storms differ from what is expected. There is significant variability in snow intensity, temperature, wind speed, duration, and extent of the area affected from storm to storm. Trucks have greater flexibility to treat roads as needed when they are assigned to specific trunks and branches on the route tree. For example, during a heavy snowstorm when A routes warrant greater attention, trucks can easily continue to treat their assigned A routes before proceeding to routes with lower priority.

Final Results

Ultimately, the Study Advisory Committee settled on the revised methodology to pursue route modeling. The modified methodology was applied to four counties: Clark County and Montgomery County in District 7, and Grant County and Pendleton County in District 6. These county pairs were selected by the Study Advisory Committee in the hope that combining these counties together in the analysis would produce further optimizations.

For this exercise, Network Analyst was used to identify, tabulate, and monitor time required and salt usage as the routes were compiled. Although Network Analyst did not ultimately generate the results, it was useful for determining which combinations of A, B, and C routes were appropriate for each truck route.

A different metric was used to gauge the success of the second round of modeling. Instead of tracking the total time required to treat all the routes on schedule in a county, the team tracked the total number of trucks needed per county to treat all the routes. This was a more useful and applicable metric due to the KYTC Study Advisory Committee's stated preference for eliminating contract truck routes where possible. To this end, the model attempted to maximize the number of roadway miles treated by state trucks before including contract trucks in the modeling scheme.

The revised modeling indicated that KYTC can treat all routes on schedule and use fewer trucks in the process. Because fewer contract trucks are needed, KYTC can potentially realize significant cost savings. By combining counties, the model also found the potential for route optimization by letting trucks cross county lines. This type of intercounty route represents a new approach for KYTC, which has previously operated the SNIC program within single counties.

For Grant and Pendleton Counties (District 6), the model demonstrated the potential to eliminate six contract trucks, resulting in a cost saving of US$25,000 per truck per year. These results are modeled in Figure 3. Different colored routes indicate different trucks. Another pair of counties exhibited similar results, with the ability to eliminate three contract trucks between the two counties. Therefore, in the four counties modeled thus far, KYTC could save up to US$225,000 per year.

Figure 4 presents another way to view the information shown in Figure 3. This chart shows the route breakdown of one truck and the data that is obtained from mapping. It includes starting and ending times along each salt "delivery" point, the amount of salt used, the name of the road, and the road's priority designation.

Conclusion

Throughout this project's development, KTC researchers adhered to KYTC expectations and maintained familiarity with the SNIC process in order to maximize improvement. In doing so, researchers were able to present a successful way to optimize snowplow routing. As the results of this project are expanded throughout the state, programs should be aware of what researchers learned throughout the project. The following list describes these lessons learned throughout the process that could be beneficial to practitioners employing similar optimization.

- Taking advantage of local knowledge when developing snowplow routes is critical. Route analysts must know what a district's current practices are, its goals, and what kinds of routes drivers are and are not willing to do. Talking with district officials and snowplow drivers was critical for this project and helped researchers build a model that was responsive to the needs of district officials and drivers.

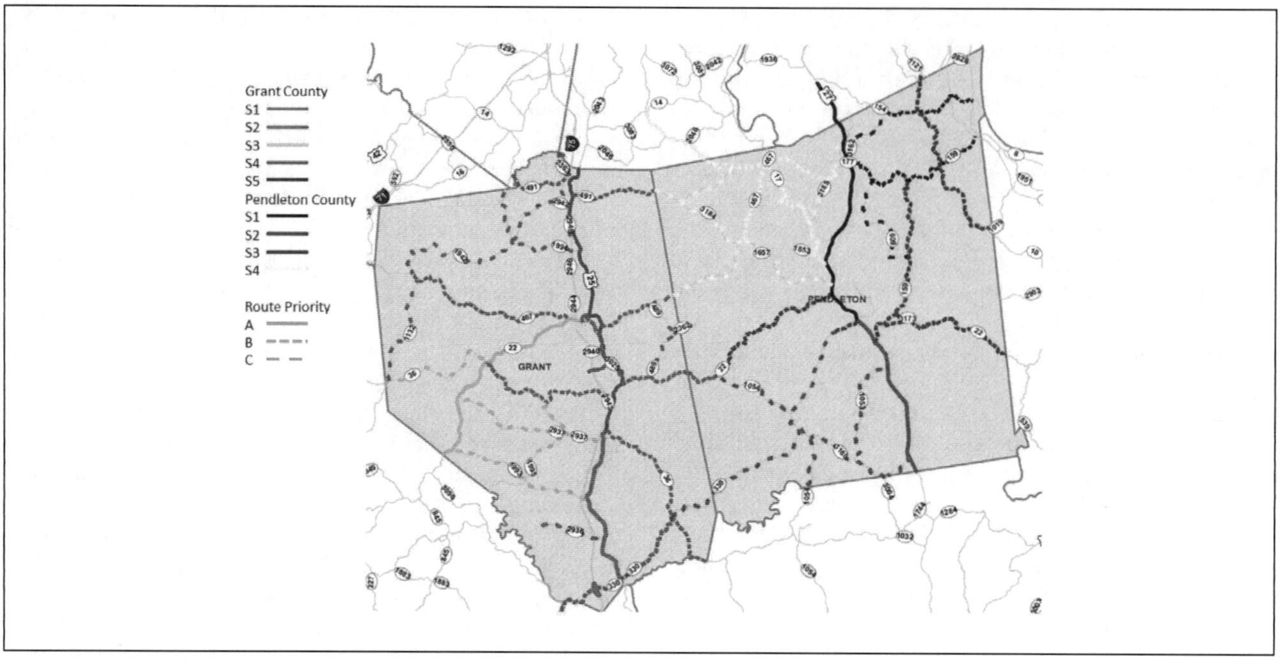

Figure 3. Grant County and Pendleton County optimized routes (excluding interstate).

- In replicating the methodology described in this report, care should be taken to simplify the network when setting up the data. Using each road's midpoint as the delivery point, rather than establishing points every 100 ft, produced a less complex network that vastly improved solver performance. Not only were the routes more straightforward, but it significantly reduced the solver's run time—from hours to seconds. A simplified network sacrifices some roadway attributes, however, so researchers and officials will need to ensure that important aspects are accounted for.
- Used alone, the VRP's effectiveness is limited. Without user input, VRP returned many routes that were clearly not the most efficient option. However, by varying constraints and trying multiple iterations, researchers developed better routes. Therefore, a combination of VRP and hands-on manipulation appears to be the best practice for developing snowplow routes.
- The network dataset should be constantly monitored and checked for quality. Even the smallest network connectivity error can lead to major problems. During this process, researchers encountered the following problems: broken connections between divided highways, incomplete lines approaching intersections, and geometric anomalies. All issues were fixed, but finding and correcting them can be time-consuming. The VRP solution depends heavily on correct data, so ensuring accuracy is very important.
- Altering constraints slightly can produce vastly different results. Some of the constraints modified by researchers included time violations, excess transit, start time, and grouping routes according to priority. Adjusting these constraints let researchers manipulate the program to generate efficient, simple, and realistic routes. Combining Network Analyst tools and researchers' knowledge to solve snowplow routing problems is a very heuristic process.
- Officials will require more preparation time to account for new routes and routing options that may be counterintuitive. Furthermore, ArcGIS can provide turn-by-turn directions. This could be especially helpful on new routes or routes that are nonintuitive. District officials should facilitate the coordination effort and understand how the optimized routes work together.
- Analysis assumed that snowplows traveled on an empty road at constant speed. Although not likely, trucks in some cities may experience delays if there is heavy traffic. Conservative estimates in other areas made up for the chance of this occurrence.
- The analysis excluded lane widths. Including this attribute creates too many segments for the model to optimize efficiently and does not affect the choice of routes.

Start	End	Minute	Salt	Salt Total	Road	Description	Priority
0:00	0:03	0:03	0	-	Transit to US 25		
0:03	0:41	0:38	6,140	6,140	US-25	from KY 36 to KY 330	A(1)
0:41	0:46	0:05	444	6,584	I-75	Exit 144, RAMP to/from KY 330	A(1)
0:46	0:59	0:13	2,283	8,867	KY-330	from US 25 to KY 607	B(1)
0:59	1:17	0:17	2,777	11,644	KY-607	from KY 1883 to KY 330	B(1)
1:17	1:22	0:05	444	12,087	I-75	Exit 144, RAMP to/from KY 330	A(2)
1:22	1:25	0:03	0	12,087	Transit back to barn		
1:25	1:40	0:15			REFILL		
1:40	1:43	0:03	0	-	Transit to US 25		
1:43	2:22	0:38	6,140	6,140	US-25	from KY 36 to KY 330	A(2)
2:22	2:27	0:05	444	6,584	I-75	Exit 144, RAMP to/from KY 330	A(3)
2:27	3:03	0:35	3,137	9,720	KY-330	from US 25 to KY 36	B(1)
3:03	3:33	0:29	2,604	12,324	KY-2936	from US 25 to End of State Maintenan	C
3:33	3:38	0:05	444	12,768	I-75	Exit 144, RAMP to/from KY 330	A(4)
3:38	3:41	0:03	0	12,768	Transit back to barn		
3:41	3:56	0:15			REFILL		
3:56	3:59	0:03	0	-	Transit to US 25		
3:59	4:38	0:38	6,140	6,140	US-25	from KY 36 to KY 330	A(3)
4:38	4:43	0:05	444	6,584	I-75	Exit 144, RAMP to/from KY 330	A(5)
4:43	4:56	0:13	2,283	8,867	KY-330	from US 25 to KY 607	B(2)
4:56	5:14	0:17	2,777	11,644	KY-607	from KY 1883 to KY 330	B(2)
5:14	5:19	0:05	444	12,087	I-75	Exit 144, RAMP to/from KY 330	A(6)
5:19	5:22	0:03	0	12,087	Transit back to barn		
5:22	5:37	0:15			REFILL		
5:37	5:40	0:03	0	-	Transit to US 25		
5:40	6:18	0:38	6,140	6,140	US-25	from KY 36 to KY 330	A(4)
6:18	6:24	0:05	444	6,584	I-75	Exit 144, RAMP to/from KY 330	A(7)
6:24	6:59	0:35	3,137	9,720	KY-330	from US 25 to KY 36	B(2)
6:59	7:04	0:05	444	10,164	I-75	Exit 144, RAMP to/from KY 330	A(8)

Figure 4. Timing, salt used, and route turn-by-turn directions for Grant County State Route 3.

Next Steps

As this research moves forward, a few additions will likely greatly improve the optimization. Preliminary results show that county-level route optimizations can further improve route by allowing trucks to cross county lines. Furthermore, optimizations are expected to yield even greater improvements when making district-level assessments.

More improvements to the optimization could be realized by moving the salt and brine facility to a more centralized location. In some cases, the salt and brine depot is located in a non-central location causing an increase in dead head time. Minimizing dead head time is a secondary measure of optimization, but significant nonetheless.

This research has only been analyzed in four counties thus far. Implementation will be tested during the next winter season in snow and ice events. It is expected that county-level route optimizations will continue to have a positive effect on cost savings for KYTC, and the rest of the state stands to benefit from the procedure.

Author Contributions

The authors confirm contribution to the paper as follows: study BB, EL, EG; data collection: BB, EL; analysis and interpretation of results: BB; draft manuscript preparation: BB, EL. All authors reviewed the results and approved the final version of the manuscript.

References

1. Sugumaran, R., M. Salim, T. Strauss, and C. Fulcher. *Web-based Implementation of a Winter Maintenance Support System Using GIS and Remote Sensing*. MTC Project 2003-05. Midwest Transportation Consortium, Ames, IA, 2005. Accessed November 2016.

2. Perrier, N., A. Langevin, and C. Amaya. Vehicle Routing for Urban Snow Plowing Operations. *Transportation Science*, Vol. 42, No. 1, 2008, pp. 44–56. http://trid.trb.org/view/851395.
3. Gang, L., Y. Ge, T. Z. Qiu, and H. R. Soleymani. *Report 13-0491: Vehicle Routing of Urban Snow Plowing Operations: Case Study for City of Edmonton, Canada.* Transportation Research Board, Washington, D.C., 2013. http://trid.trb.org/view/1240524.
4. Alfelor, R., and C. Y. D. Yang. *Managing Traffic Operations During Adverse Weather Events.* Publication FHWA-HRT-11-002. FHWA, U.S. Department of Transportation, Washington, D.C., 2011. http://www.fhwa.dot.gov/publications/publicroads/11janfeb/01.cfm.
5. Wooseung, J., J. Noble, and C. Nemmers. *Optimizing Winter/Snow Removal Operations in MoDOT St. Louis District – Includes Outcome Based Evaluation of Operations.* Cmr1200-007. University of Missouri, Columbia, MO, 2011. Accessed November 2016.
6. Hajibabai, L., S. M. Nourbakhsh, Y. Ouyang, and F. Peng. Snowplow Routing Optimization Under Resource Constraints: Formulation, Algorithm, and Decision-Support System. Presented at 92nd Annual Meeting of the Transportation Research Board, Washington, D.C., 2013.
7. Dowds, J., J. Sullivan, D. Scott, and D. Novak. *Report 2013-12: Optimization of Snow Removal in Vermont.* Transportation Research Board of the National Academies, Washington, D.C., 2013.
8. Winter Maintenance: Checking Priorities. *Roads & Bridges*, June 2013. http://www.roadsbridges.com/winter-maintenance-checking-priorities. Accessed November 2016.
9. Green, E. R., C. Blackden, and M. A. Fields. Spatial Database For Intersections. Kentucky Transportation Center Research Report. 1490. University of Kentucky, Lexington, KY, 2015. https://uknowledge.uky.edu/ktc_researchreports/1490.

The Standing Committee on Geographic Information Science and Applications (ABJ60) peer-reviewed this paper (18-05730).